U0701649

CUBE DESIGN

立方设计（下册）

深圳市立方建筑设计顾问有限公司 ／ 香港理工国际出版社 主编

中国建筑工业出版社

Project Distrbution
项目分布

Office Building, Office
写字楼、办公

Talents Garden • Shenzhen
人才园 • 深圳
Foshan News Center • Foshan
佛山新闻中心 • 佛山
Changsha Broadcasting and TV Center • Changsha
长沙广电中心 • 长沙
Shenzhen Software Industry Base • Shenzhen
深圳软件产业基地 • 深圳
China Mobile Shenzhen Branch Production Dispatching Building • Shenzhen
深圳移动生产调度大厦 • 深圳
Jintang Building • Tianjin
金唐大厦 • 天津
Liantai Building • Shenzhen
联泰大厦 • 深圳
Taitai Pharmaceutical Office Building • Shenzhen
太太药业办公楼 • 深圳
Huizhou Daily Newspaper Office • Huizhou
惠州日报社 • 惠州
Bao'an District Construction Quality Inspection Building • Shenzhen
宝安区工程质检大楼 • 深圳
Future Plaza • Shenzhen
花样香年广场 • 深圳

Commercial and Urban Complexes
商业及城市综合体

Planning & Design of Foshan Complex • Foshan
佛山综合体规划设计 • 佛山
Lsea Asia International Center • Nanning
利海亚洲国际中心 • 南宁
Central District Crystal Island • Shenzhen
中心区水晶岛 • 深圳
Emperor Star City • Shanghai
英皇明星城 • 上海
Chengdu Future Plaza • Chengdu
成都香年广场 • 成都
Maoye Baizixiang • Chongqing
贸业百子巷 • 重庆
Sheffield Land Plaza • Chengdu
谢菲尔德置地广场 • 成都
City Commercial Plaza • Guigang
城市商业广场 • 贵港

Education & Culture
文化教育

Museum of Contemporary Art and City Planning Exhibition Hall • Shenzhen
现代艺术馆与城市规划展览馆 • 深圳
Hu Yaobang Memorial Hall • Xiangtan
胡耀邦纪念馆 • 湘潭
Art Gallery at the Guanlan Engraving Base • Shenzhen
观澜版画基地美术馆 • 深圳
Guangxi Kettledrum Museum • Nanning
广西铜鼓博物馆 • 南宁
Venice Bridge Museum • Venice (Italy)
威尼斯桥博物馆 • 威尼斯（意大利）
New Campus of Xidian University • Xi'an
西安电子科技大学新校区 • 西安
University Town International Conference Center • Shenzhen (Supplementary Project)
大学城国际会议中心 • 深圳
Ezhou No. 8 Middle School • Ezhou
鄂州第八中学 • 鄂州
Longgang Pingshan Gymnasium • Shenzhen
龙岗坪山体育馆 • 深圳
Guangxi Art Gallery • Nanning
广西美术馆 • 南宁
Beirut Culture and Arts Center • Beirut (Lebanon)
贝鲁特文化艺术中心 • 贝鲁特（黎巴嫩）
Xinhai Revolution Monument (Tower) • Wuhan
辛亥革命纪念碑（塔）• 武汉
Guangxi City Planning & Construction Exhibition Hall • Nanning
广西城市规划建设展览馆 • 南宁
Shenzhen Production, Teaching and Research Base of Huazhong University of Science and Technology • Shenzhen
华中科技大学深圳产学研基地 • 深圳
Foshan Polytechnic College • Foshan
佛山职业技术学院 • 佛山
Shenzhen Bay Middle School • Shenzhen
深圳湾中学 • 深圳

Hotels
酒店

Egyptian Hotel • Cairo (Egypt)
埃及酒店 • 开罗（埃及）
Luneng Sanya Hotel • Sanya
鲁能三亚酒店 • 三亚
Guanling Hotel, Beihai, Guangxi • Beihai
广西北海冠岭酒店 • 北海
Dameisha Hotel (Liantai Club) • Shenzhen
大梅沙酒店（联泰会所）• 深圳
Dongjiu Building • Yixing
东沈大厦 • 宜兴

Urban Design and Landscape Planning
城市设计与规划

Huaqiangbei Three-dimensional Street and City Design • Shenzhen
华强北立体街道城市设计 • 深圳
Overseas Chinese Town LOFT • Shenzhen
华侨城LOFT • 深圳
Eastern Square of Nanning Conference & Exhibition Center • Nanning
南宁会展中心东广场 • 南宁
Design for the Region around the Sci-Tech Park in New Guangming District • Shenzhen
深圳光明新区科技公园周边地区城市设计 • 深圳

Modern Fashion
现代时尚

Sunny Shangeli Villa Garden • Beihai
阳光香格里别墅花园 • 北海
Tian'an Golf Sea View Garden • Shenzhen
天安高尔夫海景花园 • 深圳
Liantai Dameisha Villa • Shenzhen
联泰大梅沙别墅 • 深圳
Dongguan 0769 • Dongguan
东莞0769 • 东莞
Futong in Bao'an 96 District (Phase IV, 2006) • Shenzhen
宝安96区富通4期 • 深圳
Wei Zhen Fu • Shenzhen
唯珍府 • 深圳
Overseas Chinese Town Pure Waterfront • Chengdu
华侨城纯水岸 • 成都
Overseas Chinese Town Portofino Pure Waterfront (Phase VII) • Shenzhen
华侨城波托菲诺纯水岸七期 • 深圳
Cheng Han Man Cheng (Phase IV) • Shenzhen
承翰慢城四期 • 深圳
Bi Hai Yun Tian • Shenzhen
碧海云天 • 深圳
Nice Homestead • Wuhan
美好家园 • 武汉
Renheng Apartment • Lanzhou
仁恒住宅 • 兰州
Junyue River Paradise • Changsha
君悦香邸 • 长沙
Hua Hao Yuan Garden • Shenzhen
华浩源景园 • 深圳
Xiang Fu Jia Cheng • Changsha
湖南湘府嘉城 • 长沙
Vanke City Garden • Wuhan
万科城市花园 • 武汉
Jia Ri Lan Wan • Qinhuangdao
假日蓝湾 • 秦皇岛
Liantai Dameisha Apartment • Shenzhen
联泰大梅沙公寓 • 深圳
Overseas Chinese Town Singles Apartment • Shenzhen
华侨城单身公寓 • 深圳
Fantasia Blooming Town • Chengdu
花样年花样城 • 成都
Ao Yun Garden • Shanghai
澳韵花园 • 上海
Tao Hua Ling • Yichang
桃花岭 • 宜昌
80 Hou Jie • Shenzhen
80后街 • 深圳
Vanke Jin Yu Lan Wan • Zhuhai
万科金域蓝湾 • 珠海

Traditional and Classic
传统经典

Excellence Repulse Bay • Shenzhen
卓越浅水湾 • 深圳
Horoy Xi Yuan • Shenzhen
鸿荣源熙园 • 深圳
Zhi Di Chun Feng Ju • Shenzhen
置地春风居 • 深圳
Horoy Park Land • Shenzhen
鸿荣源公园大地 • 深圳
Hua Sheng Ling Yu • Shenzhen
华盛领域 • 深圳
China Overseas Property -Long Gang Olympic New Town • Shenzhen
中海龙岗奥体新城 • 深圳
Nan'ao Kai Xuan Bay Garden • Shenzhen
南澳凯旋湾 • 深圳
Shi Dai Tian Jiao • Wuhan
时代天娇 • 武汉
Grace Royal Apartment • Changshu
华府世家 • 常熟
China Overseas Property -Banyan Coast • Chengdu
中海翠屏湾 • 成都
China Overseas Property - International Community • Suzhou
中海国际社区 • 苏州
China Overseas Property - Xu Jiang Project • Suzhou
中海胥江项目 • 苏州

Regional Style
地域风情

Portofino Swan Castle • Shenzhen
波托菲诺天鹅堡 • 深圳
Zhong Cheng Li Jing Xiang Shan • Changsha
中城丽景香山 • 长沙
Li Jing Shan Zhuang • Shenzhen
荔景山庄 • 深圳
Xiayang Villa District • Sanya
下洋别墅区 • 三亚
Qipan Mountain Villa District • Shenyang
棋盘山别墅区 • 沈阳
Bi Lin Wan • Shanghai
碧林湾 • 上海
Du Shi Yi Jia • Shanghai
都市宜家 • 上海
Sunshine Palm Garden (Phase III) • Shenzhen
阳光棕榈院三期 • 深圳
Tonghe - Story of South-bank • Hangzhou
通和 - 南岸故事 • 杭州
Lian Tai Mangrove Bay • Shenzhen
联泰红树湾 • 深圳

写字楼、办公 Office Building, Office

人才园·深圳 Talents Garden·Shenzhen ...5
佛山新闻中心·佛山 Foshan News Center·Foshan...13
长沙广电中心·长沙 Changsha Broadcasting and TV Center·Changsha...27
深圳软件产业基地·深圳 Shenzhen Software Industry Base·Shenzhen...35
深圳移动生产调度大厦·深圳 China Mobile Shenzhen Branch Production Dispatching Building·Shenzhen...43
核电办公培训设施·阳江 Nuclear Power Office Clerk Training Complex·Yangjiang...49
金唐大厦·天津 Jintang Building·Tianjin...57
联泰大厦·深圳 Liantai Building·Shenzhen...65
太太药业办公楼·深圳 Taitai Pharmaceutical Office Building·Shenzhen...71
惠州日报社·惠州 Huizhou Daily Newspaper Office·Huizhou...77
宝安区工程质检大楼·深圳 Baoan District Construction Quality Inspection Building·Shenzhen...85

Contents 目录

立方设计（上册）·公共建筑

花样香年广场·深圳 Future Plaza·Shenzhen...91

商业及城市综合体 Commercial and Urban Complexes

佛山综合体规划设计·佛山 Planning & Design of Foshan Complex·Foshan...103

利海亚洲国际中心·南宁 Lsea Asia International Center·Nanning...113

中心区水晶岛·深圳 Central District Crystal Island·Shenzhen...119

英皇明星城·上海 Emperor Star City·Shanghai...129

成都香年广场·成都 Chengdu Future Plaza·Chengdu...135

贸业百子巷·重庆 Maoye Baizixiang·Chongqing...145

谢菲尔德置地广场·成都 Sheffield Land Plaza·Chengdu...149

城市商业广场·贵港 City Commercial Plaza·Guigang...157

教育文化 Education & Culture

现代艺术馆与城市规划展览馆·深圳 Museum of Contemporary Art and City Planning Exhibition Hall·Shenzhen...167

胡耀邦纪念馆·湘潭　Hu Yaobang Memorial Hall·Xiangtan...175
观澜版画基地美术馆·深圳　Art Gallery at the Guanlan Engraving Base·Shenzhen...183
广西铜鼓博物馆·南宁　Guangxi Kettledrum Museum·Nanning...189
威尼斯桥博物馆·威尼斯（意大利）　Venice Bridge Museum·Venice (Italy)...199
西安电子科技大学新校区·西安　New Campus of Xidian University·Xi'an...205
大学城国际会议中心·深圳　University Town International Conference Center·Shenzhen (Supplementary Project)...219
鄂州第八中学·鄂州　Ezhou No. 8 Middle School·Ezhou...229
龙岗坪山体育馆·深圳　Longgang Pingshan Gymnasium·Shenzhen...237
广西美术馆·南宁　Guangxi Art Gallery·Nanning...247
贝鲁特文化艺术中心·贝鲁特（黎巴嫩）　Beirut Culture and Arts Center·Beirut (Lebanon)...255
辛亥革命纪念碑（塔）·武汉　Xinhai Revolution Monument (Tower)·Wuhan...261
广西城市规划建设展览馆·南宁　Guangxi City Planning & Construction Exhibition Hall·Nanning...269
华中科技大学深圳产学基地·深圳　Shenzhen Production, Teaching and Research Base of Huazhong University of Science and Technology·Shenzhen...275

佛山职业技术学院·佛山 Foshan Polytechnic College·Foshan...281
深圳湾中学·深圳 Shenzhen Bay Middle School·Shenzhen...289

酒店 Hotels
埃及酒店·开罗（埃及）Egyptian Hotel·Cairo (Egypt)...297
鲁能三亚酒店·三亚 Luneng Sanya Hotel·Sanya...305
广西北海冠岭酒店·北海 Guanling Hotel, Beihai, Guangxi·Beihai...313
大梅沙酒店（联泰会所）·深圳 Dameisha Hotel (Liantai Club)·Shenzhen...319
东氿大厦·宜兴 Dongjiu Building·Yixing...325

城市设计与景观规划 Urban Design and Landscape Planning
华强北立体街道城市设计·深圳 Huaqiangbei Three-dimensional Street and City Design·Shenzhen...335
华侨城LOFT·深圳 Overseas Chinese Town LOFT·Shenzhen...341
南宁会展中心东广场·南宁 Eastern Square of Nanning Conference & Exhibition Center·Nanning...347
深圳光明新区科技公园周边地区城市设计·深圳 Design for the Region around the Sci-Tech Park in New Guangming District·Shenzhen...353

现代时尚 Modern Fashion

阳光香格里别墅花园·北海	Sunny Shangeli Villa Garden·Beihai...366
天安高尔夫海景花园·深圳	Tian'an Golf Sea View Garden·Shenzhen...376
联泰大梅沙别墅·深圳	Liantai Dameisha Villa·Shenzhen...386
东莞0769·东莞	Dongguan 0769·Dongguan...394
宝安96区富通4期·深圳	Futong in Bao'an 96 District (Phase IV, 2006)·Shenzhen...404
唯珍府·深圳	Wei Zhen Fu·Shenzhen...412
华侨城纯水岸·成都	Overseas Chinese Town Pure Waterfron·Chengdu...422
华侨城波托菲诺纯水岸七期·深圳	Overseas Chinese Town Portofino Pure Waterfront (Phase VII)·Shenzhen...434
承翰慢城四期·深圳	Cheng Han Man Cheng (Phase IV)·Shenzhen...444
碧海云天·深圳	Bi Hai Yun Tian·Shenzhen...450
美好家园·武汉	Nice Homestead·Wuhan...456
仁恒住宅·兰州	Renheng Apartment·Lanzhou...462
君悦香邸·长沙	Junyue River Paradise·Changsha...468

Contents 目录

立方设计（下册）· 创新住宅

华浩源景园 · 深圳 Hua Hao Yuan - Jing Yuan · Shenzhen...474
湖南湘府嘉城 · 长沙 Xiang Fu Jia Cheng · Changsha...478
万科城市花园 · 武汉 Vanke City Garden · Wuhan...484
假日蓝湾 · 秦皇岛 Jia Ri Lan Wan · Qinhuangdao...490
联泰大梅沙公寓 · 深圳 Liantai Dameisha Apartment · Shenzhen...496
华侨城单身公寓 · 深圳 Overseas Chinese Town Singles Apartment · Shenzhen...500
花样年花样城 · 成都 Fantasia Blooming Town · Chengdu...504
澳韵花园 · 上海 Ao Yun Garden · Shanghai...510
桃花岭 · 宜昌 Tao Hua Ling · Yichang...514
80后街 · 深圳 80Hou Jie · Shenzhen...518
万科金域蓝湾 · 珠海 Vanke Jin Yu Lan Wan · Zhuhai...524

传统经典 Traditional and Classic

卓越浅水湾 · 深圳 Excellence Repulse Bay · Shenzhen...530
鸿荣源熙园 · 深圳 Horoy Xi Yuan · Shenzhen...536

置地春风居·深圳 Zhi Di Chun Feng Ju·Shenzhen...542
鸿荣源公园大地·深圳 Horoy Park Land·Shenzhen...548
华盛领域·深圳 Hua Sheng Ling Yu·Shenzhen...558
中海龙岗奥体新城·深圳 China Overseas Property -Long Gang Olympic New Town·Shenzhen...560
嘉旺阅山华府·深圳 Jiawang Mountain Mansion·Shenzhen...568
南澳凯旋湾·深圳 Nan'ao Kai Xuan Bay Garden·Shenzhen...574
时代天骄·武汉 Shi Dai Tian Jiao·Wuhan...580
华府世家·常熟 Grace Royal Apartment·Changshu...584
中海翠屏湾·成都 China Overseas Property -Banyan Coast·Chengdu...590
中海国际社区·苏州 China Overseas Property - International Community·Suzhou...594
中海胥江项目·苏州 China Overseas Property - Xu Jiang Project·Suzhou...598

地域风情 Regional Styles

波托菲诺天鹅堡·深圳 Portofino Swan Castle·Shenzhen...604
中城丽景香山·长沙 Zhong Cheng Li Jing Xiang Shan·Changsha...612
荔景山庄·深圳 Li Jing Shan Zhuang·Shenzhen...618
下洋别墅区·三亚 Xiayang Villa District·Sanya...624
棋盘山别墅区·沈阳 Qipan Mountain Villa District·Shenyang...630
碧林湾·上海 Bi Lin Wan·Shanghai...634
都市宜家·上海 Du Shi Yi Jia·Shanghai...638
阳光棕榈院三期·深圳 Sunshine Palm Garden (Phase III)·Shenzhen...644
通和 - 南岸故事·杭州 Tonghe - Story of South-bank·Hangzhou...646
联泰红树湾·深圳 Lian Tai Mangrove Bay·Shenzhen...650

Residence section

We classify residences in this way because we think they are an embodiment of global information in nature. It is funny to label our design types in this manner according to design clothes patterns if we get away from background of the times. However, today China is facing a tide of urbanization and we need to construct buildings that can be completed in nearly two hundred years within twenty years. Thus all things seem redundant. There are too many buildings, too many types, too many ideological trends, aesthetic ways, and styles. The time needed seems too little and everything goes out instantly. Due to this manner, our architectural form will never develop into a consecutive result that undergoes sedimentation and growth. Its development shows more senses of segment, skip, and embarrassment exposed under the unprecedented ideological impact. Our design team also participates in the development process. We have designed various residences. At the same time when we stick to the idea of "space is the essence of a building" in the text book, we make use of difference, a representational form, to summarize our work in process. We also find these form-based labels are being imperceptibly influenced and new different habitation forms and lifestyles are being created. This range is defined as division of modern times, classicalism, and customs.

将住宅按这种方式分类，我们认为其本质是全球信息化的体现。如果脱离我们所处的时代背景，以这种方式依据所穿衣服的样式为我们的设计类型作标签是有些滑稽的。然而，在中国城市化高速发展的今天，我们几乎要在二十年内完成近二百年才能完成的建筑量，所有事物都显得太多了。太多的建筑数量，太多的种类，太多的思潮、审美和风格，而可以用的时间却显得太少了。一切都瞬间涌了过来。这种方式令我们的建筑形式发展不可能成为一个连续的、经过沉淀和生长的结果。它的发展体现出更多的片段感、跳跃性与面对前所未有思潮冲击下所显露出来的无耐。我们的设计团队参与其中，设计了各式各样的住宅。我们在坚持教科书上空间是建筑实质理念的同时，通过作为形式表象类型的差异归纳我们正在进行的工作。我们发现也正是这些基于形式的标签，正在潜移默化中塑造着人们不同以往的新的居住形式和生活方式。这一范畴被界定为现代、古典和风情的划分。

Modern Fashion

现代时尚

Nowadays modern style is a wide aesthetic category. According to our understanding, this style should be an embodiment of modern technology, material, and construction technology in form. Therefore the change of modern forms will follow development of technology and building materials all the time. Different from modernism of 1930s and 1940s, we also think modern styles are not a break with culture and tradition but pursue a completely-pure expression. They are mixed with different users and designers' partiality and a brand of land condition. Therefore our design is also a result generated due to mutual influence of lots of factors and is not a repeat of single styles. In the practice of design, we always stick to a belief of showing features and logic of materials used and follow the principle of beauty of architectural forms.

如今，现代风格也是一类较广的审美范畴。就我们的理解，这种风格应是当代技术、材料、施工工艺反映在形式上的体现，因此，现代形式的变化也会一直追随着技术与建筑材料的发展。同时，与20世纪三四十年代现代主义不同的是，我们认为现代风格并非同文化、传统等等的决裂，纯粹追求完全纯净的表达，其中也掺有不同使用人群，设计者的喜好和用地条件的烙印。因此，我们在这里所展现的设计是由这些众多因素相互影响的成果，并非单一风格的重复。在设计实践之中，我们一直坚守着这样的信条，即体现出所运用材料的特性和逻辑，并遵循建筑形式美的原则。

Sunny Shangeli Villa Garden•Beihai
阳光香格里别墅花园·北海

项目地点：中国，广西，北海市
项目时间：2002年
设计规模：3.5万m²
设计阶段：方案设计
项目现状：已建

Project Location: Beihai
Project Date: 2002
Project Scale: 35,000 m²
Design Phase: conceptual design
Project Status: construction completed

We bathe in the local sunshine, and as time go by throughout a day, the ever-changing white walls shift and reflect with a kaleidoscope of colors.

我们被当地阳光所感染，一天之中白色的墙体随着时间呈现出色彩的千变万化。

Project Analysis

The congregated dwelling houses (apartments) provide a living space that conforms to the living habits of the public, while villas offer various lifestyles expressing different personalities. The highest merit of the villas is an organic combination of outdoor spaces, indoor spaces and the surroundings. Its inner beauty touches the heart of the people, and bestows their fascinations.

项目分析

集合住宅（公寓）提供的是适应大众生活习惯的空间，而别墅则是能提供不同生活方式，彰显个性的居住形态。核心是别墅户外空间、内部空间与环境的总和。设计内涵给人予感动，动人的建筑会带来迷人生活。

一层平面　　　　立面图

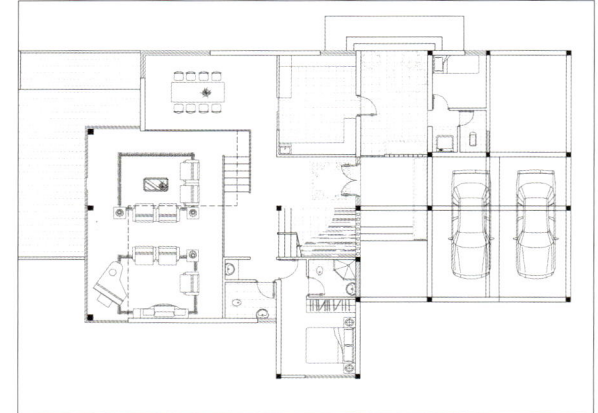

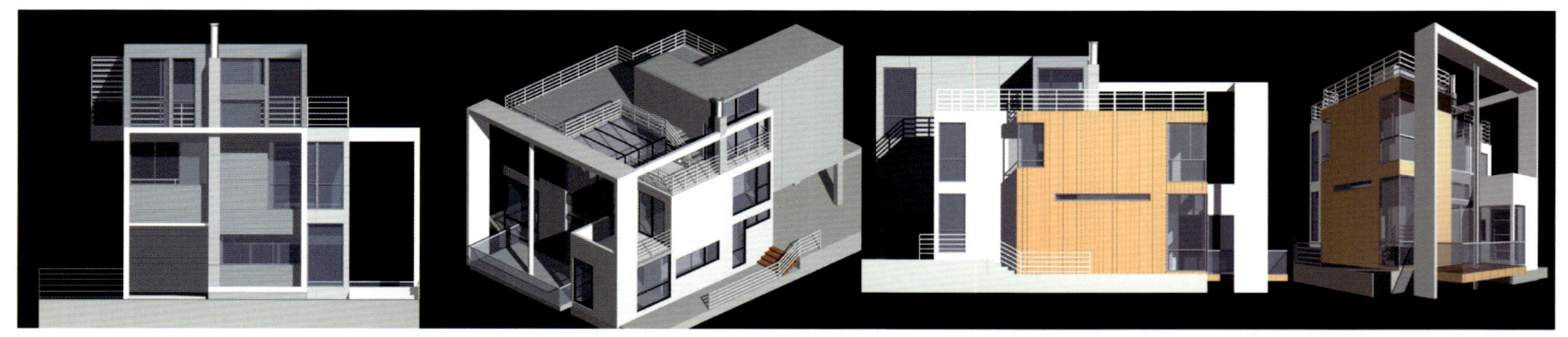

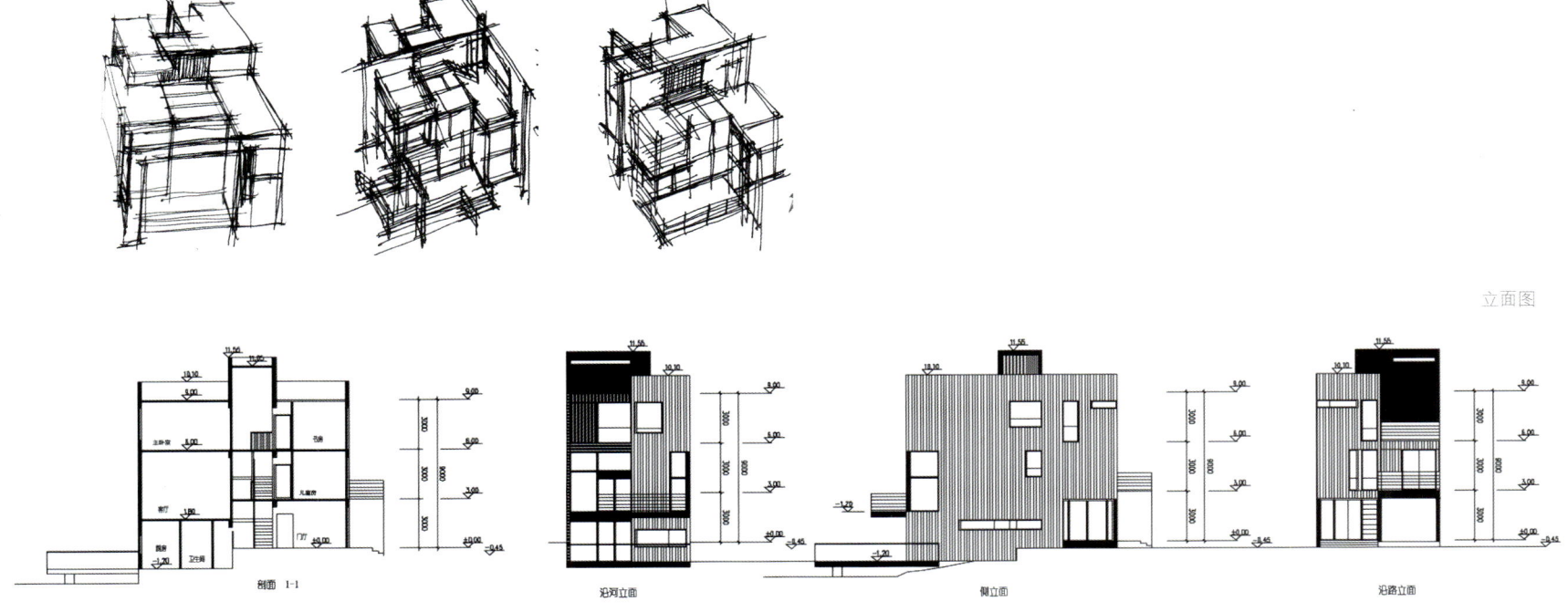

立面图

独立别墅

1、别墅引入群厅概念，将门厅、过厅、起居室、餐厅以及书房乃至户外水榭平台、层次平台、内院都作为建筑的一群开放空间处理。

2、卧室、起居空间位于建筑南北两端，以求得最好的景观和采光通风。服务定位于中部，却距离设备管道最短，经济且方便。

3、内部空间弱化楼层概念，楼层间保持良好的视线沟通；建筑内部之间的竖向联系依靠开放的竖向交通，横向联系则依靠跨跃群厅的"桥"。

4、内外空间设计充分利用了近水、临水的特点，形成丰富的由水榭平台、庭院组成的私家外部空间。

Detached Villa

1. The concept of group halls is introduced in the design of the villa. Various spaces, including the entrance hall, hallway, living and dining rooms and study, as well as the outdoor waterside pavilion, multi-leveled platforms, and the inner courtyards are taken as a group of open spaces in the building.

2. The bedrooms and living spaces are placed at the south and north ends of the building in order to obtain the best view, ventilation and lighting conditions. The service area is located in the center, which is economical and convenient since it is adjacent to facilities and plumbing.

3. As for the interior spaces, the perception of multiple stories is minimized, so people in different stories can see each other; inside the building, the vertical connections rely on the open vertical transportations, whereas the transverse connections rely on the "bridge" which spanning over the group of spaces.

4. The design of interior and exterior spaces makes full use of the surrounding water recourses, creating an outdoor private space composed of the waterside pavilions, platforms and courtyards.

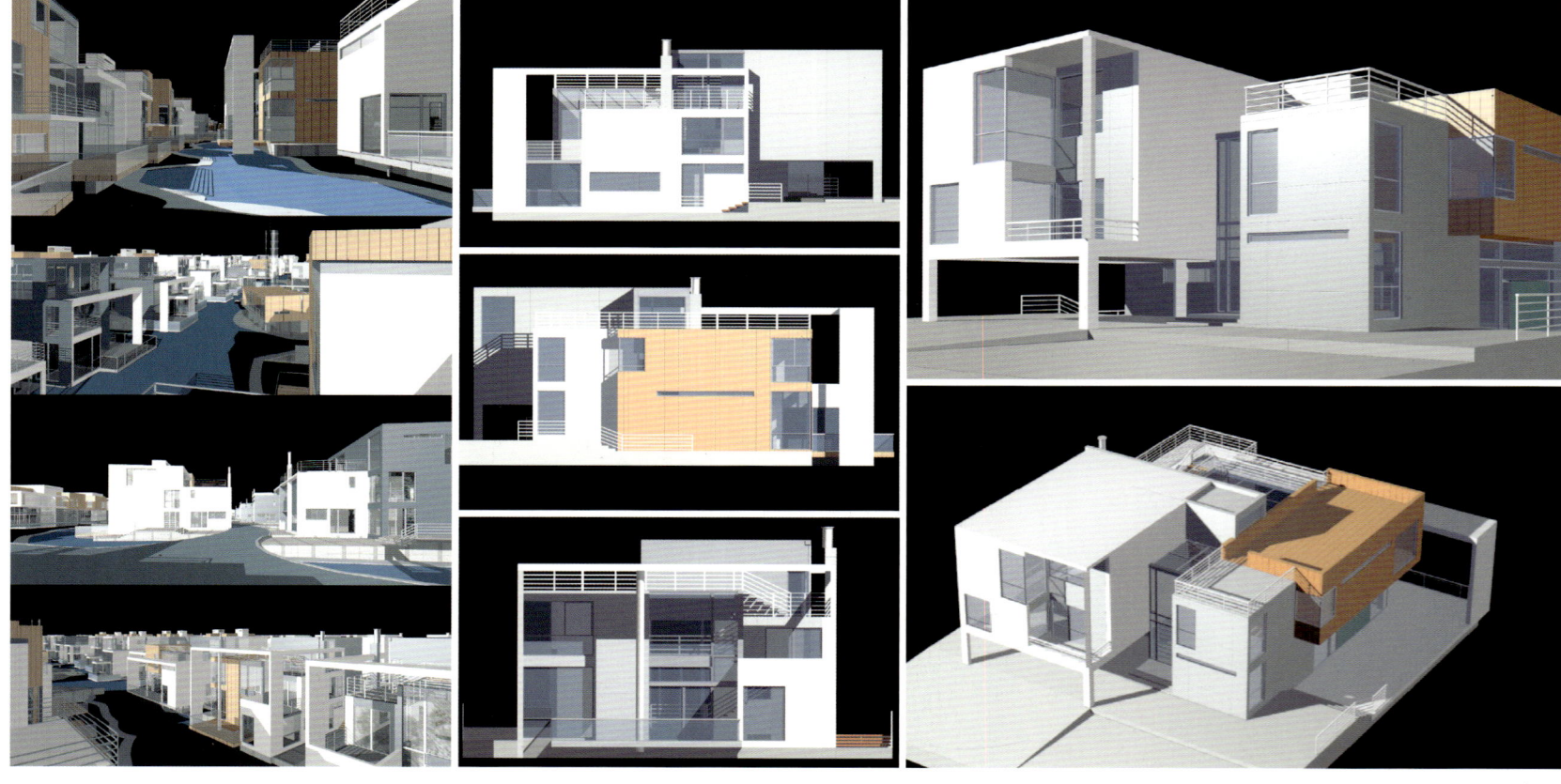

Tian'an Golf Sea View Garden•Shenzhen
天安高尔夫海景花园·深圳

项目地点：中国，广东，深圳市
项目时间：2004年
设计规模：11.2万 m²
设计阶段：方案设计
项目现状：未建

Project Location: Shenzhen
Project Date: 2004
Project Scale: 112,000 m²
Design Phase: conceptual design
Project Status: construction not started

Although we were not granted the design rights in the end, however during the process we were given the opportunity to exercise a variety of solutions.

虽然最终没能得到项目设计权，但这个过程让我们为这块用地尝试了各种不同的解决方案。

方案一

立面图

项目总览

天安高尔夫海景花园三期项目位于天安数码城内，处于其Ⅰ期、Ⅱ期之间，交通方便，景观资源极其优越。东向拥有与高尔夫最大的接触面和最佳观景角度。其南、北向视线开阔，南向高处可看海景。

本项目试图在以下三者之间寻求最佳的结合点：
1. 最大化地享受基地拥有的景观资源；
2. 在城市区域空间中扮演积极的角色；
3. 提供满足现代人居生活需求的空间。

以立体化人居空间的空中联排别墅单元为基本构成模块，以景观导向排列组合，创造具有度假氛围的城市休闲住宅，是最初的方案，也是比较激进的方案，最终因其风险较大而放弃。

Project Overview

With accessible transportation and rich landscape resources, Tian'an Golf Sea View Garden (phase III), located in the Tian'an Cyber Park, is between phase I and phase II buildings of the same project. Eastward it has an optimum angle for a panoramic view of the golf course, northward and southward it boasts an open view. From a high point toward south, one can appreciate the seascape.

This project aims to achieve an optimum combination of the following three aspects:
1. Maximize the gratification of the the site's landscape resources;
2. Playing a positive role in the regional communities of the city;
3. Providing spaces to meet the living requirements of modern residence.

The initial yet radical scheme aimed to create urban leisure housing with an atmosphere of holiday resort, with a landscape-oriented layout and taking the aerial town house units featuring a three-dimensional living space as the basic structure modules, however it was aborted due of its huge risk.

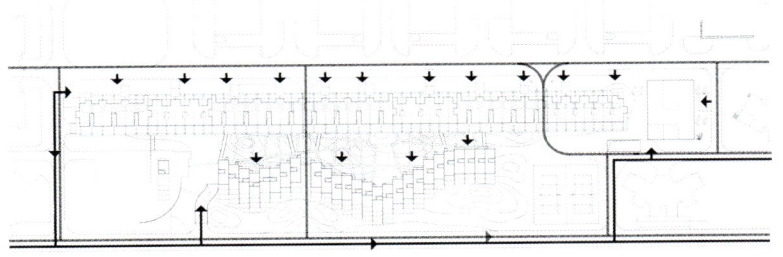

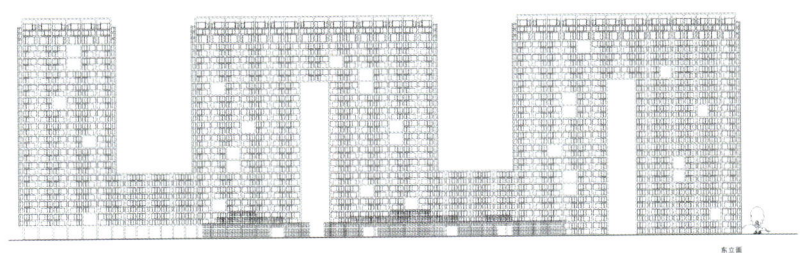

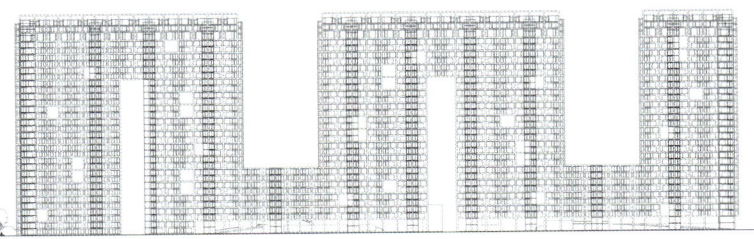

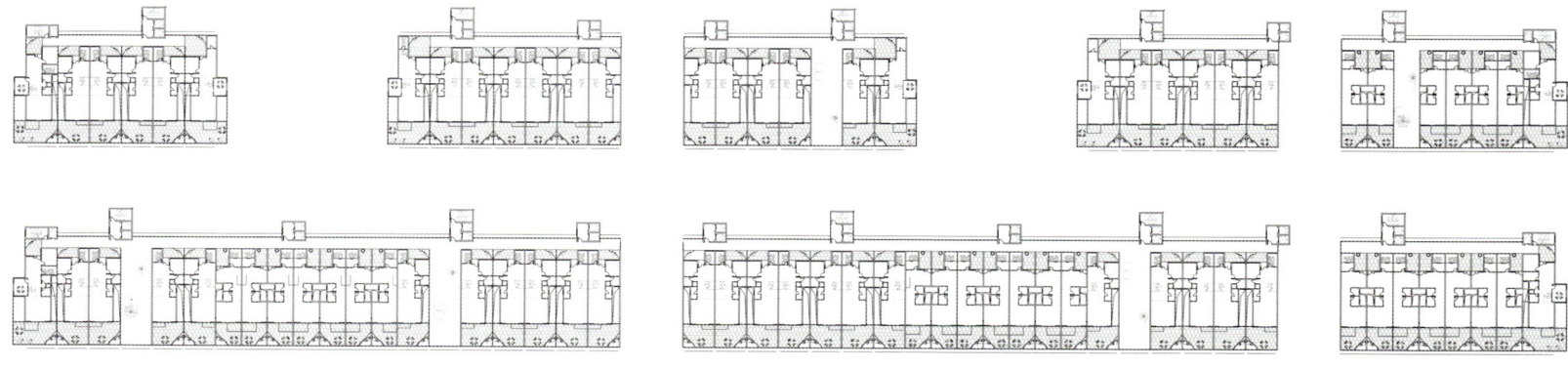

平面图

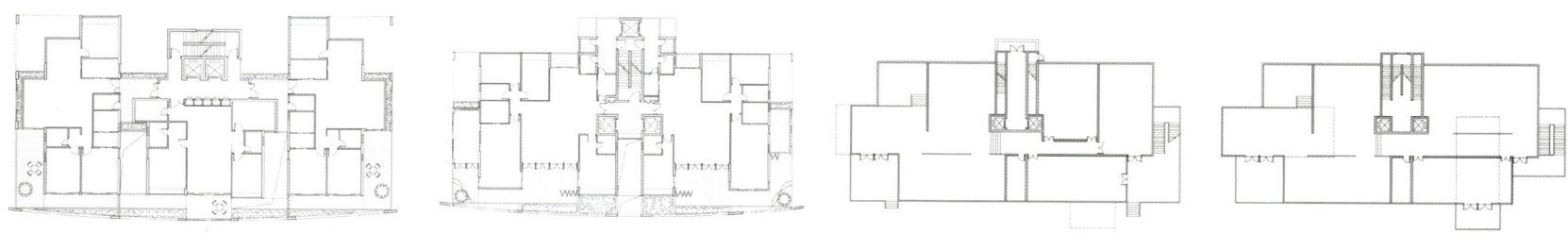

平面图

项目分析

传统建筑空间的灵活分隔给我们以启示，界面的可变性使内外沟通，浑然一体。以一梯二户的大面宽花园合院住宅单元为主要构成模块，综合考虑景观、城市空间、朝向等因素，形成板式高层的点式布局，这也是相对比较中庸的选择。

Project Analysis

The flexible division of spaces in traditional buildings inspires us to make full use of variability of the interfaces so as to obtain a smooth transition between the interior and exterior in a coherent way. It is a middle-of-the-road scheme to form a point-type layout of slab-type high-rise buildings, with units having two households on the same floor and an encircled large garden courtyard as the main modules. And the scheme is versatile as it integrates elements such as landscape, urban space, and orientation.

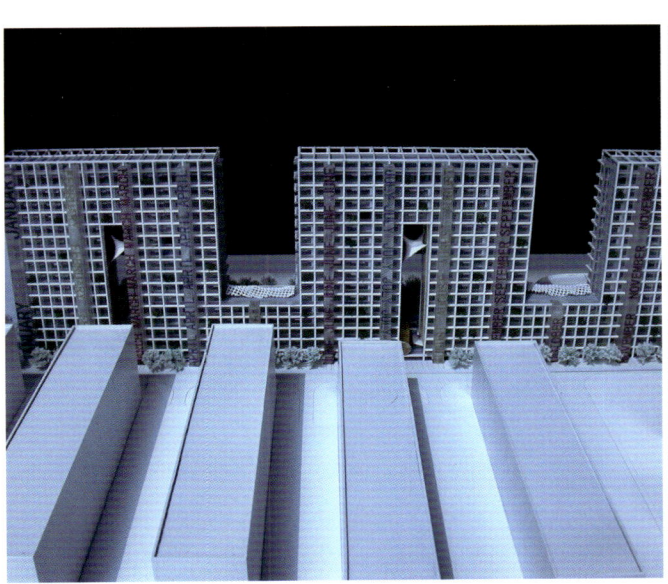
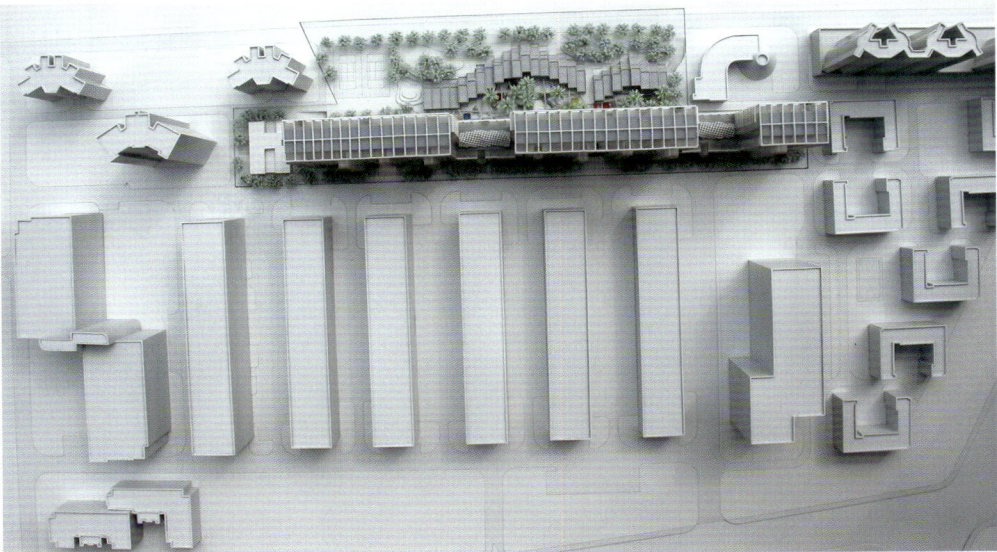

建筑设计

1. 中国民居以"合院"为显著特征。其特点为内向性强、安全、亲切、拥有公共活动空间。"合院"可以连续形成院落、街巷。这种特征和文化内涵,完全可以在现代住宅设计中通过多种设计手法得以延续。

2. 同时,这也是花园住宅的构思。现代住宅中越来越多的人们希望摆脱窗、墙的禁锢,向往更自然的生活。

200m² 以上的户型采用一梯二户合院式花园住宅,将不同功能通过院子结合在一起。而朝向院子的"墙"采用玻璃和百叶,可开启、闭合、遮挡视线、透风,形成极为灵活的内外交融空间,并在两端设置超值的两层高露台空间,与超大合院连成一体,给空中的高层住宅提供"坚实"的地面庭院。另外几个超高层的住宅方案,也是这一设计思想的延续。

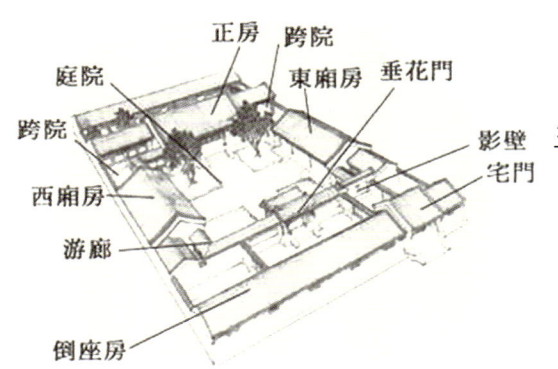

Building Design

1. A Chinese folk house is distinguished by an "encircled courtyard", featuring strong privacy, safety, geniality, and common space for activities. The "encircled courtyard" concept may be further expressed by forms of connective yards, streets and alleys. These characteristics and cultural connotation can be fully embodied in modern housing design by a variety of design measures.

2. Meanwhile, this is the concept for garden housings. In modern times, more and more people hope to break away from the strangle holds by windows and walls, and enjoy a natural lifestyle.

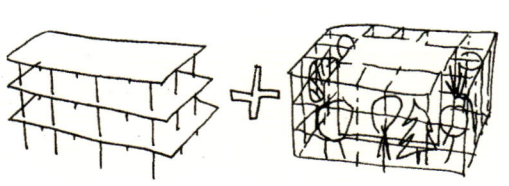

For those apartments each with more than 200 square meters, there are only two households (apartments) on each floor in the same unit, and combining different functions with the courtyard. The "walls" facing toward the courtyard are made of glass and shutters that convenient to open and close, serving as barriers and allowing ventilation so as to create a highly flexible space that integrates the interior and exterior. Meanwhile, the aerial apartments are provided with "solid" ground courtyards by setting up a two-story high terrace of great value at both sides which naturally connects with the over-sized courtyard. This concept of design is also integrated in the design of the other high-rise buildings.

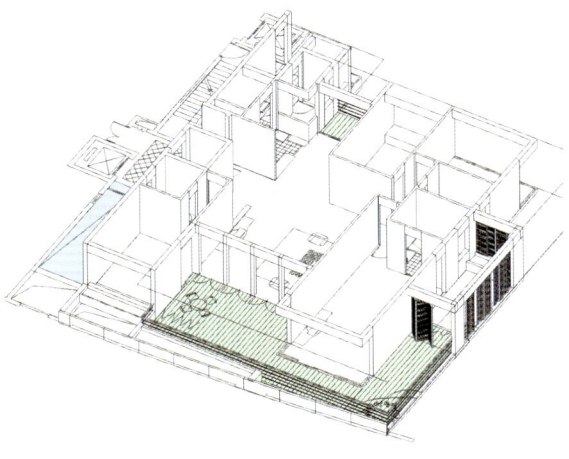
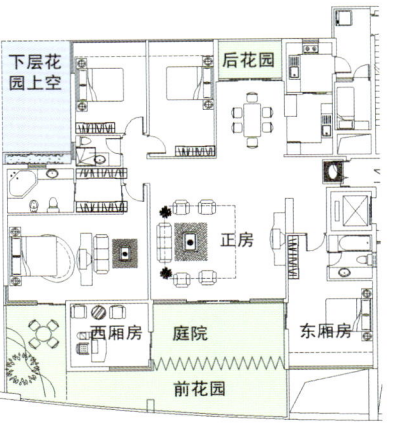

平面图

拥有两层高花园的住宅

所有户型除面向高尔夫场设置的景观阳台外，均设有超大6m高空花园。功能餐厨区、起居区、卧室区，分区明确。独有的入户双流线设计使主、辅功能分流。每户都拥有两个以上的景观面和穿堂风道，主卧保证两个不同位置的窗，关门时能保证空气对流，景观多向。900mm高窗台及窗外花池设计，既绿化建筑，同时也有效隔绝了上层对下层花园的干扰。

Residences with a two-storey-high garden

In addition to a sightseeing balcony facing a golf course, all flats each have a super-large 6m-high aerial garden. Their dinner area, living room, and bedroom are divided definitely. Their main functions are separated from auxiliary functions due to unique entrance double-line design. Each house has at least two sightseeing sides as well as a draught passage. The master bedroom has windows at different locations and thus air convection is guaranteed even though the door is closed and landscapes can be seen from multiple directions. 900mm-high windowsill and flower pool outside the window brings greenness to the building and also effectively prevent the disturbance to gardens of lower floor.

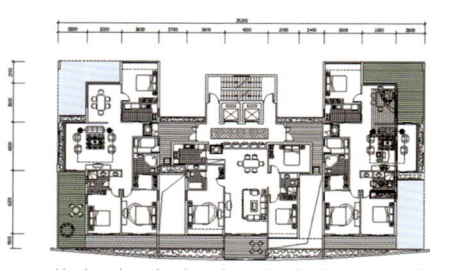
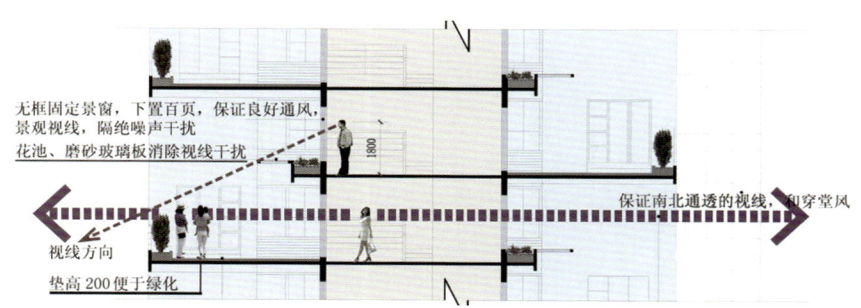

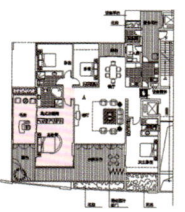
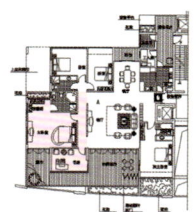

灵活的内部空间设计

灵活的分隔，以及功能的复合可变是传统建筑空间的又一特点。活动的百叶围护出亦内亦外的阳台空间，同时内部空间多样组合，可根据居住人口、功能的变化调整格局。

Flexible internal space design

Flexible partition and functional combination & changeability are another feature of traditional architectural space. Movable shutters create a large balcony space. Internal space may be assembled in a diversified way according to the number of resident and functional changes.

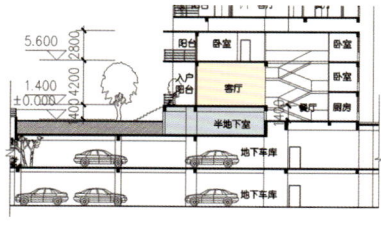

底层联排住宅

底层联排复式住宅利用架空空间设置内部错层空间，打破传统复式空间中明星"层"的界限。
根据不同空间需要设计不同层高，4.2m的客厅尺度更加宜人，结合地下室的顶板标高设计，附送私家半地下室多功能空间。

Townhouses on the ground floor

Townhouses on the ground floor have an internal split-level space, breaking the "level" boundary of traditional compound space.
Different storey heights are designed according to different needs for space. A 4.2m-high living room is more comfortable. In combination with basement roofing height design, a private semi-basement multifunctional space is offered free of charge.

在高层住宅的底层架空层里布置的是拥有私家花园的连排住宅。

② 会所
会所在南端塔楼底部架空层设置，保持塔楼落地的挺拔感，与现有会所靠近，功能上可以有很好的衔接。

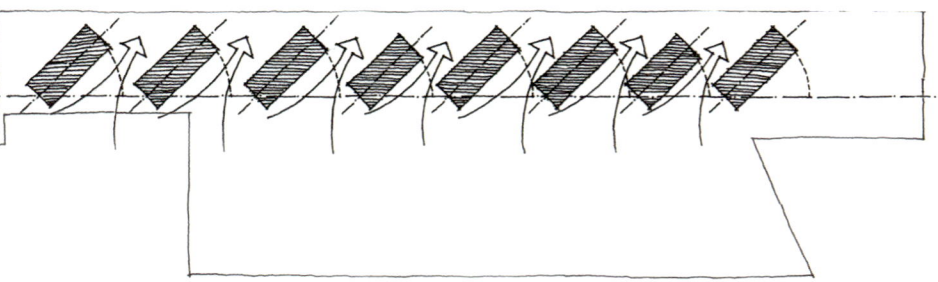

1 旋转，对城市空间的开放，对日照要求的平衡

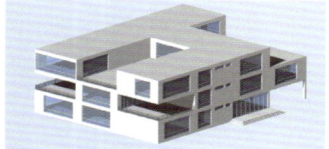

③ 幼儿园
幼儿园设在基地北端，高尔夫观景视线最差的地方，相对安静、独立。

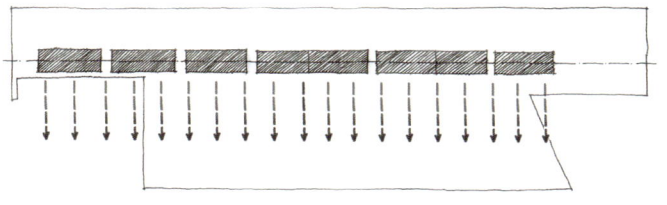

2 最佳景观取向的布局

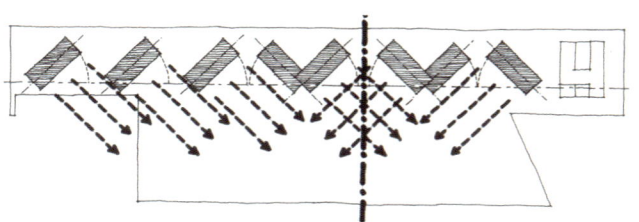

3 高度与方向调整，呼应场地及环境，尽量利用资源

■ 构思
在直接面向高尔夫板式布局的基础上，通过分离、平移、错动，形成了通透，用足景观资源，增加南北朝向的总体布局。

1 以最佳景观取向布局；

3 平移，更通透。更多层次个性空间和公共空间；

2 分离，形成通透的城市空间和天际线；

4 错动。
城市更通透，拥有多方向的视线通道。
令东西向的住宅单位却拥有了南北朝向，
景观面更长，且不产生对视。

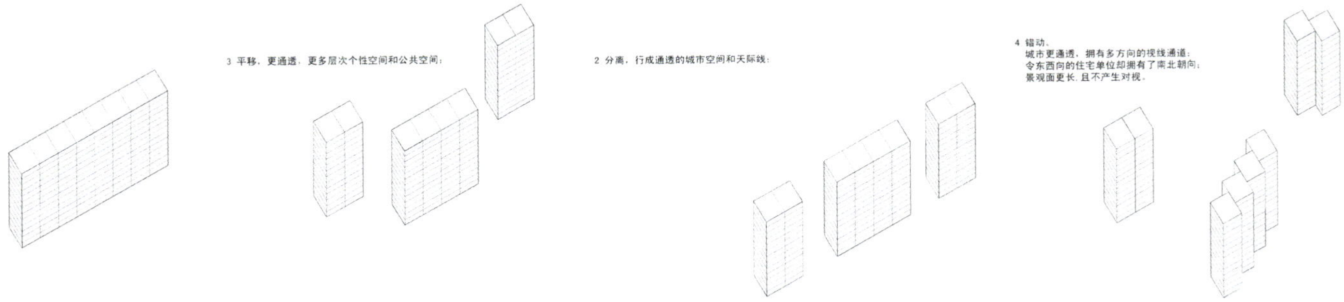

总平面

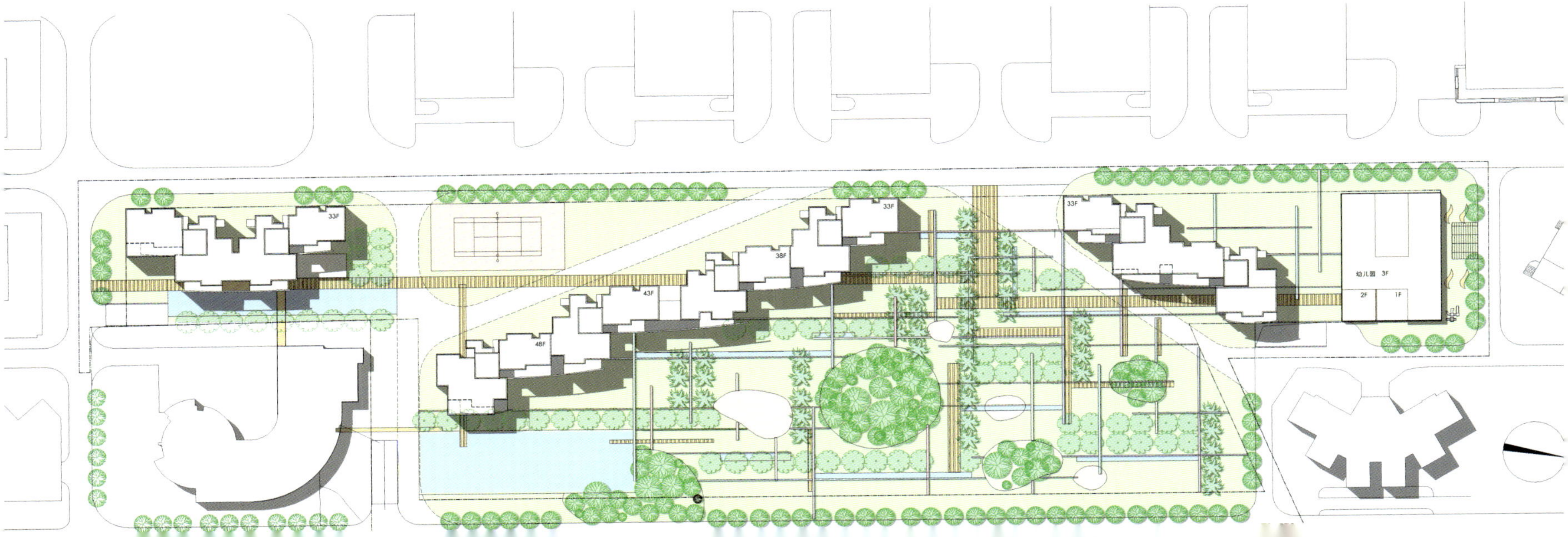

方案三

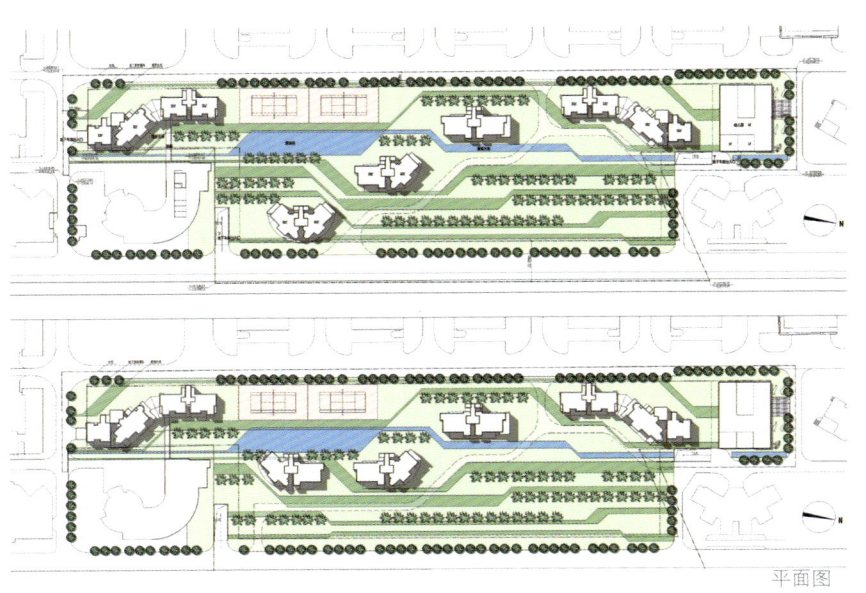

平面图

Liantai Dameisha Villa•Shenzhen
联泰大梅沙别墅·深圳

项目地点：中国，广东，深圳市
项目时间：2005年
设计规模：2.2万 m²
设计阶段：规划设计，方案设计
项目现状：在建

Project Location: Shenzhen
Project Date: 2005
Project Scale: 22,000 m²
Design Phase: layout planning, conceptual design
Project Status: construction in progress

Seascapes, seascapes and more seascapes! One building, but is separated by two groups of space – interior and exterior.

海景、海景、还是海景。一个建筑，内外两组空间。

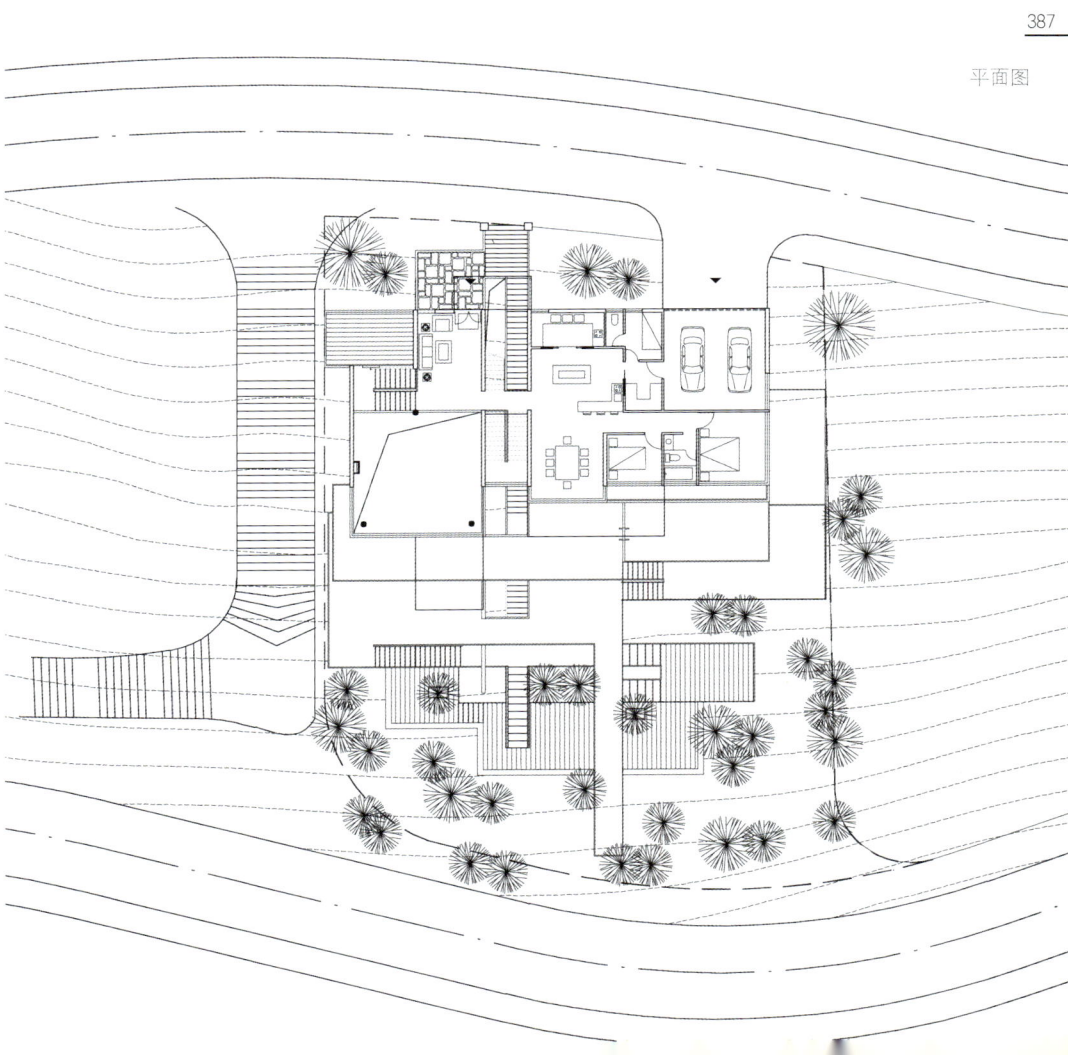

平面图

项目分析

本项目以最大限度利用海景资源为设计目的，功能分区依据海景分布。平面上，将所有厅房空间均面向海景面展开，服务空间向内；竖向上，结合景观种植及场地落差创造多层次的海景。

Project Analysis

The design of this project aims to make full use of the seascape resources on which the division of functional areas is based. Transversely, all living rooms and bedrooms are designed to face the seascape yet the service area is located in the inner part; vertically, the seascape that can be viewed at multiple levels are created by combining landscaping with height difference of the levels.

建筑设计采用了"菱形法则"。平面上,在保证退距的基础上,以菱形的方式将可视海景面宽最大化;立面上,有效地利用场地落差,上下层逐渐缩小面宽,同时相互错开,避免干扰。

The "diamond rule" is used in the building design. Transversely, on the basis of ensuring a setback from the road, the width of the facade facing seascape is maximized by means of a diamond building shape; vertically, height difference between the levels is efficiently utilized, the width of the facades of the connecting stories are gradually narrowed and the elevations are staggered to avoid interference.

别墅的上部可见区域为主要功能部分,暴露在树丛中,吸取阳光。下部则隐蔽于树丛里,参差不齐的柱子就像树根深深埋进土中,叠落的木栈道也使人更贴近自然。

The visible upper part of the villa is the main functional area above a group of trees and exposed to sunshine. The lower part hides in the shady branches; the irregular pillars look like roots buried deep in soil, the downward wooden corridor make peoples feel much closer to the nature.

整个建筑站立于斜坡上，尽量不改变现存的自然地平高度。垂直于等高线布置的这些横墙可使树木穿越与之相平行的平台，同时确定了别墅的边界。户外的楼梯使各个不同标高之间的平台形成一个相互连接的空间，将环境围绕、紧密结合，使建筑和环境融为一体。

The whole building is erected on a slope, basically making no change to the existing natural elevation. The partition walls vertical to the contour line can make trees pass over the platform parallel to them; meanwhile they fix the boundary of the villa. The outdoor staircases make the platforms with different elevations shape a space in which they are connected each other, coherently integrating the building with the surroundings.

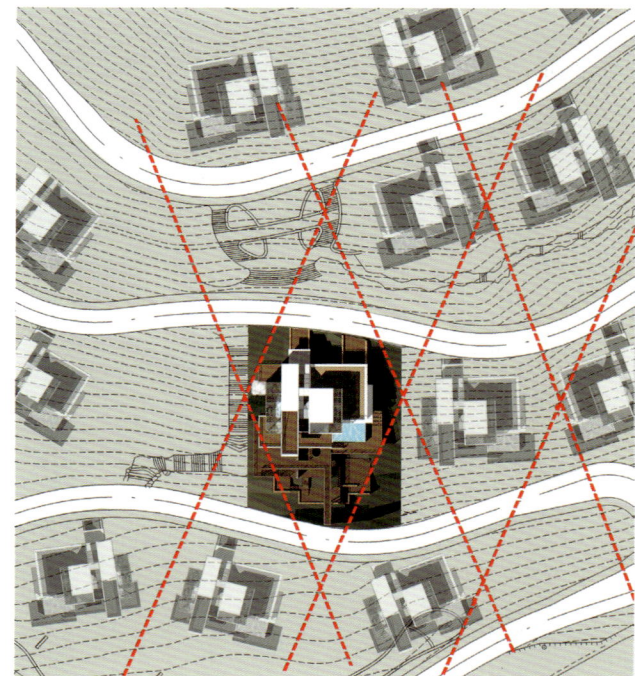

平面图

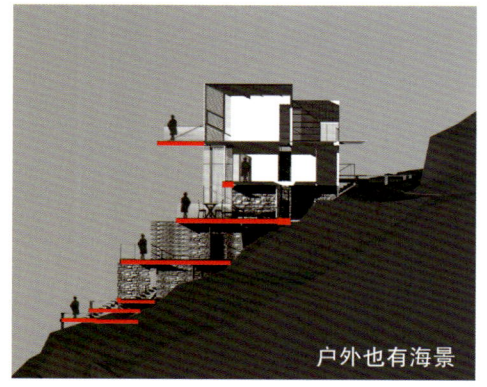

户外也有海景

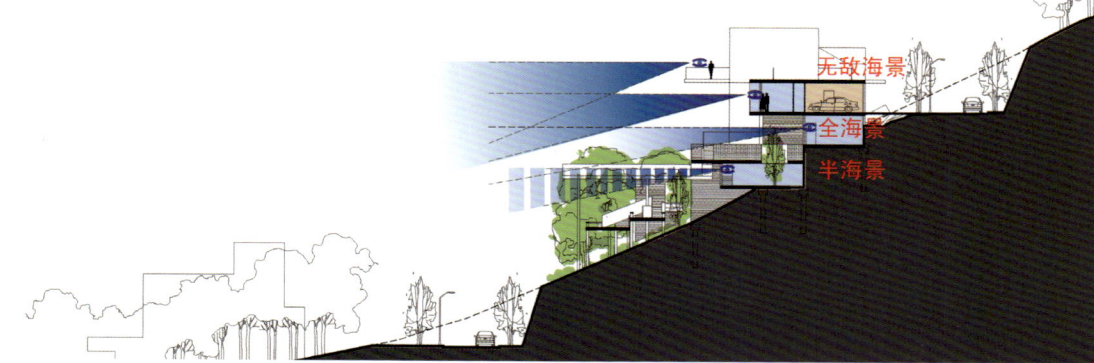

无敌海景
全海景
半海景

立面图

为海景而设计

A、采取大而宽，展开式的设计。建筑首先要让自己拥有最大限度的海景资源，身处其间，让人有美的享受。
B、建筑的功能分区依据海景分布。平面上，将所有厅房空间均面向海景面展开，侍服空间向内；竖向上，结合景观种植及场地落差创造多层次的海景。
C、结合每层空间，户外平面同样拥有海景，使人更能体验海的魅力。
D、"菱形法则"——不要挡住别人视线的解决方案。平面上，在保证退距的基础上，以菱形的方式将可视海景面宽最大化。立面上，有效地利用场地落差，上下层逐渐缩小面宽，同时相互错开，避免干扰。

Design for seascapes

A. Large, wide, and expandable: the buildings share maximum seascape resources. Staying in the buildings, you will have a kind of aesthetic enjoyment.
B. Architectural functions are distributed according to location of seascapes. All spaces spread out horizontally toward the direction of seascape. Multiple levels of seascapes are created vertically by means of landscape planting and site height difference.
C. Outdoor planes also share seascapes. Thus people can experience more charms of the sea.
D. "Rhombic rule" – a solution for unblocking other's view. Horizontally, a rhombic manner may maximize visible seascapes. Vertically, site height difference is utilized to narrow down the width of upper and lower floors and stagger them to avoid disturbance.

有效的利用地形——横墙+平台

地形、植被和周围景观的特色使我们产生了建筑由横墙承重的构想，整个建筑站立于斜坡上，而尽量不改变现存的自然地平高度。垂直于等高线布置的这些横墙和与之相平行的平台可使树木穿越，同时也确定了别墅的边界。户外的楼梯使各个不同标高之间的平台形成一个相互连接的空间，将环境环绕、紧密结合。使建筑和环境融为一体。

平台　　户外的看台，顺着山势向下叠落，加强了与山体的咬合。
横墙　　保持基地地貌，不破坏植被，插入山体，起承重作用。
植被　　利用原有植被，使其与建筑相互交织，相互渗透，相互辉映，建筑并不是孤立的，它也是山体的一部分。
泳池　　别墅中的水是稀缺的，边际化的泳池使其与海面连为一体，几层叠泉意向性地将水汇入大海，好似连成一体，海天一色。

Utilizing landform effectively – horizontal wall + platform

Landform, vegetation and surrounding landscape features make us decide that the horizontal wall bears the load of the building. The whole building stands on the slope and existing ground attitude is not changed. Horizontal walls, which are vertical to the contour line, and platforms, which are distributed in parallel with horizontal walls, make trees cross them and also define the boundary of villas. Outdoor stairs make platforms with different elevations form an interconnected space and encircle the environment. As a result, the buildings become an integral part of the environment.

Platform: outdoor platform, which goes down along the hill and is closely linked with the hill.
Horizontal wall: a wall bearing the load of buildings, which does not damage landform and vegetation or goes into the hill.
Vegetation: former vegetation blends with, permeate, and reflect buildings. Buildings are not isolated but an integral part of the hill.
Swimming pool: water is rare in villas. The marginal swimming pool integrates with the sea. Several layers of spring water flow into the sea. As if they are merged.

Dongguan 0769 • Dongguan
东莞0769 • 东莞

项目地点：中国，广西，东莞市
项目时间：2006年
设计规模：11万 m²
设计阶段：方案设计
项目现状：已建

Project Location: Dongguan
Project Date: 2006
Project Scale: 110,000 m²
Design Phase: conceptual design
Project Status: construction completed

The design hopes to connect the existing urban space, and bring a sense of freshness to the community at the same time.

设计希望在承接已建成区域的同时，给人带来一些珍贵的新鲜感。

项目总览

本案是0769小区的最后一期开发项目，我们力求在空间上与前期方案相呼应，同时有效地提升整个大社区的空间品质。

Project Overview

This project is the last phase of 0769 estate project; therefore, we aim to make it spatially compatible with the buildings of the previous phases and to improve the spatial quality of the whole community efficiently.

总平面

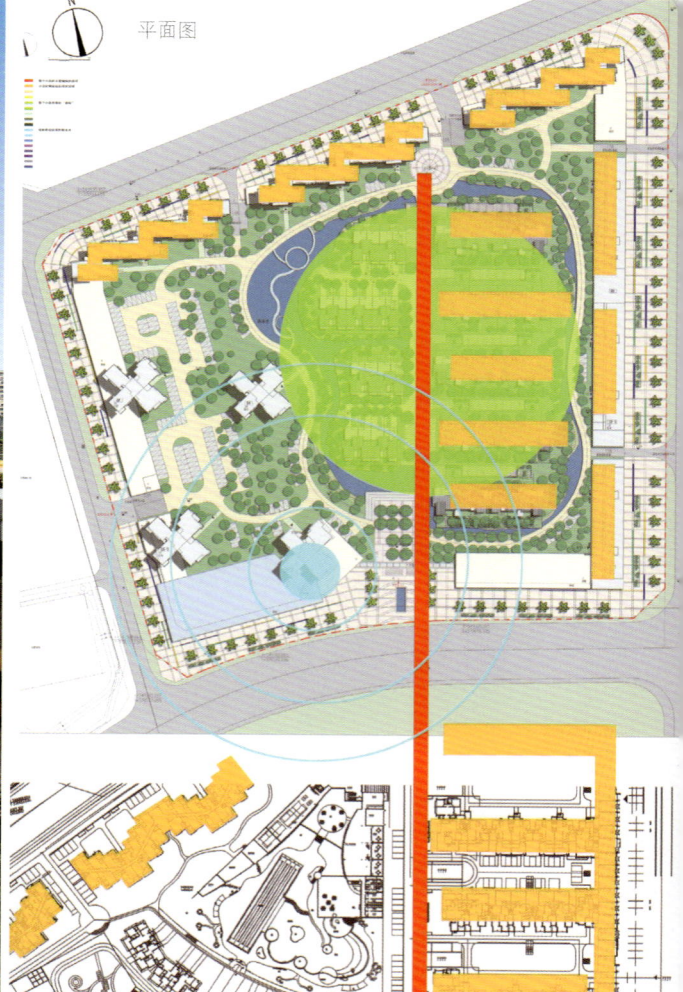

平面图

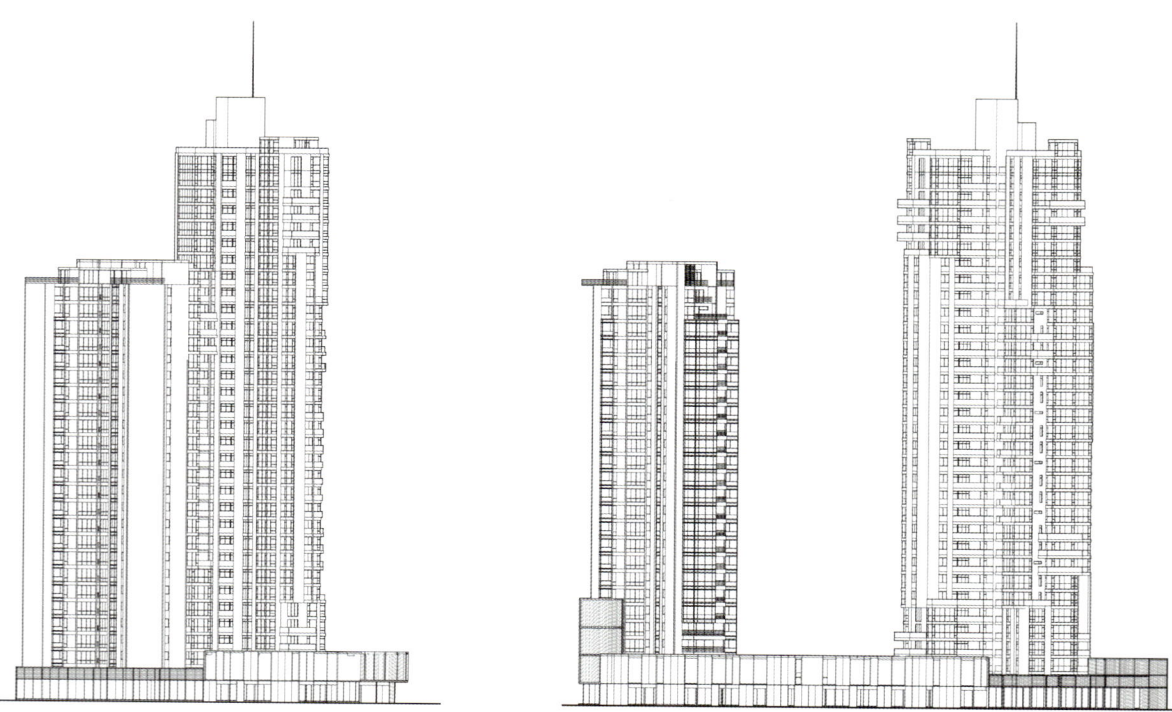

立面图 立面图

项目分析

1. 创造最大化的花园并形成相对独立的高品质社区，在原规划主轴上修改制高点来优化完善整个社区空间。

2. 打破原来单一的中低档户型模式，规划多样的产品形式，在高层半围合大庭院中布置高端产品入联排别墅、情景洋房，提升社区的整体居住品质。在东侧临街设置时尚复式公寓产品，形成有强烈标识性的城市界面。

3. 高层建筑以点、线形态布局，尽量保证住宅视野开阔，减少对视。点式建筑间距的最大化，线型建筑的边缘化，让布局显得宽松的同时营造出活泼动感的空间。

Project Analysis

1. To maximize the area of the garden and build a relatively independent high quality community; also to optimize and perfect the whole community space by changing the elevation of the highest point on the principle axis of the original planning.

2. To get rid of the low and medium monotonous flat layouts and develop buildings with various types of flat layouts; specifically, improve the residential quality for the whole community by constructing high-end buildings such as townhouses and low-rise stylish apartments in the large courtyard half enclosed by the high-rise apartments. Along the street in the east side, fashionable duplex apartments are built as a landmark to connect to the urban neighborhood.

3. To design point and linear layouts for high-rise buildings, maximize the filed of vision for occupants in the buildings and reduce the possibility that occupants in different flats can see each other. The maximization of spacing between point type buildings and marginal position of line type buildings ensure uncrowded layout while creating lively and dynamic spaces.

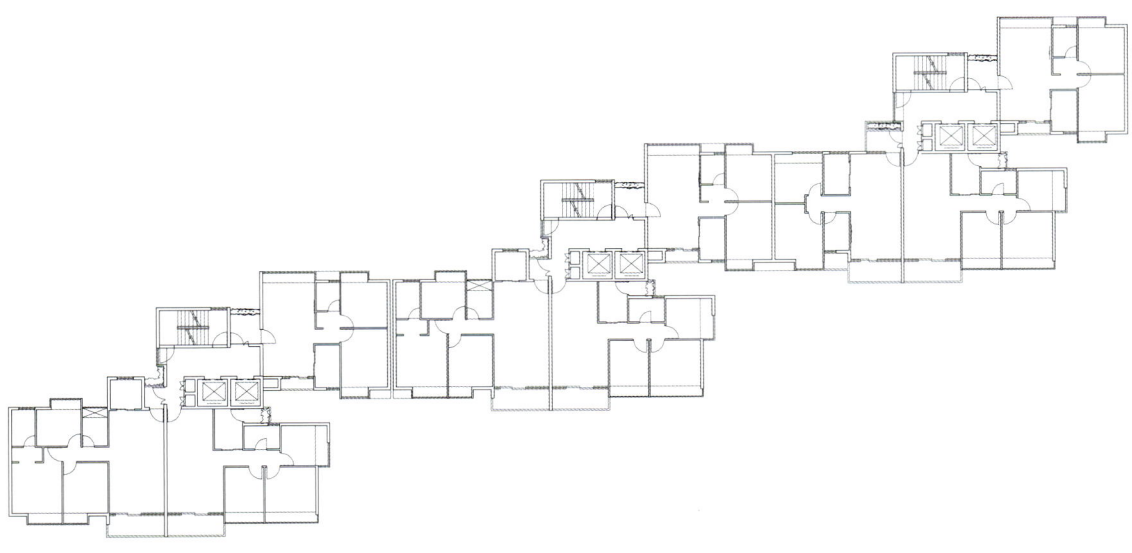

平面图

Futong in Bao'an 96 District (Phase IV, 2006) • Shenzhen
宝安96区富通4期·深圳

项目地点：中国，广东，深圳市
项目时间：2006年
设计规模：3万m²
设计阶段：方案设计
项目现状：未建

Project Location: Shenzhen
Project Date: 2006
Project Scale: 30,000 m²
Design Phase: conceptual design
Project Status: construction not started

We applied a number of methods aiming at bringing freshness and energy to this gloomy and monotonous area.

我们运用了多种的方式，希望为这一沉闷、单调的区域带来新鲜的活力。

总平面

项目总览

通过对这个地块的分析以及对周边环境的评估，我们认为本项目应具备生动、丰富以及另类的形象，以成为此区域的标志性建筑；居住形态上，我们将其定位为城市中生态、高尚、现代化风情的居住区，充分体现人性化的居住思想，坚持以人为本，创造人与环境和谐发展的空间，为业主创造良好的生活与休闲环境。我们力求在空间上做到与周围片区相互呼应的同时拥有自身生动、另类的形象，有效提升整个区域的空间品质。

Project Overview

After analysis of this plot and evaluation of the neighboring environment, we suggest that this estate should become a regional landmark presenting an ecological, noble and stylish image. This estate is designed to be a humanistic one, being people oriented, and providing an excellent living and leisure environment for the occupants through creating a harmonious relationship between people and environment. We endeavor to make this estate to be spatially compatible with the surrounding estates, and to improve the spatial qualities of the whole region by means of the lively and exceptional image of the estate.

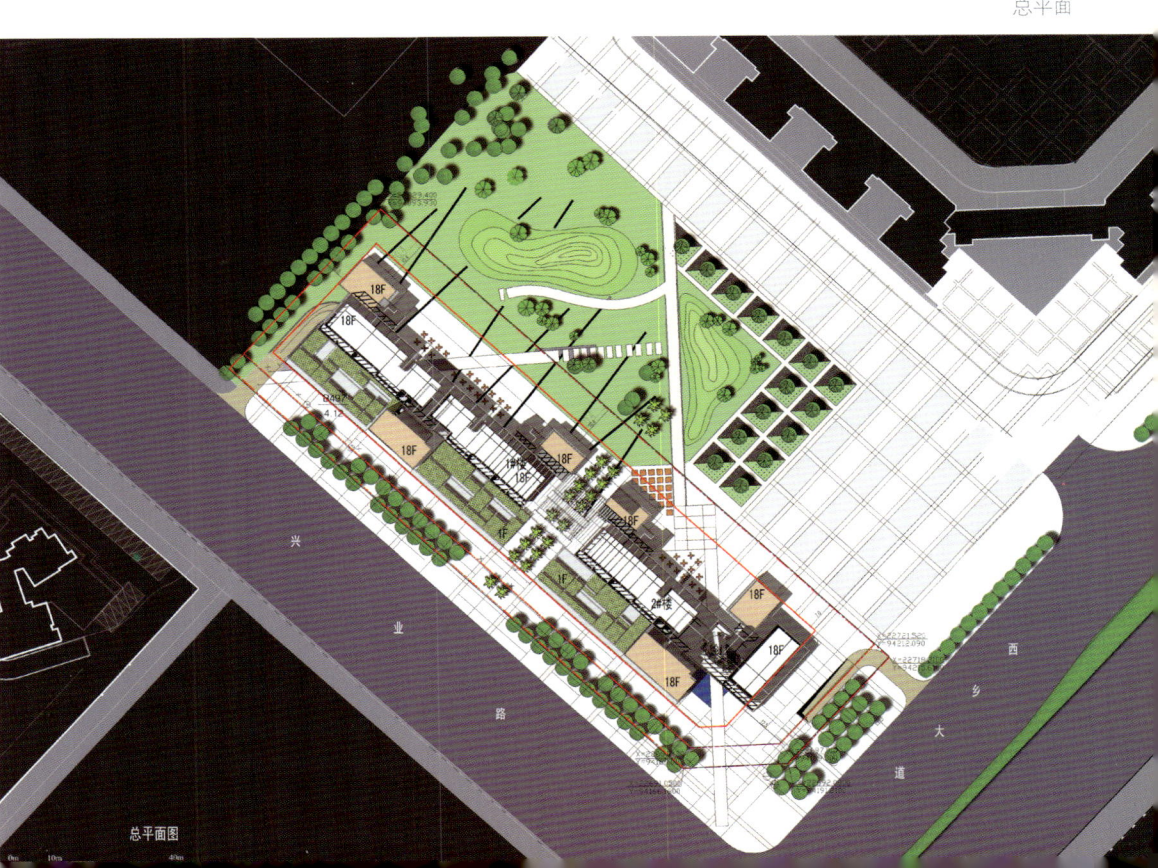

总平面图

总体布局

建筑将用地分为前后两个界面，内静外动。临街面构建了集中的商业风情街，在用地中部靠北侧构建了商业小广场，以呼应西北向一街之隔的富城三期商业区。用地东侧为相对较安静的部分，建筑围合出一个小区的内部庭院，形成半私密的小区公共活动空间。

在建筑设计上，我们前后通过多方案的尝试和比较，希望展示建筑在这一区域所承担的"明星"角色。

Overall Layout

The buildings are partitions of the plot to create static inside and dynamic outside. Commercial stylish streets are built outside the buildings; a small central commercial square is built at the north part, which is near to the Fucheng Phase III commercial area on the opposite side of a street in the northwest. At the east part, there is a relatively quiet area where a courtyard is enclosed by the buildings, being a semi-private activity space for the occupants.

After many tries and comparisons, we choose this architectural design in a hope that this estate can be a "star" one in this region.

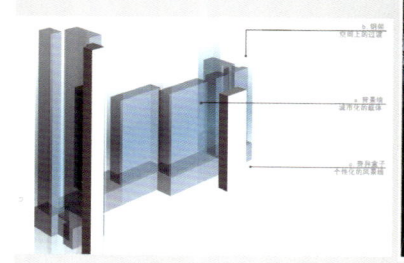

立面造型

点板结合,形成强烈对比,
钢架嵌入实体，飘动的盒子被母体吸引。

标准层平面

Wei Zhen Fu • Shenzhen

唯珍府 · 深圳

项目地点：中国，广东，深圳市
项目时间：2006年
设计规模：2.4万 m²
设计阶段：方案设计
项目现状：已建

Project Location: Shenzhen
Project Date: 2006
Project Scale: 24,000 m²
Design Phase: conceptual design
Project Status: construction completed

The architects use simple design elements to pursue intricate treatment of details, thus creating a high-end residential quarter for people.

设计师运用简约的设计元素，寻求细节上的精确处理，为人们打造高档住宅区。

项目总览

本项目位于深圳市福田区景田北路。用地西南侧隔景田北七街毗邻福田外国语学校，西北侧为片区内的开放式公园，东北侧用地为福田区网球中心，西南方向远眺香蜜湖景区。用地方正平整，景观开阔，环境幽静。

Project Overview

This project is located at Jing Tian Bei Road in Fu Tian District of Shenzhen. In the southwest of the plot is Shenzhen Futian Foreign Language School on the opposite side of Jing Tian Bei No. 7 street; in the northwest is a public park in this area; in the northeast is the Tennis Center of Futian District; from here one can see the Xiangmi Lake Scenic Spot somewhat far in the southwest. The plot is square and level, with a wide view of landscape and a peaceful and secluded environment.

十七层平面图

项目分析

从用地周围环境和居住者视觉、感觉和心理角度出发，在整体布局上充分利用环境优势，使建筑与环境相得益彰，营造出完美、和谐的居住氛围。

1. 用地大体上分两个区域，沿景田北街形成外向的商业街，以及18层的高层住宅与低层联排住宅。

Project Analysis

The surroundings and the visual, sensual and psychological perceptions of the residents are fully considered in the design. As for the overall layout, the environmental advantages are made full use of so as to make the buildings match the surroundings, creating a perfect and harmonious living atmosphere.

1. The plot is roughly divided into two areas; one is the commercial street facing Jing Tian Bei Street, another is the area where the 18-stotey high-rise buildings and low-rise townhouses are built.

2. 高层住宅和联排住宅之间为小区中心绿化带,构建了亲水特征的景观水体为住户提供了一处近距离接触自然的场所。同时,中心景观区将高层住宅与联排住宅、商业区自然分开,避免相互干扰,使住户拥有良好的生活环境。

3. 低层联排住宅用地于西南、西北方向沿景田北七街和景田北五街呈"L"型布置。不仅可充分利用景观和日照优势,还能在沿街面上形成亲切尺度。

4. 板式高层住宅布置在基地东北方向邻近福田网球中心的一侧。在东北、西北以及西南三个方向上都拥有开敞的视野和优美的景观,并可俯瞰小区中心绿化带。

2. A central green belt of the estate is built in the area between the high-rise buildings and townhouses, where there is a landscape water body, providing a place for the residents to be closer to nature. Meanwhile, the central landscape area naturally divides the high-rise buildings, townhouses, and commercial area to prevent mutual interference, allowing the residents to enjoy a good living environment.

3. The plot for low-rise townhouses is of L shape, adjacent to Jing Tian Bei No.7 Street in the southwest and to Jing Tian Bei No. 5 Street in the northwest. The design can not only take full advantage of the landscape and sunlight, but also create an amiable impression on the scale that seems to be reduced along the street frontages.

4. The slab-type high-rise buildings are located at the site's northeast part near the Tennis Center of Futian District. The northeast, northwest, and southwest orientations boast best open views and wonderful landscape, and one can, toward these directions, overlook the estate's central green belt.

Overseas Chinese Town Pure Waterfront · Chengdu
华侨城纯水岸·成都

项目地点：中国，四川，成都市
项目时间：2007年
设计规模：15万 m²
设计阶段：方案设计
项目现状：已建

Project Location: Chengdu
Project Date: 2007
Project Scale: 150,000 m²
Design Phase: conceptual design
Project Status: construction completed

We face two challenges in this project: one is to offer diversified types of residential buildings which can complement each other well; another one is how to make the buildings present a definite and clear image under the dim and foggy weather conditions.

这一项目让我们面对两大挑战：其一是提供多样类型的住宅，使其相互之间产生有益的作用；其二是在朦胧多雾的气候条件下，如何让建筑保持肯定和清晰的姿态。

总平面

总体构思及布局特点

整个小区延续了总体规划的肌理,建筑形体规划布局以南低北高的形式布置,越往北环,建筑越高。在小区内部创造了最大景观花园,既减少了北环对小区内的噪声影响,又保证了高层间视线距离的最大化。房型大小与景观质量相匹配,使得尽可能多的户型能够看到一辉花园、中央教科所深圳南山附属学校及侨城高尔夫,强调了小区景观的均好性。整体规划设计构思立足于空间美学和设计规划的基础上,深刻剖析居民的心理、生理以及社会要求。

Overall Design Concept and Layout Characteristics

The whole estate carries on the textures of the overall planning. The layout for building structures features high in the north and low in the south, with the northernmost part nearest to Bei Huan Road as the highest one. The internal garden not only helps decrease the noise from Bei Huan Road, but also maximize the length of sight lines between occupants in high-rise buildings. The size of flats matches with the landscape scale so as to ensure occupants in most flats can see Yihui garden, Nanshan School affiliated to China National Institute for Educational Research, and OCT Golf Course, attaching a great importance to ensure that the flats all have good view of landscape. Based on spatial esthetics and design planning, the overall planning is carried out after careful analysis of the psychological, physical, and social needs of the occupants.

424

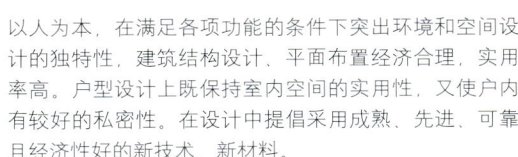

以人为本，在满足各项功能的条件下突出环境和空间设计的独特性，建筑结构设计、平面布置经济合理，实用率高。户型设计上既保持室内空间的实用性，又使户内有较好的私密性。在设计中提倡采用成熟、先进、可靠且经济性好的新技术、新材料。

住宅设计保证户型的高实用率，户户视野开阔，明厨明卫，将终级人性化的关怀带给每一位业主。

Conforming to the principles of people orientation and all functions are fulfilled, the uniqueness in environmental and spatial designs is emphasized. Building structure design and plane layout are economic and rational with a high efficiency rate. In regard to internal layout design of flats, both utility and privacy of the internal spaces are considered. Mature, advanced, reliable, and economic new technologies and materials are introduced in the design.

In the design of the internal layout of flats, it aims to achieve a high efficiency rate of flats, offer the best field of vision for the occupants in the flats, and ensure that each kitchen and each bathroom have windows overlooking the outside environment, providing humanistic care for each owner.

立面图

立面图

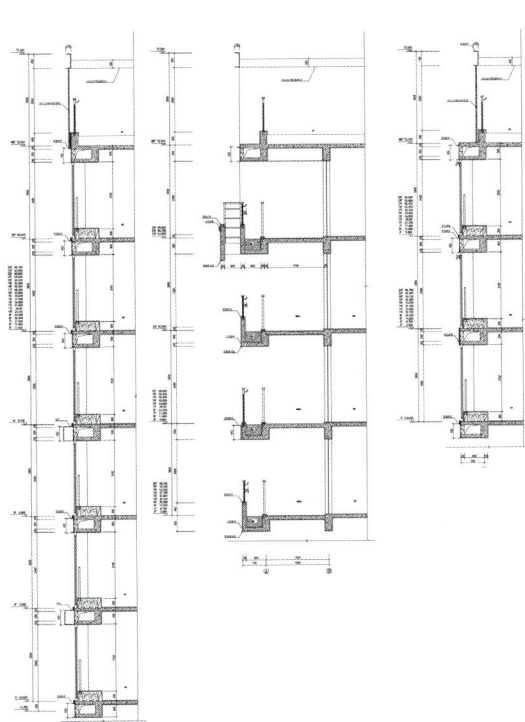
立面图

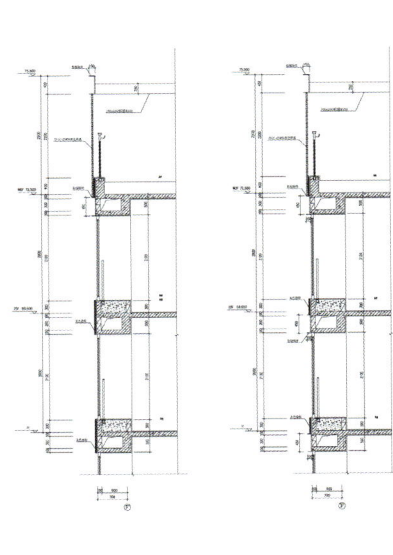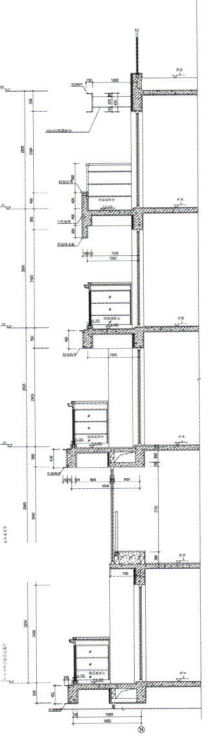
立面图

立面图

Overseas Chinese Town Portofino Pure Waterfront (Phase VII) • Shenzhen
华侨城波托菲诺纯水岸七期·深圳

项目地点：中国，广东，深圳市
项目时间：2007年
设计规模：6.5万m²
设计阶段：方案设计
项目现状：已建

Project Location: Shenzhen
Project Date: 2007
Project Scale: 65,000 m²
Design Phase: conceptual design
Project Status: construction in progress

Compared with previous phases, this phase needs architectural forms with innovation while traditional cultural elements should be included.

与前六期相比，我们需要提供既尊重文脉又有创新的建筑形式。

立面图

总平面

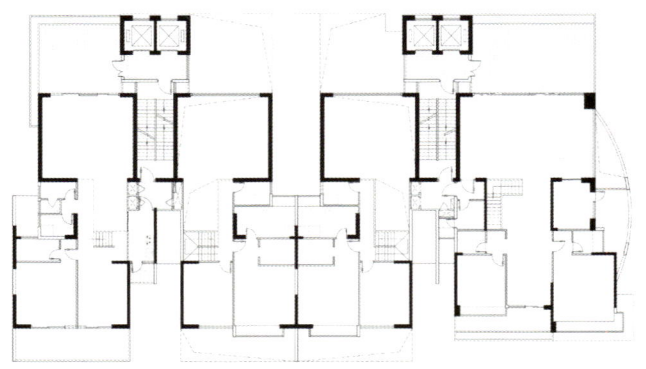

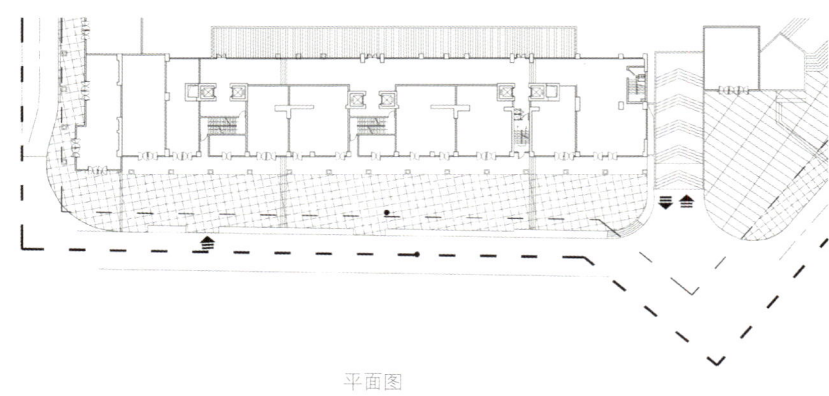

平面图

总体构思及布局特点

小区内部在创造最大景观花园的同时,又保证了高层间视线距离的最大化,使尽可能多的户型看到高尔夫球场。

Overall Design Concept and Layout Characteristics

A very large landscape garden is built in the estate, while the line-of-sight distance between high-rise buildings is maximized so as to ensure most households can enjoy the view of the golf course.

设计说明

我们尝试在高层建筑中引入别墅式空间设计，实现从高层公寓到空中别墅的跃升；同时，户内外及每户"前院"、"后院"的引入，也实现了高层建筑的立体绿化。

建筑在充分尊重前几期建筑文脉的色彩和造型语汇基础上，运用现代的设计手法及建筑材料从另一角度阐述了尊崇闲适的生活情调。

Design Description

The designers try to introduce villa-style space design for high-rise buildings, implying an upgrade from high-rise buildings to penthouses; meanwhile, the introduction of "forecourt" and "backyard" and the division of indoor and outdoor for each household also allows tridimensional greening for the high-rise buildings.

On the basis of matching the color and shape of the buildings of previous phases, modern design methods and construction materials are used to express a noble and leisured life style from another perspective.

Cheng Han Man Cheng (Phase IV) • Shenzhen
承翰慢城四期·深圳

项目地点：中国，广东，深圳市
项目时间：2008年
设计规模：15万 m²
设计阶段：方案设计
项目现状：未建

Project Location: Shenzhen
Project Date: 2008
Project Scale: 15,000 m²
Design Phase: conceptual design
Project Status: construction not started

In this design, we demonstrate the use of common materials through exquisite design to create noble sense of quality.

普通的材质通过精致的设计营造出尊贵的品质。

慢生活态度——对目前生活节奏的批判。在深圳的快节奏中，生活和工作之间需要缓冲，物质和精神缝隙需要留白。我们理应崇尚多元化，拒绝单一的文化生活。

Appreciation of slow life——It is a criticism on rapid pace of modern life in Shenzhen; rest is needed after work, and there should be a compromise between material and spiritual needs. Therefore, a varied, instead of monotonic cultural life, is desirable.

慢城中不乏自然景观，却缺乏人文景观。从无到有，从社区到街区，强调社区的弹性空间。动态居住到静态居住，低层居住与高层居住，人文居住与生态居住，人文精神才是社区的真正精神。

In Man Cheng, there is no lack of natural landscape, but a lack of human landscape. What is emphasized is the flexibility of the living spaces in the community which is developed out of nothing and connects the outside blocks. From static living space to dynamic living space, from low-rise buildings to high-rise buildings, and from humanistic living environment to ecological living environment, humanistic spirit is a central theme and true value of the community.

卫星、交通实景图

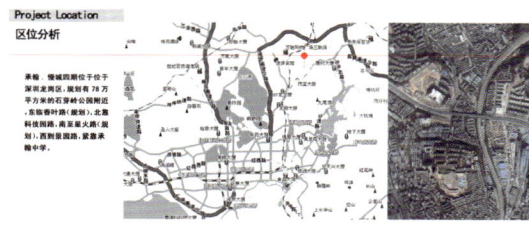

区位分析

承翰慢城四期位于深圳龙岗区，规划于78万m²的石芽岭公园附近，东临香叶路（规划），北靠科技园路，南至星火路（规划），西到景园路，紧靠承翰中学。

Projection Location

Cheng Han Man Cheng (Phase IV) is located in Longgang District, Shenzhen, close to the Shiyaling Park, which covers 780,000 square meters according to the planning. It borders Xiangye Road (planned) to the east, neighbors Kejiyuan Road to the north, and reaches Xinghuo Road (planed) to the south and Jingyuan Road to the west. It is very close to the Cheng Han Middle School.

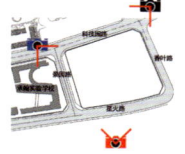

第一阶段：我们的尝试
从本地块出发，我们尝试了多种可能性。可能性的围合方式，南低北高，为了中心花园的最大化，建筑全部围绕则后用地的四周。

草案 优点：楼与楼之间视线错开，中心花园最大化
　　　缺点：空间比较封闭，单一，了无生趣

第二阶段：慢城精神的提炼

从无到有
从社区到街区
强调社区的弹性空间
动态居住到静态居住
底层居住于高层居住
人文居住与生态居住
人文精神才是社区真正的精神

规划结构
与前三期的关系：对比之后，对自身的围合感比较强，空间相对比较孤立，不能很好地与前三期融为一个社区。

第三阶段
1. 商业价值的最大化
商业呈现一个街区，向慢城前三期敞开充分利用整个社区的商业价值，提升街区文化的活力。
2. 健身步道
为慢城引入生态居住，充分利用78hm² 的石芽岭公园生态片区资源，与石芽岭生态公园保持5分钟步行的距离。
3. 规划结构
东边形成一个面，西南边的两点式结构与前三期形成一个很好的呼应。

1st period: our attempt
We have tried several possibilities for this plot. According to the possible enclosing way and the fact that southern part is higher than northern part, all buildings are built around the rear plot so as to build the largest central garden.
Strong points of the design: the line of vision among buildings is staggered; the largest central garden may be built
Weak points: the space is closed and dull.

2nd period: refinement of Slow City spirit
Growing out of nothing
From community to block
A community's flexible space is emphasized
From dynamic habitation to static habitation
From ground-floor habitation to high-rise habitation
From humanistic habitation to ecological habitation
Only humanistic spirit is a real community spirit.

Planning structure
Relations to former three projects: through comparison, this project presents a stronger sense of enclosure but the space is relatively isolated. It cannot make up a community with former three projects.

3rd period:
1. Maximization of commercial value
A commercial block is built to attract residents of former three projects. Thus the commercial value of the whole community is utilized fully and the vitality of block culture is improved.
2. Keep-fit walkway
Residents of the Slow City project can share the 780,000-m2 Shiyaling Park's ecological resources. They may take five minutes to reach the park by foot.
Planning structure
The planar structure in the east and two-point structure in the southwest respond to former three projects well.

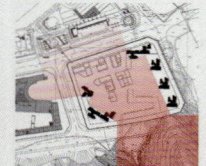

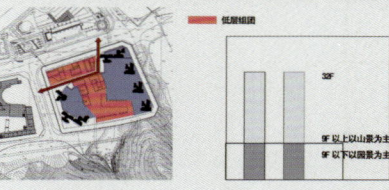
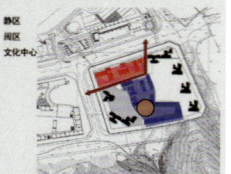

产品分区

慢城的设计目标：

1 成为最早和国际接轨的社区
2 一个具有文化艺术气息的社区
3 多样性的一个国际住宅公园

Bi Hai Yun Tian • Shenzhen
碧海云天·深圳

项目地点：中国，广东，深圳市，南山滨海居住区
项目时间：2001年
设计规模：29万 m²
设计阶段：方案设计，初步设计
项目现状：已建

Project Location: coastal residential area, Nanshan, Shenzhen
Project Date: 2001
Project Scale: 290,000 m²
Design Phase: conceptual design, preliminary design
Project Status: construction completed

Interestingly, the site neighbors a miniature landscape park. Due to its special location, a good transition from the large architectural volumes of this estate to the miniaturized scale of the landscape should be established.

很有趣的是该项目与微缩景观公园为邻，因其特殊的区位需解决巨大的建筑体量和"邻居"微缩景观尺度之间的过渡。

项目总览

用地的周边景观具有极强的不平衡性，南临填海区，西邻锦绣中华的微缩景观公园。方案为寻求景观自身优势与尊重城市公共资源之间的平衡点，自西向东采用了逐步递增高度的布局方式，由4层向30层渐变。

Project Overview

The surrounding landscape of the site features great imbalance. The site is adjacent to a reclamation area in the south and to the miniature scenic park of the Splendid China in the west. An balance is reached in the design between making use of the landscape and matching the planning of the city's public facilities. From west to east, the elevation of the buildings is increased one by one, with the height increasing from 4 stories to 30 stories in succession.

立面图

立面图

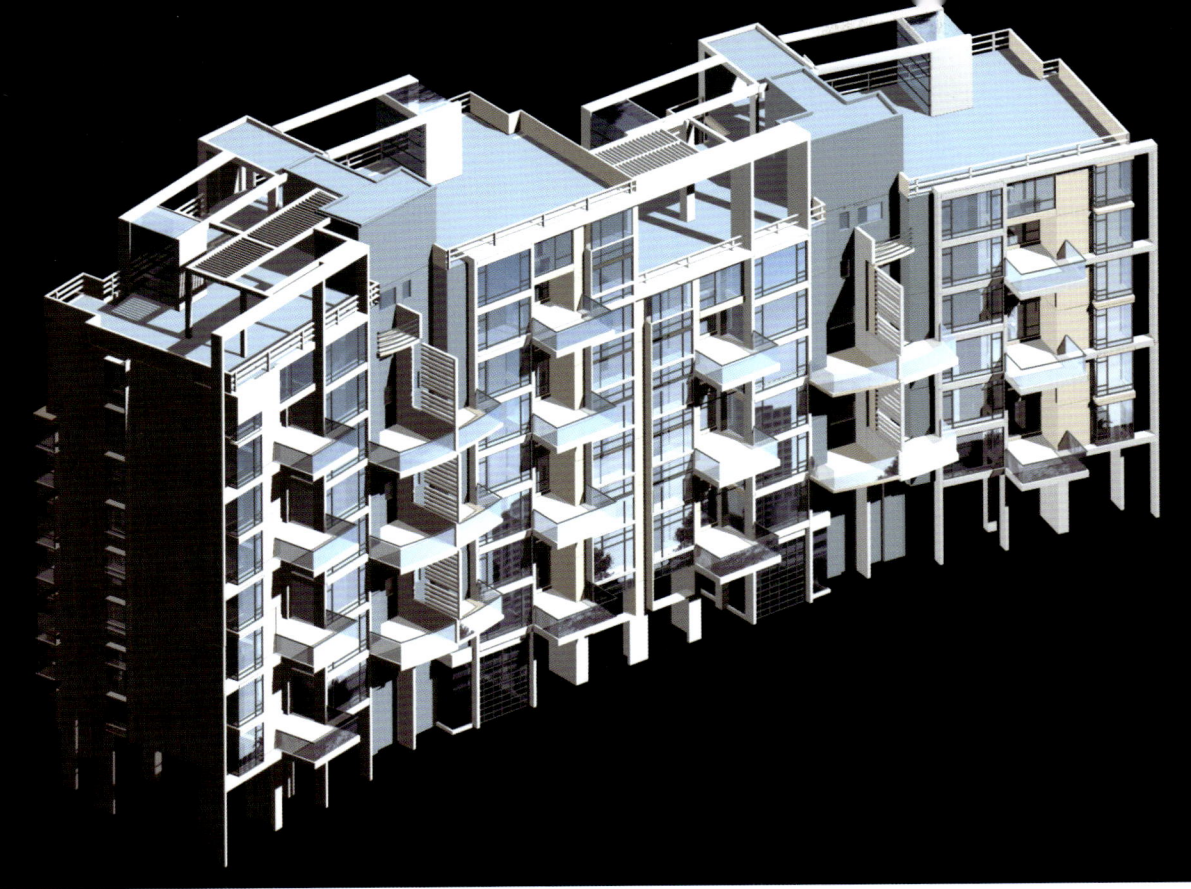
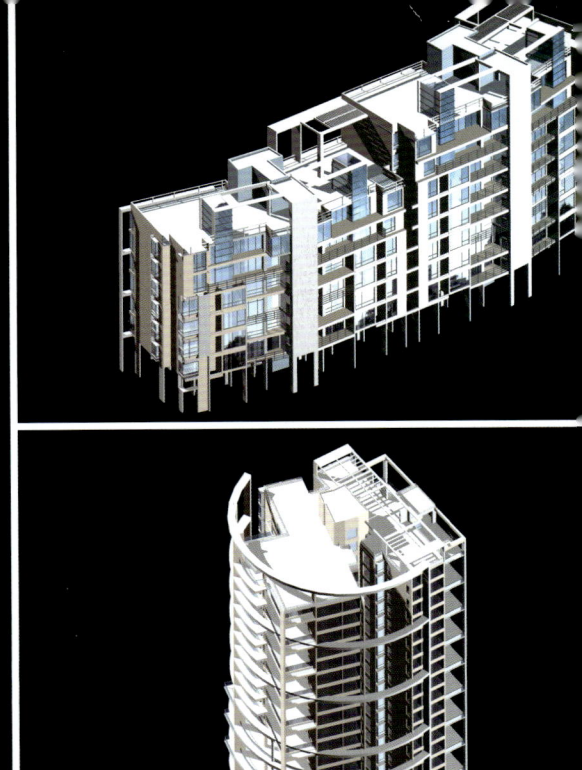

本项目率先在深圳地产项目的高层户型中引入两层空中花园，将"天人合一"的中国园林空间思想引入到高层建筑，并且在高密度的开发下提供了高质量的居住环境，让住户在家中能同时享受户内与户外的生活，实现高层建筑从"鸽子笼"向人性化居住的转变。建筑的立面设计则采用另一种方式，即一处建筑群在社区的不同位置设置不同的建筑色彩与特性，同时又保持社区的整体感和协调感。

For the first time, two storeys hanging gardens are built in the high-rise buildings in a real estate project in Shenzhen; the philosophy of "oneness between heaven and man" in spatial design of traditional Chinese garden is reflected in the design of the high-rise buildings. So in the urban area with dense buildings, a high quality living environment is created to allow the residents to enjoy indoor and outdoor life, no longer living in small flats like "pigeonholes" but enjoying humanistic residence. An unique method is used for design of building facades, in which various architectural styles and characteristics are demonstrated in different positions in the community, while a sense of wholeness and harmony is kept in the community.

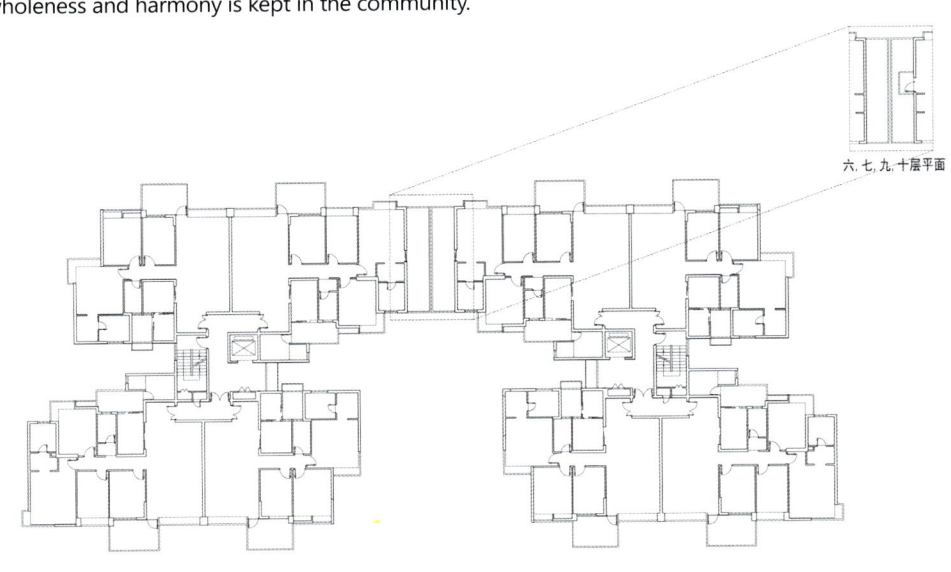

六、七、九、十层平面

Nice Homestead · Wuhan
美好家园·武汉

项目地点：中国，湖北，武汉市
项目时间：2002年
设计规模：7万 m²
设计阶段：方案设计
项目现状：已建

Project Location: Wuhan
Project Date: 2002
Project Scale: 70,000 m²
Design Phase: conceptual design
Project Status: construction completed

The whole design process is the logical analysis of land and project.

整个设计过程是对用地和项目逻辑分析的解题过程。

项目总览

项目用地位于汉口江汉区新华下路八古墩，地块成规则的长方形，临街面距离较短，进深较大，用地平整。

Project Overview

This project site is located at Ba Gu Dun in Xin Hua Xia Road in Jianghan District of Hankou. The plot is level and rectangular, with a relatively short side at the frontage and a long depth.

设计说明

采用行列式布局，小高层与多层相结合，并通过加减法取得足够的日照间距，实现户户南北通透，营造丰富多变的外部空间。建筑被绿地所环绕，使整个规划理性而不乏灵动，同时巧妙利用建筑不同高度的特点，合理规划布置以提高电梯利用率，解决经济适用性问题。

Design Description

A row-style layout is adopted in this design, in which medium high-rise buildings and multi-storey buildings are arranged in rows; enough distance is kept between buildings allowing sunlight through plus-minus method; every flat has good ventilation as there are windows in the south and in the north; rich and diversified outdoor spaces are created. The buildings are surrounded by greenbelts, which makes the whole layout rational and dynamic. Meanwhile, the difference in building heights is skillfully utilized to create a feasible planning, thus the efficiency of elevators is increased and the flats are economic and sound for living.

平面图

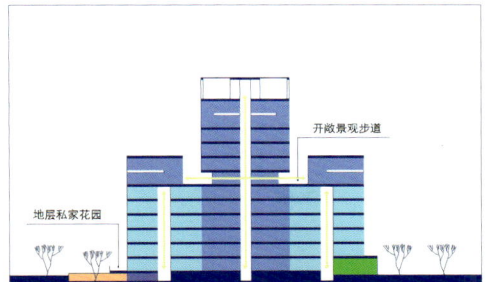

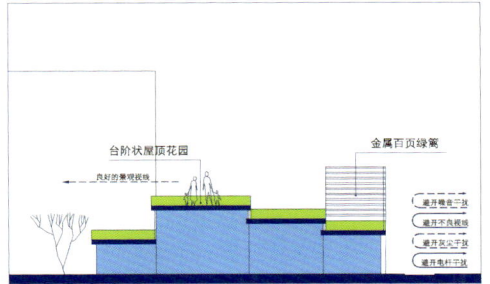

建筑设计的特点

住宅建筑设置半地下停车及储藏，首层用户有较大花园；每栋住宅均设有双层共享式入口门厅，并通过开敞的景观步道入户；二期住宅多高层混排，多层住宅顶层用户可利用高层住宅的电梯；所有户型均为明厨明厕；部分小高层户型带有绿化；住宅立面采用现代造型，每栋住宅带有不同的色彩倾向，并随机排列组合，形成鲜明、独特、丰富的建筑格调及小区形象，从而收到与许多一次性建设小区千篇一律的风格完全不同的效果。

商业建筑简洁明快，形成轻松愉悦的购物环境；建筑模块化，并加入变化，使得大进深的商业空间依然具有良好的自然光环境；商业建筑的屋顶绿化成为整个小区环境的一部分。顶部外置一个金属百页隔墙，供绿色植物攀爬，从而成为一个保护小区景观，隔离噪声、灰尘以及不良视线的绿色屏障。

根据不同特点，提供不同的开发概念。

Architectural design features

Residences have a half-underground parking space and store room. Ground-floor residents have a large garden. Each residence has a sharable two-storey-high entrance vestibule with a broad landscaped passageway. Most residences of phase-II project are high-rise mixed houses. Top-storey residents of multi-storey residences may use elevators in the high-rise residences. All flats have a bright kitchen and toilet. Some flats in medium high-rise buildings present a green color. Residence façade is of modern shape. Each residence has different color orientations, which are arranged at random, to form a bright, particular and rich architectural style and residential quarter image. Thus this project presents a completely different style compared with many other residential quarters.

Business buildings look concise and bright, with a pleasant shopping environment. Since buildings are modularized and added with changes, the deep commercial space is still a good environment with natural light. Roof greening of business buildings becomes an integral part of the whole residential quarter environment. A metal shutter partition is set outside the roof so that green plants may climb up. It is a green barrier to protect landscapes of the residential quarter and insulate noises, dust and bad line of vision.

Different development concepts are offered according to different features.

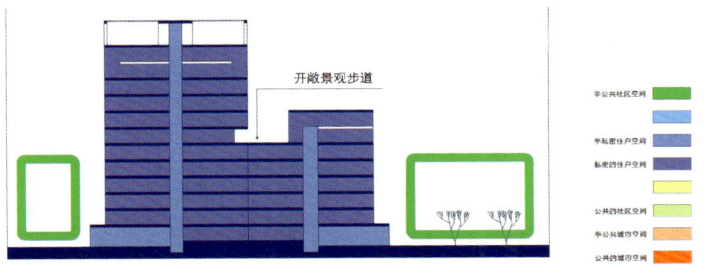

沿新华下路形成连续的商业步行空间，并在入口形成景观步行街的半公共化场所，将小区入口至于步行街的一个尽端；每栋住宅设置两层高的共享式入口门厅；实现了从公共化的城市空间——半公共化的城市空间——公共的社区空间——半公共社区空间——半私密的住户空间——私密的住户空间的层层转化，同时也实现了功能上由商业到居住的转化，由城市到社区的转化。

Continuous pedestrian spaces form along Xinhua Xialu Road. A half-public area comes into being at the entrance to the landscaped pedestrian street. Entrance of the residential quarter is located at one end of the pedestrian street. Each residence has a sharable two-storey-high entrance vestibule, realizing a conversion from public urban space, half-public urban space, public community space, half-public community space to half-private resident space and private resident space. It also realizes a functional conversion from business to residence and from city to community.

建筑与环境设计体现了年轻、现代、艺术等新风格,并实现全方位的立体绿化。建筑采用现代风格,利用色彩增强其可识别性、亲和力及艺术感染力,并将小区环境作为一个整体来处理,不但层次丰富,而且能形成一个连续的生活、休闲景观序列。

The designs for buildings and the environment express a new style featuring juvenility, modernization and art, aiming to achieve overall tridimensional greening. The buildings are of modern styles; the identifiability, affinity, and artistic appeal are enhanced by colors. Treating the environment of the estate as a whole in the design can not only get multi-level spaces, but also create continuous landscape for living and leisure.

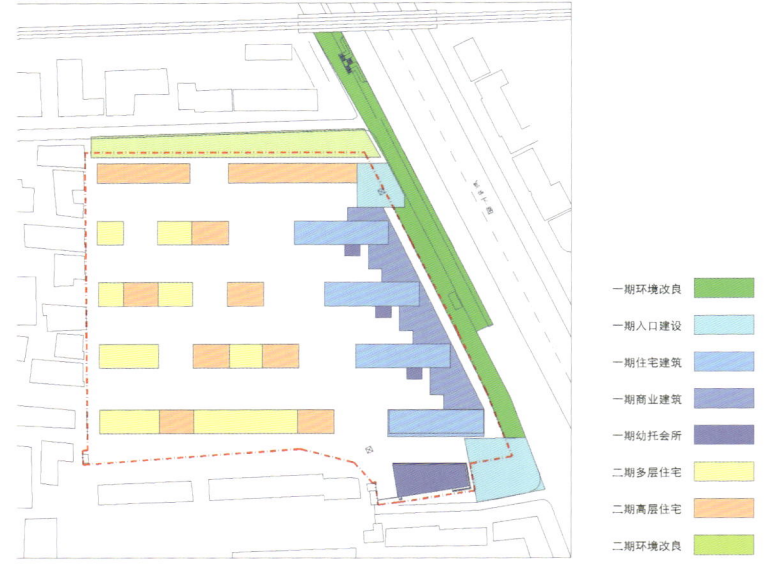

461

Renheng Apartment • Lanzhou
仁恒住宅·兰州

项目地点：中国，甘肃，兰州市
项目时间：2003年
设计规模：20万 m²
设计阶段：方案投标
项目现状：未建

Project Location: Lanzhou
Project Date: 2003
Project Scale: 200,000 m²
Design Phase: conceptual design
Project Status: construction not started

Situated in a culturally rich city of Lanzhou, a modern and straightforward way of building is used to express the architectural language.

在富有文化底蕴的兰州，用现代直率的方式演绎全新的建筑。

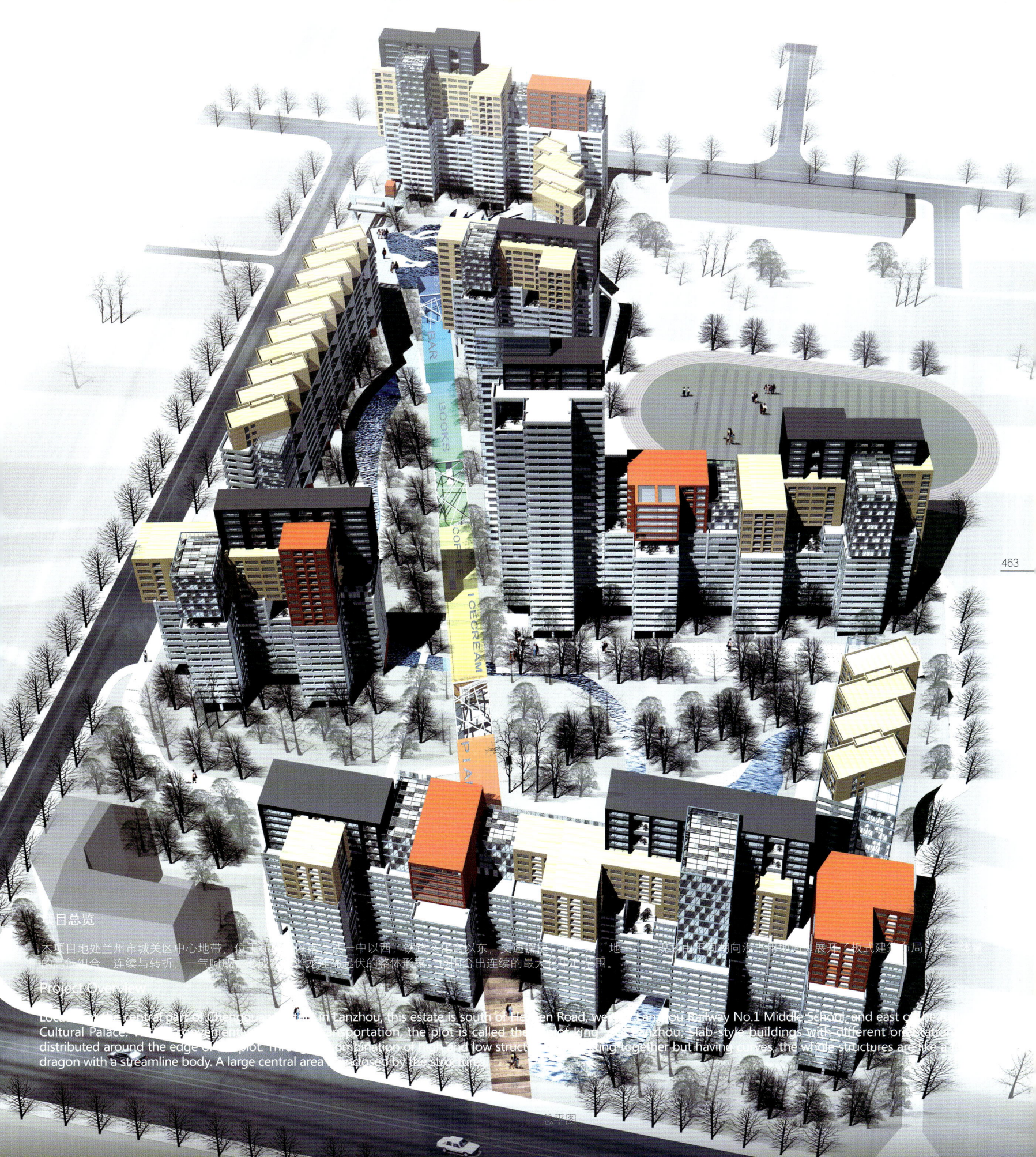

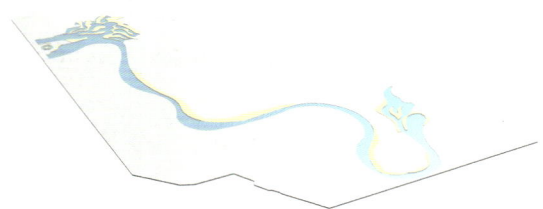
"龙、黄河"

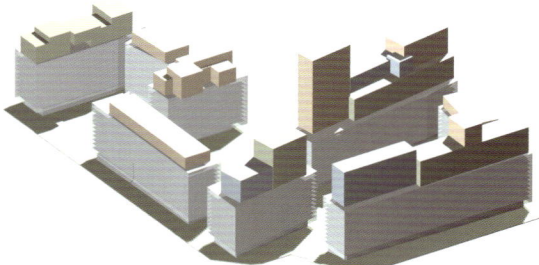
"长城"

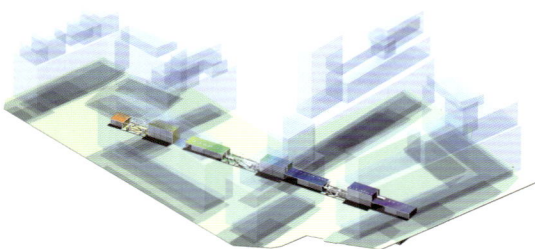
"丝绸之路"

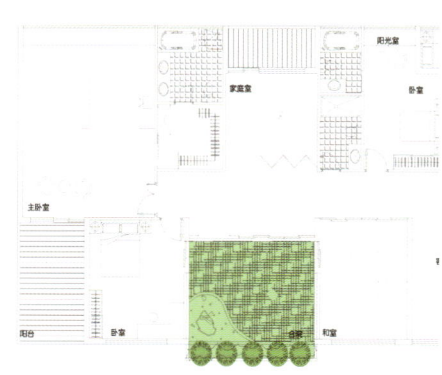
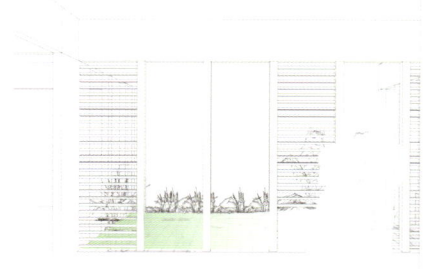

现代三合院-玻璃+百叶

现代三合院-休闲之所

花园及合院

本户型试图从中国古代民居及兰州民居中提取精华，应用到现代居住建筑中。以"合院"的形式，将不同的功能通过院子结合在一起，而朝向院子的"墙"则用一些玻璃及百叶代替，可开可合，可遮视线也可透风，形成极为灵活的内外交融空间。

Garden and Encircled Courtyard

In the design of the internal layout of the flats, the excellent elements of ancient Chinese houses and Langzhou folk houses are introduced. The encircled courtyard connects different functional areas, however, the "walls" facing the courtyard are built with glass and equipped with shutters, which can be opened and closed, serving as vision barriers and ventilation openings, and creating highly flexible spaces interpenetrating between the internal and external.

天际线

如果说长城是宇航员在太空中唯一能看到的人类活动建筑物，黄河就是三台阁上看到的兰州最富"动感"的线条，那么也许这个小区就是兰州带有一定标识性的居住区。

居住庭院，建筑层数不超过12层时，是一个比较宜人的尺度。我们的住宅不可能全部做成12层（容积率的要求），但可以在心理上给人12层的感觉：首先，建筑底下12层全部处理成简洁的白色，即形成了12层的白色"街墙"；然后，在白色"街墙"上面放置不同层数、不同颜色的体块，有蓝色、黄色、红色、深灰色，高低不同，错落有致。这样，一年有不同的季节，不同的天气，一天有不同的时辰，对应有不同颜色的天空。当清晨日出的时候，天空呈朦胧的黄色，"街墙"上面黄色的体块溶入到天空中；当晴天的中午，天空呈蓝色，蓝色的体块溶入到天空中；当日落的时候，红色体块融入到晚霞中；当晚上的时候，深灰色体块消失于天际。不同的时候都会给人12层的感觉，同时产生不同的建筑轮廓线，即天际线是变化的。

Skyline

If the Great Wall is the only human building that can be seen by astronauts in the space, the Yellow River is the most dynamic line in Lanzhou that can be seen at San Tai Ge, then this residential quarter can bring an identifiable residential area to Lanzhou.

A residential courtyard is a pleasant environment for residents when it has no more than 12 floors. It is impossible that all our residences have 12 floors (plot ratio requirement). However we may give people a psychological feeling of 12 floors. First we paint the first 12 floors white. Thus they become a white 12-storey "street wall". Then different colors (blue, yellow, red, dark grey, etc.) of blocks are placed on different floors of the white "street wall". In this way, different seasons and weathers in a year and different time of a day correspond to the sky of different colors. As the sun rises in the morning, the sky looks yellow and the yellow block on the "street wall" blends with the sky. At sunny noon, the sky looks blue and the blue block on the "street wall" blends with the sky. At the time of sunset, the red block blends with sunset glows. In the evening, the dark grey disappears. People will have a feeling of 12 floors at different times. At the same time, different building contour lines appear. That is, the skyline changes.

会所设计

会所引喻"丝绸之路",从南到北两个小区人行贯穿,其间功能均匀分布,方便小区住户共享资源。"巨构"式设计既可作为小区的形象指引,又可营造人文气氛。丝绸之路上的各国文字和特征赋予会所以内涵,形成有特色的"文化带"。

Club Design

In allusion to the Silk Road, the clubhouse has a passage running through the north and south areas, in which the functions are evenly distributed for convenient sharing by the residents. With a "mega-structure" design, the clubhouse can not only enhance the image of the estate, but create a humanistic atmosphere. Various characters and letters of the countries as well as specialties of the countries by which the Silk Road pass are used in the decoration, which endow the clubhouse with cultivation and create a "cultural belt" with distinctive features.

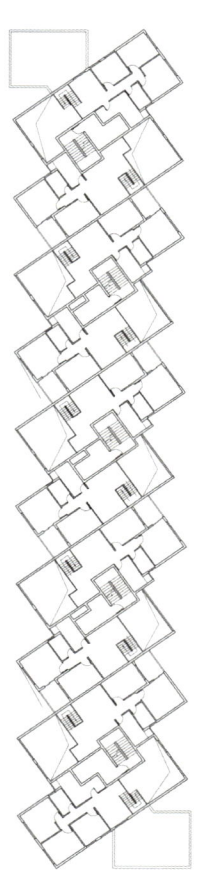

平面图

立面设计

住宅立面在12层高度设计阳台形成的街墙，构成小区统一流畅的白色立面，有效减小了小区的竖向尺度，使其更亲切、宜人。顶端的"盒子设计"反映了地域特征和人文特色。带形的自然状态立面反映了内在生活状态和人为功能。人们在装点自己的同时也装点了社区。

Facade Design

The same smooth white facades with balconies located at the height equivalent to 12th floor are designed as frontage wall in the estate. This facade efficiently reduces the vertical scale, which expresses a feeling of hospitality and amenity. On the "frontage wall", blocks with different layers, height and colors, including blue, yellow, red, and charcoal grey, are embedded so as to produce an uneven yet well-arranged image. The "box design" reflects local characteristics and humanistic features. The natural facade with the stripe decorations embodies natural lifestyles and artificial functions. The whole community is decorated and the people can enjoy a colorful environment.

立面图

户型分布

户型以资源的"均好性"原则来组合布置。每户面积的大小、层数的高低、有无占天占地、有无私家花园，这是实实在在的差别。在资源分配上是"杀富济贫"、缩小差距的"平均主义"呢，还是"加大差别"？我们的选择是：根据对资源的占有率大小来分配户型，同时尽量避免不利因素。

"均好性"就是：好景观+大面积+好朝向+……+高的价格=好的享受、偏的景观+小面积+适中的朝向+……+低的价格=舒适的享受。层数低、密度小的小区资源好坏分布在平面上展开；高密度、高层数的小区资源好坏分布应在坚向上分析，层数越高、日照越充分、视野越开阔、通风越好……

Flats Distribution

Different types of flats are distributed on the principle of equality. All flats are different in respect of area, floor, spatial location, and availability of private garden. We may allocate resources based on equalitarianism or increase the gap. Our choice is to allocate flats according to the occupancy rate of resources and try to avoid unfavorable conditions.

Well-balanced layout means that good landscape + large area + good orientation + ...+ high price = good enjoyment; houses sharing few landscapes + small area + moderate orientation + ...+ low price = comfortable enjoyment. For residential districts with low-rise buildings and low-density, the superior and inferior resources are distributed horizontally. For residential districts with high-density and high-rise buildings, the superior and inferior resources are distributed vertically, that is, the higher a floor, the more sunlight, the broader field of vision, and the better ventilating condition...

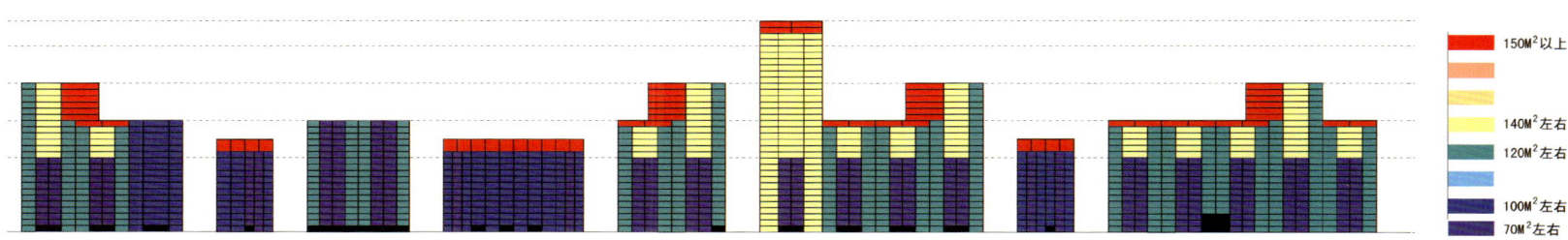

Junyue River Paradise • Changsha
君悦香邸·长沙

项目地点：中国，湖南，长沙市
项目时间：2005年
设计规模：32万 m²
设计阶段：方案设计
项目现状：已建

Project Location: Changsha
Project Date: 2005
Project Scale: 320,000 m²
Design Phase: conceptual design
Project Status: construction completed

This project presents a vivid architectural image, simple architectural style, and reasonable layout.

鲜明的建筑外衣，简洁的的建筑风格，合理的建筑布局。

设计策略

1. 提高建筑层数，降低建筑密度，做到庭院最大化，并通过底层架空等方法将地面更多的公共空间提供给住户；

2. 利用地形高度差设置生态车库；

3. 通过建筑体量的跌落，塑造丰富的城市形象。在住区的周边引入小规模的商业街，为住区提供配套设施；

4. 户型上力求做到个性化，板式一梯多户相互错位排列，达到南北通透，户户朝阳，并在高层中创新性地引入空中庭院的概念，将别墅化的生活品质提供给住户；

5. 建筑造型简洁现代，色彩鲜明，确立建筑群在城市中的形象；

6. 模数化的整体设计理念，采用标准化配置，提高施工的精确性，缩短施工周期，降低建设成本。

Design Strategy

1. The courtyards are maximized by increasing buildings stories and reducing building density. Meanwhile, more public spaces are provided for the residents by making the first floor as empty space.

2. Ecological garages are built by making use of the elevation difference of the plot surface.

3. With a cascade layout of buildings, a luxuriant urban image is built for the estate. Small-scale commercial streets are set in surrounding areas as amenities for the residents.

南立面　　　　　　　　　西立面

4. As for the internal layout of flats, unique features are emphasized; in each unit of the slab-type buildings there are multiple flats in the same floor, with the units arranged in a staggered way so that each flat is thoroughly ventilated through windows in the north and south walls and can be exposed to the sunlight. The concept of sky courtyard is introduced in the design of high-rise buildings, providing the living quality of villas for the residents.

5. The shape of buildings is uncomplicated and modern and the colors are bright, which help building an outstanding image of the complex in the city.

6. Overall modular design concept is introduced and standardized configuration is used, so that the accuracy of construction is improved, the construction period is shortened, and the construction costs are reduced.

北立面

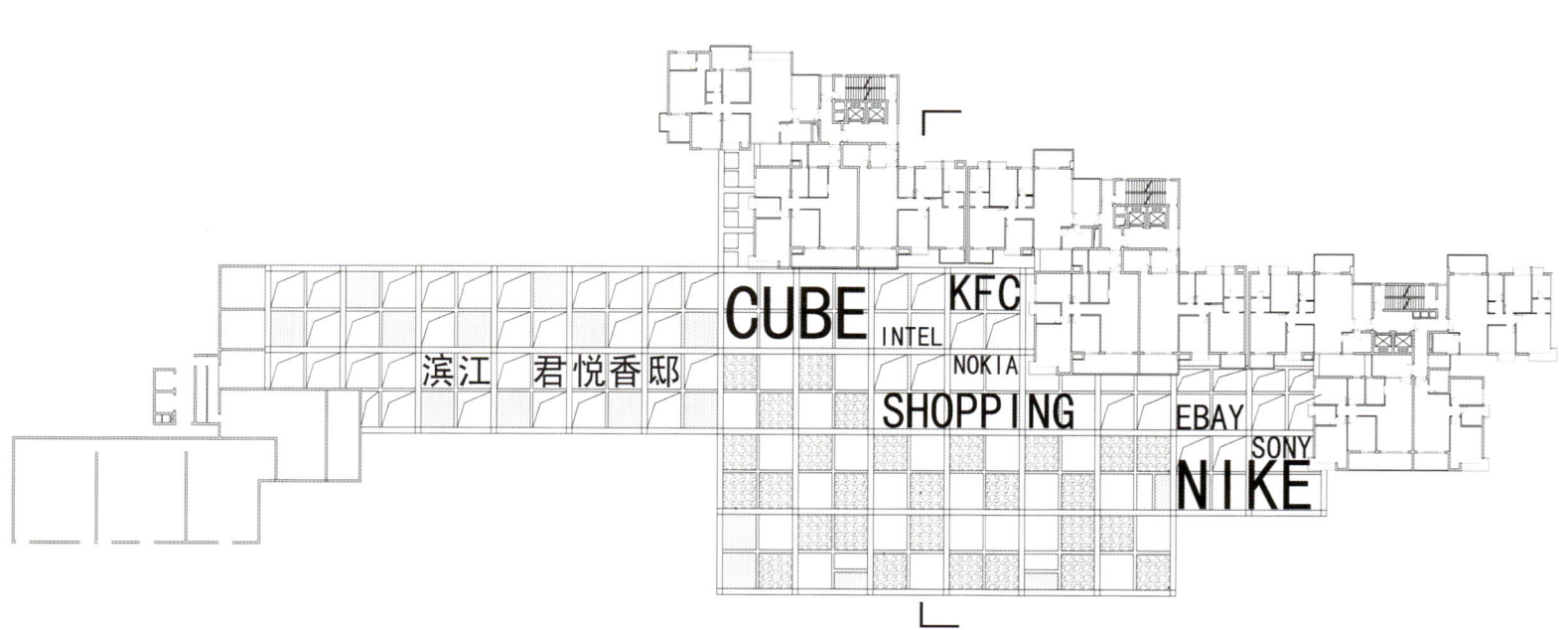

三层平面

Hua Hao Yuan Garden • Shenzhen
华浩源景园 · 深圳

项目地点：中国，广东，深圳市
项目时间：2005年
设计规模：8万 m²
设计阶段：方案设计
项目现状：已建

Project Location: Shenzhen
Project Date: 2005
Project Scale: 80,000 m²
Design Phase: conceptual design
Project Status: construction completed

Like playing a game, the thick and solid point-type tower is cleverly changed into a thin and transparent slab-type high-rise building with good ventilation and improved vantage point.

如同做一个游戏，将厚而实的点式塔楼变成薄而透的板式高层住宅，巧妙地赋予它优良的通风条件和景观优势。

项目总览

本项目用地位于深圳市布吉街道办石芽岭生态居住区内,南临科技园路,东壤景芬路,北接京九铁路线。基地呈带状,A区北边红线内有一凹地,高差达15m,规划将其建成一个以体育、休闲为主的生态公园。本项目用地面积为65800m²,与已建成的华浩源一、二、三期遥遥相望。

Project Overview

The project site is located in Shiyaling eco-residential community in Buji Subdistrict of Shenzhen, adjacent to High-tech Zone road in the south, to Jing Fen Road in the east, and to the Beijing-Kowloon railway line in the north. The plot is of a belt shape, with a hollow zone within the boundary line in the north which has an elevation difference of 15 meters and is to be built as an eco-park with the themes of sports and leisure. The site has an area totaling 65,800 m2, and is adjacent to the completed Huahaoyuan phase I, II, and III estates.

总平面

设计特点

1. 尊重用地现状，使建筑、自然环境相互交融，与已建成的二、三期住宅形成一体；

2. 每户均带有入户花园和3层高的观景阳台，拥有良好的景观面。户型设计上为两房双卫或一卫双用，三房均有阳光浴室；

3. 地下车库不仅与住宅中庭有交错的竖向空间设计，还把体育公园的设计概念很好地运用到地下车库中去，这样由室外到地下车库，再由地下车库到住宅中庭，形成丰富的三度空间；

4. 建筑体形相互交错，有高有低，流畅的曲线设计与周边山脉相呼应，形成有一定韵律感的天际线。

Design Features

1. The existing conditions of the plot is made full use of to make the buildings match with natural environment, and integrate with the completed phase II and III estates.

2. Every flat has a home garden and a three-storey-high balcony overlooking landscape, enjoying a wide field of view. The internal layout of the flats is designed as two bedrooms with two washrooms or one washroom, or as three bedrooms all with a colorfully painted bathroom with ample sunlight.

3. The underground garage not only has a vertical space interlaced with the atrium, but also is designed by introducing the concept of a sports park. Thus, an ample tridimensional space is created from outside to the underground garage and again to the atrium.

4. The buildings are staggered, some high and some low, with smooth curve matching the surrounding mountain ranges, forming a skyline with some rhythm.

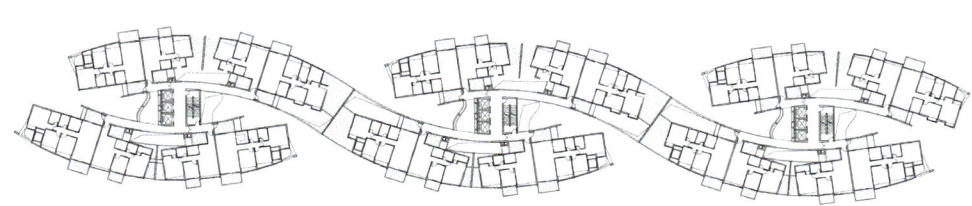
栋标准层平面图

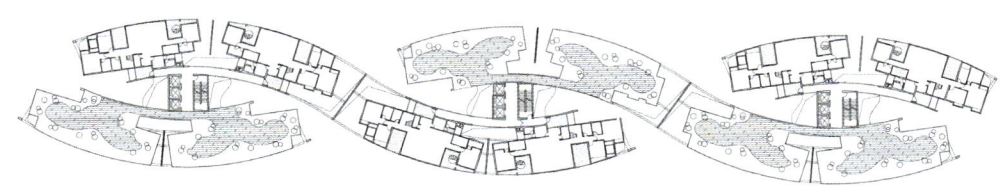
栋复式一层平面图

Xiang Fu Jia Cheng • Changsha

湘府嘉城 · 长沙

项目地点：中国，湖南，长沙市
项目时间：2009年
设计规模：34.2万m²
设计阶段：方案设计
项目现状：未建

Project Location: Changsha
Project Date: 2009
Project Scale: 342,000 m²
Design Phase: conceptual design
Project Status: construction not started

As an interface for the provincial government plaza, it opens and extends the central courtyard.

作为省府中心广场的一个界面，开放和引入让其成为中心广场的延续。

基地位置

湘府佳城位于长沙市天心生态新城的核心区内，与湖南省政府机关大院相距仅300m。具体位置在芙蓉南路以东，杉木冲路以南。

Site Location

Located in the core area of Tianxin Ecological New Town, Changshan, the Xiangfu Jiacheng project is only 300m away from the residential quarter for employees of institutions of Hunan Provincial Government. It is exactly situated in the east of South Furong Road and south of Shanmuchong Road.

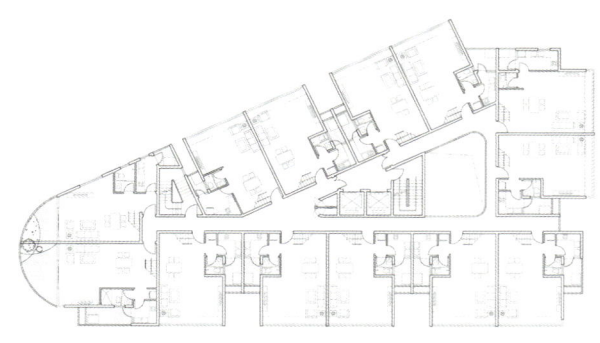

平面图

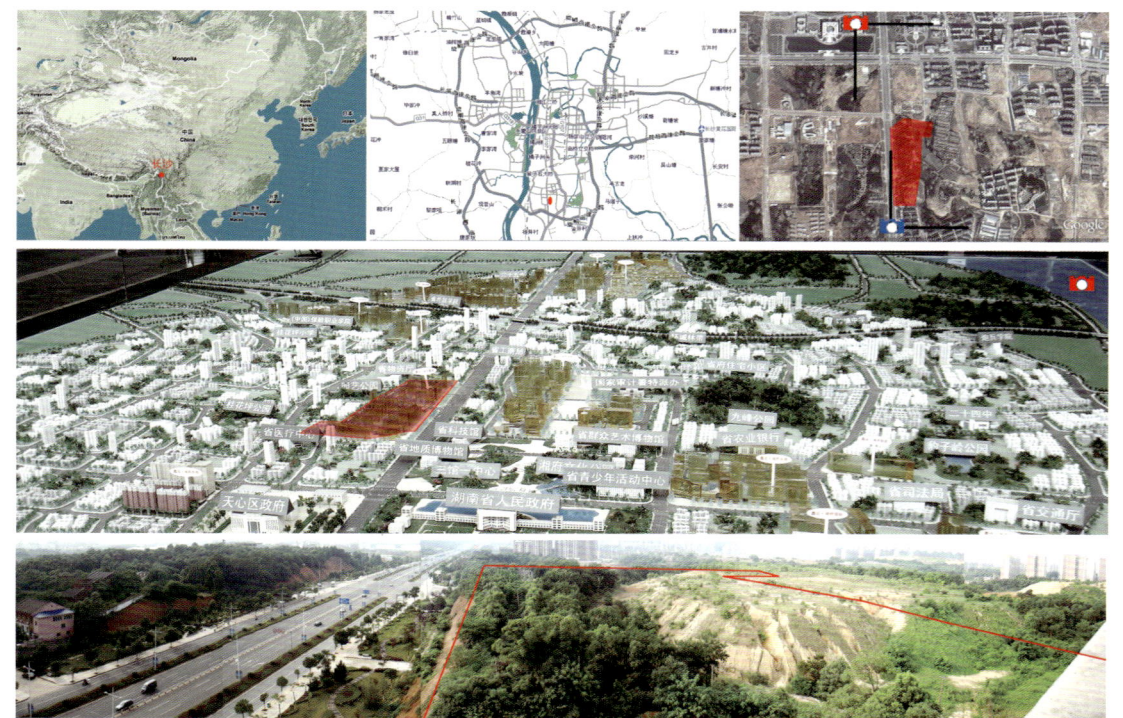

基地照片

地块内部有大量的原生植被，以香樟为主，树龄较长的大香樟估计在20年以上。同时，有山坡、洼地，地块高低起伏；地块已完成七通，但未完成拆迁及土地平整工作。

Site Photo

There are lots of vegetations on the plot and most are camphor trees. Some trees are 20 years old or older. Besides, there are slopes and low-lying grounds on the plot. "Seven supplies" is available on the plot but removal and leveling have not been completed.

Plan Analysis

个性化产品增值策略

艺术坊之FREEHOUSE 平面及分析：

- 自由多样的分隔方式
- 每户均赠送地下室或屋顶花园
- 内天井增加采光

改造方式一：用简单的单片隔墙对FREEHOUSE原有大空间进行简单的分隔成接待、展示、工作、休息等空间，成为艺术坊中的"艺术之家"。

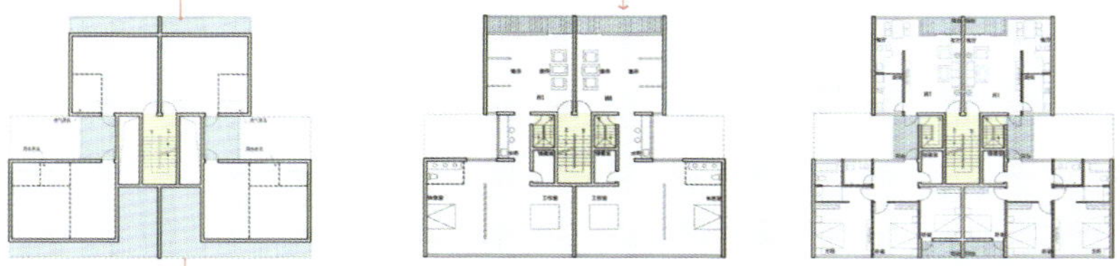

改造方式二：根据FREEHOUSE中的燃气开关和用水开关的位置，我们可以将其改造成厨房和卫生间，其余空间可相应的改成厅和房间，最终成为三房两厅两卫的"幸福之家"。

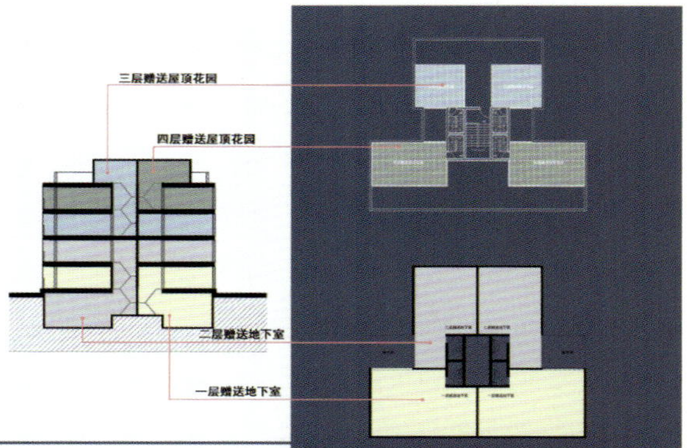

三层赠送屋顶花园
四层赠送屋顶花园
二层赠送地下室
一层赠送地下室

大院山居之叠TOWNHOUSE 平面及分析：

- 两层高客厅
- 每户赠送地下室或屋顶花园
- 坐拥园艺公园山景

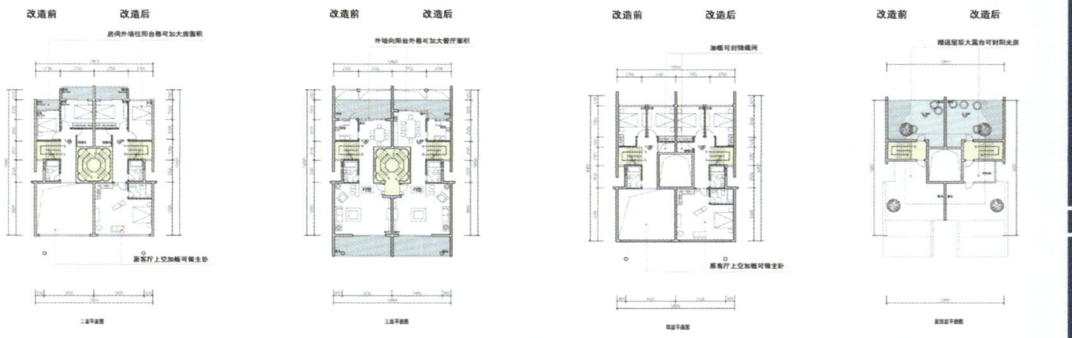

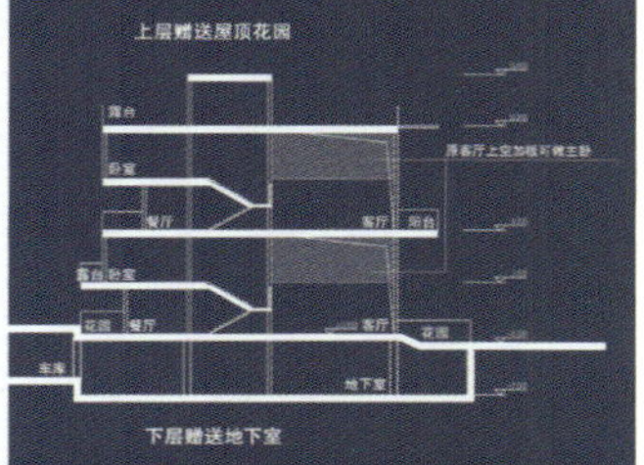

改造前　改造后

上层赠送屋顶花园
下层赠送地下室

大院山居之楼王 平面及分析：

- 两层高客厅
- 270度全景阳台
- 正对园艺公园、视野开阔

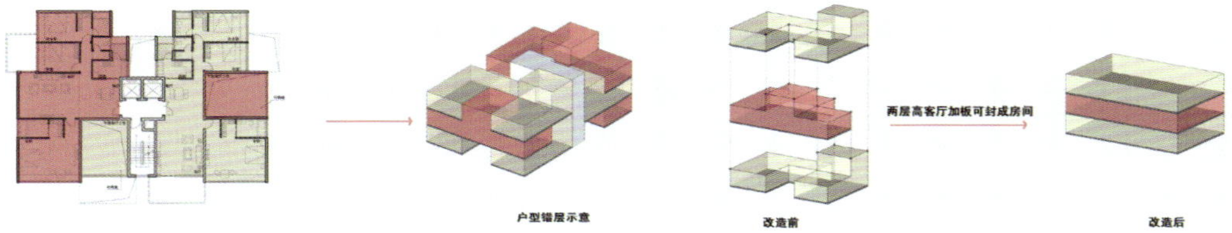

户型错层示意　　改造前　　两层高客厅加板可封成房间　　改造后

慕城　　Yo-Town

本项目在40m限高和2.4容积率的双重压力下，力图打造以年轻的白领人群为客户群的时尚"Yo-town"。项目规划具有四个特点：
1. 充分尊重城市资源、城市空间；
2. 尝试去捕捉城市新生活的丰富性与复合性；
3. 资源的集约分配巧妙解决了容积率与限高的压力；
4. 立体化、个性化的居住产品配置提供多样化、有层次的选择。

In spite of the double stresses from a height limit of 40m and a floor area ratio of 2.4, the design aims at building a fashionable "Yo-town" targeting white-collar youths. The project design features:
① Compatible with the city's public resources and spaces;
② Trying to follow the rich and diversified new lifestyles in the city.
③ Skillful arrangement of resources to create compact design satisfying the requirement on floor area ratio and height limit.
④ Three-dimensional and personalized configuration of architectural structures provides multiple choices for people with different interests and preferences.

一条保留城市视线的通道连通了区域文化中心与用地南的城市公园，在用地内贯穿南北，依次形成"天街秀场"、"艺术坊城"、"大院山居"三大分区。"天街秀场"将城市公寓与城市休闲风情街结合，提供居家型、办公型及多变的loft分离产品。"艺术坊城"由两个大小高层院落和半开放的4层"艺术之家"形成的街坊组成。"大院山居"将山体公园引入，形成独立MINI HOUSE区域，小高层沿边开放布局，体现了高品质生活居住。

A passageway allowing sight lines for city view connects the regional cultural center and the city park south of the estate. The passageway runs through the estate from north to south, along it three sub-communities are built, known as "Heaven Show Square", "Artistic Lane Town" and "Mountain Dwelling with Large Courtyard". With a combination of urban apartments and leisure style streets, the "Heaven Show Square" provides apartments, offices, and large indoor spaces for flexible partition. "Artistic House City" consists of two courtyards encircled by high-rise buildings and medium high-rise buildings, and a semi-open "Arts House" of four stories which forms a block. The concept of mountain park is introduced in the design of the "Mountain Dwelling with Large Courtyard", forming a detached MINI HOUSE area, with the medium high-rise buildings distributed along the sides, expressing a high quality lifestyle.

Vanke City Garden · Wuhan
万科城市花园·武汉

项目地点：中国，湖北，武汉市
项目时间：2004年
设计规模：12万 m²
设计阶段：方案设计
项目现状：已建

Project Location: Wuhan
Project Date: 2004
Project Scale: 120,000 m²
Design Phase: conceptual design
Project Status: construction completed

This project is in fact a uniquely designed "outlander" inside a systematic, industrious, and repetitious residential area.

这一项目事实上是在一片型制化、工业化重复单调的居住区中镶嵌一块"特别设计"的异质体。

设计原则

针对这一项目的特点，我们力求解决：
1. 构筑一群独立、特点突出的和谐建筑；
2. 强调整个大规划的整体感和对内围合的空间；
3. 保护原始地形，重视绿脊公园的延伸和渗透；
4. 尊重城市环境、小区内部建筑和地形的关系，解决建筑沿街面和转角的形象问题。

Design Principle

On the basis of the features of this project, the following objectives are set up:
1. To construct a group of detached and harmonious buildings with distinctive characteristics.
2. To emphasize a sense of wholeness as well as the internally enclosed spaces by means of the overall planning.
3. To preserve the original landform and attach importance to extension and penetration of the green ridge park.
4. To establish a harmonious relationship between city environment, buildings inside the residential community and existing landform, and establish a good image for the frontages and turning corners.

项目总览

武汉万科城市花园非标区方案设计位于武汉万科城市花园规划小区中部，用地呈不规则形状，内部环抱小区山脊公园，西面为已建成的一期住宅和公建，北面与规划中的体育公园相隔三环路辅道，属于城市花园二期建设项目。

Project Overview

The non-standard plot of Wuhan Vanke City Garden is located at the center of planed area of the garden; the plot features an irregular shape, in which a mountain ridge park is embraced. The completed phase I residential and public buildings are in the west; the sports park which is under planning will be located on the other side of the relief road of Third Ring Avenue in the north. This is the phase II construction project.

标准层平面

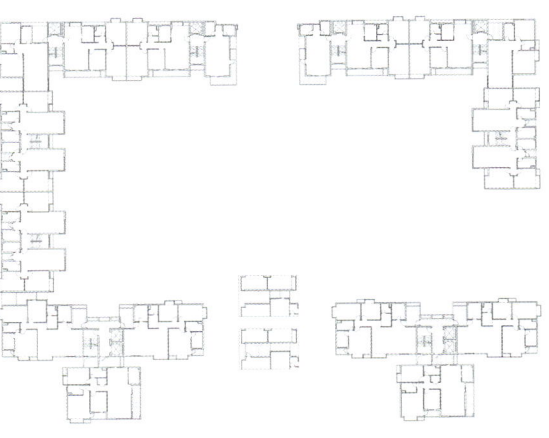

立面图

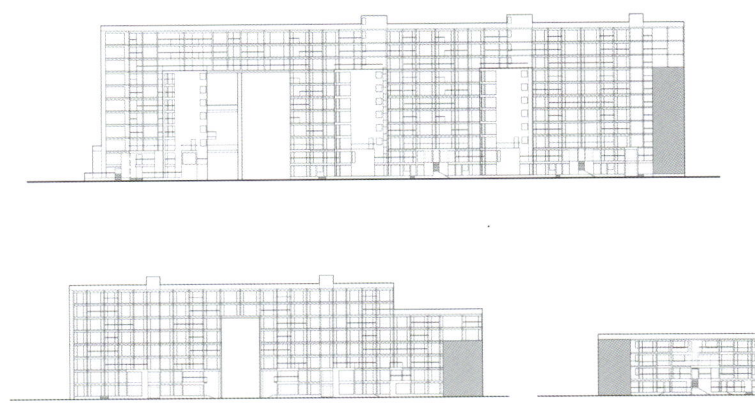

Jia Ri Lan Wan • Qinhuangdao
假日蓝湾·秦皇岛

项目地点：中国，河北，秦皇岛市
项目时间：2005年
设计规模：8.3万 m²
设计阶段：方案设计
项目现状：已建

Project Location: Qinhuangdao
Project Date: 2005
Project Scale: 83,000 m²
Design Phase: conceptual design
Project Status: construction completed

This site is situated in the vicinity of the estuary of the Great Wall. We hope our design can ignite people's imaginations.

地块位于长城入海口附近，我们希望设计能让人产生丰富的联想。

项目总览

秦皇岛假日蓝湾项目位于秦皇岛市中心区的南部，地块东接河南路，南临渤海湾，西依凯莱度假村，北面为平整空地，与地块西边相隔一公里处则是国家重点旅游区——老龙头。

Project Overview

This project site is located in the southern part of the downtown Qinhuangdao, east to He'nan Road, south to Bohai Bay, west to the Gloria Holiday Villas, in the north there is a level vacant plot. A kilometer away from the site, there is a national key scenic spot known as Laolongtou.

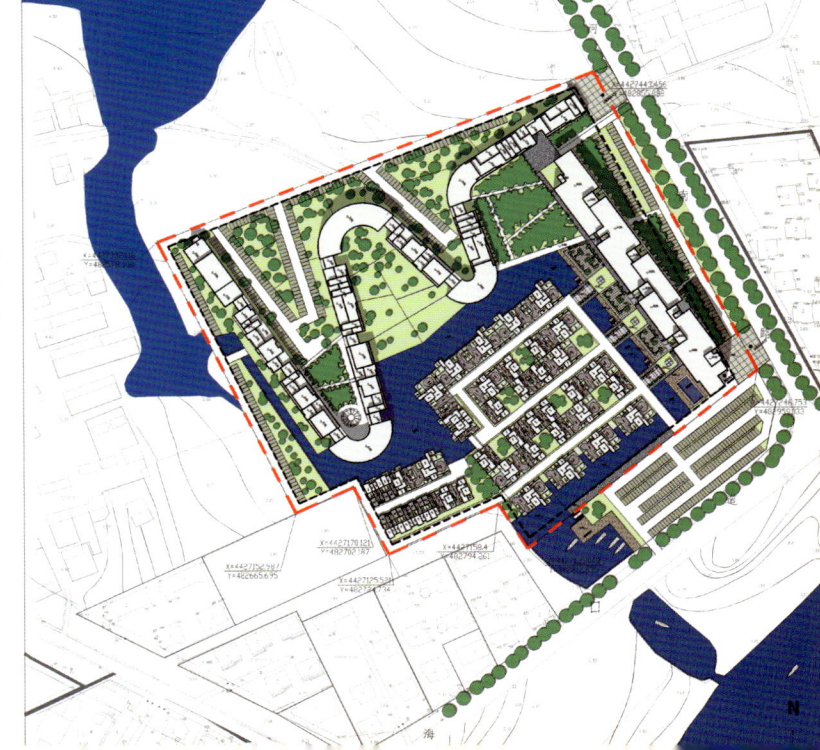

总平面

立面图

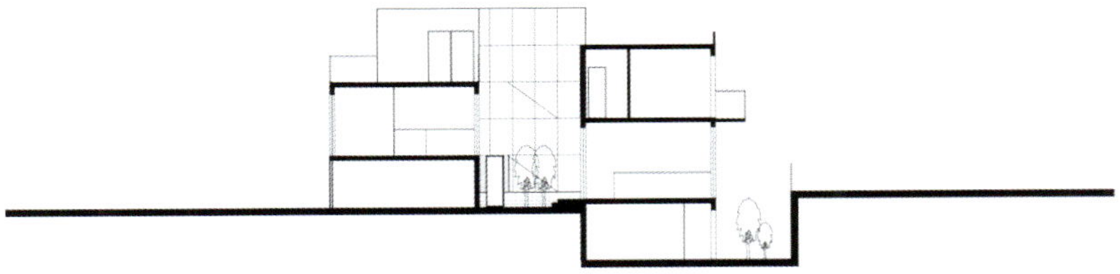

剖面示意A-A

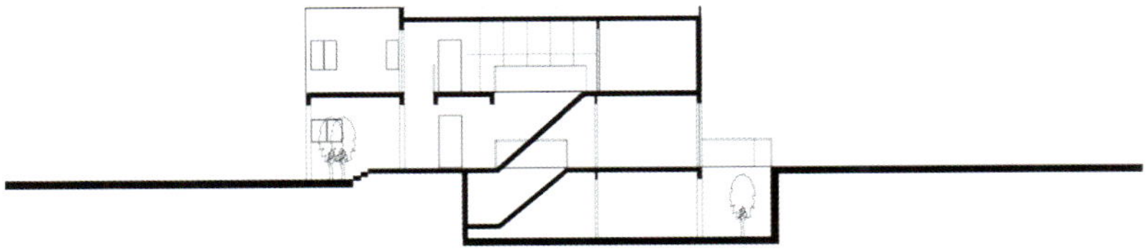

剖面示意B-B

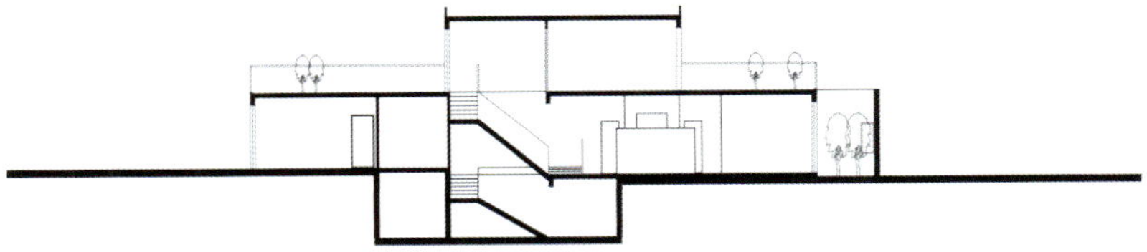

剖面示意C-C

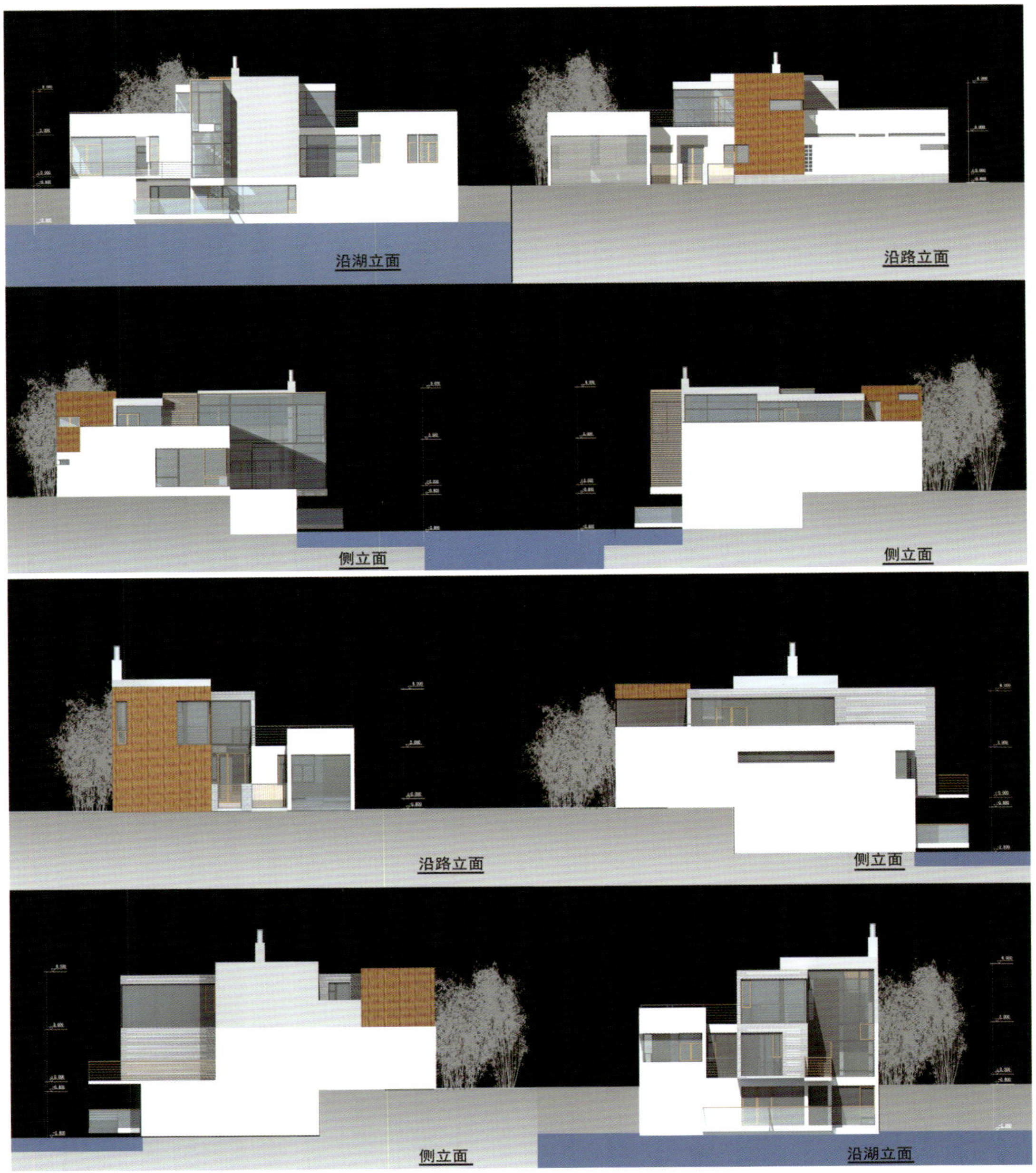

沿湖立面　沿路立面

侧立面　侧立面

沿路立面　侧立面

侧立面　沿湖立面

建筑设计

1. 项目拥有全新的居住空间，丰富且个性化的建筑风格。

2. "龙"形公寓建筑单体具有多景观朝向性的优点，每户都拥有两个或以上的景观面，成功达到了无对视的状态。

3. 设置绿色竖向的观海平台及观海大堂。在丰富居住空间层次的同时，也增加了竖向空间的绿化。

4. 考虑到秦皇岛典型的海洋性气候，项目采用"穿堂风"设计，别墅和公寓都享有自然的穿堂风，充分享受沿海带来的便利。

5. 整个小区立面采用简洁的处理手法，以灰、白、木色体块三种元素贯穿每一个建筑单体，体现简洁质朴的海滨城市建筑特征。

Building Design

1. This project boasts brand-new living spaces and rich and unique building styles.

2. Each unit of the "dragon" shape apartment has good orientations overlooking landscape. Each flat has two or more landscape-overlooking sides, and occupants in different flats can not see each other.

3. Green vertical platforms and lobbies overlooking seascape are set up so as to increase vertical spatial greening while enriching levels of living space.

4. After considering typical marine climate in Qinhuangdao, "cross ventilation" design is adopted in this project; both villas and apartments have draught flowing through the rooms, fully benefiting from the climate.

5. Simple and brief treatment methods are adopted for all the facades in the estate, every single building is painted with three colors including grey, white, and wood color, expressing a sea town's simple and unadorned architectural characteristics.

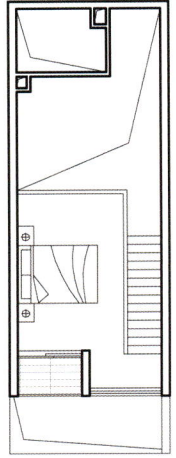 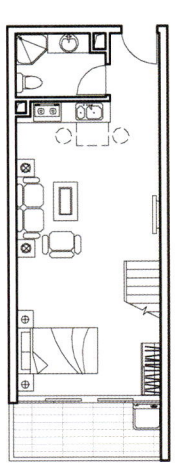 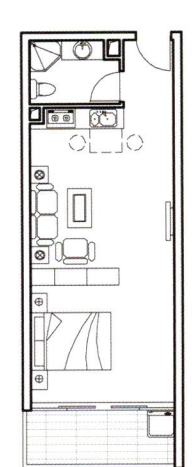 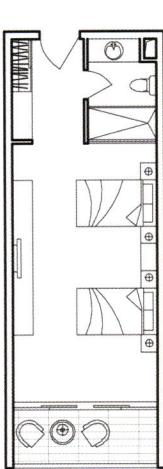

平面图

Liantai Dameisha Apartment • Shenzhen
联泰大梅沙公寓·深圳

项目地点：中国，广东，深圳市
项目时间：2006年
设计规模：6 000m²
设计阶段：方案设计
项目现状：在建

Project Location: Shenzhen
Project Date: 2006
Project Scale: 6,000 m²
Design Phase: conceptual design
Project Status: construction in progress

Under the premise of minimal cover rate, we experimented with a variety of design methods.

在这里，我们在最小覆盖率的前提下尝试了多种不同的处理方式。

总平面

方位图

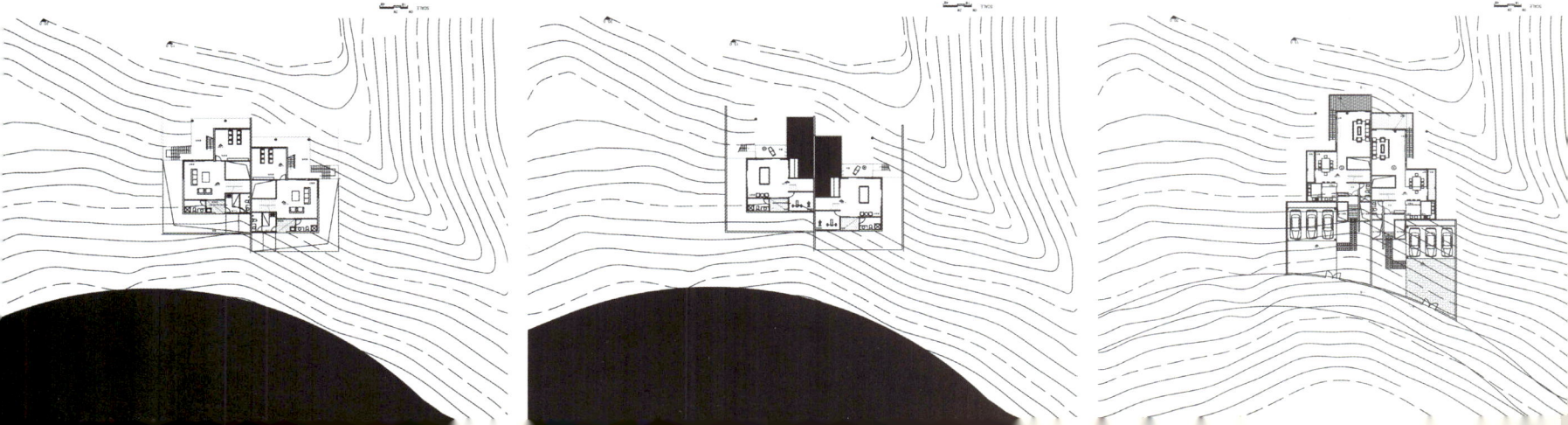

项目分析

根据项目良好的区位及地形、地貌特点,我们确定如下的总体设计原则:

1、充分利用基地本身的唯一看山通道,我们将主体建筑设计成"一"字形。东朝南偏转8度,标准层朝东为厅、房,朝西向山体为辅助交通空间,如此的设计保证了所有房间均享有一览无遗的海景资源。

2、现有山涧小溪从主体建筑公共平台层穿过,稍加改造成小水潭,设汀步、木栈桥,形成第一亲水场所。小溪潺潺而下,在主入口处汇成较大水潭,使回家的人们从踏上入口的那一刻起就能听到溪水的声音,感受水的灵性。

3、为了减少对山体的破坏,我们采用"干"式的建筑形式,将建筑大部分底层架空,仅以结构柱与地面接触,架空层下则尽量恢复原有植被。

4、为了减少建筑体量过长带来的突兀,立面大量使用木纹板、毛石墙等自然材料,少量外墙为深灰色,深浅与周围山体颜色一致,从而使整个建筑融入山体中。

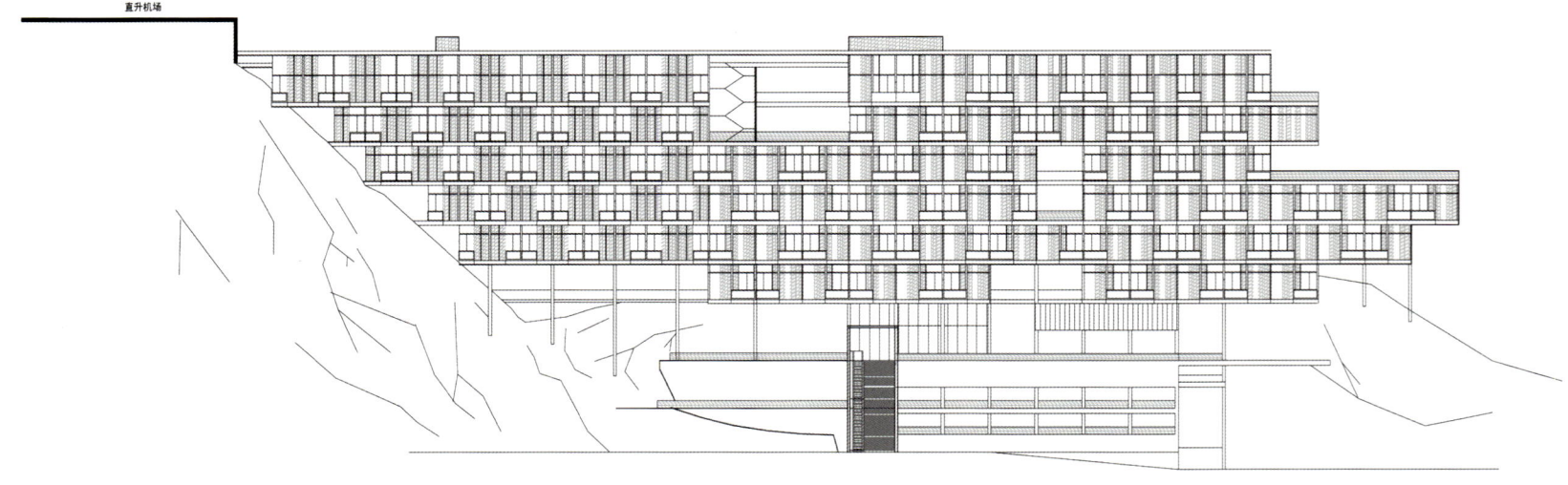

立面分析图

Project Analysis

On the basis of the favorable location and landform of the plot, the following general principles is adhered to in the design:

1. By making full use of the only channel allowing sight lines to view the mountain, the main building is designed with a linear shape, deflecting from the east to south by 8 degree. In standard stories, living rooms and bedrooms are located in the east, auxiliary transportation space is in the west facing the mountain. Such design ensures all rooms overlook seascape.

2. The existing mountain stream runs across the public platform of the main building. After small rebuilding, a small pool with stepping stones and a timber trestle bridge can be built as water landscapee. The babbling stream run down to the main entrance where a larger pool forms. The residents returning home can hear the running stream and enjoy the water landscape at the entrance.

3. In order to minimize impairment to the hill bought by the construction, the "stilt style" is adopted for the buildings, of which the most first floors are built as empty space, only the structural columns contact the ground; the vegetation on the ground is restored to the original state to the greatest extent.

4. In order to minimize the unfavorable feeling brought by the wide face width of the building, plenty of natural materials such as wood grain boards, and rough walls are used for constructing the facades, and some outer walls is of charcoal grey, which shade conform to the colors of surrounding hills, so that the buildings are integrated with the hills.

Overseas Chinese Town Singles Apartment · Shenzhen
华侨城单身公寓·深圳

项目地点：中国，广东，深圳市
项目时间：2006年
设计规模：3.9万m²
设计阶段：方案设计
项目现状：已建

Project Location: Shenzhen
Project Date: 2006
Project Scale: 39,000 m²
Design Phase: conceptual design
Project Status: construction completed

Three-dimensional and diverse urban lifestyle is created inside a small area.

在一个很小的区域内营造立体而多样的城市生活。

设计原则

1. 充分利用基地的景观资源；
2. 在有限的用地内创造必要的公共活动空间；
3. 尊重城市关系，有机配置建筑体量，使建筑与环境完美结合，相得益彰；
4. 户型经济、实用、精致。

Design Principle

1. To make full use of the site's landscape resources;
2. To create necessary public activity spaces in a plot with limited area;
3. Conform to the city planning and arrange the building structures in an organic way, so as to achieve a perfect match between the buildings and the environment;
4. To create economic, functional, and fine internal layout for the flats.

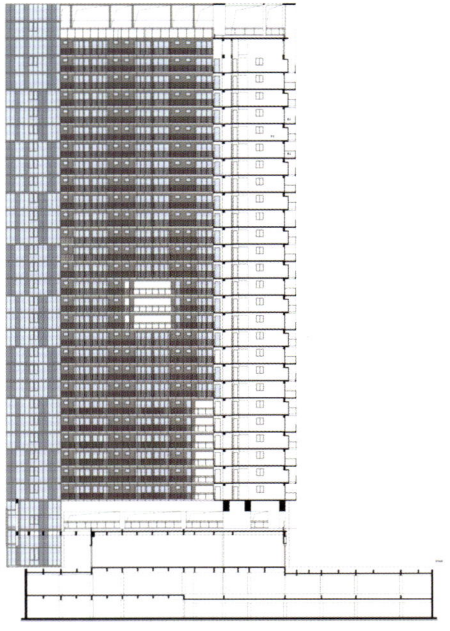

立面图

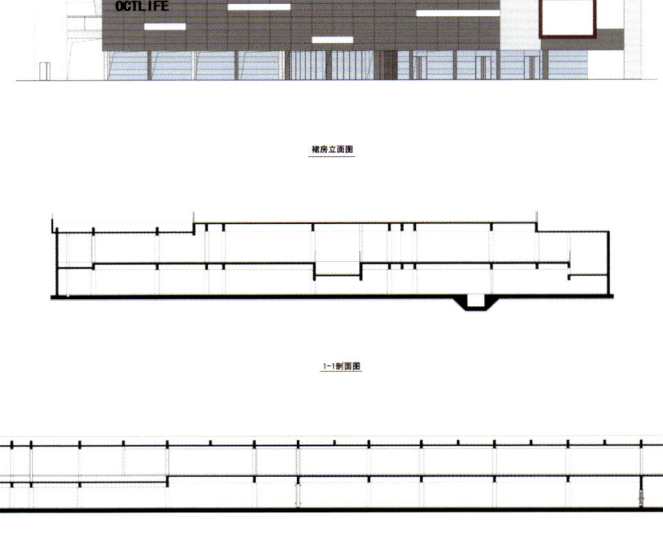

剖面图

立面图

立面设计

1. 立面设计以景观为导向，以大幅落地景窗和阳台作为主要造型元素；

2. 以黑白相间的格构架为底，透明玻璃落地窗和阳台栏板形成的体量为中心点，营造出整体大气、洗炼的视觉效果。

Facade Design

1. The facade design is landscape-oriented, taking large-scale French landscape windows and balconies as the major shaping elements.

2. An overall grand and clear visual effect is created by taking a black and white lattice framework as background and with the structure composed of transparent glass French windows and balcony balustrades as the center.

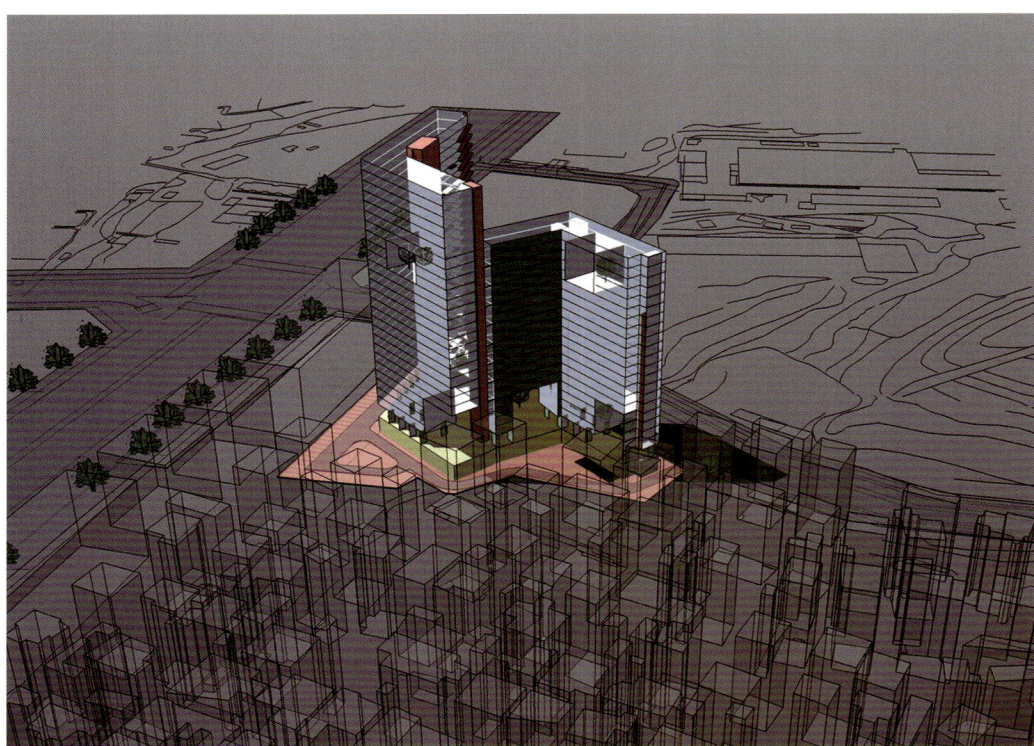
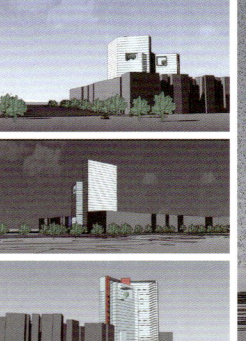

建筑布局

建筑主体由相连的分别为31层、29层、28层、26层的几座公寓形成，沿基地西南面和南面展开布局，一层为商业用房，二层为架空平台层，管理用房设于三层。户型基本为单廊式布置的单身公寓，并对西晒采取措施。展开式布置的建筑体量，使建筑可以获得最大的通风采光面，同时与城市建筑群及道路相协调并呼应。

Buildings Layout

The main building consists of four connected apartment towers distributed in the southwest and south parts of the plot, which numbers of stories are 31, 29, 28, and 26 respectively. The first floors are used for commercial purpose, the second floors are platforms of empty space, and the third floors are management offices. The basic internal layout of the flats features single room with a outside common corridor for all flats, and some measures are taken for preventing western sunlight. The building structures are spread for the maximum ventilation and lighting while creating a harmonious relationship with neighboring buildings and roads in the city.

Fantasia Blooming Town • Chengdu
花样年花样城·成都

项目地点：中国，四川，成都市
项目时间：2007年
设计规模：63.2万 m²
设计阶段：方案设计
项目现状：已建

Project Location: Chengdu
Project Date: 2007
Project Scale: 632,000 m²
Design Phase: conceptual design
Project Status: construction completed

While facing densely distributed buildings, our design allows us to experience maximum visual distance through integration of external spaces.

面对高密度的居住形式，通过整合外部空间求得相对最大的视距。

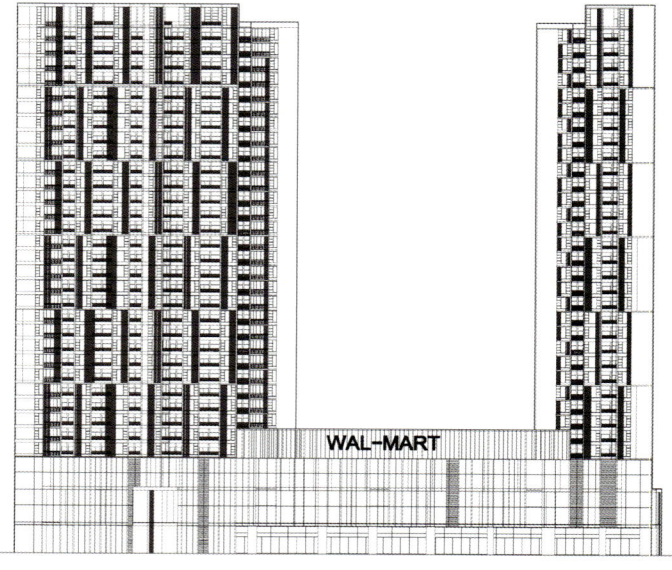

立面图

立面图

规划布局

整个小区采用建筑形体高低结合的布局方式，于小区中间布置6栋100m以下的高层住宅。东面和西北侧为5栋25层的高层住宅和3栋100m以下的高层住宅，南侧为5栋12层小高层住宅，西南角则布置2栋100m以下的高层住宅，形成中间高、四周低，且城市节点又拔高的空间形式。此布局有利于形成更加丰富的天际线，在小区内部创造更大的中心花园。同时也使得尽可能多的户型有良好的景观视线。

Planning Layout

The whole estate is designed with a layout with a combination of high-rise and low-rise buildings. Six high-rise buildings with height no more than 100m are built in the central area of the estate; five 25-storey buildings and three high-rise buildings with height no more than 100m are built at the east and northwest parts; five 12-storey medium high-rise buildings are located at the south part and two high-rise buildings lower than 100m are constructed at the southwest corner. The overall spatial layout features a high center and four low sides while the urban nodes (public places) have somewhat higher elevation. This layout is favorable for creating more luxuriant skylines, building a larger central garden in the estate, and for creating good landscape view for the occupants.

规划结构

小区以一个中心绿地把内部各组团院落连接起来，沿途串连了小区内部活动主广场和沿途各个景观节点，是小区人文、自然活动最精彩之处。

Planning Framework

All internal courtyards are connected by the central green area in the estate. Around the greenbelt, there are the main activity square and some small scenic spots, which is the most wonderful place full of humanistic and natural attractions.

竖向设计

小区内部车库采用半地下式设计，四个组团形成一个大的台地。中心绿地与小区商业街等城市半公共空间为同一个标高，形成中间低、四周高的半环绕式盆地。而在小区的主入口以及一些节点做了一些小高差作为区域的空间过渡。在各栋楼宇的架空层布置配套的商业、会所以及生态车库的通风采光口，不但解决了车库的通风、采光等问题，而且能达到节约能源的目的。

Vertical Design

The garages in the estate are designed as half underground. The platform-type central green area consisting four structures has the same elevation as that of the public spaces such as the commercial street, so that a semi-encircled basin with a low center and high sides around it is created. Small elevation difference is designed at the main entrance and some connecting nodes to guide the transition between spaces. The empty space of the first floors of all buildings are used for stores and pan clubs, as well as for ventilating and lighting openings of the eco-garages, which not only solve the problems of ventilation and lighting for the garages but also can save energy.

Ao Yun Garden • Shanghai
澳韵花园·上海

项目地点：中国，上海市
项目时间：2002年
设计规模：3.5万 m²
设计阶段：方案设计
项目现状：已建

Project Location: Shanghai
Project Date: 2002
Project Scale: 35,000 m²
Design Phase: conceptual design
Project Status: construction completed

For this low density project, richness and diversity of the community can be easily integrated into the city.

在这一低开发程度的项目中，社区的丰富性、多样化使其非常容易地融入城市。

项目总览

上海澳韵花园位于上海市闵行区七莘路与华茂路交汇口的西南部，为万泰花园的三期项目。项目定位为具有异域休闲情调的联排别墅和精品城市住宅。

Project Overview

This garden is located southwest of the intersection between Qi Xin Road and Hua Mao Road in Minhang District of Shanghai; it is the third phase project of Wan Tai Garden. The garden is positioned as town houses and premium urban residence with an aura of exotica and leisure.

立面图

联排别墅造型充分发挥屋顶"第五立面"的造型作用，通过屋顶单坡、双坡、双向单坡组合变化及立面基本造型变化设计成4种相同平面功能的外部造型单元，再通过单元的不同拼接方式及色彩的变化形成丰富的联排别墅立面形象。立面材质富于变化，涂料、金属百叶、玻璃、石材、木材等天然质朴的材质组合也准确地表达出项目的定位。

In the design of townhouse shape, the roof is treated as "the fifth facade". Four kinds of external shaping units with the same planar function are designed by means of combinations of pentroofs, double-pitch roofs, and pentroofs in two directions, as well as varied shapes of facades. Furthermore, luxuriant images of the townhouse facades are created through using different ways to connect the units and using different colors. The combination of various materials and textures used for the facades, including coatings, metal shutters, glass, stones, timber and other natural unadorned materials, precisely express the position of this garden estate.

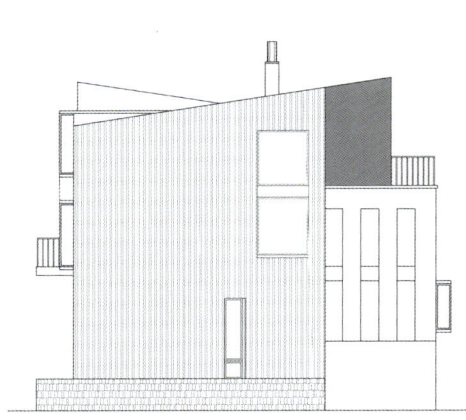

东立面图

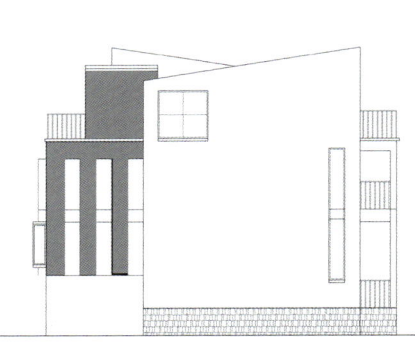

西立面图

Tao Hua Ling • Yichang
桃花岭·宜昌

项目地点：中国，湖北，宜昌市
项目时间：2004年
设计规模：5.15万 m²
设计阶段：方案设计
项目现状：在建

Project Location: Yichang
Project Date: 2004
Project Scale: 51,500 m²
Design Phase: conceptual design
Project Status: construction in progress

This is a reconstruction and expansion project. The difficulty in the design is how to bring new vitality to the buildings.

该项目是一个改建和加建项目，其难度更多地体现在如何重新让其迸发新的活力。

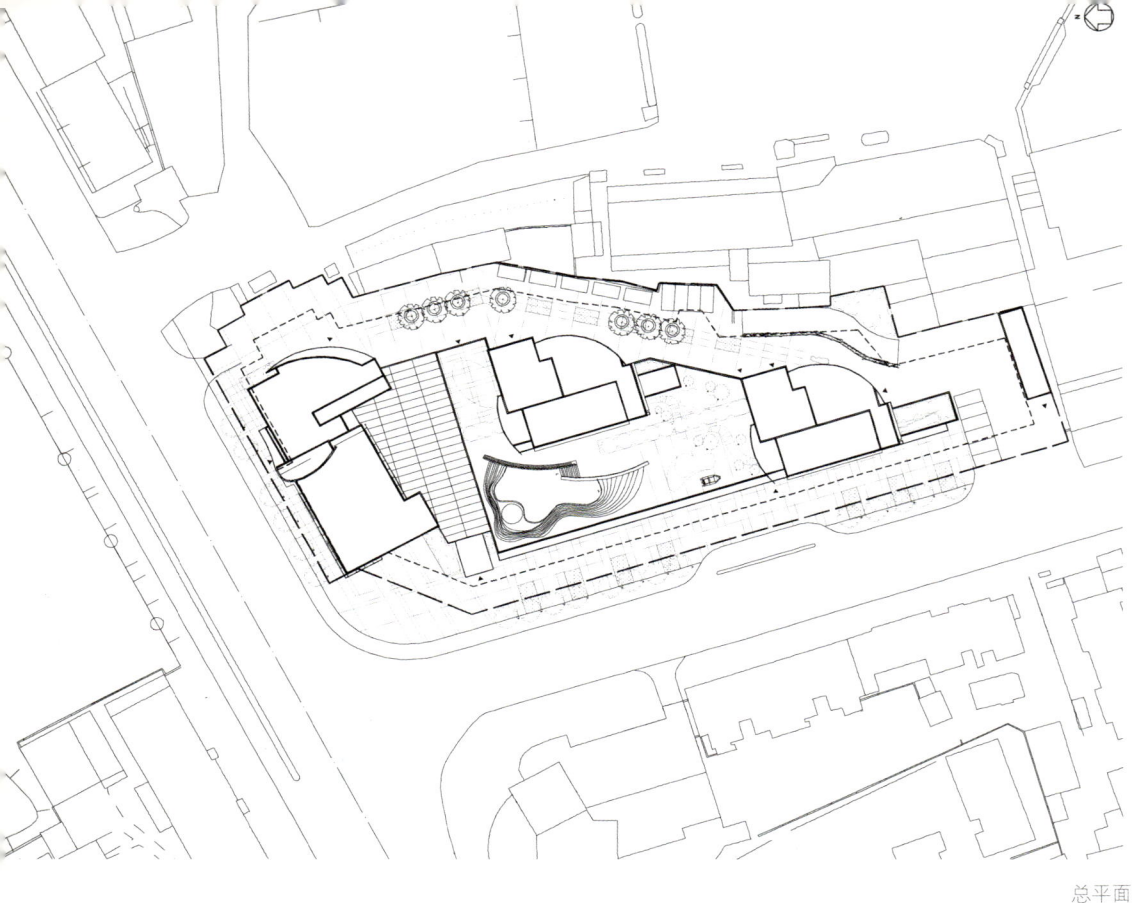

项目总览

项目用地狭小不规则，且北部有一幢烂尾高层建筑，通过合理布置商业、公寓、酒店三个功能体块的位置，以及有机处理三者之间的相互关系，实现了具有城市角色的综合建筑群。

Project Overview

The plot is narrow and irregular and there is abandoned unfinished high-rise building in the north. Therefore a complex with complete urban functions are designed, in which the three functional volumes of commerce, apartment, and hotel are well arranged in terms of position and an organic relationship among the three volumes is achieved.

总平面

1. 商业：商业是城市的"粘合剂"。商业的设计关注发掘那些伴随的、潜在的、丰富的、更人性化的要求，令简短的商业交换活动变成可受时间参与和享用的休闲活动。1-5层的复合商业包括商业街、超市、酒楼和健身会所，结合通高中庭设计，使商业空间更具有可视性和可达性。

2. 广场：广场作为城市的客厅，具有集散、展示、休闲等多种功能。本项目退让城市道路空间形成商业广场，吸引更多的消费者，同时也为宜昌市民提供娱乐、休闲的场所，达到双赢效果。

3. 酒店：酒店临近云集路，作为城市的对外形象窗口，可提升城市品味。将原烂尾楼改造成为酒店的一部分。酒店以标准间为主，部分单间为辅，与桃花岭饭店相互补充。

4. 公寓：公寓布置在用地中部和南部，平面布局紧凑，以尽可能缩小体量。

1. Commerce: commerce is the "adhesive" of a city. In the design of commercial structure, the exploration of incidental, potential, and more humanistic needs is emphasized, so that the simple and temporal commercial activities can become enjoyable and recreational ones. On the first floor to the fifth floor of the commercial center, there are commercial streets, supermarkets, restaurants, and fitness clubs; with a central atrium reaching the roof, the commercial center has spaces easily seen and accessible.

2. Square: as the "living room" of a city, a square has many functions such as distribution, exhibition, and recreation. A space is left between the building and the city road to build a commercial square, with the purpose of attracting more customers and providing a place for the recreation and leisure-time activities of the citizens in Yichang, creating a win-win situation.

3. Hotel: the hotel is close to Yun Ji Road. As a window to the outside world, the hotel help to enhance the image of the city. The existing abandoned unfinished building is rebuilt as a part of the hotel. Most rooms are standard rooms, and others are single rooms. The hotel and the Tao Hua Ling restaurant together offer service for the guests.

4. Apartment: it is located in the central and south parts of the plot; the plane layout is compact so as to reduce the volume as much as possible.

80 Hou Jie • Shenzhen
80后街·深圳

项目地点：中国，广东，深圳市
项目时间：2004年
设计规模：4.4万 m²
设计阶段：方案投标
项目现状：已建

Project Location: Shenzhen
Project Date: 2004
Project Scale: 44,000 m²
Design Phase: competition
Project Status: construction completed

The site is located in a cluttered old industrial district. We strive to create a vivid image of the estate by breathing a new life into it.

该项目处在混杂的旧工业区中，我们力求创造出如同蜕皮新生的生动效果。

立面设计

以景观为导向的立面设计,以大幅景窗和有韵律感的阳台作为主要造型元素。

以红灰相间的网状肌理为底,绿色玻璃落地窗和阳台栏板形成的体量为中心点,营造出翡翠镶嵌般的视觉效果。

Facade Design

This design is landscape-oriented, with the large-scale landscape-overlooking windows and the balconies with a sense of rhythm as major shaping elements.

Red and grey mesh texture is used as background and the volume composed of green glass French windows and balcony balustrades is located in the center, creating an visual effect that as if the whole structure is inlaid with emerald.

住宅设计

标准层采用有自然通风与采光的内廊式布局。

户型布局尽量将厅、房置于外沿，厨厕窗开于凹槽。A、B座标准层分别为1梯14户和1梯12户。将大套型（大2房）分布在标准层的端部（东南、西南、西北角、东北角）。

Residence Design

The standard floors are designed with a gallery style layout with natural ventilating and lighting conditions.

In the internal layout of flats, the living rooms and bedrooms are generally placed in outer edge of the buildings, and the windows of washroom and kitchens are in the recessed parts of the building. The standard floors of Unit A and Unit B have 14 flats and 12 flats respectively on the same floor. Larger flats (with two large bedrooms) are located at the ends of the standard floors (southeast, southwest, northwest and northeast corners).

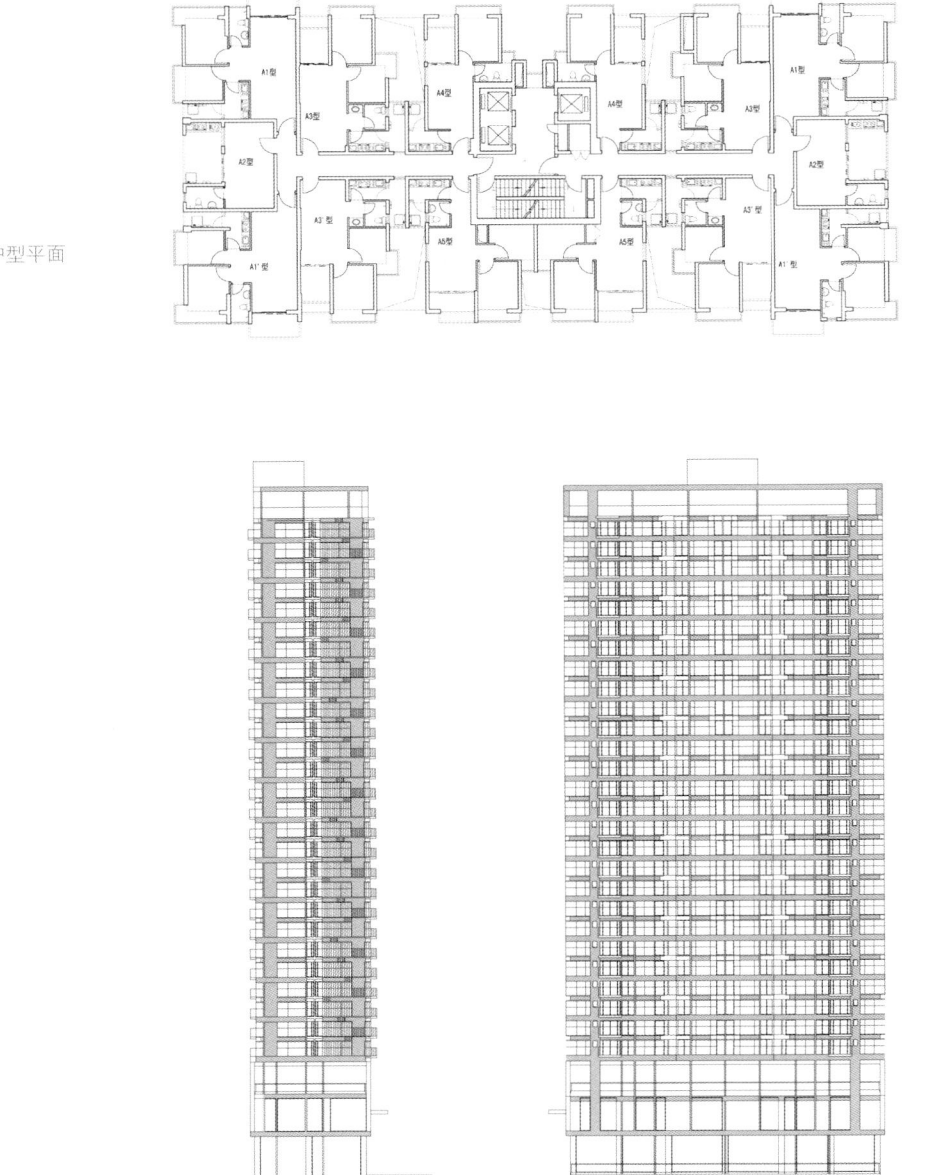

户型平面

立面图

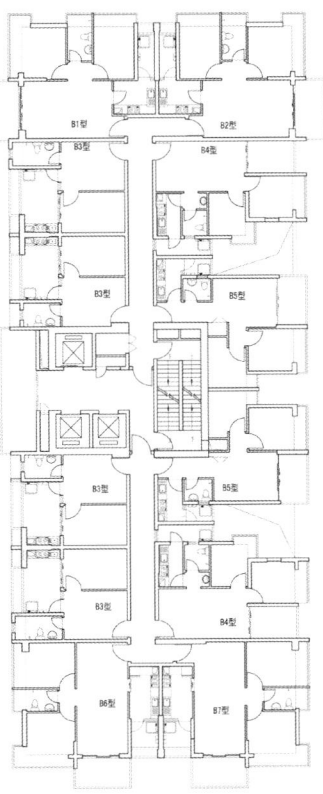

Vanke Jin Yu Lan Wan · Zhuhai
万科金域蓝湾·珠海

项目地点：中国，广东，珠海市
项目时间：2006年
设计规模：5.4万m²
设计阶段：方案设计
项目现状：未建

Project Location: Zhuhai
Project Date: 2006
Project Scale: 54,000 m²
Design Phase: conceptual design
Project Status: construction not started

The dynamic and elegant facade design creates a dialogue between the buildings and the sea.

流动、飘逸的立面设计是与"海"的对话。

设计说明

1. 建筑立面以深色的底，白色的面呈现海边建筑的阳光感和轻盈柔和的氛围。同时，一栋160m的超高层配以三栋100m的高层作为背景，造型强调体块与水平线条的穿插，醒目而突出。三栋高层相应低调而淡雅，与超高层的设计策略相似，但更强调水平面。三栋高层与一栋超高层互为图底关系。

2. Townhouse：Townhouse正面以暖色体块穿插，形成沿街面热闹而有人气的商业氛围；而背面与屋面的高层呼应，相似以黑、白、灰为主色调，体块简化处理。

Design Description

1. The facades of the buildings have a deep color background, and the white planes presenting a sense of sunshine and lighthearted and mild atmosphere of the seaside buildings. A skyscraper with the height of 160 meters is set off with three 100-meter-high high-rise buildings. The shape of the buildings is highlighted by horizontal lines inserted between volumes. The three high-rise buildings are of low-profile style and are quietly elegant, which design strategy is like that of the skyscraper, laying stress on planes. The relationship between the skyscraper and the three high-rise buildings is a figure-ground one.

2. Townhouse: the frontage facade of the townhouse are dotted with warm color blocks so as to create a bustling and booming commercial atmosphere along the street; the back facade is painted with the colors of black, white and grey, matching the high-rise buildings with similar colors. The shape of volumes is simplified.

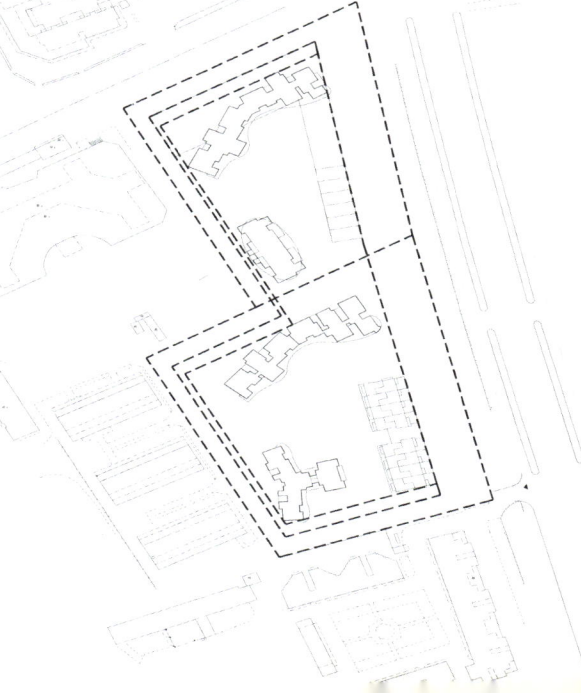

总平面

一层平面　　　　　　　　　　　　　二层平面

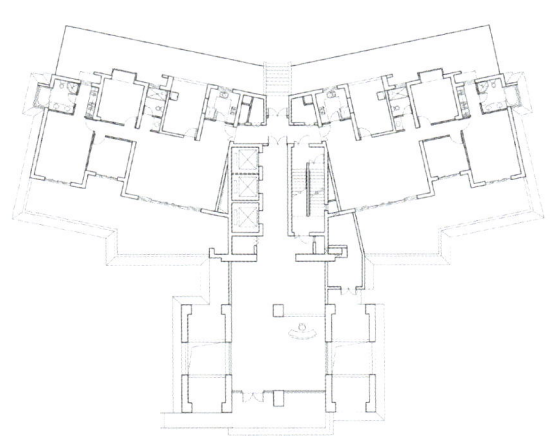

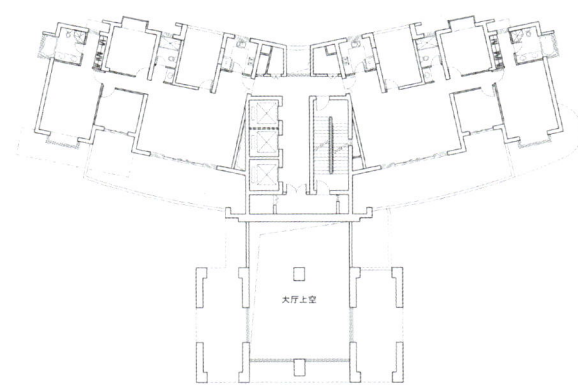

Traditional and Classic......

Traditional and Classic

传统经典

Classics and modern styles are always perpetual themes of architectural designs. We think traditional and classic aesthetics are a basic human psychological requirement and are nostalgia for current society. Therefore we should not treat them by imitation and reappearance but focus on and make a study of their changes and development under modern technical conditions and modern architectural use and function requirements.

Traditions and their nationality under different cultural backgrounds, e.g. oriental and western characters, not only provide a new subject for our design but also bring a new challenge to us.

古典与现代一直是建筑设计永恒的主题,对于传统和古典的审美,我们认为是人基本的心理要求,是基于当今社会的怀旧心态。因此,我们对待它绝对不是机械的模仿再现,而应关注研究其在现代技术条件以及当代建筑使用功能要求下的变化和发展。

传统,加之其属于不同文化背景的民族性,如东方、西方等等,既为我们的设计工作提供了新课题,同时也是新的挑战。

Excellence Repulse Bay • Shenzhen
卓越浅水湾·深圳

项目地点：中国，广东，深圳市
项目时间：2007年
设计规模：6万m²
设计阶段：立面方案设计
项目现状：已建

Project Location: Shenzhen
Project Date: 2007
Project Scale: 60,000 m²
Design Phase: facade design
Project Status: construction completed

We combine common and simple materials to recreate a style featuring the combination of nobleness, grandness, exquisiteness, and elegance.

我们将普通平实的材料通过重新组合实现了高贵大气与精致典雅的交融。

项目总览

本项目位于深圳市南山区高新科技园南区，南临滨海大道，交通便利，处于城市重要景观带上，西南以及正南方向视野开阔，可观赏到红树湾海景，地理位置优越。

立面外观采用了现代风格的设计，简洁挺拔，却又不失亲切、稳重，突显新一代高尚住宅区的清新品质。

Project Overview

The estate is located in the south district of the Hi-tech Park in Nanshan District of Shenzhen. Adjacent to Bin Hai Avenue in the south, the estate is conveniently accessible. With the position in a major landscape belt of the city, it boasts an excellent geographical location with wide field of view in southwest and due south. The occupants in the estate can appreciate seascape of the Mangrove Bay.

The facades are designed with a modern style, straight and forceful, while genial and steady, highlighting the fresh quality of the new generation of noble estate.

建筑设计

1. 利用连续的空间围合出充满生活气息的中庭大花园,并将自然通风、采光的半地下停车场巧妙设计为首层观景大露台;
2. 充分挖掘景观潜力和滨海建筑的审美特征,融入丰富的滨海文化内涵,以新古典主义的手法,通过具有厚重体量感的实墙面和通透的玻璃面所形成的对比,赋予建筑典雅而不失时尚的观感;
3. 建筑实墙面材采用极具质感的砂岩板材以及仿砂岩板的涂料和面砖,营造出尊贵、典雅的怡人居住环境。

Architectural Design

1. An large atrium garden full of liveliness is encircled by the continuous spaces, and the half underground garage with natural ventilation and lighting is skillfully designed as a large landscape-viewing balcony on the first floor;

2. The potentialities of making use of the landscape resources are fully exploited and aesthetic features of coastal buildings are fully considered in the design. Rich coastal culture elements are introduced into the design. With a neoclassicism style, the buildings have blank walls with thick and heavy volume and transparent glass facade which create a striking contrast and an elegant yet fashionable effect;

3. The solid wall facades are covered with sandstone boards with sound texture and artificial face tiles and coatings with finish like sandstone, creating a dignified, elegant, and pleasant living environment.

平面图

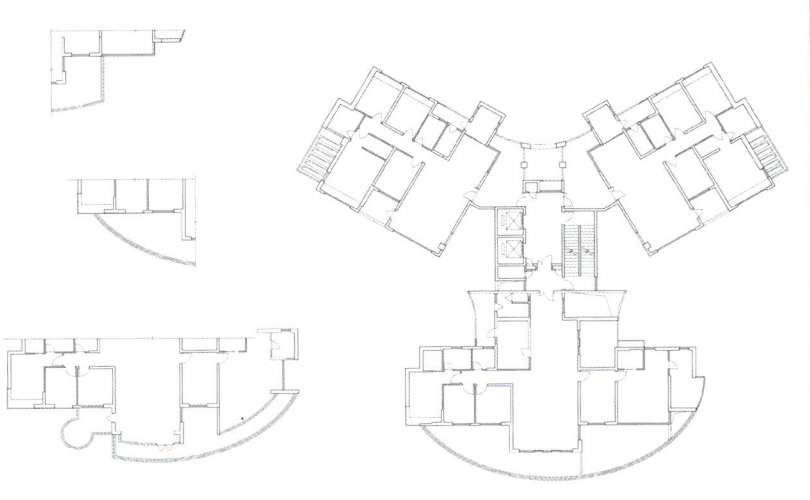

Horoy Xi Yuan • Shenzhen

鸿荣源熙园·深圳

项目地点：中国，广东，深圳市，香蜜湖片区
项目时间：2001年
设计规模：14.3万 m²
设计阶段：方案设计 初步设计
项目现状：已建

Project Location: Xiangmihu Area, Shenzhen
Project Date: 2001
Project Scale: 143,000 m²
Design Phase: conceptual design, preliminary design
Project Status: construction completed

By means of highly efficient utilization of land in three-dimensions, variety of different types of residential organization with mixed greening on a three-dimensional level, this allows a new perspective on the concept of "luxury housing".

通过土地高效立体的利用、多样化不同类型住宅建筑的组织，以及建筑同立体绿化之间的结合，让人们对"豪宅"有了重新的认知。

项目总览

鸿荣源熙园位于深圳市福田区有中心区后花园"绿核"之称的香蜜湖片区用地,自然环境得天独厚,旅游景观近在咫尺。

Project Overview

Located in the Honey Lake region, which is called as the "Green Core" in the backyard garden of the CBD in Futian District of Shenzhen, Horoy Xi Yuan enjoys a charming environment endowed by nature and is only a stone's throw away from the resorts with attractive scenery.

建筑造型采用简约古典的处理手法，通过坡屋顶简化檐口的处理、合适的材质及细部搭配，形成简化的三段式造型，力求体现大气、尊贵典雅的形象气质。

A tidy and classical style is adopted in the design of the building shape. Through the sloping roof, simplified cornice, appropriate materials and perfect matching between details, the simplified three-section design is created to present an image of magnificence, nobility and elegance.

规划布局

社区活力是评价一个社区的重要因素，而这种活力来源于居住人群的类型、空间与建筑风格的多样化以及本社区同周边相邻环境的交融关系。设计考虑到其1.5左右的容积率要求，将联排别墅、多层集合住宅、小高层住宅、板式高层到点式高层住宅等多种不同居住类型建筑进行组合，通过良好的组织，在低层社区实现了良好的小尺度空间感，高层社区实现了低密度的优质景观感受。规划上照应城市规划采用北面板式高层，南面点式塔楼，中间低层的布局方式，沟通东西向景观，最终使各户充分享受景观资源并获得良好朝向。同时，以建筑划分场景，使区内生活宜人且富有情调。户型设计摒弃了传统阵列式的设计，利用高差创造出富有个性的集合住宅。

Planning and Layout

The vitality of a community is one of the most important factors used to evaluate this community. Vitality is generated from the type of people living here, the diversity of the spaces and the architectural styles as well as the harmonious relationship between this community and the neighborhoods. Due to the requirement of a plot ratio of about 1.5, different types of residential buildings are designed to form a delightful combination, including townhouses, multi-storey buildings, medium high-rise buildings, slab-type high-rise buildings and point-type high-rise buildings. Through perfect organization, a sound experience of space is created in the lower building area and an impression of low density and exquisite scenery is generated in the high-rise building area. To match the surrounding city planning, the layout of this community is designed with the slab-type high-rise buildings located in the north, the point-type towers in the south and the lower buildings in between, so that the east-west belt of landscape can be fully viewed and all the apartments can enjoy the scenery while all getting good orientations. In addition, the buildings are utilized to separate different garden views so that the life in the community can be more pleasant with romantic atmosphere. As for the layout of apartments, the traditional array design is abandoned and height difference is utilized to create congregated dwelling houses with strong personality.

一层平面

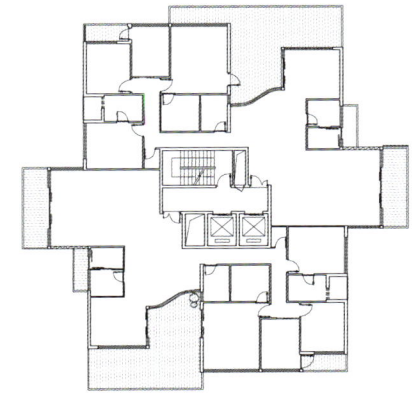

二层平面

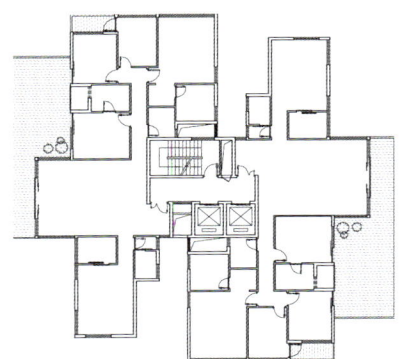

三层平面

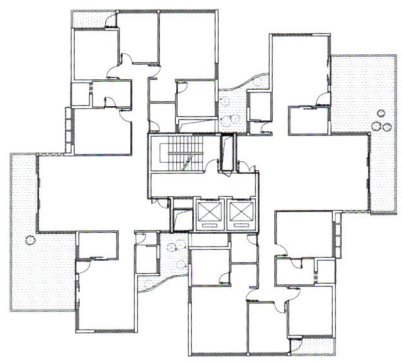

立面图

平面图

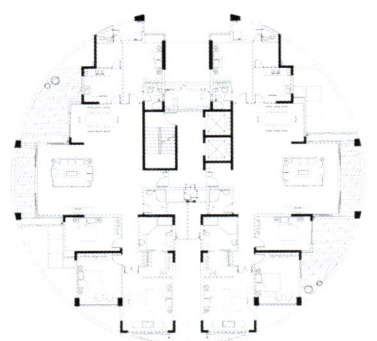
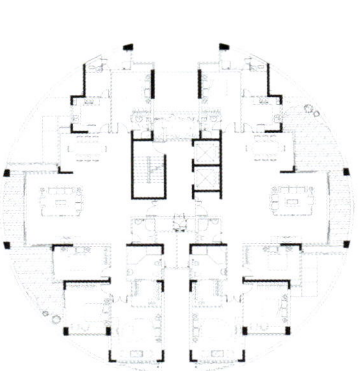

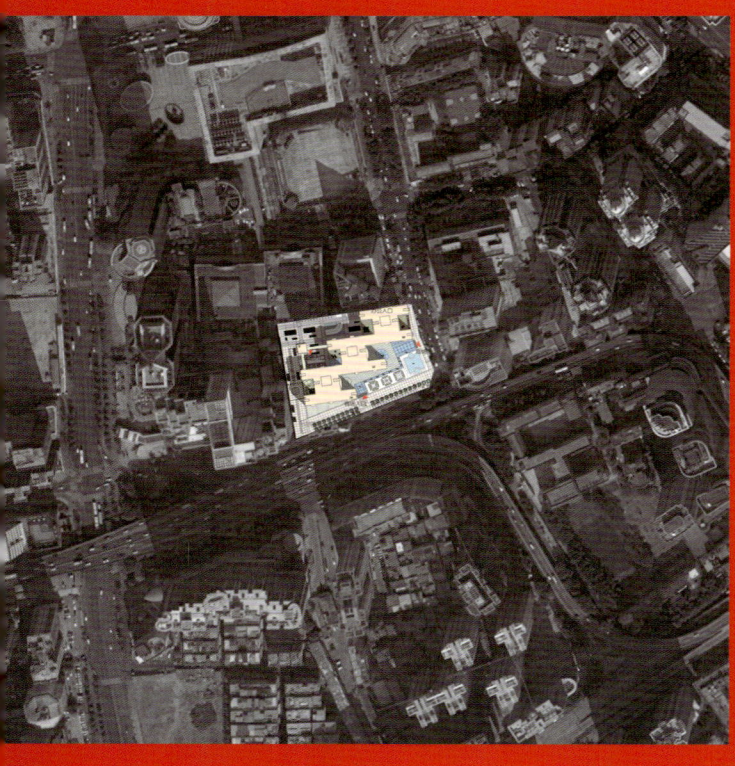

Zhi Di Chun Feng Ju • Shenzhen
置地春风居·深圳

项目地点：中国，广东，深圳市，罗湖区，春风路
项目时间：2003年
设计规模：8.1万 m²
设计阶段：方案设计
项目现状：已建

Project Location: Chunfeng Road, Luohu District, Shenzhen
Project Date: 2003
Project Scale: 81,000 m²
Design Phase: conceptual design
Project Status: construction completed

This project provides an open urban area that can be traversed to reconnect urban segments which were isolated before. The efficient and compact layout ensures high-quality living spaces.

提供的可穿越性城市开放空间，将以前割裂的城市片段重新连接起来；高效、紧凑的平面设计保证了居住空间的优良品质。

Horoy Park Land •Shenzhen
鸿荣源公园大地·深圳

项目地点：中国，广东，深圳市，龙岗龙城29区
项目时间：2004年
设计规模：60.7万 m²
设计阶段：方案设计
项目现状：已建

Project Location: Longcheng No. 29 District, Longgang, Shenzhen
Project Date: 2004
Project Scale: 607,000 m²
Design Phase: conceptual design
Project Status: construction completed

This project is in fact an additional development for the previously completed "New Asia" project. The main focus of the design is to deal with the relationship between the existing buildings and the surrounding urban park.

该项目其实为已建成"新亚洲"项目的后续开发，如何处理同已建成部分以及作为背景的城市森林公园的关系是设计的重点。

543

立面图

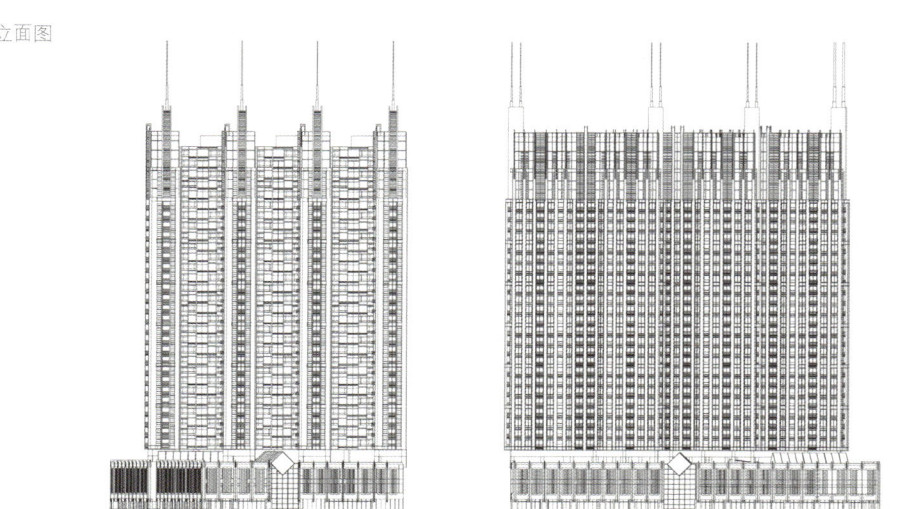

革命性的哥特式造型设计

建筑造型根据小户型面积小、竖向感强的特点，结合哥特式建筑风格创造出具有强烈视觉震撼力、直率而不造作的建筑形象，在罗湖口岸略带西洋古风的建筑环境中既统一和谐，又独树一帜。

Revolutionary Gothic Style Design

As the buildings consist of small flats with small area and a strong vertical sense, an architectural image with strong visual impact and inartificial straightness is created in the building shape design by introducing the style Gothic architectures. Agreeing with the western archaistic style of other buildings around Luohu border crossing, the estate also has its own unique style.

剖面图

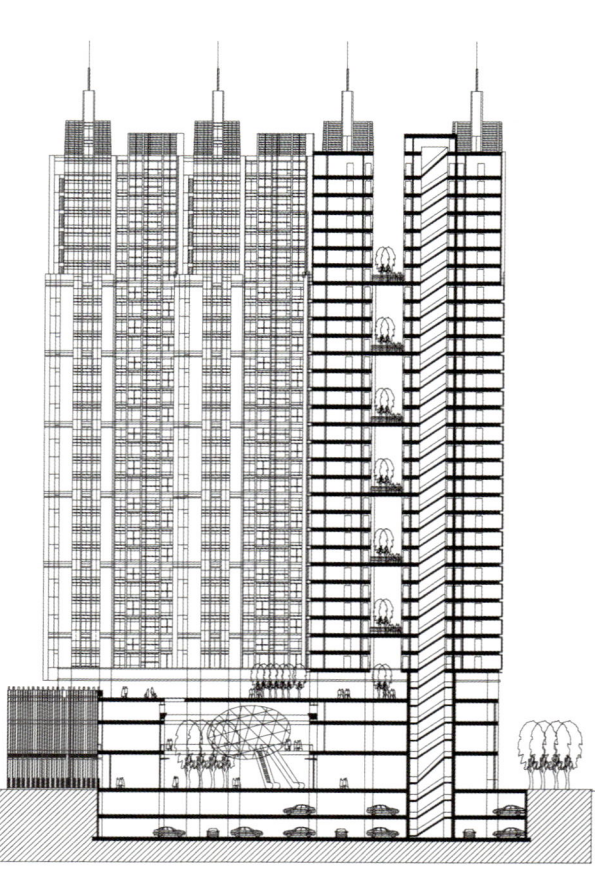

项目分析

项目地理位置优越，商业价值极高，距罗湖火车站不足700米，地处国贸、发展中心商业圈辐射范围，具有典型的港居口岸物业特征，定位为罗湖口岸国际服务公寓。

Project Analysis

Enjoying a favorable location, the estate has high commercial value as it is less than 700 meters away from Luohu Railway Station. With the position within the business circle radiating from the Shenzhen International Trade Building and Shenzhen Development Center, the estate is favored by Hong Kong residents as it is near the border with Hong Kong, so this estate is position as an international service apartment near Luohu border crossing.

设计原则

1. 尊重城市
——寻求与现有城市脉络的契合关系，并积极互动，以求共荣。
2. 作为城市活力元素，活力的商业空间
——以促进现有城市商圈的提升为前提，使本项目的商业成为区域城市活力的新发生器。
3. 城市公寓型的居住类型
——在城市环境中最大化地争取资源：日照、通风、景观、视线并考虑功能的灵活可变性。

Design principles

1. Respect the city
   -- To strive to blend with urban pulses and interact with them so as to realize joint prosperity.
2. Act as an active urban element and dynamic commercial space
   -- To make this project's business facilities become a new generator for regional urban vitality on the premise of promoting the development of existing urban business circles.
3. A habitation type of urban apartment
   -- To seek maximum resources, such day-lighting, ventilation, landscape, and line of vision, in the urban environment and take into account functional flexibility and changeability.

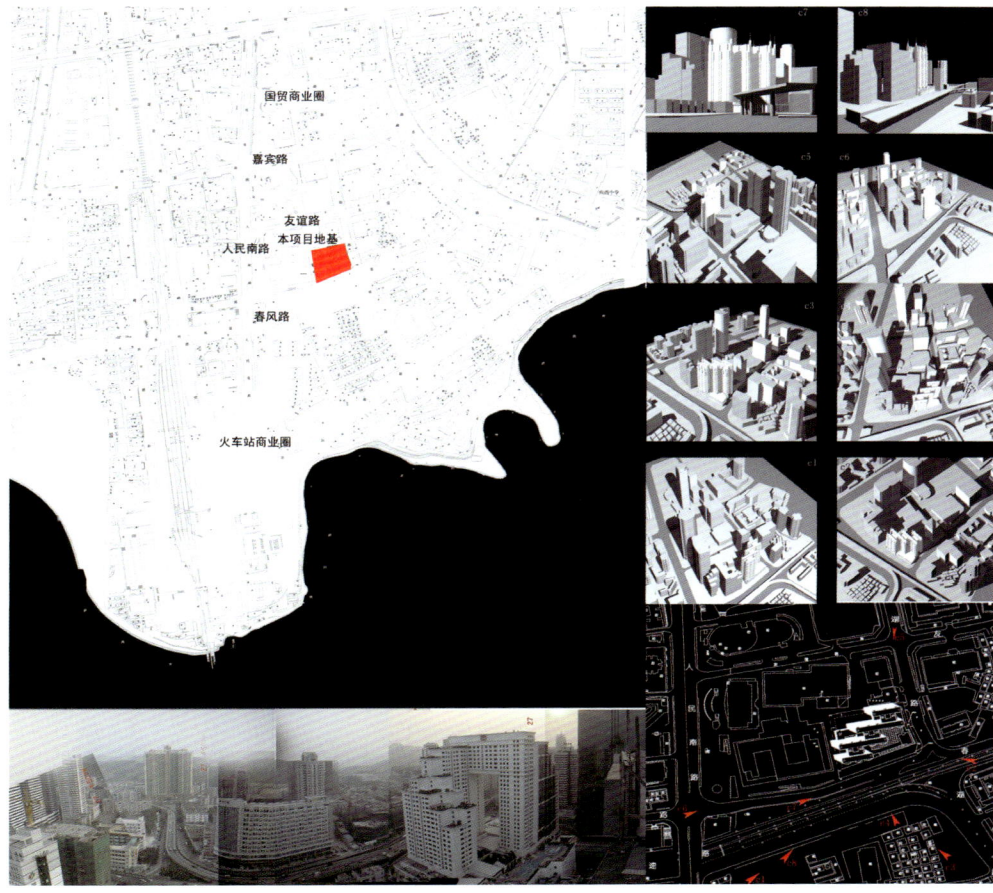

模型、方位图

分析图

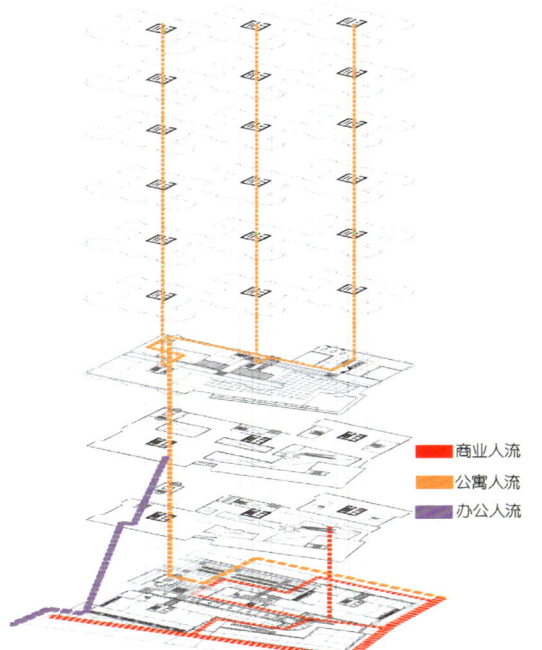

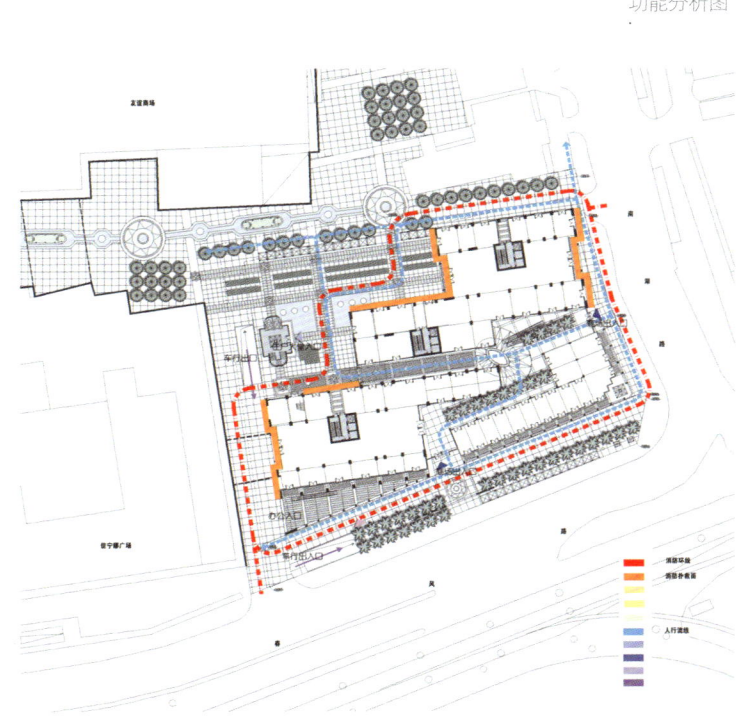

功能分析图

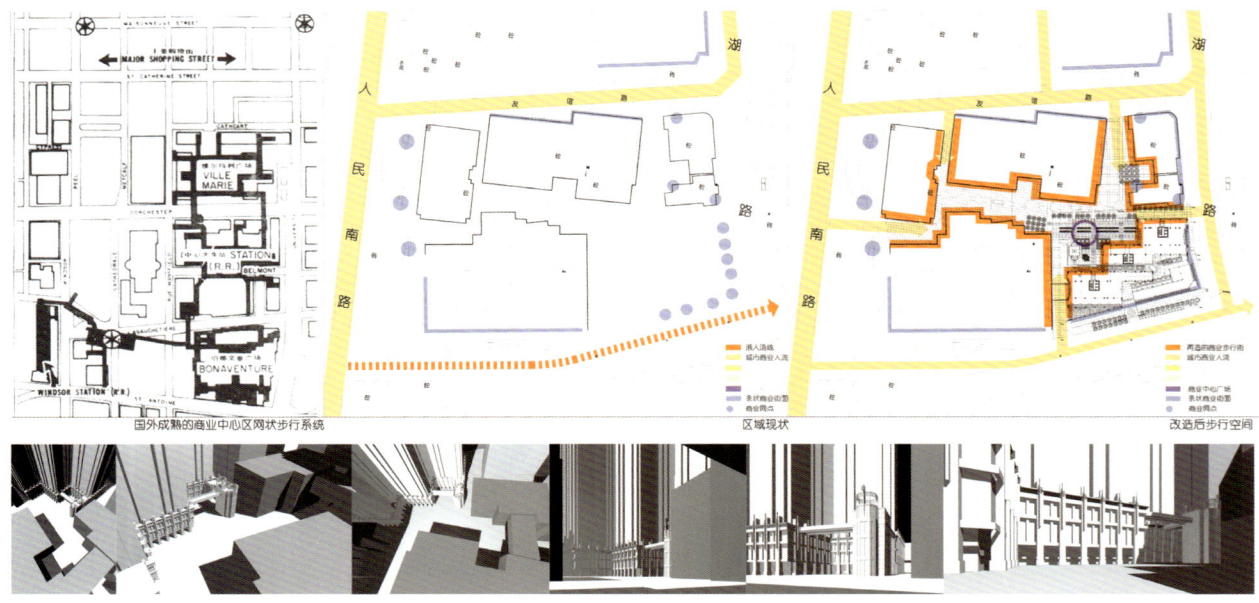

"网络"状商业街区空间布局

——"后街":商业价值的最大化。
商业裙房沿南部春风路布局,沿用地北部设置"后街",使底层商业形成环状街铺,同时,在空间上与周边的佳宁娜、友谊城贯穿起来,成为可通达人民南路、友谊路的次一级网状不行商业带,为区域商业氛围增添活力。

——顺应主楼体量错列有序而成的"广场"、"街道"等室外空间形态,收放有致,更增加了底层街铺的长度,同时与建筑体量形成良好的图底反转关系,使建筑内外共同作用,互相交流,成为积极的商业空间。

——南向布局的商业裙房也为主体住宅提供了南向可沐浴阳光的屋顶花园空间。

Network-like spatial layout of commercial blocks

-- "Back street": maximization of commercial value
Commercial skirt buildings are allocated along Chunfeng Road in the south of the plot. A "back street" is set in the north of the plot. Ground-floor shops take on a ring shape. Meanwhile, they are linked up with surrounding CARRIANNA and Friendship Department Store. Thus a sub-grade-I network-shape pedestrian commercial zone with an access to South Renmin Road and Youyi Road comes into being, bringing a new vitality to regional commercial atmosphere.

-- Such outdoor spatial forms as "plaza" and "street" are arranged orderly, increasing the length of ground-floor shops. They form a good relationship with buildings. Thus outside and inside of buildings interact and communicate with each other to form an active commercial space.

-- Southward commercial skirt buildings also provide a southward roof garden space with plentiful sunlight to main residences.

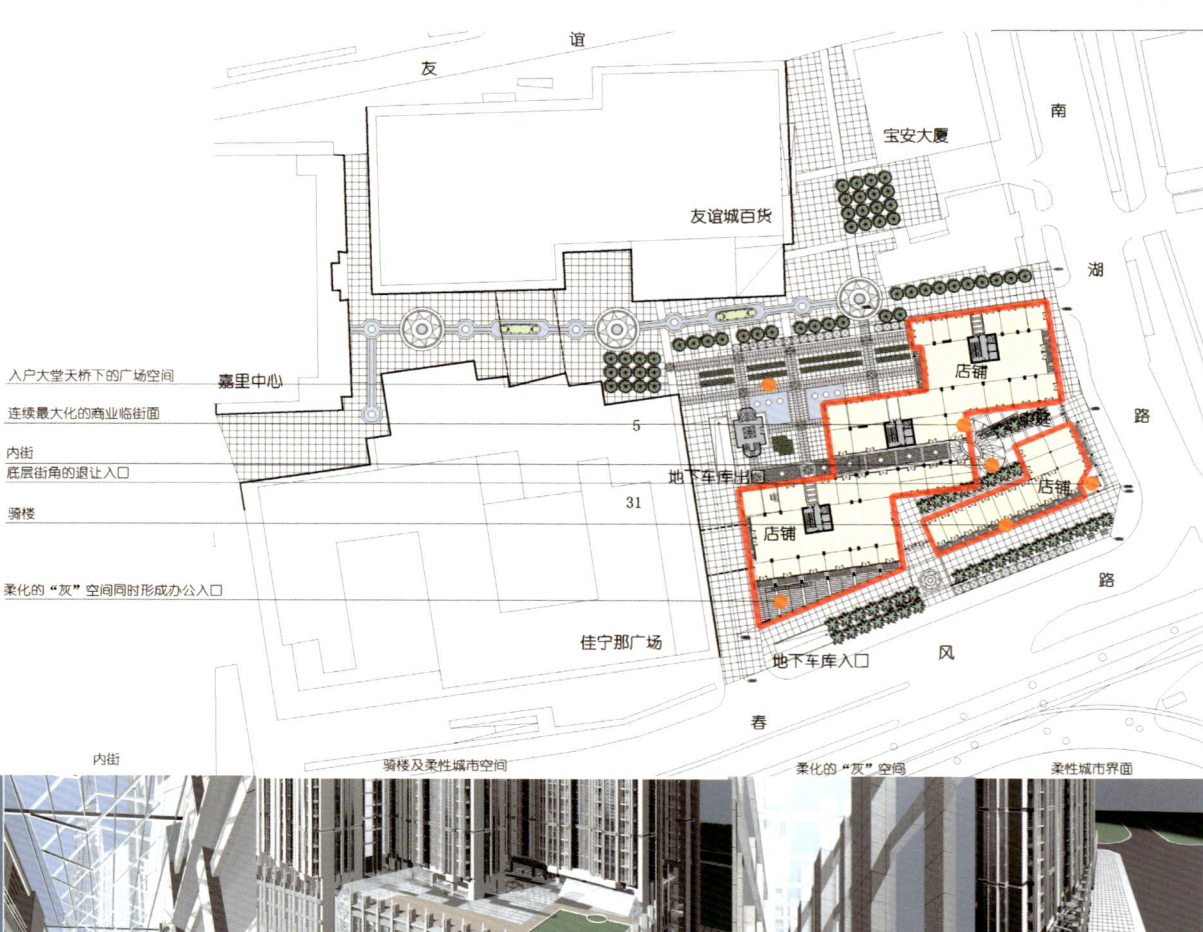

方位平面

项目设计

以促进现有城市商圈的发展为前提，尊重城市关系，寻求与现有城市脉络的契合，打造城市公寓型的居住类型，在城市环境中最大化地争取资源。主楼由三幢"工"字单元错位相连，保证了高层小户型视野的开阔，又巧妙地契合地形的特点，实现了对城市街角的避让和人流视线的引导作用。网状的步行商业街使建筑没有背立面，商业价值也全方位得到发掘。

Project Design

Aiming at promoting the development of the existing business circle of the city, the design creates a good relationship between the estate and the surroundings and strives to match the planning style of the city. The estate is built as urban apartments, utilizing as mush resources as possible in the urban environment. The main building consisting of three"工"(Chinese letter) shape units connected in a staggered way, which ensures a broad field of vision for occupants in small flats of high-rise buildings. The topography of plot is considered in the design to avoid creating street corners and well guide the sight lines of people. As there is a network of pedestrian commercial streets around the building, so the building does not have back facade, which greatly enhance the commercial value of it.

Horoy Park Land • Shenzhen
鸿荣源公园大地 · 深圳

项目地点：中国，广东，深圳市，龙岗龙城29区
项目时间：2004年
设计规模：60.7万㎡
设计阶段：方案设计
项目现状：已建

Project Location: Longcheng No. 29 District, Longgang, Shenzhen
Project Date: 2004
Project Scale: 607,000 m²
Design Phase: conceptual design
Project Status: construction completed

This project is in fact an additional development for the previously completed "New Asia" project. The main focus of the design is to deal with the relationship between the existing buildings and the surrounding urban park.

该项目其实为已建成"新亚洲"项目的后续开发，如何处理同已建成部分以及作为背景的城市森林公园的关系是设计的重点。

项目总览

本项目的规划设计旨在通过"山、水、院、街、镇"等并列元素的设计和整合,实现一种自然、生态、阳光、开放、丰富、和谐的人居生活环境和居住的舒适。

Project Overview

The design aims at creating a natural, ecological, sunny, open, luxuriant and harmonious living environment with great comfort through the design and combination of the parallel elements such as "hill, water, courtyard, street and town".

总平面

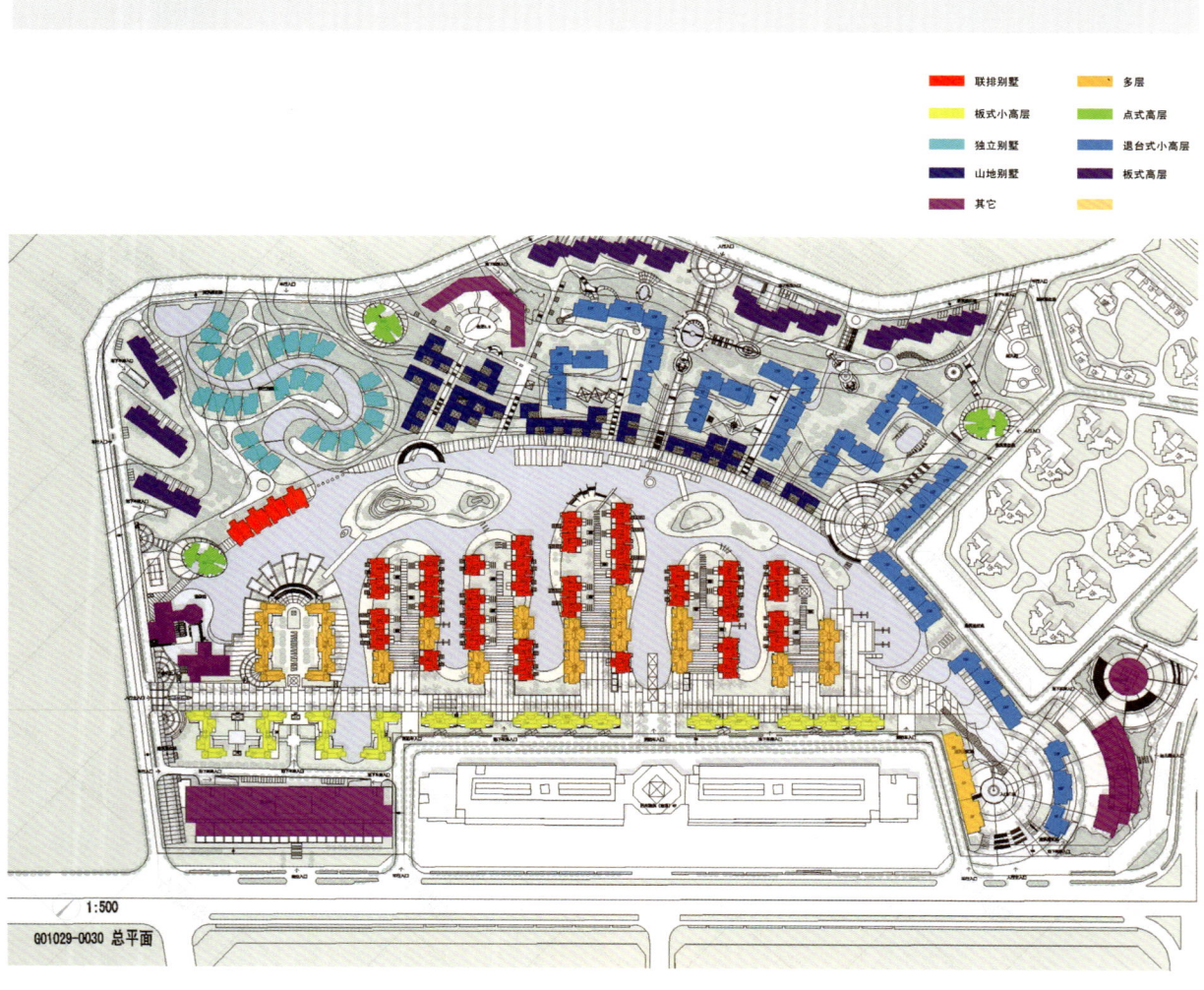

总平面

立面图

项目规划

"山"：充分利用北面佘山公园优越的景观资源，北高南低的地形地势营造出叠落的建筑群落，形成高低起伏的自然山地组团；同时保留多条看山的景观视线通道，让青山的绿色能渗透到社区内，渗透到城市中，形成山的主题社区。

"水"：北高南低的地势让社区的南北轴自然成为天然的汇水空间。在此基础上衍生出人工湖和连串溪流，求得了最多的社区临水面。广场、游船码头、近水平台油然而生，形成水的主题社区。

"院"：既是中国人居住空间的核心和精髓，同时也是组成城市和社区的单元和细胞；从微观层次上看，平地处设计大小相宜的前庭后院，与立体高层中每户拥有的空中露台，共同形成一个"院之社区"。

"街"：本项目由两条贯穿社区南北的"水主题街"和"步行主题街"成为主线索串联起东北向的"山巷"和"水巷"；加之龙翔大道口的社区"商业主题街"，共同为社区带来充满闲适、舒缓的生活节律和氛围。

"镇"：所代表的居住社区是人本的、人性的。由小区到小镇的变化，是一维到多维的进步，是由关注室内到同时关注室内外的质变，是由物质到精神的升华。

Project Planning

"Hill": by making full use of the wonderful landscape resources of Sheshan Park in the north and the landform featuring high in the north and low in the south, a cascade of buildings are built in the hilly natural environment; many channels allowing sightlines to view the hills are kept to make the green landscape become part of the community and the city. Thus, a hill-themed community is created.

"Water": the landform featuring high in the north and low in the south makes the south-north axis of the community become a natural water-gathering channel. Based on this channel, an artificial lake and plenty of streams are derived, and a large area of water body is created. Naturally, squares, marina, waterside platforms are designed to create a water-themed community.

"Courtyard": courtyard is the core and essence of residential space for Chinese; meanwhile, it is also the basic unit and cell to construct cities and communities. From a microcosmic view, a "community of courtyards" is created by designing forecourts and backyards of appropriate sizes on ground and an aerial gazebo for each flat in tridimensional high-rise buildings.

"Street": the "water theme street" and "pedestrian theme street" are two main streets running through the community from south to north and linked with the "mountain lane" and "water lane" in the northeast; these streets and lanes, together with the "business theme street" in the community at the entrance of Long Xiang Avenue, create leisurely and comfortable life rhythm and atmosphere for the whole community.

"Town": it represents a people-oriented and humanistic residence community. The change from a residential quarter to a town is an increase from one dimension to multiple dimensions, a qualitative change from focusing on indoor environment to both the indoor and outdoor environment, and the sublimation from materiality to spirituality.

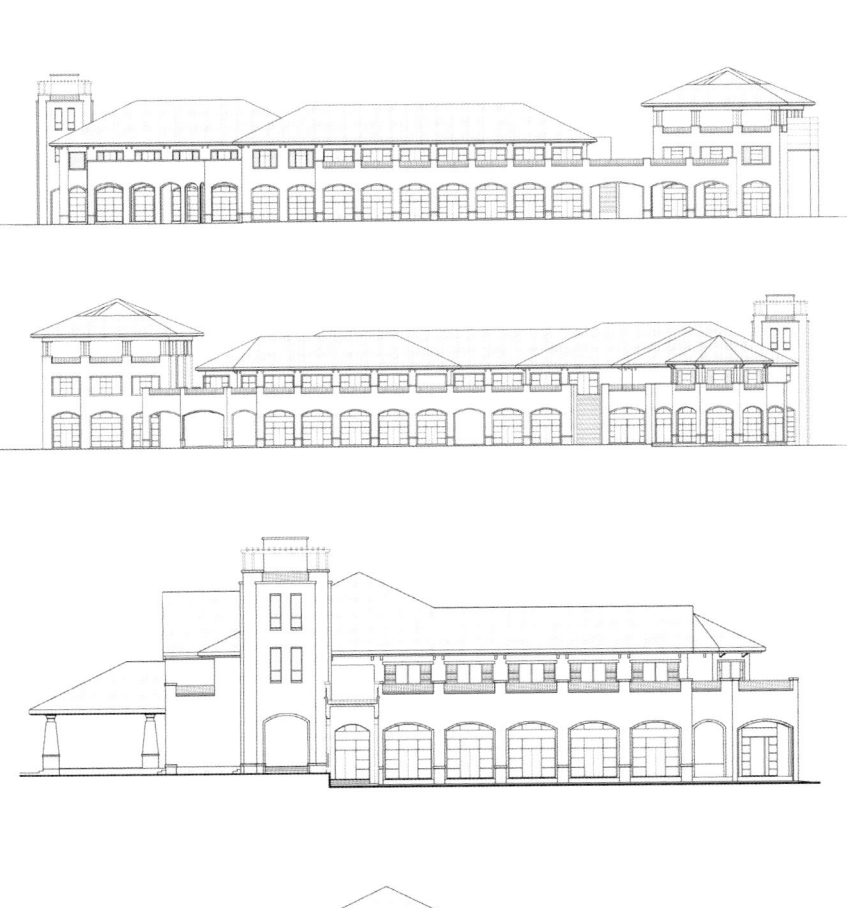

立面图

一层平面

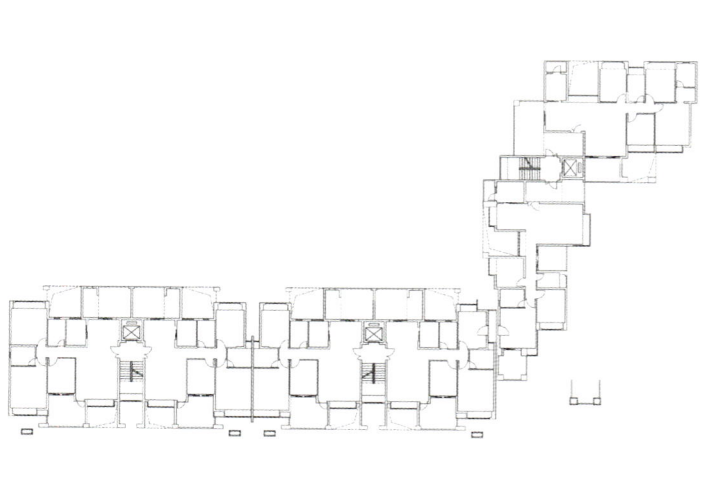
二层平面

Hua Sheng Ling Yu • Shenzhen
华盛领域·深圳

项目地点：中国，广东，深圳市
项目时间：2005年
设计规模：3万 m²
设计阶段：方案设计
项目现状：已建

Project Location: Shenzhen
Project Date: 2005
Project Scale: 30,000 m²
Design Phase: conceptual design
Project Status: construction completed

For the design of the small residential units with limited investment, a remarkable result is achieved through reasonable allocation of resource.

作为小户型的住宅设计，在投资严格控制的条件下，通过合理配置资源达到意想不到的效果。

项目总览

本项目位于深圳市景田北景田七街,北临北环辅道,是该地区唯一的小户型公寓楼。

Project Overview

This estate is located in Jing Tian No. 7 street in Shenzhen and is adjacent to the relief road of Bei Huan road. It is the only small flat apartment in this area.

项目分析

尊重城市关系,充分利用基地的景观资源,在有限的用地范围内创造最大的公共活动空间和居住空间。户型设计经济、实用,立面精致、时尚。

Project Analysis

In the design of the estate, a good relationship between it and public spaces outside this estate is established, the landscape resource in the site is made full use of, and maximum spaces for public activities and residence are created in a limited area. The internal layout of flats is designed to be economic and practical, and the facades are fine and fashionable.

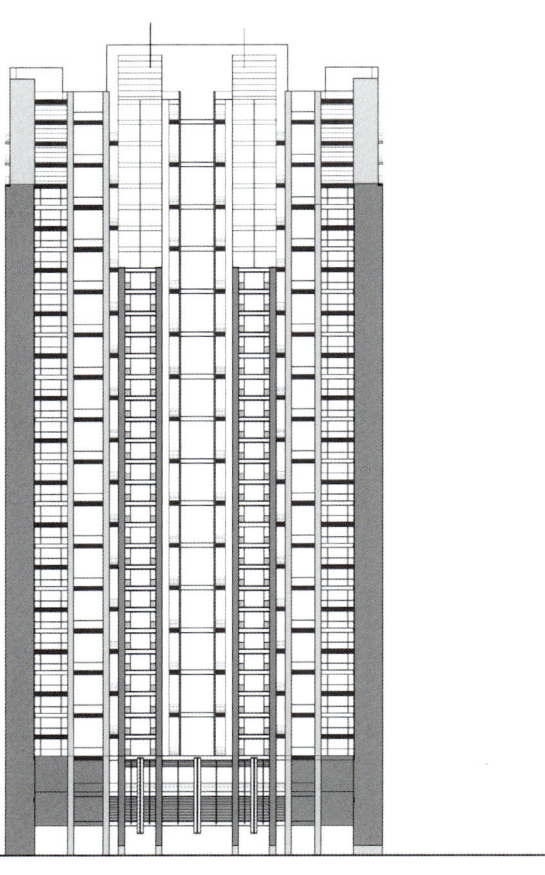
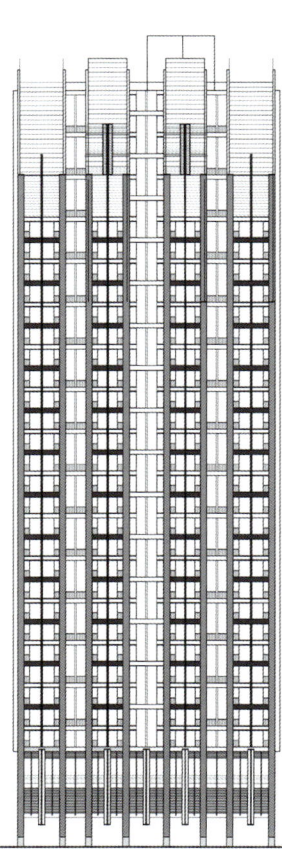

立面图

China Overseas Property - Long Gang Olympic New Town · Shenzhen
中海龙岗奥体新城·深圳

项目地点：中国，广东，深圳市
项目时间：2006年
设计规模：45万㎡
设计阶段：方案设计
项目现状：已建

Project Location: Shenzhen
Project Date: 2006
Project Scale: 450,000 m²
Design Phase: conceptual design
Project Status: construction completed

For such a project with densely distributed buildings, we strive to maximize various values by designing a combination of different types of houses.

对于如此密集的居住类型，我们通过不同类型的组合，实现各类价值取向的最大化。

规划构思

本项目用地规划巨大，交通便利，自然环境优美，具有形成高品质大型社区的先决条件。

1. 在整体布局上，采用高层、小高层围合大院子的手法，打造资源型高品质楼盘，在空间尺度、视野、公共资源上形成领先市场大部分楼盘的优势。小高层等高端产品直接朝向中心庭院地面景观，外围高层则享有庭院全景和城市远景。

2. 在楼栋布置上，通过错拼的方式组织各楼栋，使每户都能拥有超大尺度的景观视野，充分发掘地块内及地块外的景观价值。

Planning Concept

This project site is of a large area plot, conveniently accessible and with picturesque surroundings, which boasts the conditions for building a large high quality community.

1. As for the overall layout, a big courtyard is encircled by high-rise buildings and medium high-rise buildings, aiming to build high quality buildings through quality resources. The spatial scale, field of vision, and public resources of this estate outperformed most estates in the real estate market. The high-end buildings such as medium high-rise buildings are directly orientated toward the landscape of the central courtyard, whereas the outer high-rise buildings enjoy the whole landscape of the courtyard and the faraway view of the city.

2. As for the building layout, all buildings are staggered so that the occupants in each flat can have a very wide field of vision for viewing landscape. Therefore the value of the landscape inside and outside the plot is maximized.

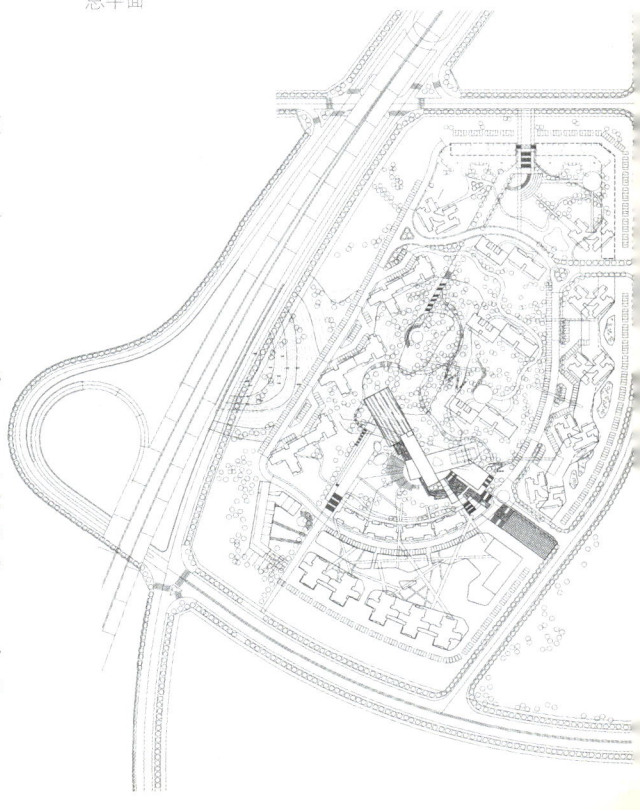

总平面

3．同时错拼的方式使中心庭园与周边自然山体之间相互渗透。

4．在整体空间布局上，开放西南角，使住区与远景山体形成良好的对话关系，同时也形成有节奏的城市天际线。

3.he buildings are staggered so that the central courtyard and the surrounding natural hills form an integrated landscape.

4.As for the overall spatial layout, the southwest corner is open so as to allow the faraway mountains can be seen and a rhythmic city skyline is created.

户型特色

高端户型特色

1. 将别墅级的立体生活空间带入高层住宅中，创造有特色的户内生活空间；
2. 提供高品质无干扰的居家生活方式；
3. 因地制宜，使住户能够全方位使用基地内外的景观资源。

经济性户型特色

1. 户型灵活多变，提供多样性的家庭生活方式，适应力强；
2. 户内空间与户外空间相互交融，提高经济户型的舒适性；
3. 体现人性化的关怀，为住户提供最佳尺度的生活空间。

造型分析

1. 整体采用简洁明快的风格，与整个奥体公园的风格相协调；
2. 整体形成几大分区：地面尺度采用较为丰富的颜色，给人以现代、时尚的感受；中低区采用温和、中性的色调；高层区采用简洁、直率的现代风格，体现出高品质楼盘的大气、豪华。

Features of Flats

Features of High-end Flats

1. Villa-level tridimensional living spaces are brought into the high-rise building to create distinctive indoor living spaces.
2. High quality and interference-free living style is provided.
3. Perfect adaption to the site's actual situations, so that the landscape resources inside and outside the site are made full use for the occupants.

Features of Economic Flats

1. Flexible and adaptable internal layouts provide diverse life styles.
2. A perfect integration between indoor spaces and outdoor spaces improves comfortableness of economic flats.
3. Humanistic care is expressed and living spaces of optimum scale are provided for the occupants.

Shaping Analysis

1. The shape of all buildings is of a simple and vivid style in line with that of the whole Olympic Park.
2. The different parts have different styles: the ground has rich colors which express a sense of modernity and fashion; the medium and low-rise areas adopt moderate and neutral colors; the high-rise areas adopt straight and modern style which expresses grandnesss and luxury of the high quality buildings.

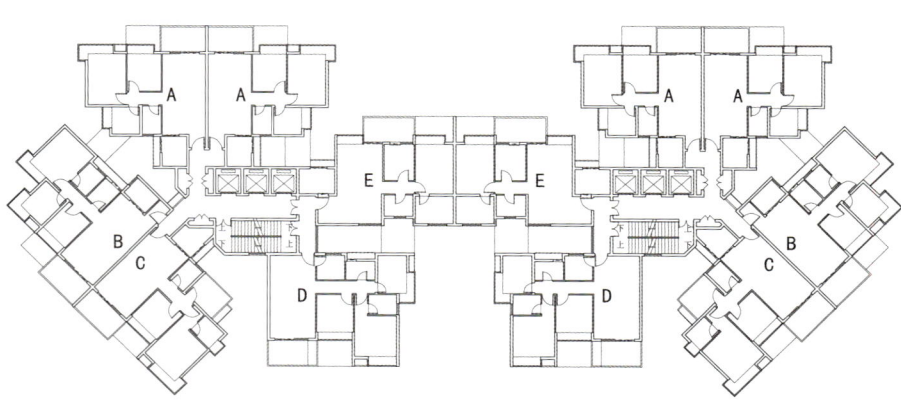

平面图

生态车库及山体保护

项目用地地势复杂，大量的土方开发必然会对自然山体产生巨大影响，同时在经济上也造成浪费。所以，我们在设置地下车库时，要同时注重以下问题：

1. 结合山体的坡度变化，层层叠落地布置了地毯式车库，用短坡道连接，与山体的自然坡度形成完美契合，节约了大量的土方开挖，同时提高了停车效率；

2. 利用坡地的高差变化，巧妙设置车库，将周边自然环境延伸进车库。

Eco-garage and Hill Landform Preservation

The plot has a complicated topography, and excavation of a large amount of earth will unavoidably brings significant impact on the hill landform and may result in extra costs; therefore, to the following issues are considered when designing the underground garage:

1. Terraced garages are distributed on the slope and are connected with short slope passage, so as to perfectly match the natural gradient of the hill slope. A lot of earthwork is saved while the parking efficiency is raised.

2. The garages are positioned skillfully by making use of the elevation difference of the slope land and the natural environment integrates with the garages.

Jiawang Mountain Mansion · Shenzhen
嘉旺阅山华府·深圳

项目地点：中国，广东，深圳市
项目时间：2007年
设计规模：15.5万 m²
设计阶段：方案设计
项目现状：在建

Project Location: Shenzhen
Project Date: 2007
Project Scale: 155,000 m²
Design Phase: conceptual design
Project Status: construction in progress

Common materials are used to create quality through exceptional design.

普通的材质，通过精致的设计营造尊贵的品质。

总体布局

项目利用地形结合高层点板，南北错位，延边布局，形成了对城市街道和北侧相邻地块的适度退让，保证了开放的庭院景观和高层建筑的开阔视野。

Overall Layout

By making use of the landform, a combination of point-type and slab-type high-rise buildings is designed. The buildings are staggered from south to north and are positioned along with the sides of the plot, so there is a appropriate setback from the city streets and the neighboring plot in the north. Such layout ensures open courtyard landscape and wide field of vision for occupants in the high-rise buildings.

总平面

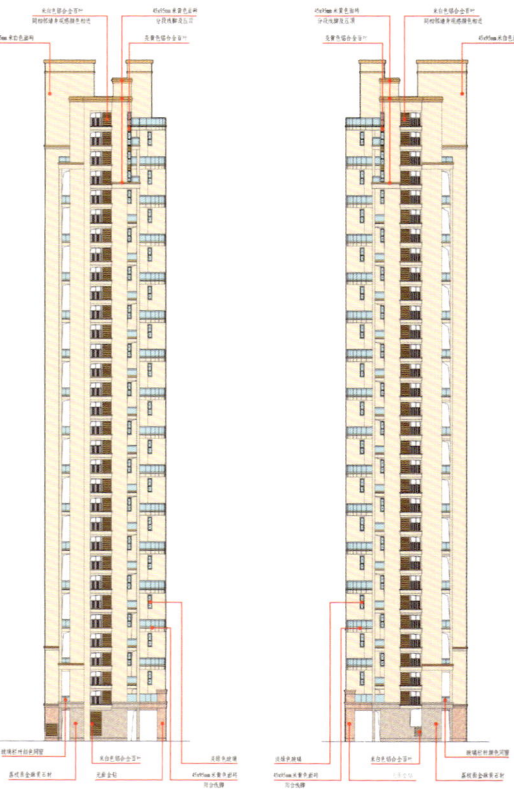

平面图

低层产品布置在用地中部地质条件较差的微风化区域，同时与北面小区外别墅相呼应。低层产品旋转45°，在有限的基地范围内争取日照等自然资源，避免与高层单元的直接对视。商业区临街面设置，并在西边设置集中式商业区。

设计充分利用高台地形气势，创造挺拔的建筑形象和尺度震撼的商业与会所空间。造型强调品质感，色彩与材质搭配典雅、细节精致，体现出项目的尊贵性。

The low-rise buildings are positioned in the central area where the geological condition features slight weathering. From here the villas in the north outside the estate can be seen. The low-rise buildings are oriented with a turn of 45 degree, so that natural resources such as sunlight are easily available in the limited site area, and occupants in the low-rise buildings will not see those in the high-rise buildings. The commercial facilities are set up along frontages and there is a central commercial area in the west.

The topology and terrain of high terrace are successfully utilized in the design so as to create an upstanding architectural image and scale astonishment as well as commercial and clubhouse spaces. The quality of the shape is emphasized, and the colors and materials go well with each other to create elegant appearance and exquisite details, expressing the dignity of the estate.

立面图

布局
总体布局采用周边高层围绕中心庭院、点板结合的布置方式。
1. 南面临城市街道是点式布局，以减少对城市的压迫感；
2. 北面临深业项目是点+板式的布局方式，尽可能避开多层住宅，并利用开敞部位，以保证北边深业项目的日照要求；
3. 多层住宅布置在地质条件较差的微风化区域，同时与北面别墅相呼应、沟通，浑然一体。

高差处理
有效利用基地落差大的现有地形条件，在沿街面设置商业以对应城市的人流，而住宅部分则分布在商业和车库的屋面之上。

空间与景观视线
高层部分沿基地长边方向交错布置，尽量做到视线无遮挡。同时，利用高层围合自然形成小区服务设施+东面中心庭院+湖面的景观轴线，以及西面中心庭院+景观道路+东面中心庭院的又一条景观轴，这两条轴共同构成了社区的景观"面"。而多层住宅内部，又是自成体系的线性景观。

功能布局
周围高层住宅交错布置，保证庭院空间不封闭地质条件差处错位布置多层住宅产品，形成韵律感，扩大了建筑间距。建筑旋转45°，在有限的基地范围内有效争取了日照，避免了与高层单元的直接对视，并且与旋转后的楼形成呼应。商业临街面设置，并在西边设置集中式商业。

1. Layout
General layout: central courtyards encircled by surrounding high-rise buildings; points combined with areas.
a. Point-type layout is used in the south side facing urban streets so as to reduce the feeling of oppression to the city;
b. "Point + area" layout is used in the north side facing the Shenye project. Efforts will be made to keep away from multi-storey residences. The open part will be utilized to guarantee the day-lighting requirement of the north Shenye project.
c. Multi-storey residences are arranged in the slightly weathered area where geological conditions are bad. They respond to and communicate with northern villas to make up a whole.

2. Height Difference Treatment
In consideration of obvious height difference of the site, business facilities are built along streets to attract urban population while residences are arranged above business facilities and garages.

3. Space and View of Landscape
High-rise buildings are arranged along the direction of longer side of the site. The line of sight will not be blocked. Meanwhile, the enclosure of high-rise buildings forms two landscape axles, one covering service facilities, eastern central courtyards, and lake while the other covering western central courtyards, landscaped roads, and eastern central courtyards. The two landscape axles make up a landscape area of the community. In addition, there are systematical linear landscapes inside multi-storey residences.

4. Functional Layout
Surrounding high-rise residences are staggered so as to ensure that courtyard space is not closed. Multi-storey residences are arranged at the place where geological conditions are bad. Thus a sense of rhythm comes into being to expand building interval. Buildings rotate 45 degrees to get sunlight as much as possible, avoid direct facing high-rise buildings, and respond to 3# building. Business buildings are set to face streets and concentrated in the west side.

Nan'ao Kai Xuan Bay Garden • Shenzhen
南澳凯旋湾·深圳

项目地点：中国，广东，深圳市
项目时间：2007年
设计规模：3.2万 m²
设计阶段：方案设计
项目现状：在建

Project Location: Shenzhen
Project Date: 2007
Project Scale: 32,000 m²
Design Phase: conceptual design
Project Status: construction in progress

The concept of group is used in the design of this villa.

以群体的概念来看待该别墅区的设计。

项目总览

整体用地背山面海，自然条件优越，是不可多得的可开发滨海别墅区。

Project Overview

Backed by hills and facing the sea, this plot enjoys favorable natural conditions, and it is a rare land suitable to develop seaside villas.

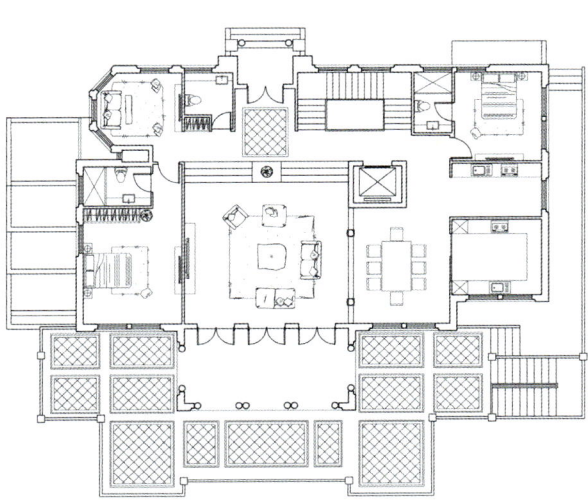

一层平面图

576

总体规划

在基本尊重原有地貌态势的基础上，对地形作了合理的整体改造，从而将海景资源最大化，保证大部分住宅享受海景资源。根据地块的不同标高和海景等级，南北两边（进深较短）分作三排，中间（进深稍长）分作四排。其次，借助路网格局将项目分作三个自然组团，有助于分期开发。

Overall Layout

The landform is transformed in an appropriate way without making great change to the original state, so as to maximize the field of vision for viewing seascape and ensure occupants in most buildings can overlook seascape. On the basis of the elevation difference in the plot and the levels of seascape, three rows of buildings are built in the north and south sides (with shorter depth), and four rows of buildings are built in the in-between area (with longer depth). In addition, on the basis of the road network layout, the plot is divided into three natural subareas, facilitating development phase by phase.

单体设计

会所作为项目的主要配套设施，设置在主入口正对面，依山就势，还可远眺海景，具有极佳的景观视野。同时，奢侈豪华的顶级会所也提升了小区的整体档次。小区主打品牌以豪华型住宅为主，小区的建筑风格定位为古典风格，在体现高贵气质的基础上偏重度假风情。

Single Building Design

As the one of the main amenities, the clubhouse is built in the place right opposite the main entrance. The clubhouse is backed by a hill and enjoys a very wide field of vision for viewing seascape. The luxurious clubhouse helps enhance the quality and image of the whole estate. Most buildings in the estate are designed as luxurious ones with classical style, laying particular stress on creating a holiday theme on the basis of an atmosphere of nobleness and dignity.

充分考虑海滨地区的气候与环境特征，立面材料以质感涂料搭配石材，应用丰富的细节设计使建筑远近都极具品味。

In the design, the characteristics of the weather and environment of the seaside area are fully considered; the facades are covered with texture coatings and stone materials, with plenty of well-designed details. The buildings look very noble from both a near and far position.

Shi Dai Tian Jiao • Wuhan
时代天娇·武汉

项目地点：中国，湖北，武汉市
项目时间：2001年
设计规模：4.2万 m²
设计阶段：方案设计
项目现状：已建

Project Location: Wuhan
Project Date: 2001
Project Scale: 42,000 m²
Design Phase: conceptual design
Project Status: construction completed

The reconstruction of the old city area with traditional Chinese style does not only sustain original urban life but also creates new high-quality open urban spaces.

富有中国特色的旧城改建既保持了原来城市的生活线索，同时也更新为积极高质量的城市开放空间。

总体布局

本项目同时面临容积率及限高的压力，采用均好性布局，沿街展开式布置，每栋住宅建筑尽量错开且间距在18m以上，争取户户朝南及得到更多的远景和园景。公寓与住宅分开，24层公寓设在地形短临近解放大道处，与东侧商发大楼交相辉应。三栋16层风车形住宅，沿后街营房布置，避开城市主干道的噪声，与公寓相独立，避免相互干扰。

General Layout

There are strict requirements on floor area ratio and height limit in the project. Therefore, a well-balance layout is used with the buildings built along the streets; they are staggered with a space of more than 18m between them, so that each flat has a southern orientation and enjoys more faraway landscape and garden view. The high apartment and the lower styled residential buildings are separated; the 24-storey apartments erected in a narrow area near Jie Fang Avenue, a well match with the Shang Fa Building in the east. The three 16-storey windmill-style residential buildings are positioned beside the back street houses so as to keep away from the noise from city arterial roads. They are also separated from the apartment so that mutual interference is avoided.

总平面

建筑造型

建筑造型力图表现建筑设计的内部功能变化，创造清新明亮的现代建筑形象。宽大的露台、绿荫葱郁的构架结合有着丰富立面颜色且通透明亮的景窗，丰富的弧形形成明亮的变化，成为解放路上具有现代风格的新地标。

Building Modelling

The architectural shape aims to demonstrate the transition between internal functional areas and to create a fresh and bright image of modern building. The estate becomes a new landmark in Jie Fang Road for their spacious balconies, verdant frameworks, rich colors on facades, clear and transparent sightseeing windows, rich arc shapes, and reasonable changes.

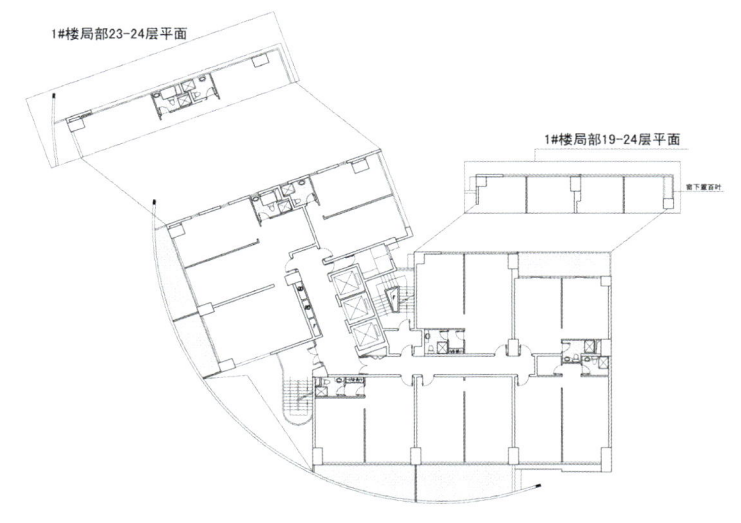

平面图

Grace Royal Apartment · Changshu
华府世家 · 常熟

项目地点：中国，江苏，常熟市
项目时间：2004年
设计规模：7.5万m²
设计阶段：方案设计
项目现状：已建

Project Location: Changshu, Jiangsu
Project Date: 2004
Project Scale: 75,000 m²
Design Phase: conceptual design
Project Status: construction completed

The project cleverly meets the tough guidelines while assuring that every residential unit meets the daylight requirement, and at the same time maximizing the view distance.

项目巧妙地在高强度开发要求下保证了每户住宅的日照要求，同时提供了最大的视野间距。

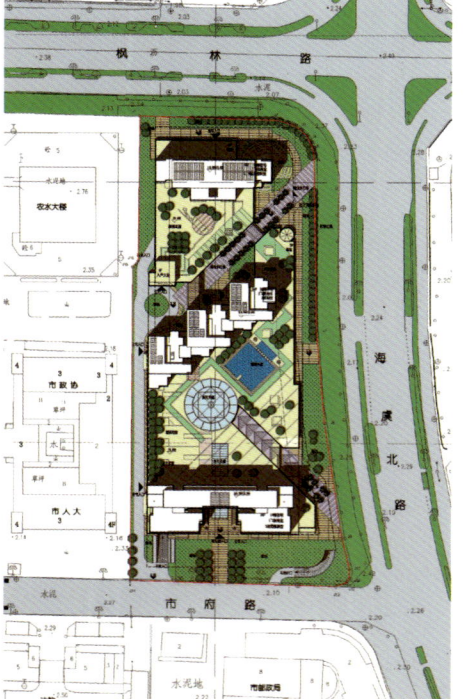

总平面

设计原则

以发挥城市中心的积极作用为目的，以城市为依托，通过对该区域城市空间的呼应与提升创造该区域的活力节点，与周边的城市空间积极互动。依据当地的相关规划及设计规范，在满足日照、通风等条件的前提下解决高容积率、商业面积大的设计难点。充分利用现有景观资源，通过设计求得西向景观与南北向通风、采光兼而有之的统一。通过设计创新，提升高尚居住的生活品质。

Design Concept

To bring the role of downtown into full play; create a vitality node for this area and make it interact with surrounding urban spaces by echoing and promoting the urban space in this area; overcome the design difficulty of high plot ratio and large business facility area according to local planning and design codes under the premise of meeting day-lighting and ventilating conditions; make fully use of existing landscape resources to realize unification of westward landscapes and south-north ventilation and day-lighting; to improve life quality of high-end residences by means of design innovation.

设计构思

项目难度在于6.5的容积率和满足日照条件之间的矛盾，以及居住和商业功能的叠合，设计过程充满了一种求唯一解的乐趣。在设计出中间的小面宽错列户型单元后，规划形态也随之成立，所有问题一瞬间迎刃而解。

1. 规划

"南北两排＋中间斜向错列板式布局"，充分考虑到城市结构布局、道路及周边建筑关系。中间的住宅采用斜向错列板式布局，形成整体建筑群体量，产生步移景异的空间感，削弱对城市道路的压迫，并使自身获得东南向的日照、通风和更大的虞山景观面。同时也为三排的住宅开阔了视野，减少了相互间的视线干扰。

2. 商业

顺应主楼体量，在裙房内错列有序的设置"街道"、"广场"等室外空间形态，增加了底层街铺的临街面。同时，与建筑体量形成良好的图底关系，使商业内部空间丰富有趣，形成积极的商业空间。

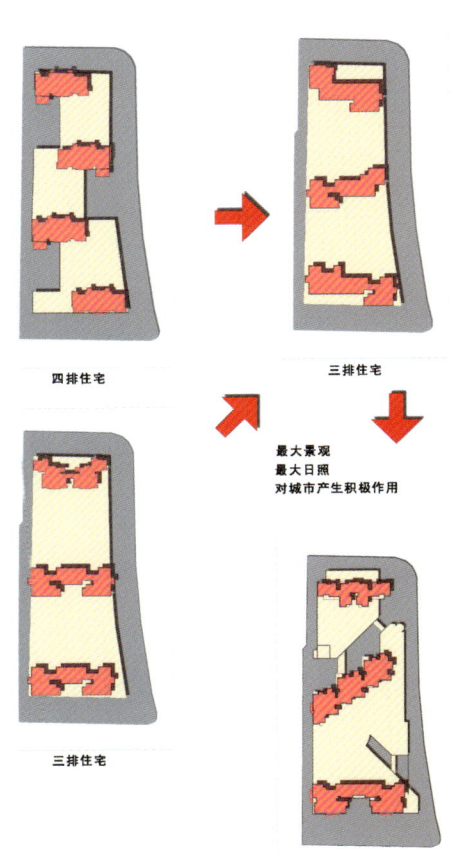

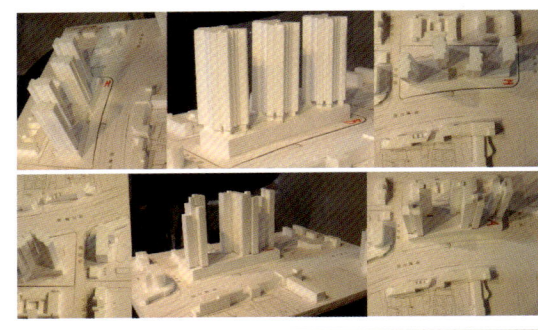

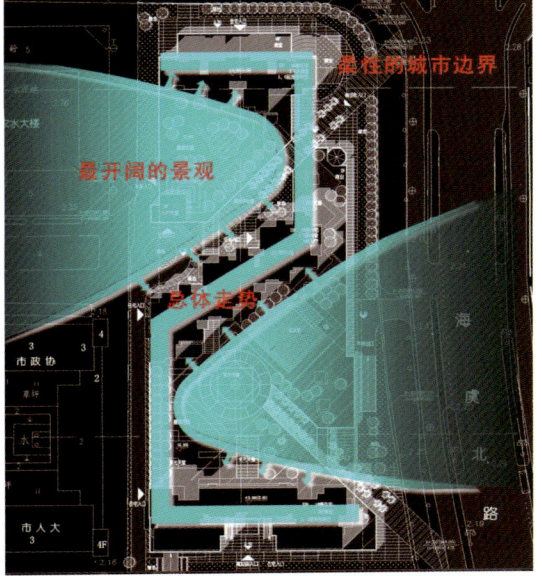

功能分析图

Design Concept

The difficulty in designing this estate lies in the contradiction between satisfying the floor area ratio of 6.5 and meeting the sunlight acquiring conditions, as well as in the combination of residential and commercial functions. The process of design is full of delight as an unique solution is found. After an in-between oblique row of slab-type buildings arranged in a staggered way is designed, the planning comes into being and all problems are readily solved.

1. Planning

The layout features "one row of buildings in the south and one row of buildings in the north plus a in-between oblique row of slab-type buildings arranged in a staggered way". The city planning and the relationship between the estate with city roads and the surrounding building are fully considered in the design. The in-between row of slab-type buildings are arranged in a staggered way, which form a large volume of buildings, a sense of moving scenery as one walks beside the buildings, and a sense of pressure toward the city roads is weakened. Such layout also allows the in-between row of buildings to receive sunlight from the southeast, obtain better ventilation and get wider field of vision to view Yu Mountain. An open field of vision is achieved for the three rows of buildings and the meeting of sightlines between occupants in different rows of buildings is avoided by a large extent.

2. Commerce

In harmony with the main buildings' volumes, outdoor spatial forms such as "street", "square" are introduced in the skirt building in a well-organized way. This not only increases the frontage area of stores on the first floors, but also creates a good figure-ground relation between the commercial areas and the building volumes. Therefore the internal commercial spaces are diversified and active.

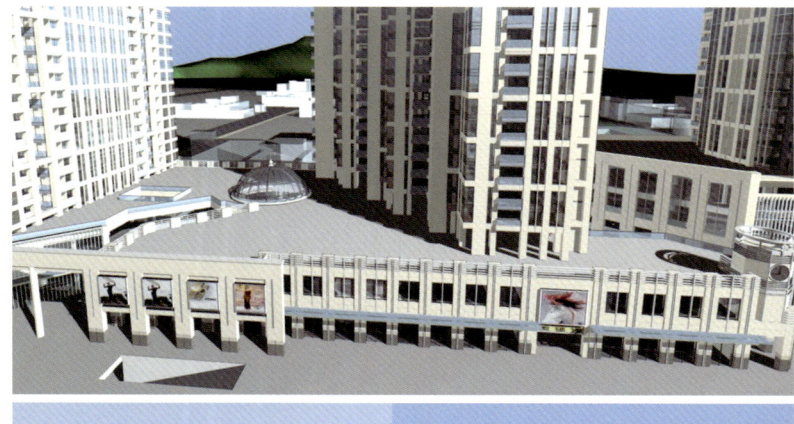

造型设计

精致、典雅

石材、玻璃、深灰色钢材、面砖等都为立面造型元素，直接体现建筑精致、典雅的特点，符合高层次住宅的性格特质，成为城市的亮点。

Shaping Design

Delicateness and Classical Elegance
Materials such as stone, glass, charcoal grey steel and face tile are all used as the shaping elements of the facades, directly expressing the delicateness and classical elegance of the buildings. The style fits the high-end residential buildings, being a bright spot in the city.

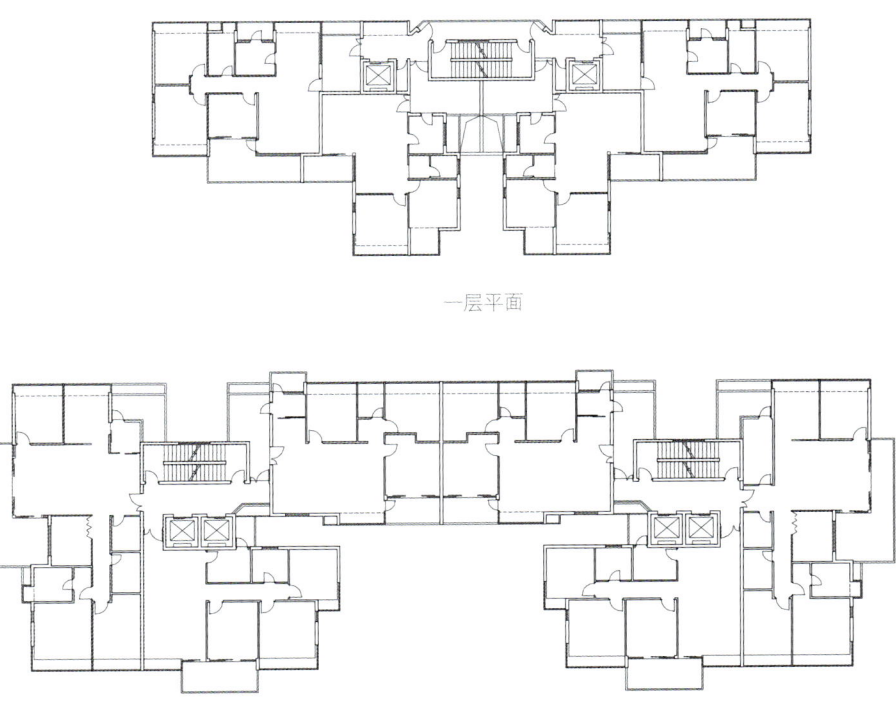

一层平面

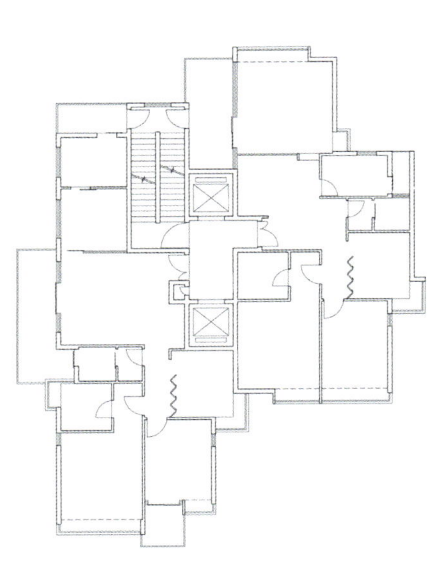

二层平面

三层平面

China Overseas Property -Banyan Coast • Chengdu
中海翠屏湾·成都

项目地点：中国，四川，成都市
项目时间：2007年
设计规模：23万 m²
设计阶段：方案设计
项目现状：未建

Project Location: Chengdu
Project Date: 2007
Project Scale: 230,000 m²
Design Phase: conceptual design
Project Status: construction not started

Appropriate plan layout, quality of materials, and affordability are the highlights of the design.

户型均好性、建筑的质感、造价的经济性是设计的重点。

设计构思

我们提出三个规划目标：

1. 最大化利用用地东南西三面的景观资源，提高住户的景观品质。

2. 打破原来单一的住宅户型，规划多样的产品形式，通过公寓与商业的结合引入商业区，有效提升社区的整体品质。

3. 最大化小区花园，为小区内良好的景观品质提供条件。

Design Concept

Three planning objectives are proposed:

1. To make full use of the landscape resource in the east, south, and west areas of the site, so as to improve the landscape quality for the occupants.

2. Unlike traditional monotonous internal layout of flats, diversified internal layouts styles are introduced. In the commercial areas, apartment and commercial spaces are combined to greatly improve the overall quality of the community.

3. To maximize the area of the garden in the estate, establishing a sound basis for offering quality landscape in the estate.

总体布局

总体布局采用高层建筑围合布局，并根据地形、地势、组织各功能块位置，西北角侧布置为综合商业区，由三栋公寓及其底层商业休闲广场组成，南侧为小高层围合的大花园社区，有利于形成丰富的围合空间，在小区内部创造丰富的花园空间，同时保证高层间视线距离的最大化。

Overall Layout

The overall layout features a courtyard encircled by high-rise buildings, and various functional areas are organized in accordance with the landform and topology. A comprehensive commercial area is set up at the northwest corner, consisting of 3 apartments with the first floors as commercial and relaxation squares. In the south part is a grand garden encircled by medium high-rise buildings. This layout is good for creating luxuriant encircled spaces and garden area and maximizing line-of-sight distance between occupants in high-rise buildings.

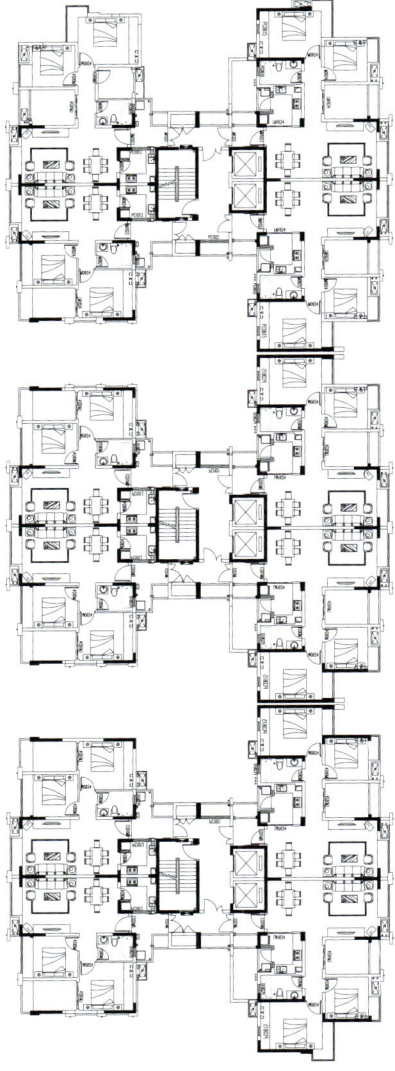

平面图

China Overseas Property - International Community · Suzhou

中海国际社区·苏州

项目地点：中国，江苏，苏州市
项目时间：2007年
设计规模：5276.4万 m²
设计阶段：方案设计
项目现状：已建

Project Location: Suzhou
Project Date: 2007
Project Scale: 52,764,000 m²
Design Phase: conceptual design
Project Status: construction completed

Real estate development is actually a process to explore, integrate and distribute resources.

房地产开发其实是对资源的发掘、整合和分配。

基地位置

项目地理位置极为优越，周边配套设施齐全，位于苏州新工业园区金鸡湖东板块，与工业园21号地块相邻。

Site Location

The estate is located in the east block of Jing Ji Lake in Suzhou Industrial Park and is adjacent to the No. 21 plot in the Park. The location is very favorable and the amenities around the site are all in readiness.

总平面

构思草图

功能分析图

居住组团细胞

建筑——环境

本项目地处两条重要的城市轴线之间，它的建成将加强两轴线之间的联系，为城市建设添彩。"中央河畔，科技园区，国际社区"是本案的主题概念。老苏州城的历史和传统，新工业园区的现代和包容，在规划设计中已形成了多样性、开放性、包容性的多元城市文化特征。介于两条城市轴线之间的特殊地位，以方舟公园和中央河为核心，于弧形主轴内布置低层住宅，外侧依次布置11层、18层、25层住宅，取得较好的空间过渡。超高层双塔布置在湖东副横轴端点，是轴线的结束，也是该区域的地标。

The estate is located in the area between two major city axes. It enhances the connection between the two axes and is a wonderful landmark in city planning. The theme of this estate is "Central River Side, Hi-tech Park, International Community". The design has multiple urban cultural traits characterized by diversity, openness, and compatibility, with the elements drawn from the history and tradition of the old Suzhou city and the modernity and compatibility of the new industrial park. As the plot is located between the two city axes, the Fang Zhou Park and Central River are deemed as the center in the layout. Low-rise buildings are built inside the curve axis of the estate. Outside the low-rise buildings, 11-storey, 18-storey, and 25-storey buildings are erected to achieve sound transition between the spaces. The twin-tower skyscraper is located at the end of the horizontal axis east of the lake, creating a landmark in this area.

China Overseas Property – Xu Jiang Project • Suzhou
中海胥江项目·苏州

项目地点：中国，江苏，苏州市
项目时间：2007年
设计规模：20万 m²
设计阶段：方案设计
项目现状：已建

Project Location: Suzhou
Project Date: 2007
Project Scale: 200,000 m²
Design Phase: conceptual design
Project Status: construction completed

The strong contrast between the tall and low buildings creates a unique result.

高层和低层风格的迥异营造出奇特的效果。

项目总览

由于周边已有建筑的日照要求，使得高层建筑体量必须尽可能靠北海临胥江。通过选择点式和短板体量，并进行角度的不同扭转，错开视线和空出一定缝隙，在减少对胥江压迫感的同时，部分建筑采用退台处理，形成临桐泾路富有亲和力的城市界面。在日照影响范围内布置低层多进院联排别墅产品，提升项目价值。

Project Overview

The high-rise buildings should be built as near to Bei Hai and Xu River as possible because enough space should be left to allow the surrounding buildings accept sunlight. Point type and short plate volumes are designed and they are positioned with different torsion angles. The channels allowing sight lines are staggered and some gaps are designed. While a sense of pressure on Xu River is reduced through design, some buildings are also designed with setback so that the frontage side along Tong Jing Road seems friendly to the public. Within the area to which sunlight may be blocked, low-rise townhouses with multiple courtyards arranged in series are built to increase the value of this project.

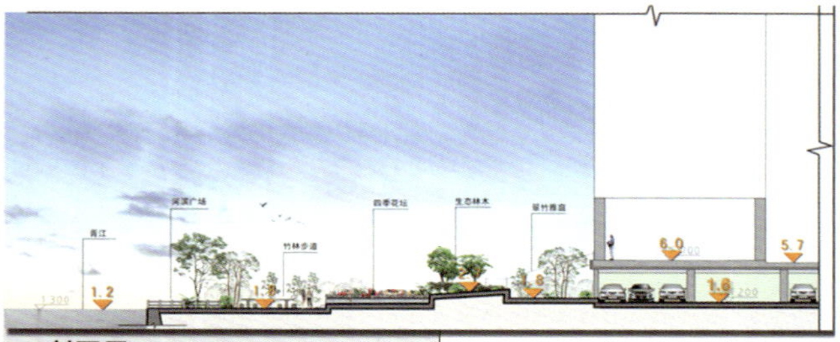

A-A剖面图

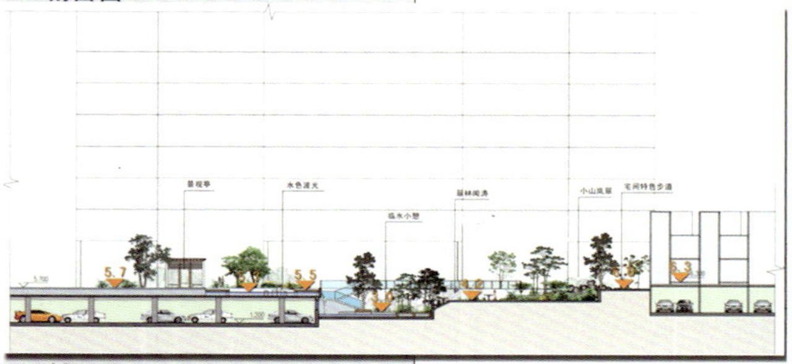

B-B剖面图

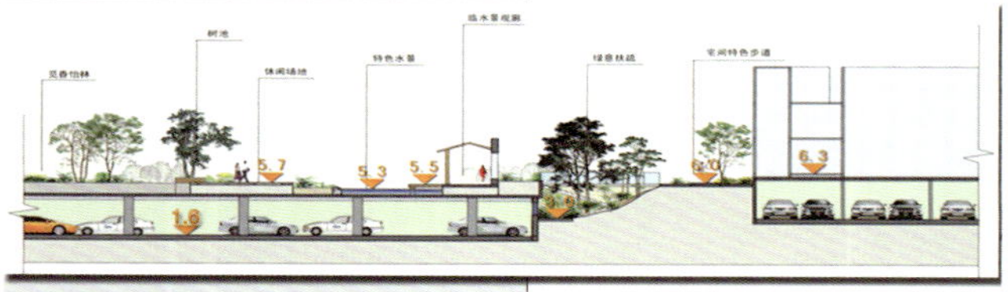

C-C剖面图

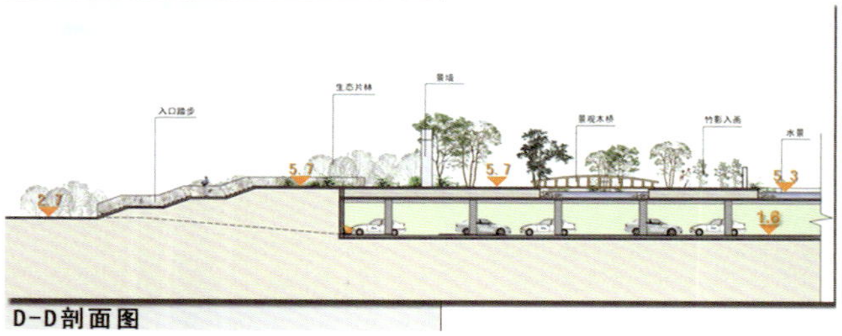

D-D剖面图

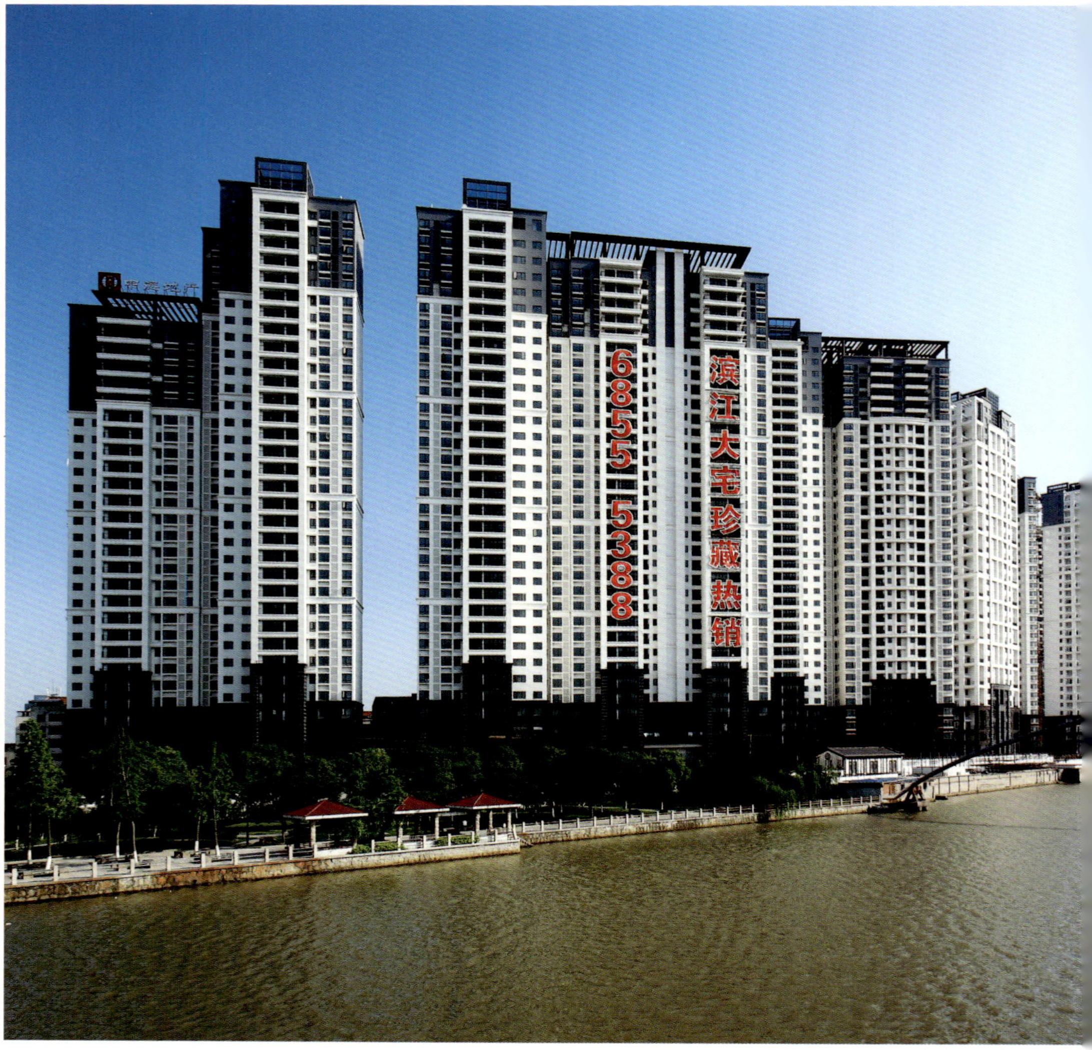

高层造型在注重品质的同时运用灰白色彩，低层联排采用现代中式风格，体现从建筑形象与环境主题上对传统文脉的呼应。

In the design of the shape and appearance of the high-rise buildings, greyish white is used and the quality is emphasize; the low-rise town houses are of modern Chinese style, full of traditional culture elements in terms of architectural image and environmental theme.

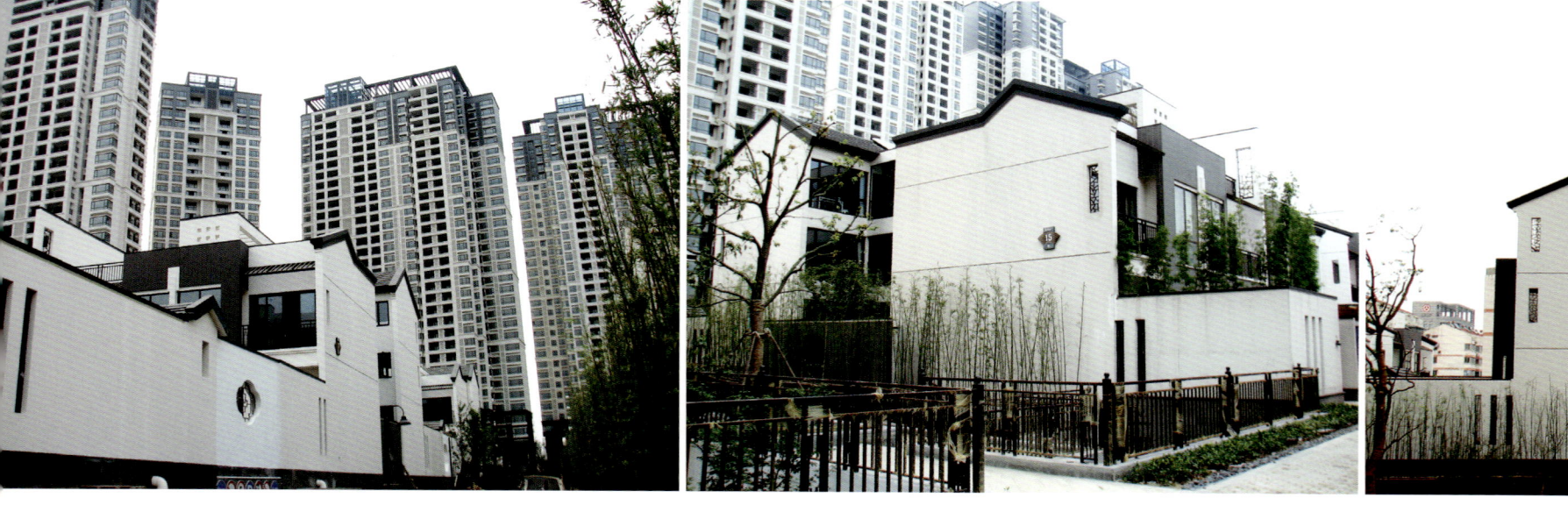

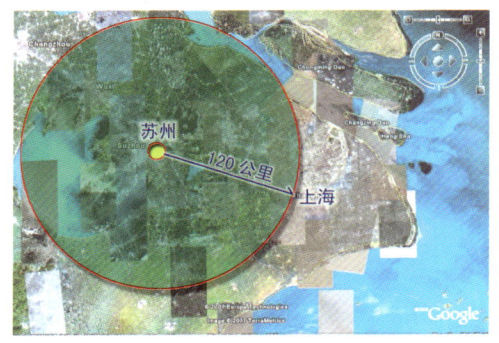

区位分析

苏州位于中国华东地区，距上海核心区120km。本项目位于苏州市内环线内，靠近老城区西南角，在新区和新加坡工业园区的中间位置，毗邻胥江，拥有非常良好的区位优势。

Location Analysis

Located in east China, Suzhou is 120km away from the core area of Shanghai.
This project is situated inside inner ring road of Suzhou, approaching southwest corner of the old urban area. Since it is right in the middle of new urban area and Singapore Industry Park and neighbors Xujiang River, it has an advantage of a favorable location.

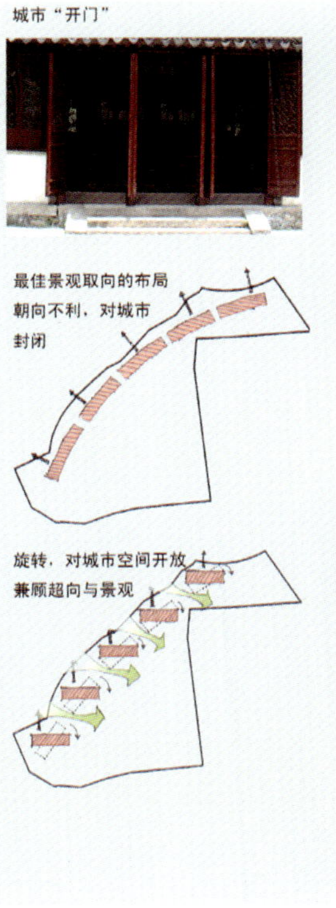

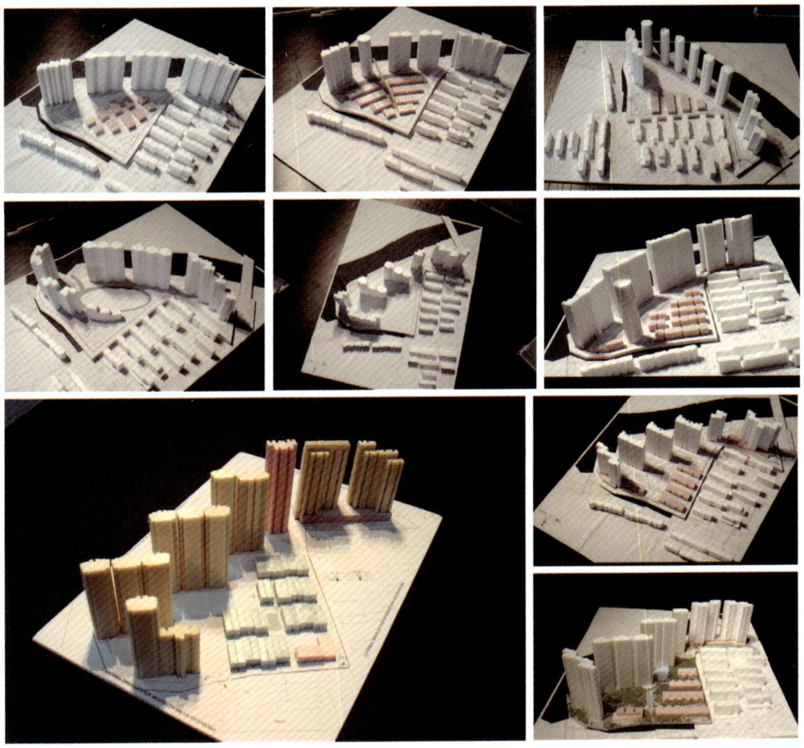

Regional Style……

Regional Style
地域风情

Due to large-area development, high-speed construction, and people's further cultural requirement for living space, many new communities are being built in the form of custom implantation. We think this is an embodiment of current "fast food culture" in our industry. Meanwhile "fast foods" show different qualities. Therefore, for these projects we focus not only on imitation of architectural images but also creation of outer space, environment, and atmosphere. We will not arrange landscapes simply like movie but solve the problem of conflict and harmony between foreign cultures, living habits and users' cultural background.

整片的开发、高速的建设,以及人们对于居住空间进一步的文化要求让很多新建的社区以这种风情移植的手法来开发。我们认为这其实是当今社会"快餐文化"在我们这一行业中的体现,但同样"快餐"也有品质的差别。因此,在这类项目之中,我们所关注的不仅是建筑形象的模仿,更应该注重外部空间、环境及氛围的综合塑造;并非简单的电影式布景,还需解决其外来文化、生活习惯与其使用人群文化背景之间的冲突和调和。

Portofino Swan Castle • Shenzhen
波托菲诺天鹅堡·深圳

项目地点：中国，广东，深圳市
项目时间：2002年
设计规模：63万 m²
设计阶段：方案设计，初步设计
项目现状：已建

Project Location: Shenzhen
Project Date: 2002
Project Scale: 630,000 m²
Design Phase: conceptual design, preliminary design
Project Status: construction completed

The design is based on the principle that the original water body, hills and passages on this plot are not to be altered due of this project.

设计希望原有地块的水系、山脉、人的路线都不会因该项目的建设而改变。

项目总览

占地30多万m²，建筑面积为63万m²的天鹅堡，作为华侨城最北侧的围合边界，集自然山水之气、海之气、人之气于一体，尽显华侨城美景。项目以其规模地域形成同西组团住宅区、东组团住宅区以及纯水岸同等重要的新兴高端居住社区。以欢乐谷和燕晗山为中心，也成为华侨城的一个副中心，具备自我社区的向心性和标志性。

Project Overview

This estate covers a land area of more than 300,000 m², and boasts a building area of 630,000m². Located at the northernmost part of Overseas Chinese Town, the estate enjoys landscape consisting of mountain, water and sea, stimulating popular enthusiasm and thoroughly showing the charm of Overseas Chinese Town. The large area estate is divided into the western residential area, eastern residential area, and the innovative high-class residential area equivalent to Pure Waterfront estate. With Happy Valley and Yanhan hill as the center, the community also becomes a sub-center of Overseas Chinese Town with its own essence and landmark.

总平面

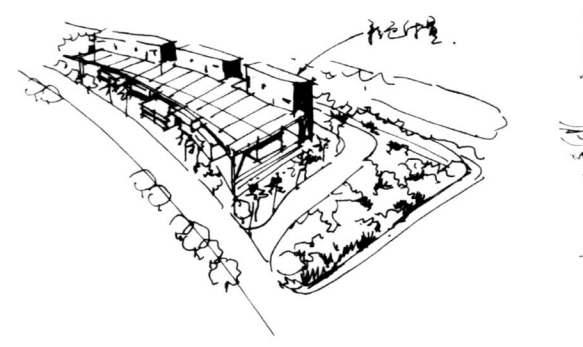
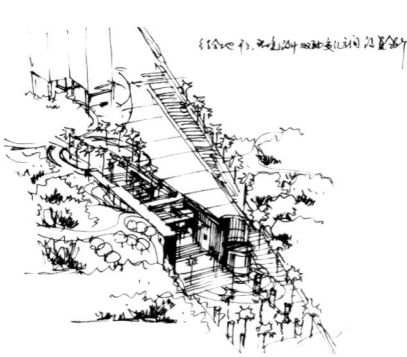

宏观层次——总体规划

规划首先考虑的是对周边城市资源的保护，保证华侨城大区域范围内城市空间的连续与完整。它自西向东保留了帝诺山、天鹅湖-燕晗山之间的视线通畅；自南向北延续了燕栖湖-纯水岸-天鹅堡，由低到高的尺度递增。

本项目是低密度高层规划的经典之作：浅弧形建筑体量围合出东西向内部庭院超大尺度的开放空间，南、北楼间距最大为150m，实现了公园内居住的体验。

Macroscopic Level——General Planning

Preservation of the resources in the surroundings is considered as the priority in the planning. This is to ensure the consistency and integration of city spaces in the large Overseas Chinese Town region. From west to east, there is no obstruction between sightlines connecting Dinuo hill, Swan Lake and Yanhan hill. From south to north, Yanxi Lake, Pure Waterfront estate and Swan Castle estate forms a continuous increase of elevations.

The layout featuring low-density high-rise buildings is a classic design. A large-scale courtyard with open sides in the east and west is surrounded by buildings with slightly curved volumes. The maximum space between the northern building and the southern building is 150 meters, so the occupants can feel like living in a park. .

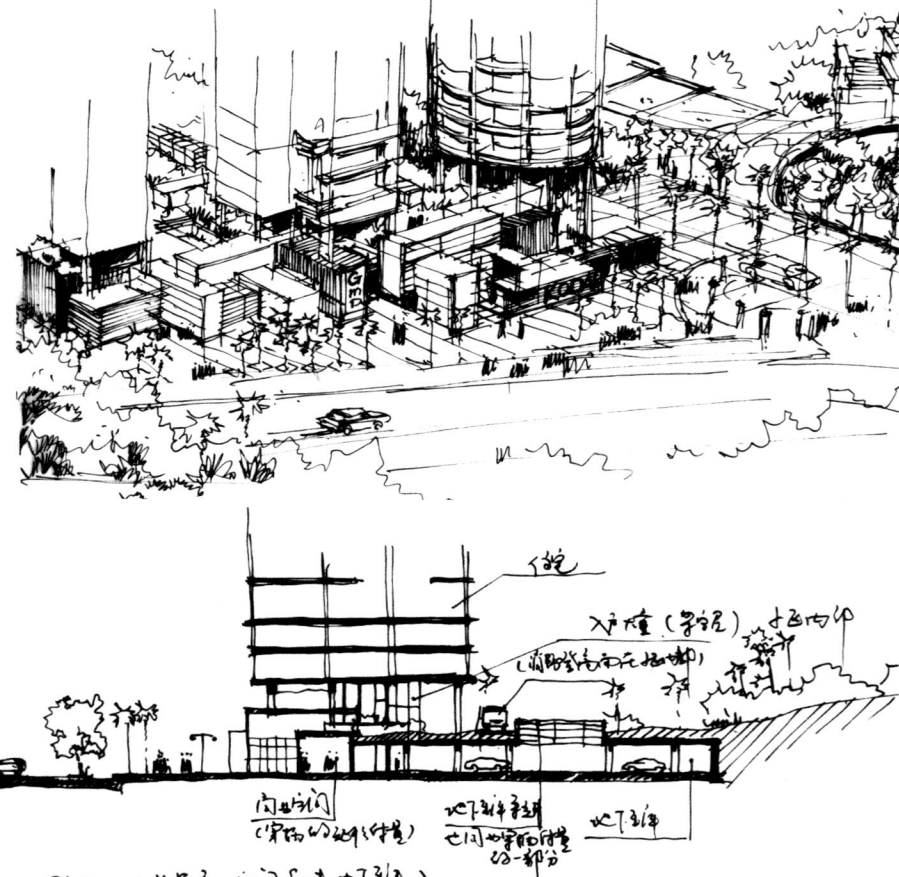

立面图

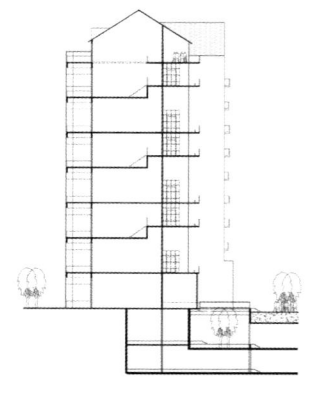
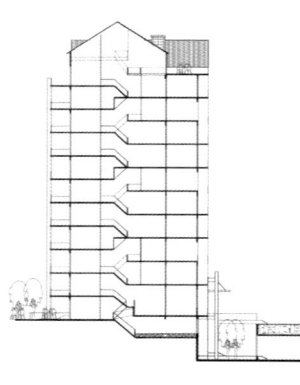

剖面图

户均面积在200m²以上，户型设计充分考虑高端客户的需求，注重公共空间的舒适度及双流线的内部设计。户型产品丰富多样，单元形态包括一层两户的小高层、一层三户的中高层及一层四户的高层单元。户型从平层大户到复式大宅以及客厅4.8m高、卧室3.2m高的跃式空中别墅产品，不断冲击传统居住观念，也不断刷新华侨城居住项目"天价"的销售记录。

The average floor area is more than 200m². The needs of high-end customers are fully considered in the design of the floor plans, emphasizing comfort in public spaces and double streamlines in the interior design. With diversified floor plans, the units include small high-rise units with two suites on the same floor, medium high-rise units with three suites on the same floor, and high-rise units of with four suites on the same floor, as well as large suites with single floor, large duplex suites, and penthouses with 4.8m-high living rooms and 3.2m-high bedrooms. These new designs are a break to the traditional residential concept, and the selling prices of suites in this estate have never been higher – exceeding the selling price of Overseas Chinese Town.

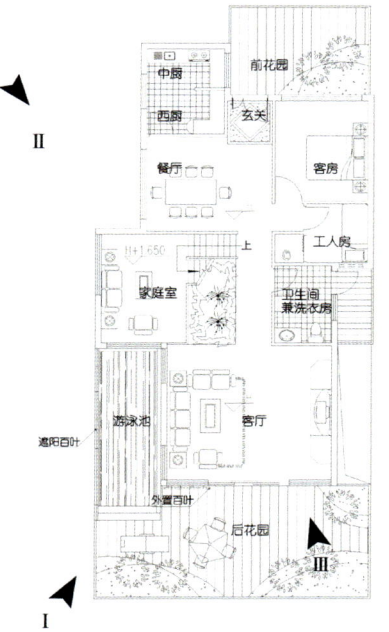
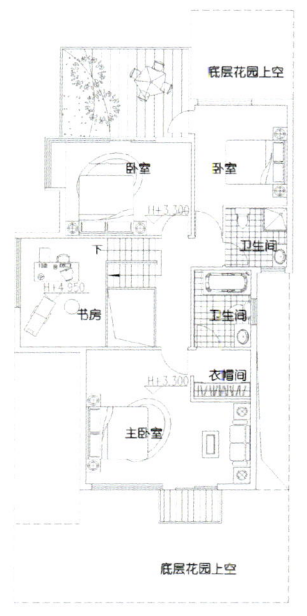
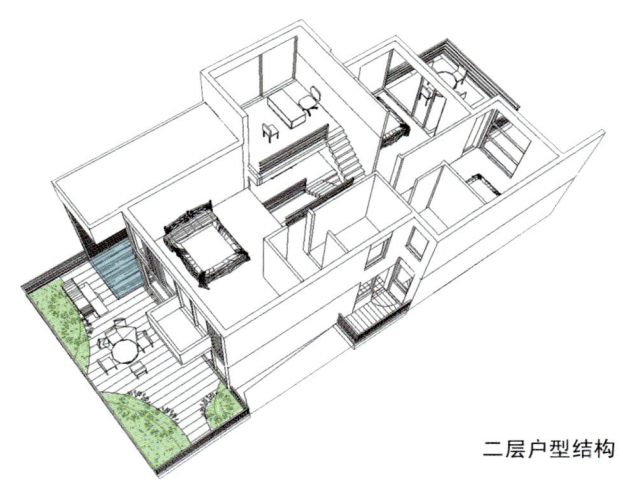

二层户型结构

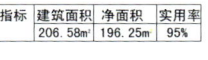

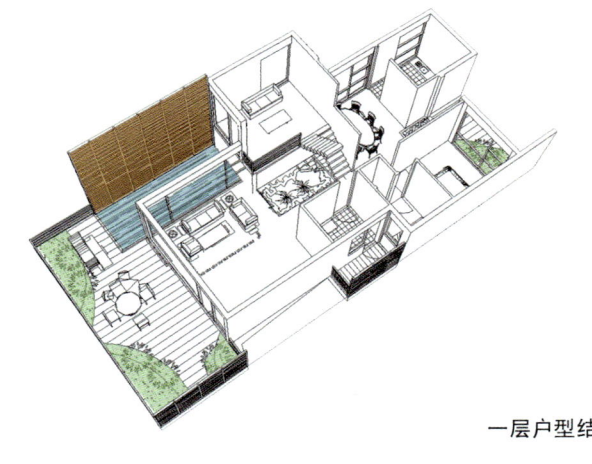

一层户型结构

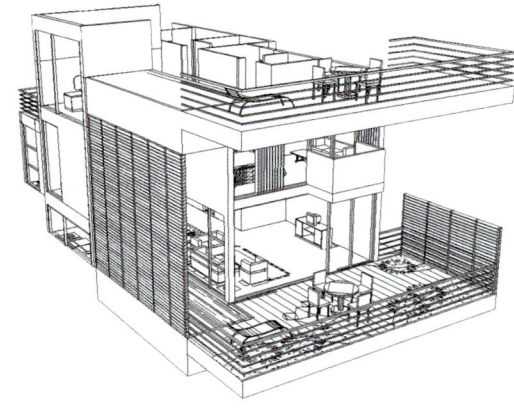

透视 I

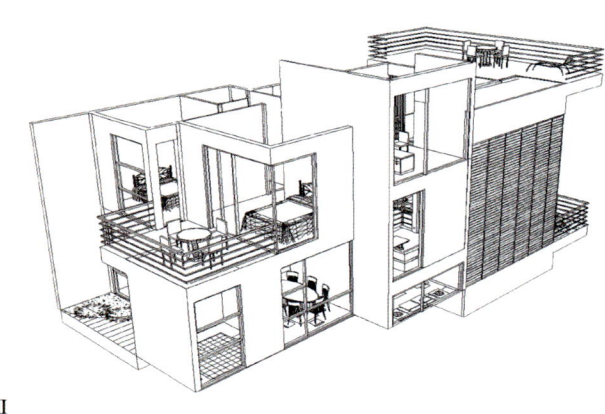

透视 II

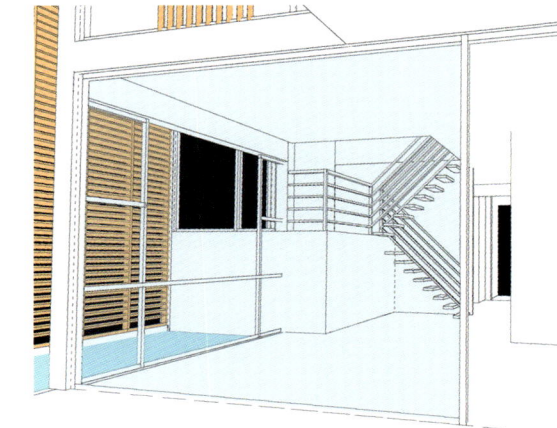

透视 III

Zhong Cheng Li Jing Xiang Shan · Changsha
中城丽景香山·长沙

项目地点：中国，湖南，长沙市，体育新城
项目时间：2004年
设计规模：39.5万m²
设计阶段：方案设计
项目现状：已建

Project Location: Changsha Sports New Town
Project Date: 2004
Project Scale: 395,000 m²
Design Phase: conceptual design
Project Status: construction completed

Different architectural forms are presented to meet the demands of different occupants. The diversified designs make this Mediterranean-style "town" appear colorful and lovely.

不同的建筑形态满足了不同使用人群的需求，多样性的设计让这个充满地中海情调的"小镇"显得丰富别致。

项目总览

本项目容积率为2.5，社区内需保留一个约9000m²的自然山体，且业主要求以11层或以下的小高层为主。为打破常规均质化、行列式小高层的规划模式，该设计引用了两种不同形式的居住形态：一为以高层景观优势为主的小高层建筑，另一个以小尺度低层高度情趣空间为重点的低层洋房。以一条贯穿社区并直达保留山体的内街为线索，将低层洋房串联起来，形成了中、小高层的"景"，并在低层洋房区构建出不同类型住宅的集合建筑形态，运用地中海建筑风格，令建筑显得丰富而可爱。

Project Overview

The floor area ratio of this project is 2.5, and a natural hill with the area of 9,000m² needs to be kept in the community. The owner requires that most buildings should be small high-rise buildings of 11 floors or less. To break the conventional layout of homogeneous and row-type small high-rise buildings, the designers introduce two different types of residential housing in this design; one is the small high-rise building with the advantage of overlooking the landscape and the other is the low-rise foreign style house with interesting spaces. The low-rise foreign style houses are built along an inner street running through the community to the hills, which becomes the "landscape" of the small high-rise buildings. Congregated dwelling houses with diversified forms are constructed in the low-rise foreign style house area; the houses with Mediterranean architectural style look colorful and lovely.

总平面图

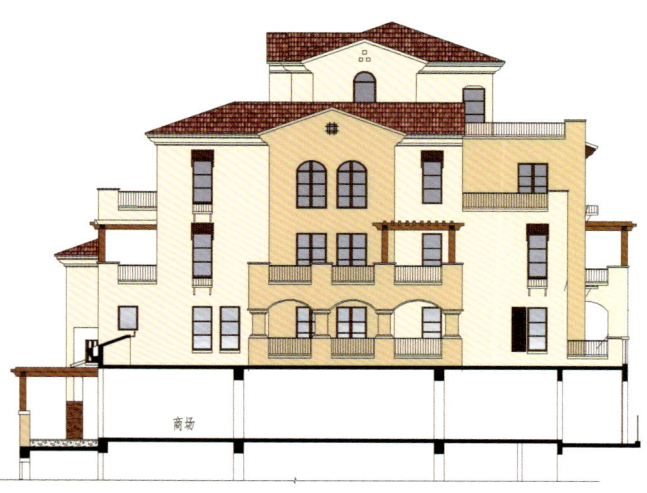

3-3剖面图 1:100

| 墙面5-1-9 | 墙面5-3-4 | 住宅线脚与墙身同色 所有外墙阳角、所有窗四周、凹凸线脚都倒成半径1cm圆角 |

会所北一街立面图 1:100

会所色号　主墙面5-1-9　肌理墙面5-3-4　线角5-1-7　住宅色号　墙面5-1-9　墙面5-3-4　住宅线脚与墙身同色 所有外墙阳角、所有窗四周、凹凸线脚都倒成半径1cm圆角

溪一街立面图 1:100

会所色号　主墙面5-1-9　肌理墙面5-3-4　线角5-1-7　住宅色号　墙面5-1-9　墙面5-3-4　住宅线脚与墙身同色 所有外墙阳角、所有窗四周、凹凸线脚都倒成半径1cm圆角

立面图

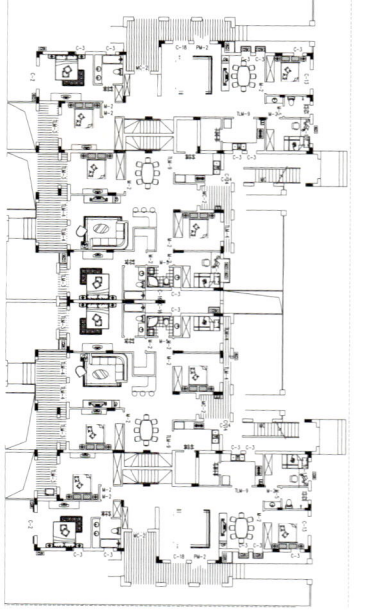

平面图

Li Jing Shan Zhuang • Shenzhen
荔景山庄·深圳

项目地点：中国，广东，深圳市
项目时间：2007年
设计规模：3万 m²
设计阶段：方案设计
项目现状：在建

Project Location: Shenzhen
Project Date: 2007
Project Scale: 30,000 m²
Design Phase: conceptual design
Project Status: construction not started

This is a reconstruction project. The greatest challenge is to resolve the conflict between the distinctively variable terrain and transportation.

这是一个改建项目，最大挑战来自于高差巨大的地形和交通之间的矛盾。

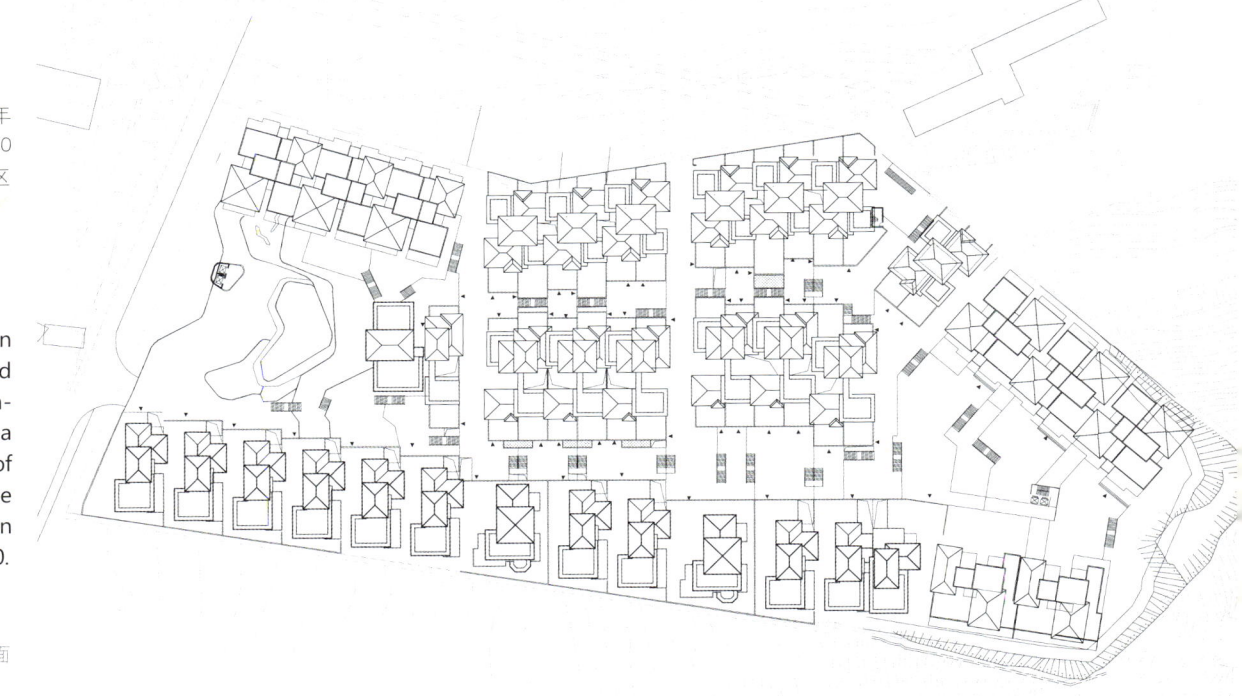

项目总览

本项目独踞大南山西坡，南北双向均为碧波绿海的百年荔枝林。用地西低东高，东西向高差约为50m，在1.0的容积率条件下，如何实现大南山的自然景观与别墅区建筑群的共荣共生是本项目的重点。

Project Overview

The estate is located on western slope of Da'nan Mountain, where there is a large area of hundred years old lychee woods spreading in the north-south direction. The west and east plot has a height difference of about 50 meters. The key of this project is to achieve harmony between the villa buildings and the natural scenery of Da'nan Mountain while satisfying the floor area ratio of 1.0.

总平面

"宫殿式"生态车库与"地衣式"别墅群

本项目原为已拆除的老别墅用地，被削成几个台地。新设计希望顺应原始山形"高之愈高"的趋势，运用"双±0.00"的概念，利用现状地形停车入户，并通过侧采光和顶采光，形成高大的阳光车库；在车库上盖形成别墅的花园休闲平台层，顺自然山势逐波升高，使别墅体量依附于山体自然生长。

所有建筑分南、北、中三排顺山而布，形成西、东两大集中绿化空间和两条线形登山景观通廊，在庭院中随时可看到自然山景。

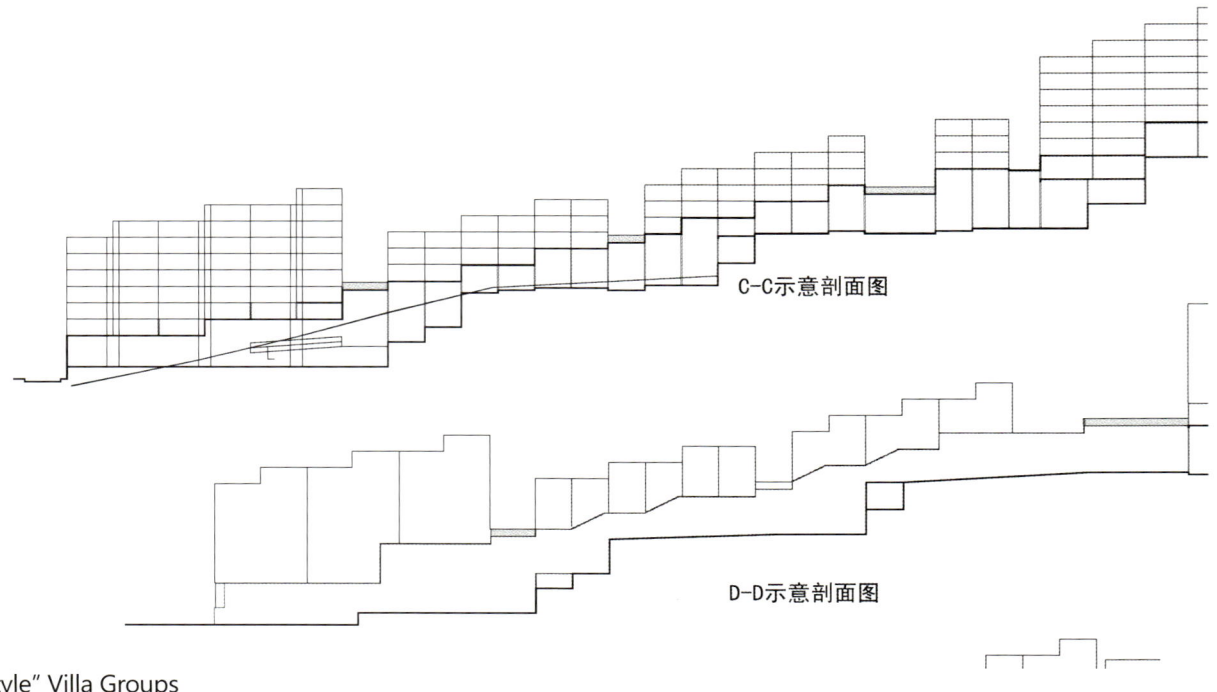

C-C示意剖面图

D-D示意剖面图

"Palace-style" Eco-garages and "Lichen-style" Villa Groups

The plot of this project was previously used for villas which have been dismantled, and it is bulldozed into some platforms. The new design aims at making use of the ascending elevation of the slope and the concept of "double ±0.00". On the basis of the existing landform, high and spacious sunny garages are achieved through various skylights and windows. On the roof of the garages, platform gardens are built; the platform gardens are distributed along the slope in a cascade way, so that the villas naturally integrate with the surrounding hills.

Conforming to the hilly landform, all buildings are arranged in three rows– north, middle, and south respectively, forming two centralized green spaces in the west and east with two linear passages for hiking and sightseeing, therefore the scenic view of the mountains can be seen at any time from the courtyard.

621

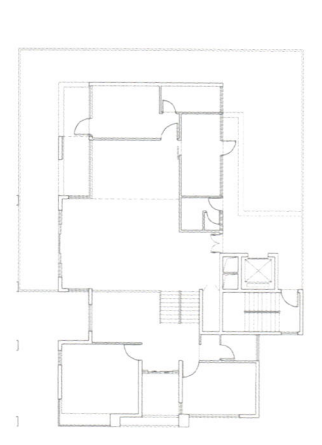

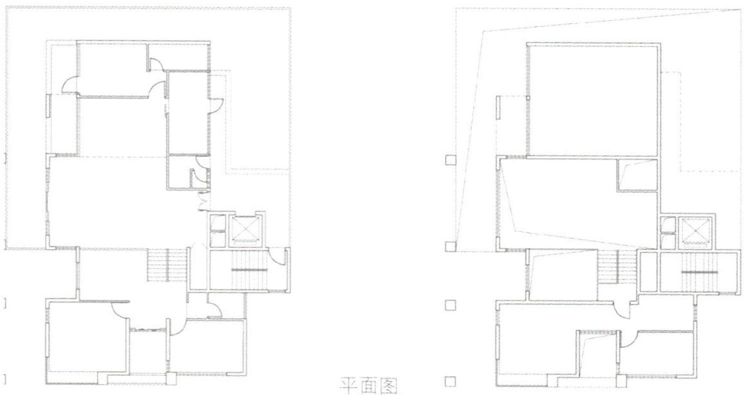

平面图

立面图

产品造型丰富，因景而设

建筑沿山势布局，南边景观最优，布置级别最高的独栋住宅；北边次之，布置联排别墅；中间没有明显的外部景观，布置内庭院较大的联排别墅；东边及西边受环境影响较大，布置相对较高的多层空中别墅。

Creative building shapes are designed on the basis of surrounding landscape.

The buildings are arranged by conforming to the hilly landform. In the southern area with best scenery, single-family houses with the highest value are built; in the northern area where the scenery is the second best, townhouses are erected; in the central area where there is no significant outdoor scenery, townhouses with larger courtyards are built. In the eastern and western areas where there is more impacts from the surroundings, and multiple-story villas are arranged.

造型设计

借鉴"草原别墅"的设计手法，强调自然材质，如石材、金属、玻璃的运用及与地形紧密结合的自然通台，创造根植于山林中的建筑形象。

Form Design

The design methods of "prairie villa" are referred to in the design of this project. The application of natural materials such as stone, metal and glass is emphasized, and natural open platforms are designed by making use of the lanscape. Therefore an image of buildings erecting from forest is created.

平面图

立面图

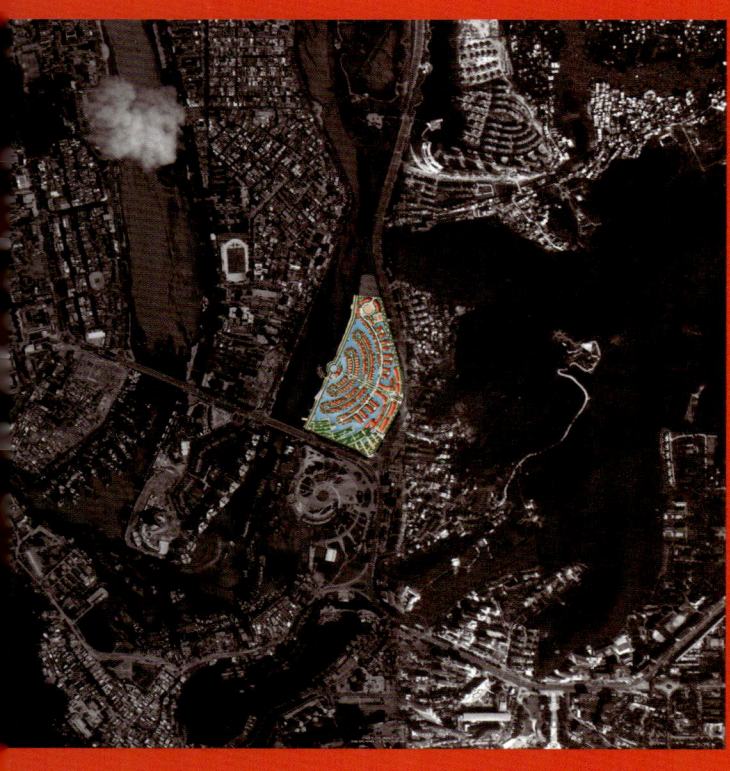

Xiayang Villa District • Sanya
下洋别墅区·三亚

项目地点：中国，海南，三亚市
项目时间：2006年
设计规模：10.5万 m²
设计阶段：方案设计
项目现状：未建

Project Location: Sanya
Project Date: 2006
Project Scale: 105,000 m²
Design Phase: conceptual design
Project Status: construction not started

While making use of the tides as a sustainable landscape, the design provides maximum contact area per household.

利用自然的潮汐实现景观的可持续，为每户提供最大的接触面。

总平面

项目总览

本项目位于临春河入海口附近，是凤凰路与临春河之间鲜有的未开发地块之一。我们力求把本项目打造成具有热带风情的地中海式水岸高档度假休闲小镇。

Project Overview

Located near the estuary of Linchun River, the plot is one of the rare undeveloped plots between Fenghuang Road and Linchun River. The design aims to build a luxurious and leisurely water-side resort town of tropical and Mediterranean elements.

涟漪

棕榈

指状

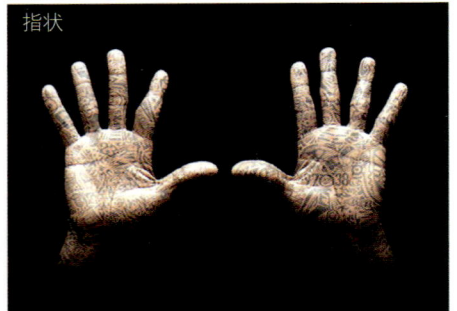

立面图

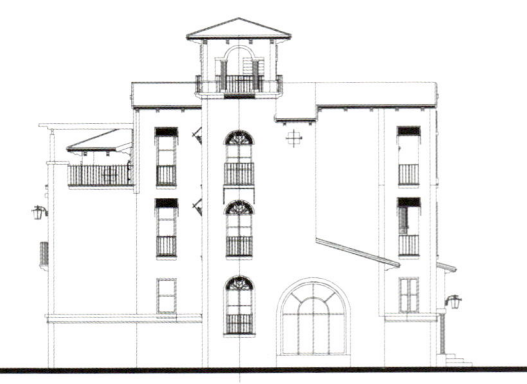

立面图

规划结构分析

主轴

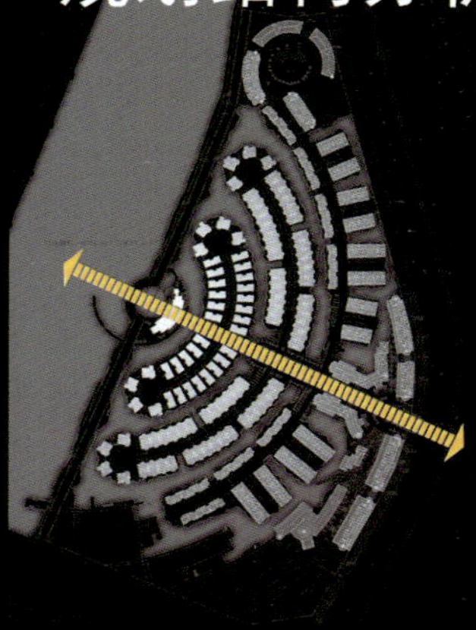

叶脉状路网

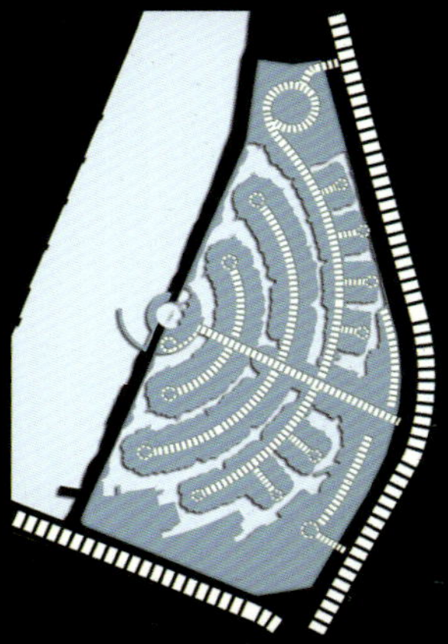

指状水岸半岛

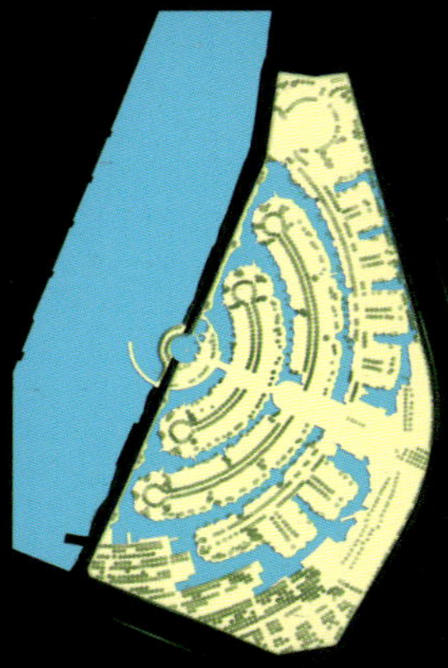

中心

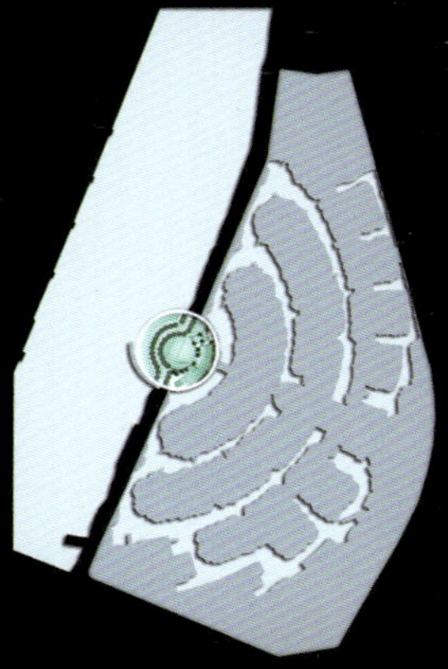

设计方案的灵感来源于椰子树树叶和海水的波纹。
项目规划结构可由几个关键词概括，即"主轴线"、"叶脉状车行路网"、"水岸半岛"、"中心点"。

1."主轴线"
主轴线从凤凰路主入口通向临春河，是整个社区的主要交通枢纽，也是社区的主要景观轴。

2."叶脉状车行路网"
本项目的车行交通系统如同叶脉，这样可以高效率地使用有限面积的道路系统。

3."水岸半岛"
每条支状路向水面沿伸，沿路两边布置住宅，形成一个个有序的半岛，每个半岛自组成团。

4."中心点"
在整个放射状社区的中心，设置一个带沙滩泳池的小型会所，使整个社区更加具有凝聚力。

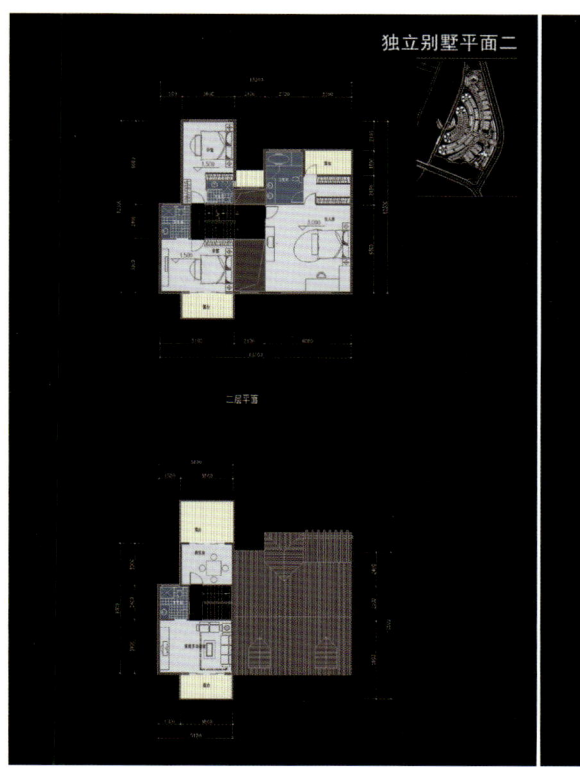

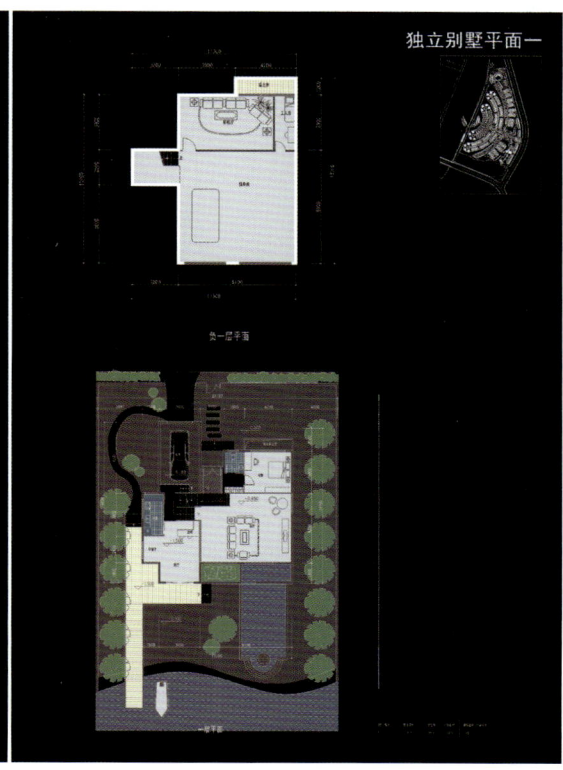

The design ideas derive from leaves of coconut palm and ripples of the sea.
The planning scheme of this project can be outlined by the following phrases: "main axis", "network of vein-like roads", "waterfront and peninsula", and "central point".

1. "Main Axis"
   The main axis extends from the main entrance in Feng Huang Road to Linchun River; it is the transportation hub as well as the main sightseeing axis of the community.

2. "Network of vein-like roads"
   The traffic system of this project is like veins of a leaf, thus the traffic system with limited area can be efficiently utilized.

3. "Waterfront and Peninsula"
   Every branch road extends to the direction of the water body, and residential buildings are distributed along the sides of the roads. Thus a series of well-arranged peninsulas are created, with each peninsula as a separate group.

4. "Central Point"
   At the center of the community with a radial pattern of roads, a small clubhouse with a beach swimming pool is erected, bringing more cohesion to the whole community.

总体布局

我们把独立别墅、联排别墅、叠拼别墅、公寓由西向东阶梯布置，达到资源的合理分配。社区的高层公寓集中放在地块的北端，高层公寓拥有河景、海景，视野开阔。低层住宅亲水，高层住宅看景，形成了"互惠互利"的双赢模式。

Master Plan

From west to east, single-houses, townhouses, multi-story townhouses, and apartments are arranged in a step-like array, achieving reasonable distribution of resources. The high-rise apartments are concentrated in the north part of the community, enjoying wide field of vision for river view and seascape. The low-rise buildings are located near water body while the high-rise buildings overlook the landscape, creating a win-win situation.

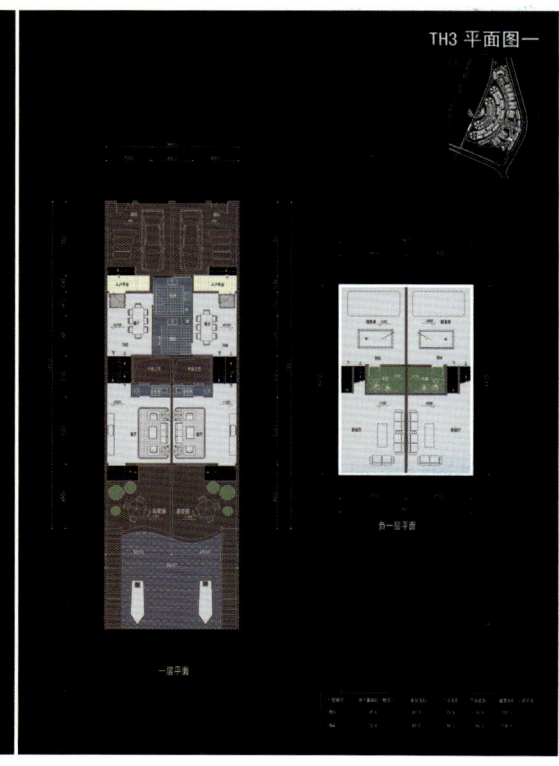

Qipan Mountain Villa District • Shenyang
棋盘山别墅区·沈阳

项目地点：中国，辽宁，沈阳市
项目时间：2008年
设计规模：14.5万 m²
设计阶段：方案设计
项目现状：未建

Project Location: Shenyang
Project Date: 2008
Project Scale: 145,000 m²
Design Phase: conceptual design
Project Status: construction not started

Bold and resolute designs and stage-like effects create dramatic changes like those in the Disney cartoons.

掠夺式的设计、布景式的手法，营造出迪斯尼式的戏剧化效果。

设计原则

设计紧扣"沈阳棋盘山欧陆旅游小镇"的主题,在镇中造山城,在山下建街镇。二期的西北地块是整个内环中最重要也是档次最高的地块,所以,"山城"将会成为整个总规划中的制高点,山城中的双拼别墅立面风格则属西班牙宫廷风格,更显高贵。

"镇中有山":在二期用地的西北地块,即价值最高的地块,利用总规划用地的挖方进行填土,形成一座山坡,将双拼别墅布置于山坡上,使高品质的双拼别墅拥有更开阔的视野。山地的营造打破了一期规划中二元次平面设计的单一性,形成了多元次的空间形态,也丰富了地块内的景观资源,使得居住和旅游品质得到提升。

Design Principle

By strictly adhering to the theme of "European style tourist town in Shenyang Qipan mountain", the design aims at building mountain dwellings in the town and building street-towns at the foot of hills. The northwest plot of phase II project is the most valuable plot within the inner circle. Therefore, the "mountain dwellings" becomes the highlight in the overall planning. The semi-detached house facade is of royal Spanish style.

"Hill in town": On the northwest plot of phase II project, which is the most valuable plot, a hillside is piled with earth excavated in the land leveling project under general planning; the semi-detached house style villas are erected along the hillside so that the occupants in the villas have wider field of vision. The hillside creates multi-dimensional spaces to enhance the landscape, avoiding the monotonous plan design in the planning of phase I; and increasing both the quality of residence and tourism.

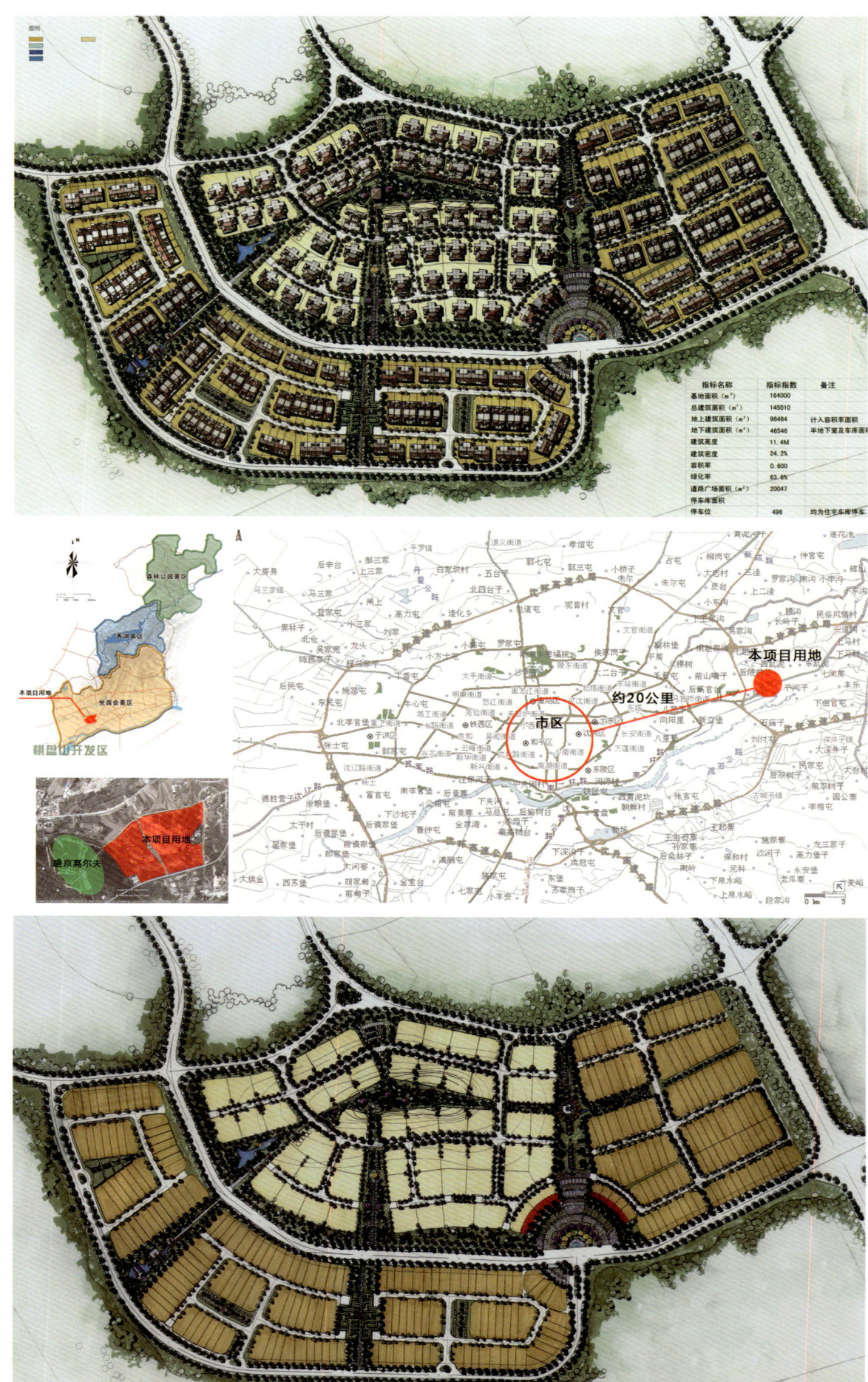

立面图

Bi Lin Wan • Shanghai
碧林湾·上海

项目地点：中国，上海市
项目时间：2002年
设计规模：26.5万 m²
设计阶段：方案设计
项目现状：已建

Project Location: Shanghai
Project Date: 2002
Project Scale: 265,000 m²
Design Phase: conceptual design
Project Status: construction completed

Using water as the main theme, "water and mountain" scenery is created for this rural residential area, providing a natural and ecological life style.

以水为主题打造山水为脉的小镇住宅区，提供一种自然生态居住的生活方式。

项目总览

碧林湾项目位于上海市闵行区七宝镇，华林路西、华茂路南、华莘港东、中谊路贯穿整个地块，西北面为已建的绿野香洲。项目定位为休闲风情山镇。

Project Overview

The estate is located in Qibao town in Minhang district of Shanghai. Four roads- West Hua Lin Road, South Hua Mao Road, East Hua Shen Port Road and Zhong Yi Road, all of which run through the plot. In the northwest lies the completed Lu Ye Xiang Zhou estate. This estate is positioned as a mountain town with recreational atmosphere.

总平面

规划构思

1. 项目设计以"湾"为主题，合理利用地块周边的水资源，设计出水广场、码头、游艇、水边Townhouse、水边住宅、临水商业街等，构建出一个符合主题的社区形象。

2. 以"湾"造"林"，在充分利用水资源的基础上，广种树木、花卉，形成条条绿脉。林、水交织，从而形成一个自然、生态的宜人社区。

3. 在小区人行主入口处设计商业步行街，向内延伸城市空间，加强小区内部城市空间意向，创造出一种小镇形象，营造一种休闲、轻松的氛围，同时也保证整个城市街区的统一连贯。

Planning Concept

1. The water resource around the plot is appropriately utilized in the design, with "bay" as the theme. Water plaza, dock, yacht, waterside townhouses, waterside residential buildings and waterside commercial streets are designed, building a communal image matching the theme.

2. Trees are planted around the bay. One the basis of making full use of the water resource, plenty of plants and flowers are planted to form green belts. The woods are interlaced with water bodies; therefore a pleasant community full of natural and ecological environment is created.

3. A commercial pedestrian street is designed at the main entrance. The street extends to the inside of the estate, bringing a sense of urban space and creating an image of small town, forming a recreational and relaxing atmosphere, at the same time, ensuring integration and consistency with the city streets.

Du Shi Yi Jia • Shanghai

都市宜家 • 上海

项目地点：中国，上海市
项目时间：2002年
设计规模：8.8万m²
设计阶段：方案设计
项目现状：已建

Project Location: Shanghai
Project Date: 2002
Project Scale: 88,000 m²
Design Phase: conceptual design
Project Status: construction completed

An intriguing street extends into the entrance of the estate, allowing people entering the estate in a leisurely way.

让一条有趣的小街成为小区的入口，让人们非常自然地进入。

总平面

项目分析

建筑设计追求具有现代感、富有异域情调的居住建筑形式。以坡顶作为统一的屋顶形式，体现居住建筑的总体特征。但设计并未简单套用欧洲传统形式，而是求新求变，运用现代设计手法重新构图，取得更为丰富多样的屋顶形式。商业街立面设计延续从稳重到活泼，古典到现代的风格走势。

Project Analysis

The design aims to create architectural forms full of modern and exotic atmosphere. All roofs are of pitched ones, a main feature of the residential buildings. However, the design does not simply copy traditional European architecture, but makes innovations and changes. Modern methods are used to design new and diversified roof shapes. The facades along the commercial street are designed with the styles featuring a gradual transition from classical to modern.

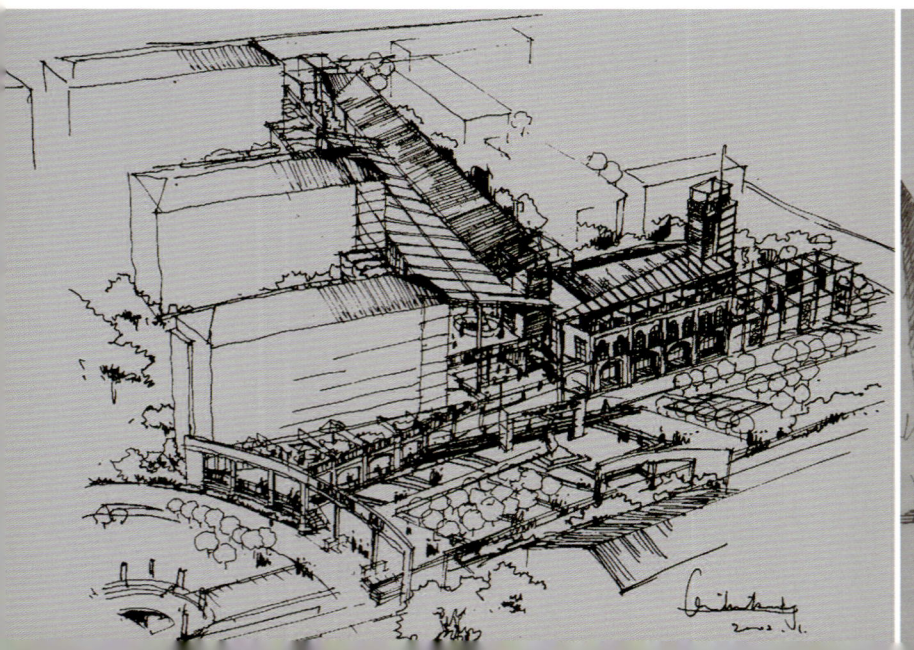

方案一

方案二

方案三

Sunshine Palm Garden (Phase III) • Shenzhen
阳光棕榈院三期·深圳

项目地点：中国，广东，深圳市，南山区，桂庙路
项目时间：2002年
设计规模：14.1万 m²
设计阶段：方案设计
项目现状：已建

Project Location: Guimiao Road, Nanshan District, Shenzhen
Project Date: 2002
Project Scale: 141,000 m²
Design Phase: conceptual design
Project Status: construction completed

This project provides an open urban area that can be traversed to reconnect urban segments which were isolated before. The efficient and compact layout guarantees quality of living spaces.

提供的可穿越的城市开放空间，将之前割裂的的城市片段重新连接了起来；高效、紧凑的平面设计保证了居住空间的优良品质。

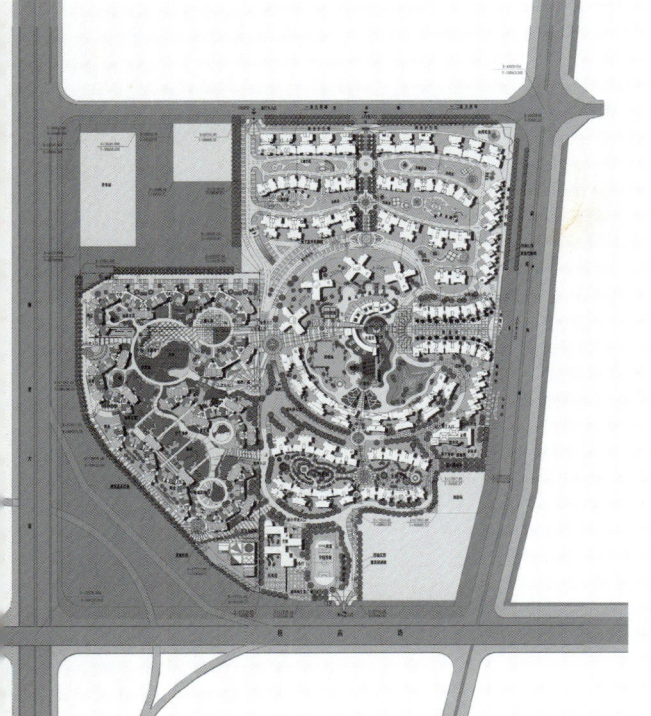

设计说明

整体布局以"Y"型小高层单元为模块，串联成半围合组团，组团间相互沟通并组合成中心大花园，形成步移景异、形态层次丰富的外部空间。

造型采用简洁的单坡屋顶，坡间正反错落形成丰富的屋顶聚落，抽象出简洁质朴的地中海小城建筑顶部特征。屋顶色彩选取米色与褐红的组合，局部变异的同色灰调系列具有阳光感，对地中海风情进行了全新的解释。

Design Description

In the overall layout, the Y-shaped small high-rise units are modules connecting together to form a semi-enclosed group. The groups share a good spatial relationship between each other and creating a centralized garden with multiple levels of outdoor spaces. The view changes as one move through the space.

The roofs are of simple pent roof, with the vertical sides of any two neighboring pent roofs placed closely. Therefore diversified roof groups are created, with the style of the simple and unvarnished building roofs in Mediterranean small towns. A combination of cream and maroon is used as the colors of the roofs, while some parts are painted with the same color in different shade to create a sense of vibrance, which is a complete embodiment of Mediterranean style.

总平面

Tonghe - Story of South-bank • Hangzhou
通和 — 南岸故事·杭州

项目地点：中国，浙江，杭州市，萧山区
项目时间：2003年
设计规模：50万 m²
设计阶段：方案投标
项目现状：未建

Project Location: Xiaoshan District, Hangzhou
Project Date: 2003
Project Scale: 500,000 m²
Design Phase: competition
Project Status: construction not started

White walls, grey tiles, ornamental perforated windows, and houseboats are well arranged with modern techniques, aiming to create a poetic imagery of "a small bridge over the flowing stream by a village house".

用现代手法演绎，白墙、灰瓦、漏窗、船屋，营造"小桥流水人家"之意境。

用地规划

项目不以形、围或大花园、大绿地的"常规"手法来设计，而是从空间、氛围上入手，围绕"带状公建街"的社区公共室外空间组织建筑和布局。从西到东为商业风情街、幼儿园、现代中国园林会所和学校。以"街区"感将项目分为亲水、近山南北两片。此南北穿插的辐射状绿轴，形成了山、水、林的对话与空间的"对接"。围绕这些绿轴又形成了多层次的户外空间，层层叠叠、相互交织，形成动态延伸、静态扩展有序列展开的"情境"空间。

Land Use Planning

This design does not follow the conventional methods emphasizing shape, enclosure, large garden, or large greenbelt, but strives to create unique spaces and atmosphere. The buildings are organized around the public outdoor spaces of the "belt-shaped street"; from west to east, commercial stylish street, kindergarten, clubhouse with the style of modern Chinese garden and school are built. The whole estate is divided into the water side area to the south and the foothill area to the north. This radial green axes extending from the south to north creates a dialogue between the hills, water, and woods and a close connection of spaces. Outdoor spaces of various levels are created around these green axes, levels upon levels and interlaced with each other, creating a peaceful yet dynamic space.

全景平面图

基地有一片水域，在处理上不是简单地将其处理成观赏层面上的"湖水公园"，而是将看水景、思水境与江南水乡联系起来，设计出一个个浮岛，形成一道道水网，构成一串串临水别墅。

There is a water area in the site. The water area is not simply modified to become a "lake park" for sightseeing. Instead, it is designed to be associated with the scenery in the south of the lower reaches of the Yangze River. A series of floating islands, a network of water body, and consequently a series of waterside villas are designed.

功能分析图

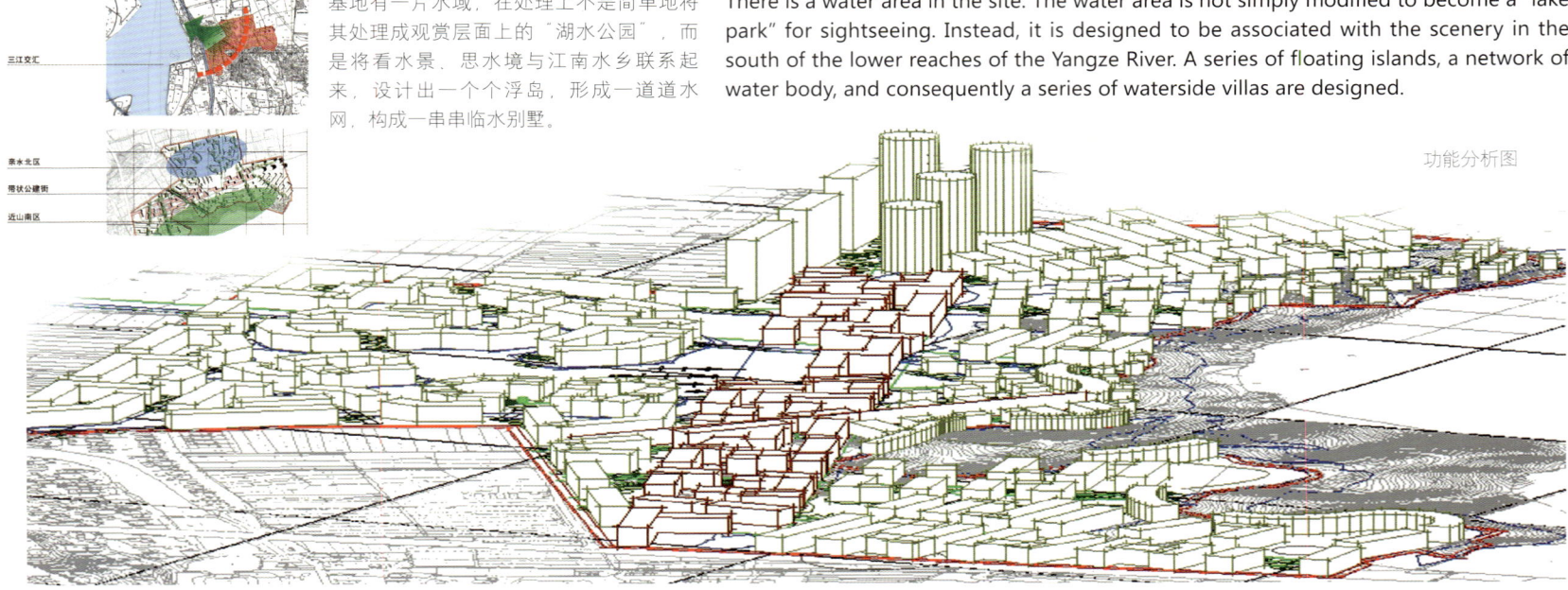

立面设计

住宅采用地方材料与玻璃、钢等相结合，力求体现阳光感、水乡感。立面由建筑线条、遮阳百叶、白墙等组成一幅生动活泼的画卷。公建带有江南水乡民居之形，用现代手法演绎，白墙、灰瓦、漏窗、船屋，力求营造"小桥流水人家"之意境。

Facade Design

The residential buildings are built with a combination of glass, steel and locally available materials, aiming to express vibrant atmosphere and water scenery. The facade, like a lovely scroll painting, is composed of the elements such as structural lines, shutters and white walls. The public buildings are designed with the shapes of folk houses in the lower reaches of the Yangze River; white walls, grey tiles, ornamental perforated windows, and houseboats are well arranged with modern techniques, aiming to create a poetic imagery of "a small bridge over the flowing stream by a village house".

Lian Tai Mangrove Bay • Shenzhen
联泰红树湾·深圳

项目地点：中国，广东，深圳市
项目时间：2007年
设计规模：5.2万 m²
设计阶段：方案设计
项目现状：在建

Project Location: Shenzhen
Project Date: 2007
Project Scale: 52,000 m²
Design Phase: conceptual design
Project Status: construction in progress

The design ideas derive from traditional southern Chinese gardens. This estate is a microcosm of the society, culture and life of southern China.

从江南园林中吸取灵感，犹如一个江南社会文化与物质生活的缩影。

工人房
UP
1.500
中厨
车库
UP
西厨
餐厅
DN
起居室
1.500
接待
±0.000
UP
DN
-1.600
下沉花园
UP
客卧
客厅
-0.400
-0.600
泳池
水面

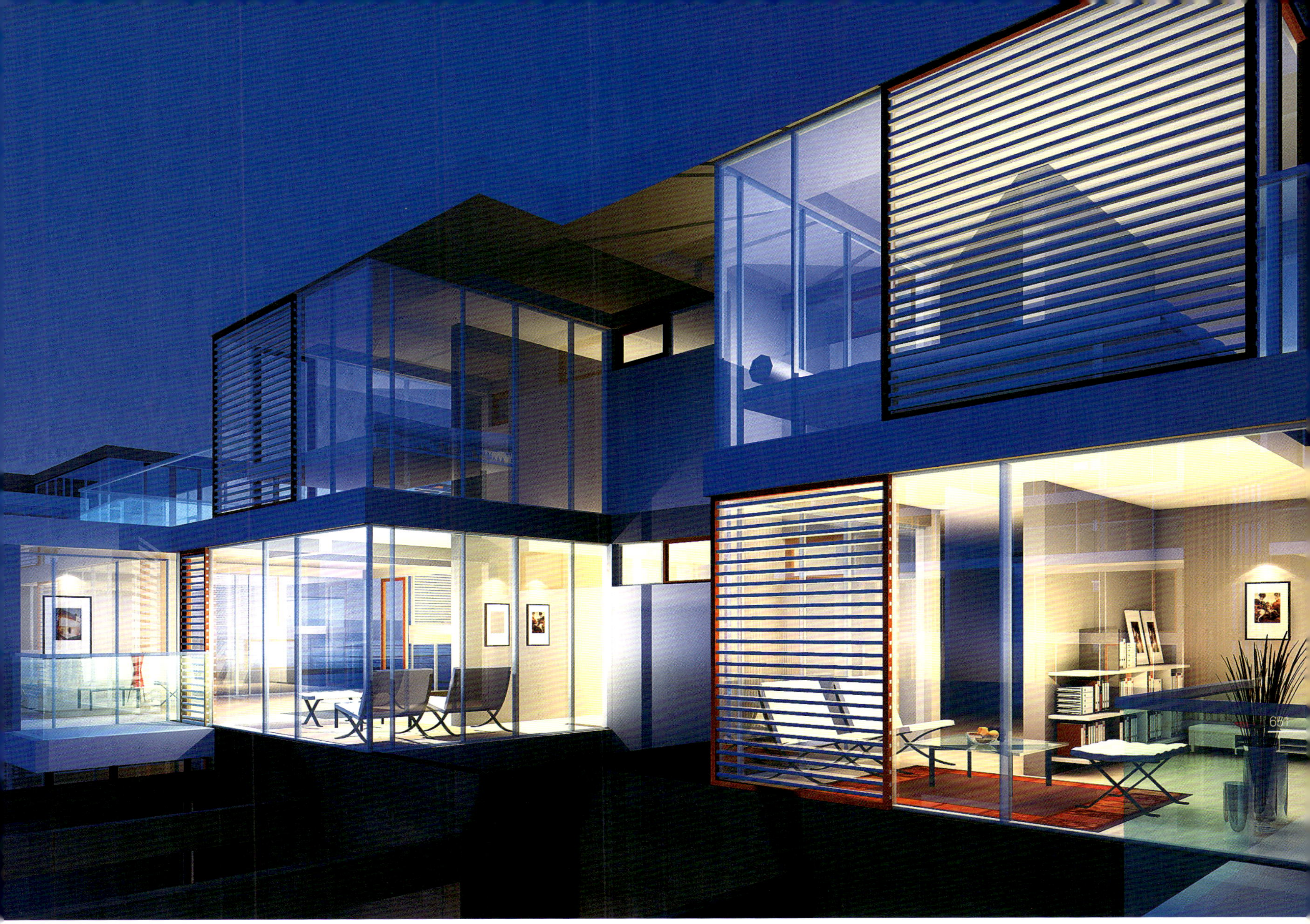

一层平面

总平面

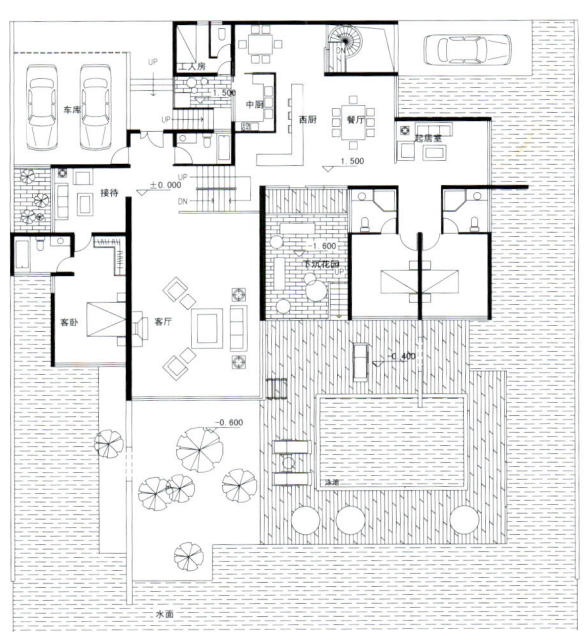

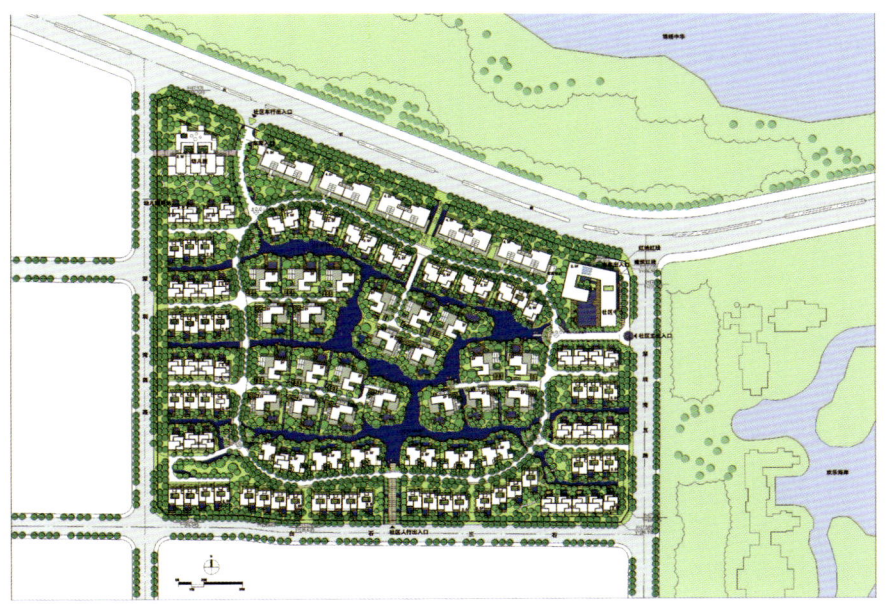

户型

1. 传承中国传统文化，借鉴中国园林中借景、对景的手法，在空间表达上含蓄、内敛，注重空间的过渡、传承关系。

2. 流动空间：公共的客厅、餐厅，半公共的厨房、过厅、家庭厅以及私密的卧房、书房之间，通过墙体的穿插，透明及半透明墙体的运用来划分空间，但不隔离空间，达到空间和视线上的流动性和连续性。

3. 户内空间与户外空间的交融。我们在园林中作建筑，在建筑内部作园林，"建筑是室内的园林，园林是室外的建筑。"

Floor Plans

1. By referring to the heritage of traditional Chinese culture, the design of floor plan of suites in this estate aims at embodying Chinese traditional culture. The spaces are arranged in a reserved and confined way, while emphasizing the transition and integral relationship between them.

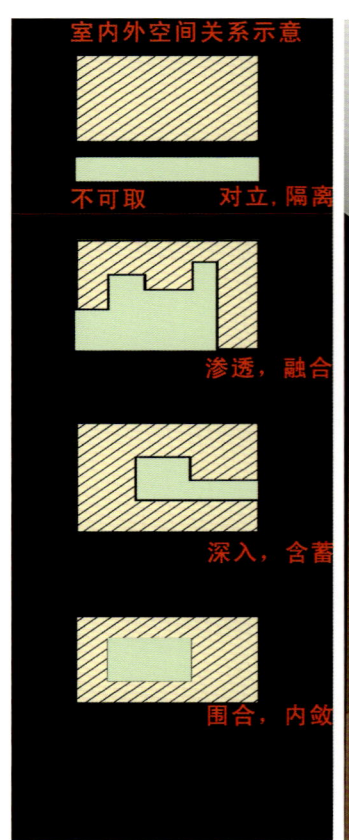

2. Flowing Spaces: through well-organized partition of walls and utilization of transparent and semi-transparent walls, the whole space is divided into open living room and dining room, semi-open kitchen, gallery, and family room, and private bedroom and study. But the divided spaces are not isolated from each other, so as to obtain the circulation and continuity of spaces and sightlines.

3. Blending of outdoor and indoor spaces: the buildings are constructed in a garden, and gardens are set up in the buildings; "building is the indoor garden, and garden is the outdoor building."

组团：别墅并不是分散布局在基地里，而是规划条理地布局成团。

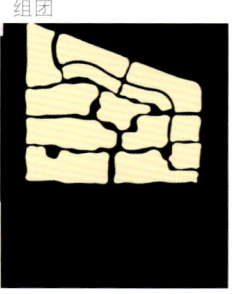

Grouping: villas do not scatter at the site but are grouped rationally and orderly.

生态水网：网状的水系是社区主要的内部景观，水网把整个地块划分成一个个组团，形成岛屿和半岛屿。

Ecological water network: the network-shaped water system is a main landscape of the community. It divides the whole plot into different groups to form islands and peninsulas.

绿环：沿着区外边的城市环道和社区内部的环道，有高大的乔木植被，形成两道绿环，绿环粗略地划分了社区的空间；内部绿环以内为高档区，以外为较高档区。

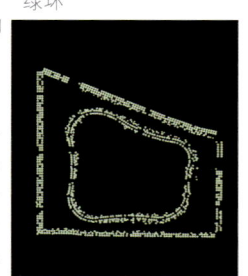

Green ring: tall arbor trees are planted along outside urban ring roads and ring roads in the community to form two green rings, which divide the space of the community roughly. Inside the green ring is a high-end residential quarter while outside the green ring is a very high-end residential quarter.

路网：路网是社区的交通系统。

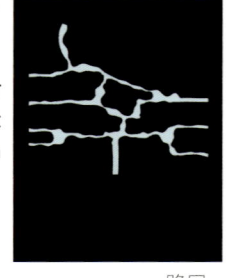

Road network: the road network is a traffic system of the community.

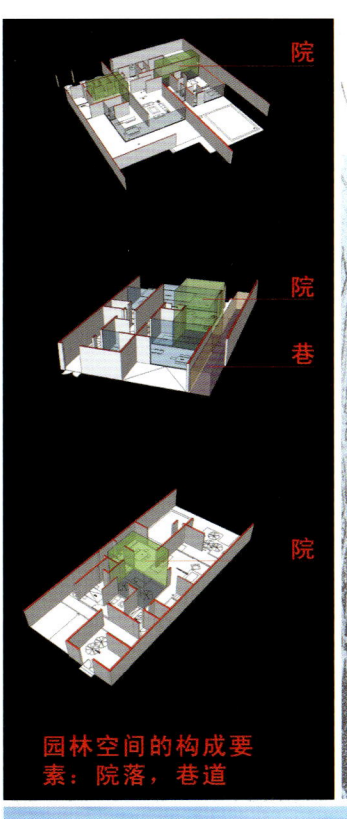

园林空间的构成要素：院落，巷道

Project
Index……

PROJECT INDEX 作品索引

项目名称	时间	规模	地点
碧海云天	2001	29万m²	广东深圳
灏景湾	2001	6万m²	广东深圳
锦上花（花样年华）	2001	5万m²	广东深圳
锦缎之滨	2001	14万m²	广东深圳
鸿荣源香蜜湖熙园	2001	15万m²	广东深圳
广西南宁华星时代广场	2001	16万m²	广西南宁
盈晖翠苑（龙宝发）	2001	4.7万m²	广东深圳
弘雅城市花园	2001	19万m²	广东深圳
宜昌时代天骄(龙苑三期)	2001	1.8万m²	湖北宜昌
山西太原恒运通商厦	2001	3万m²	山西太原
武汉时代天骄	2001	4.2万m²	湖北武汉
城建大厦	2001	7万m²	广东深圳
成都花样年	2001	1.6万m²	四川成都
上海都市宜家（彩虹堡）	2001	10万m²	上海
花样年农科中心	2001	20万m²	广东深圳
上海绿野香洲（澳韵花园）	2002	3.5万m²	上海
万兆家园入口环境	2002	0.2万m²	上海
城建花园立面方案调整	2002	9.8万m²	广东深圳
武汉美好家园	2002	7万m²	湖北武汉
北海蔚蓝香格里	2002	9万m²	广西北海
阳光棕榈园三期	2002	10万m²	广东深圳
上海彩云堡	2002	10万m²	上海
上海碧林湾	2002	30万m²	上海
三亚兰海花园二期	2002	2万m²	海南三亚
成都新希望	2002	16万m²	四川成都
贵阳监管小区	2002	40万m²	贵州贵阳
北海阳光香格里小区	2002	9万m²	广西北海
鸿威大梅沙海怡轩	2002	1.8万m²	广东深圳
侨城花园（天鹅堡）	2002	48.8万m²	广东深圳
上海万兆东一#地块	2002	10万m²	上海
上海万兆中华村	2002	7万m²	上海
侨城东方花园	2002	3万m²	广东深圳
贵港城市广场	2002	2万m²	广西贵港
深圳太太药业	2002	1.5万m²	广东深圳
成都花样年洗面桥	2002	2.7万m²	四川成都
中海横岗项目	2002	28.6万m²	广东深圳
杭州通和项目	2003	50万m²	浙江杭州
武汉紫菘枫亭	2003	11万m²	湖北武汉
万科华宇项目	2003	28万m²	广东深圳
贵港假日酒店	2003	1.5万m²	广西贵港
广州大东景项目	2003	43万m²	广东广州
集浩北海工业园	2003	50hm²	广西北海
武汉光谷硕博苑	2003	4万m²	湖北武汉
兰州仁恒住宅项目	2003	20万m²	甘肃兰州
广西南宁鸣翠谷	2003	43万m²	广西南宁
华侨城东部厂区改造	2003	5.6万m²	广东深圳
佛山市新闻中心	2003	8万m²	广东佛山
北海城市会馆	2003	3万m²	广西北海
深圳公安边防支队机关大院	2003	11.7万m²	广东深圳
置地春风居	2003	8.1万m²	广东深圳
鄂州汽车客运站	2003	11万m²	湖北鄂州
浙江台州黄岩名仕家园	2003	6.3万m²	浙江台州
三亚市丰兴隆区规划设计	2003	43万m²	海南三亚
深圳市宝安区质检站办公楼	2003	1.4万m²	广东深圳
福建泉州中总商住区	2003	18万m²	福建泉州
上海虹梅家园	2003	30万m²	上海
北海德宝国际新城	2003	63万m²	广西北海
联泰大厦	2004	4万m²	广东深圳
天安高尔夫海景花园三期	2004	11.2万m²	广东深圳
北海北部湾（长安花园）	2004	3.5万m²	广西北海
宜昌中房·桃花美邻	2004	3.6万m²	湖北宜昌
东莞华侨新村别墅区孔雀岛	2004	2万m²	广东东莞
杭州水榭花城	2004	28.6万m²	浙江杭州
武汉时代天骄二期	2004	15.8万m²	湖北武汉
重庆茂业中心（百子巷）	2004	59万m²	重庆
深中旅布吉项目	2004	4.9万m²	广东深圳
鸿荣源公园大地	2004	60万m²	广东深圳
南宁会展中心广场东广场规划设计	2004	4万m²	广西南宁
西安电子科技大学新校区二期"巨构"	2004	25万m²	陕西西安
葵涌办公楼投标	2004	1.4万m²	广东深圳
长沙格蓝康都（金盈城）	2004	7万m²	湖南长沙
龙岗住宅局人才村	2004	8万m²	广东深圳
上海碧林湾新苑	2004	13万m²	上海
重庆茂业沙坪坝项目	2004	42.7万m²	重庆
广西南宁德宝大厦（园湖路）	2004	2.3万m²	广西南宁
坪山体育中心体育场	2004	0.73万m²	广东深圳
坪山体育中心体育馆	2004	0.48万m²	广东深圳
三洲田	2004	0.13万m²	广东深圳
长沙中达丽景香山项目	2004	30万m²	湖南长沙
上海英皇明星城	2004	11万m²	上海
大梅沙酒店（联泰会所）	2004	1.1万m²	广东深圳
茂业爱华路项目	2004	20万m²	广东深圳
万科武汉项目	2004	5万m²	湖北武汉
武汉润海时代天骄二期	2004	16万m²	湖北武汉
南宁东盟国际水果蔬菜批发市场	2005	35万m²	广西南宁
胡耀邦纪念馆	2005	0.3万m²	湖南
茂业和平店立面改造	2005	5.3万m²	重庆
花样年别样城	2005	13万m²	四川成都
佛冈项目	2005	1万m²	广东清远
联泰大梅沙别墅	2005	2.2万m²	广东深圳
汕头联泰项目	2005	14万m²	广东汕头
深圳湾中学设计	2005	6万m²	广东深圳
南宁江南新天地	2005	1万m²	广西南宁
华侨城单身公寓	2005	4万m²	广东深圳
北海蔚蓝家园三期	2005	2万m²	广西北海
桂林软件城	2005	7万m²	广西桂林
龙城街道办事处综合办公楼	2005	2万m²	广东深圳
茂业成都商场改造	2005	5.3万m²	四川成都
长沙圣爵菲斯二期	2005	15.6万m²	湖南长沙
长沙滨江君悦香邸项目	2005	33万m²	湖南长沙
布吉宏达大厦改造	2005	0.2万m²	广东深圳
长沙圣爵菲斯三期规划	2005	22万m²	湖南长沙
秦皇岛假日蓝湾	2005	9万m²	河北秦皇岛
信义长安项目	2005	27万m²	广东东莞
联泰九号地块	2005	5.2万m²	广东深圳
华盛领域	2005	3万m²	广东深圳
鄂城秋林中学	2005	4万m²	湖北鄂州
鸿威惠州项目	2005	11万m²	广东惠州
英皇上海南京东路项目	2005	7万m²	上海
卓越浅水湾立面改造	2005	6万m²	广东深圳

项目名称	年份	面积	地点
恒安华浩源四期	2005	8万m²	广东深圳
茂业成都人民商场二期	2005	24.5万m²	四川成都
莞城中心区	2005	13万m²	广东东莞
联泰大梅沙公寓	2005	0.6万m²	广东深圳
华侨城纯水岸100号别墅	2005	0.1万m²	广东深圳
新光高尔夫豪园	2006	60万m²	广东深圳
长沙书院路521号	2006	36万m²	湖南长沙
中粮万福阁（达利空间）	2006	2.3万m²	广东深圳
花样年侨香路项目（香年广场）	2006	7.4万m²	广东深圳
南昌联泰	2006	115万m²	江西南昌
东莞理想0769项目	2006	13万m²	广东东莞
好景社会（景田北）项目	2006	2.1万m²	广东深圳
万科广州项目1	2006	1.4万m²	广东广州
三洲田领导别墅	2006	1.3万m²	广东深圳
万科广州项目2	2006	15万m²	广东广州
东莞天安项目	2006	62.3万m²	广东东莞
珠海万科金域蓝湾	2006	6.7万m²	广东珠海
天津梅江南项目	2006	3.4万m²	天津
威尼斯桥博物馆投标	2006	0.1万m²	意大利威尼斯
宜昌桃花岭城市广场	2006	6万m²	湖北宜昌
长沙宜居花园	2006	36万m²	湖南长沙
富通城四期投标	2006	3万m²	广东深圳
东莞福都花园	2006	3万m²	广东东莞
深圳卓越南新路改造	2006	51万m²	广东深圳
澳门鲍思高球场项目	2006	0.3万m²	澳门
花样年盐田明珠项目	2006	3万m²	广东深圳
河南郑州农科路项目	2006	7万m²	河南郑州
上海碧林湾三期	2006	2.4万m²	上海
深圳中海康城国际	2006	45万m²	广东深圳
承翰龙岗山居别院	2006	8万m²	广东深圳
长沙新簇银盆岭项目	2006	19万m²	湖南长沙
长沙中城·丽景香山二期	2006	20万m²	湖南长沙
长沙广电中心投标	2006	8万m²	湖南长沙
淮北泰富华凤凰城	2007	60万m²	安徽淮北
深业坪山项目投标	2007	11万m²	广东深圳
金光华项目投标	2007	20万m²	广东深圳
广西南宁利海项目	2007	8万m²	广西南宁
三亚下洋田项目	2007	10万m²	海南三亚
福州三迪·凯旋枫丹项目	2007	9.2万m²	福建福州
华侨城侨北苑一期	2007	11万m²	广东深圳
天津团泊高尔夫公寓	2007	9.8万m²	天津
华侨城五号地	2007	8.4万m²	广东深圳
中海苏州胥江项目	2007	20万m²	江苏苏州
天津花样年C地块	2007	48万m²	天津
天津金唐大厦（深福保）	2007	9万m²	天津
龙光汕头42街区项目	2007	34万m²	广东汕头
苏州中海国际社区	2007	120万m²	江苏苏州
深圳嘉旺城项目	2007	14万m²	广东深圳
深圳市现代艺术馆与城市规划展览馆国际竞赛	2007	8万m²	广东深圳
鄂州第八中学	2007	2.5万m²	湖北鄂州
深圳东都花园二期	2007	7.7万m²	广东深圳
泰华N7项目	2007	53.6万m²	广东深圳
长沙圣爵菲斯酒店别墅	2007	60万m²	湖南长沙
华侨城侨北苑二期	2007	8.7万m²	广东深圳
成都华侨城立面	2007	15万m²	四川成都
信贤坂田项目	2007	42万m²	广东深圳
广西钦州·体育公园	2007	50万m²	广西钦州
成都中海城南2号地项目	2007	23万m²	四川成都
无锡茂业	2007	15.3万m²	江苏无锡
成都花样年·花样城项目	2007	50万m²	四川成都
东莞卓越蔚蓝城邦	2007	41万m²	广东东莞
东莞鼎峰板岭项目	2007	20万m²	广东东莞
宝安中心区N27地块项目	2007	6万m²	广东深圳
南山·苏州项目	2007	45万m²	江苏苏州
花样年成都新津老君山项目	2007	44万m²	四川成都
马峦山	2007	40万m²	广东深圳
华侨城六号地	2007	6万m²	广东深圳
华科大产基地	2007	2万m²	广东深圳
大万世居	2007	1.6万m²	广东深圳
深圳南澳凯旋湾花园	2007	4万m²	广东深圳
福建融信宗地2007-12号项目（白马路地块）	2007	11万m²	福建福州
福建融信宗地2007-14号项目（则徐大道地块）	2007	10万m²	福建福州
南山荔景山庄	2007	4.7万m²	广东深圳
中海康城花园北地块项目	2007	10万m²	广东深圳
兆恒抚顺特钢公司公明厂房及配套	2007	1.3万m²	辽宁抚顺
长沙同瑞项目	2007	25万m²	湖南长沙
深业惠州火车站项目	2007	14万m²	广东惠州
深圳华侨城天鹅堡三期	2007	20万m²	广东深圳
深圳横岗湛宝大厦	2007	10万m²	广东深圳
武汉喻家湖项目（联峰）	2008	20万m²	湖北武汉
沈阳棋盘山投标项目	2008	8万m²	辽宁沈阳
惠州日报社项目	2008	10万m²	广东惠州
华侨城机场项目	2008	132万m²	广东深圳
光明中心区公园投标	2008	474.85hm²	广东深圳
长沙君悦香邸南地块	2008	20万m²	湖南长沙
青岛港中旅海泉湾项目	2008	20万m²	辽宁青岛
深圳金众葵涌丰树山项目	2008	17万m²	广东深圳
深圳华侨城可行性研究-燕晗花园二期	2008	3.3万m²	广东深圳
长沙地矿局项目	2008	28万m²	湖南长沙
惠州山水江南二期	2008	2.3万m²	广东惠州
长沙中城·丽景香山三期	2008	11万m²	湖南长沙
成都华侨城二期多层部分设计	2008	22万m²	四川成都
海口海湾花园项目	2008	22万m²	海南海口
佛山世界名牌店现代服务产业区规划（奥特莱斯）	2008	142万m²	广东佛山
深圳市光明新区科技公园周边地区城市设计	2008	475hm²	广东深圳
佛山市公共文化综合体建筑方案设计投标	2008	28万m²	广东佛山
龙华和平路项目（古浓滕）	2008	32.5万m²	广东深圳
新洲项目	2008	9.3万m²	广东深圳
鸿威增城项目	2008	150万m²	广东广州
深圳人才园设计及设计咨询投标	2008	4.8万m²	广东深圳
湖南宁乡春城商业广场	2008	20万m²	湖南宁乡
深圳大学城国际会议中心及中心广场概念投标	2008	1万m²	广东深圳
天安惠州市惠阳区新阳城一期	2008	6.7万m²	广东惠州
贝鲁特国际竞赛	2008	1.8万m²	贝鲁特

项目名称	年份	面积	地点	项目名称	年份	面积	地点
珠海中信湾项目	2008	5.2万m²	广东珠海	信义芜湖项目	2009	105万m²	安徽芜湖
深业巢湖项目	2008	11万m²	广东深圳	信义布吉金稻田路项目	2009	76.6万m²	广东深圳
宜兴东 大厦	2008	10.6万m²	江苏宜兴	梅花万岁城项目	2009	18万m²	四川乐山
承翰慢城四期投标	2008	15万m²	广东深圳	广西美术馆AND铜鼓博物馆投标	2009	3.6万m²	广西
华侨城燕晗组团地块（oct-03）	2008	4.4万m²	广东深圳	友盛英伦名苑四期（华润中央工程）	2009	4.7万m²	广东深圳
长沙滨江君悦香邸六栋项目	2009	32万m²	湖南长沙	中粮地产61区项目	2009	8.4万m²	广东深圳
深圳岗宏家和盛世二期	2009	3万m²	广东深圳	深圳绿景牛栏前项目	2009	6.3万m²	广东深圳
惠州宝安 山水江南二期项目	2009	1.5万m²	广东惠州	成都华韵天府	2009	21万m²	四川成都
辛亥革命纪念碑（塔）概念性设计方案	2009	0.1万m²	湖北武汉	皇庭东莞石龙项目	2009	34.8万m²	广东东莞
华强北路立体街道城市设计方案国际咨询	2009	1hm²	广东深圳	北海冠岭项目会议中心方案设计	2009	1.5万m²	广西北海
深圳市中心区水晶岛规划设计方案国际咨询	2009	6.4万m²	广东深圳	北海冠岭项目五星级酒店方案设计	2009	3.8万m²	广西北海
惠州皇庭波西塔诺项目(大亚湾)项目	2009	77万m²	广东惠州	皇庭钦州滨海新城白石湖项目	2010	42万m²	广西钦州
深圳市观澜版画基地美术馆及交易中心设计方案国际竞赛	2009	3万m²	广东深圳	信贤坂田一期项目	2010	42万m²	广东深圳
				山东鲁能三亚酒店	2010	14万m²	海南三亚
				山东鲁能领秀城A2A3商业综合体	2010	24万m²	山东鲁能
				岗宏地产龙岗项目	2010	31.6万m²	广东深圳
				蓝湾西安项目	2010	118.5万m²	西安
深圳市软件产业基地项目	2009	4.8万m²	广东深圳	创维半导体设计中心	2010	12万m²	广东深圳
佛山职业技术学院投标	2009	21.7万m²	广东佛山	天俊坪山景观项目	2010	2.6万m²	广东深圳
大运新城美域蓝湾项目投标	2009	15万m²	广东深圳	天津深福宝张家窝别墅概念规划	2010	9.4万m²	天津
深圳卓越南新路改造（南区）	2009	68.7万m²	广东深圳	北京广安贵都酒店扩建工程立面设计概念方案	2010	3.6万m²	北京
阳江安宁路住宅项目	2009	5.6万m²	广东阳江	朱古石项目概念规划设计	2010	55万m²	广东深圳
阳江核电办公培训设施概念方案设计	2009	21.8万m²	广东阳江	深福保天津工业园概念规划	2010	48万m²	天津
深圳侨城7号地	2009	6.5万m²	广东深圳	华侨城栖湖花园3期1#地	2010	4.8万m²	广东深圳
深圳市龙岗区128区改造项目	2009	14.8万m²	广东深圳	华侨城栖湖花园3期2#地	2010	12万m²	广东深圳
花样年花郡立面改造	2009	13万m²	广东深圳	华侨城栖湖花园3期3#地	2010	2万m²	广东深圳
华侨城锦绣四期	2009	17万m²	广东深圳	重庆龙湖U2项目2~5号地块	2010	89万m²	重庆
佛山ABC大楼	2009	3万m²	广东佛山	业尊公交上盖	2010	12.7万m²	广东深圳
西安华侨城项目	2009	17.4万m²	西安	花样年云南大理项目	2010	7.8万m²	云南大理
百合名门项目（龙城街道五联）	2009	27万m²	广东深圳	信义横岗酒店项目	2010	77.6万m²	广东深圳
信义横岗综合楼立面改造	2009	77.6万m²	广东深圳	中海桂林项目	2010	13万m²	广西桂林
广西钦州商厦地块项目	2009	18万m²	广西钦州	科兴生物谷二期项目	2010	6.6万m²	广东深圳
星河南沙项目	2009	40万m²	广东广州	业尊光明住宅项目	2010	12.7万m²	广东深圳
清远翡翠绿洲	2009	2.1万m²	广东清远	绿景化州橘州新城	2010	160万m²	广东化州
皇庭钦州御珑湾项目（体育公园）	2009	11.8万m²	广西钦州	华侨城整体规划研究和城市更新可行性研究	2010	133万m²	广东深圳
成都花样年 香年广场项目	2009	20万m²	四川成都	卓越下步庙项目	2010	84万m²	广东深圳
湖南湘府嘉城项目	2009	34万m²	湖南湘府	绿景珠海项目	2010	3.6万m²	广东珠海
西安曲江项目	2009	34万m²	西安曲江	卓越龙华项目	2010	14万m²	广东深圳
深圳移动生产调度大厦	2009	10万m²	广东深圳	武汉华侨城H/A地块概念规划建筑设计研究	2010	175万m²	湖北武汉
彭成东城一号项目	2009	7.1万m²	广东深圳	青岛中海 李沧河南南庄项目	2010	186万m²	山东青岛
骏泰龙岗龙东项目	2009	22万m²	广东深圳	蓝湾商务大楼项目	2010	13.9万m²	广东深圳
万泰铜陵项目	2009	35.6万m²	安徽铜陵	成都中海高层办公楼项目	2010	38万m²	四川成都
万泰杭州华丰项目	2009	6.3万m²	浙江杭州	北京房山港中旅温泉度假城项目	2010	63万m²	北京
北京国歌广场	2009	6万m²	北京	石家庄项目	2010	12万m²	河北石家庄
天津华侨城	2009	158.14万m²	天津	花样年惠州大亚湾.碧云天花园	2010	14万m²	广东惠州
花样年深圳保税区B105-30地块项目	2009	4.6万m²	广东深圳	石家庄余底村旧改项目	2010	110万m²	河北石家庄
花样年渔农村	2009	2.3万m²	广东深圳	天津花样年项目	2010	659万m²	天津
成都谢菲尔德置地广场	2009	14万m²	四川成都	苏州南山项目	2010	39.5万m²	江苏苏州
深圳樱花苑项目	2009	27.7万m²	广东深圳	惠州鼎峰项目	2010	61.8万m²	广东惠州
皇家领地酒庄项目	2009	3.5万m²	山东	苏州相城项目	2010	17.8万m²	江苏苏州
惠阳花样年别样城项目	2009	58万m²	广东惠州	台山东郊马山项目(嘉旺城国际公馆)	2010	3.9万m²	广东台山
康佳工业区改造项目	2009	24.6万m²	广东深圳	平谷大连项目	2010	6.9万m²	辽宁大连
沈阳中海国际社区2、4号地	2009	17万m²	辽宁沈阳	嘉旺惠东项目前期规划	2010	18.7万m²	广东惠州
中海济南九曲项目	2009	225万m²	山东济南				
太东花园项目	2009	18.7万m²	广东惠州				
东莞信义长安金三角	2009	24万m²	广东东莞				

项目名称	年份	面积	地点
青少年活动中心改扩建工程投标	2010	6.2万m²	广东深圳
彭城沙湖项目	2010	145万m²	广东深圳
彭城龙西项目	2010	5.8万m²	广东深圳
天津睿德信住宅项目	2010	18万m²	天津
成都汇峰中心投标	2010	12.9万m²	四川成都
烟台龙湖项目	2010	213万m²	山东烟台
华美达酒店立面施工图设计	2010	0.135万m²	湖北宜昌
南京项目（东方置地S04、N04地块）	2010	100万m²	江苏南京
成都花样年福年广场（诺亚舟）	2010	16.5万m²	四川成都
纯水岸幼儿园地块方案设计项目	2010	0.3万m²	广东深圳
中海城南1号B地块商业二期、三期	2010	26.7万m²	四川成都
嘉宏花园2期规划方案项目（天俊实业股份有限公司）	2010	18万m²	广东深圳
江苏启东崇明岛项目概念规划	2010	62.4万m²	江苏启东
港中旅（北京）长阳行政中心	2010	84.5万m²	北京
龙岗区百达厂及周边地区改造提升设计	2010	7.1万m²	广东深圳
常州市武进西太湖生态休闲区项目概念规划设计	2010	177万m²	江苏常州
海丰肉联厂项目	2010	11万m²	广东海丰
蓝湾莲花路公交上盖住宅项目概念	2010	14万m²	广东深圳

2001 CUBE DESIGN

	成都花样年	花样年农科中心	山西太原恒运通通商厦	城建大厦投标
宝安弘雅城市花园投标方案	锦缎之滨	上海都市宜家（彩虹堡）	灏景湾	锦上花（花样年华）
广西南宁华星时代广场	鸿荣源熙园	宜昌时代天骄	龙宝发盈晖翠苑	碧海云天

2002 CUBE DESIGN

武汉时代天骄		北海阳光香格里	成都花样年洗面桥	成都新希望
城建花园立面	华侨城-天鹅堡二期项目	贵港城市广场	贵阳监管小区	鸿威大梅沙海怡轩
上海澳韵花园(绿野香洲)	三亚兰海花园二期	侨城东方花园	上海碧林湾	上海彩云堡
上海万兆东一地块	上海万兆中华村	深圳市道路收费站设计方案	深圳太太药业	万兆家园入口设计

2003 CUBE DESIGN

武汉美好家园	阳光棕榈园	中海横岗项目		鄂州汽车客运站
宝安弘雅城市花园投标方案	福建泉州中总商住区	广州大东景项目	贵港假日酒店	杭州通和项目
华侨城东部厂区改造	集浩北海工业园	兰州仁恒住宅项目	三亚市丰兴隆区规划设计	深圳公安边防支队机关大院
深圳市宝安区质检站办公楼	万科华宇项目	武汉光谷硕博苑	武汉紫菘枫亭	浙江台州黄岩名仕家园

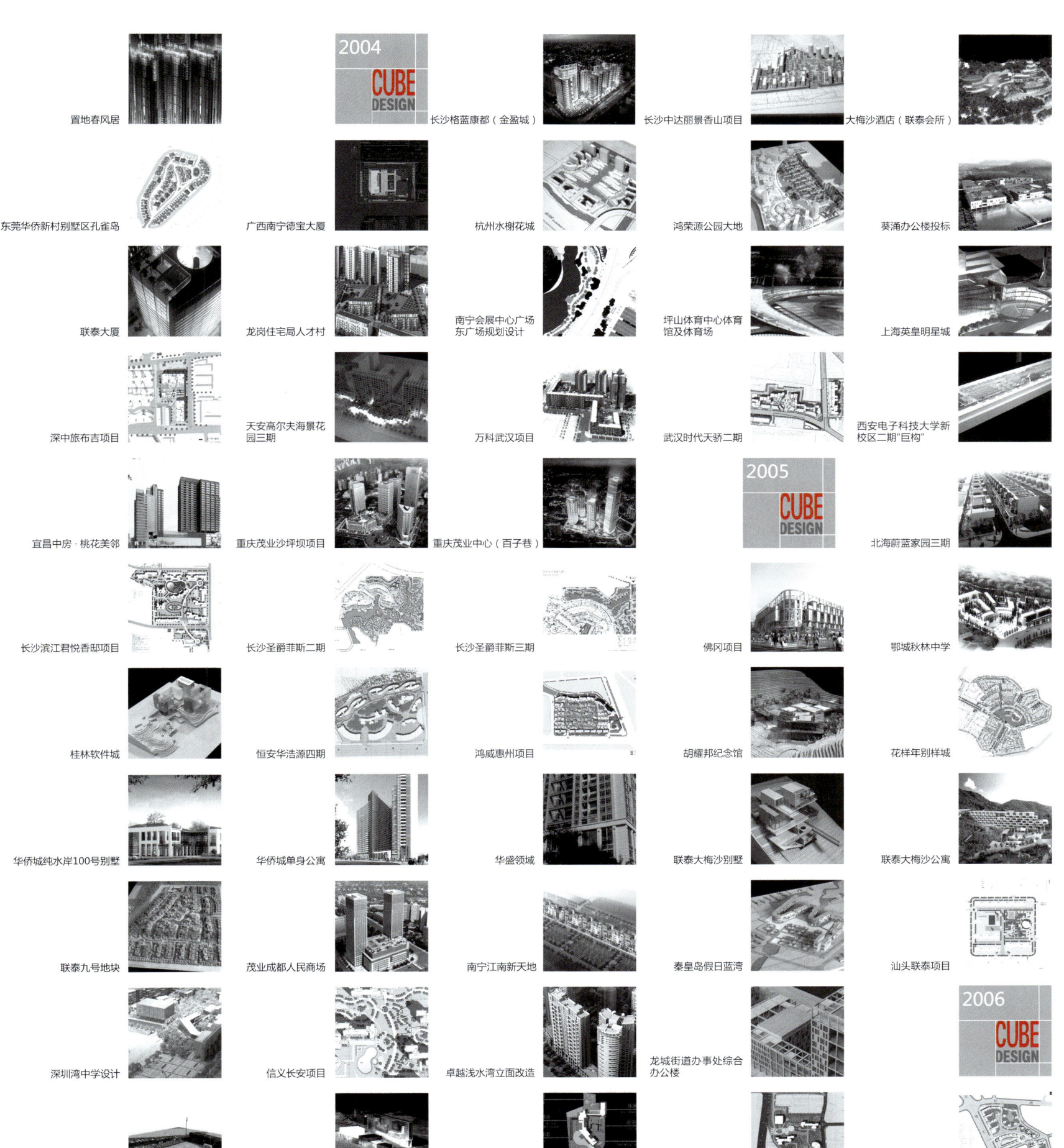

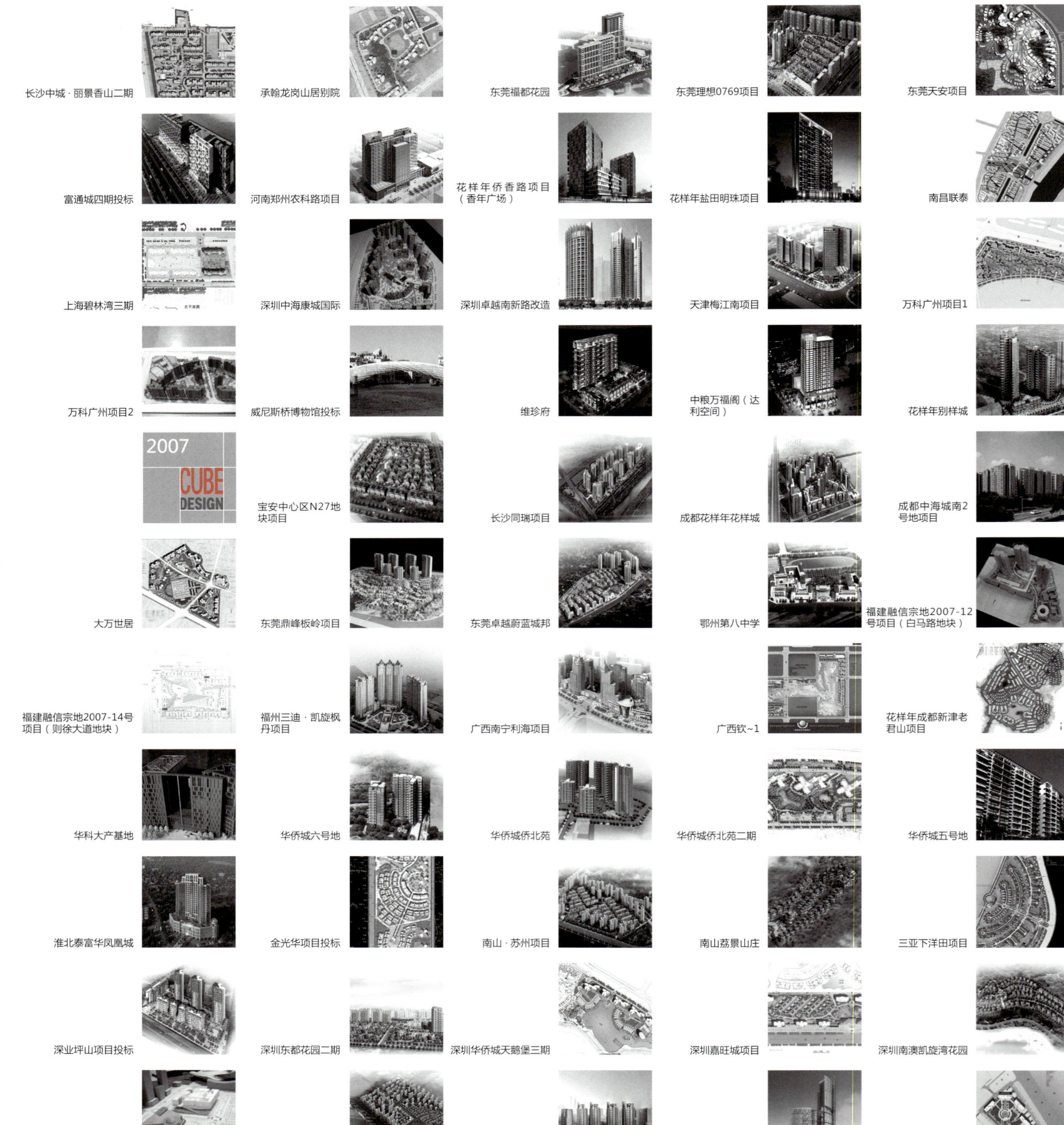

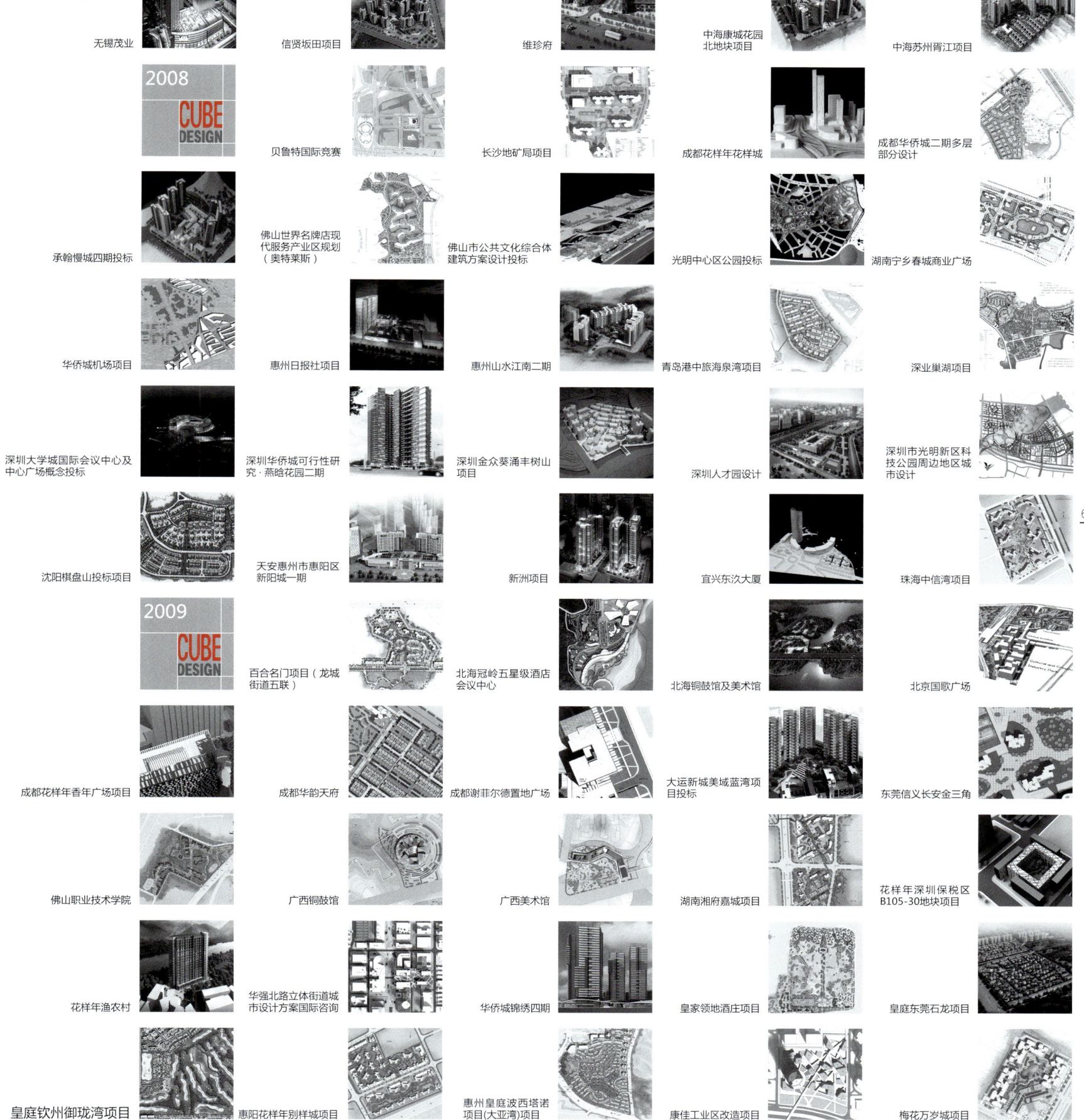

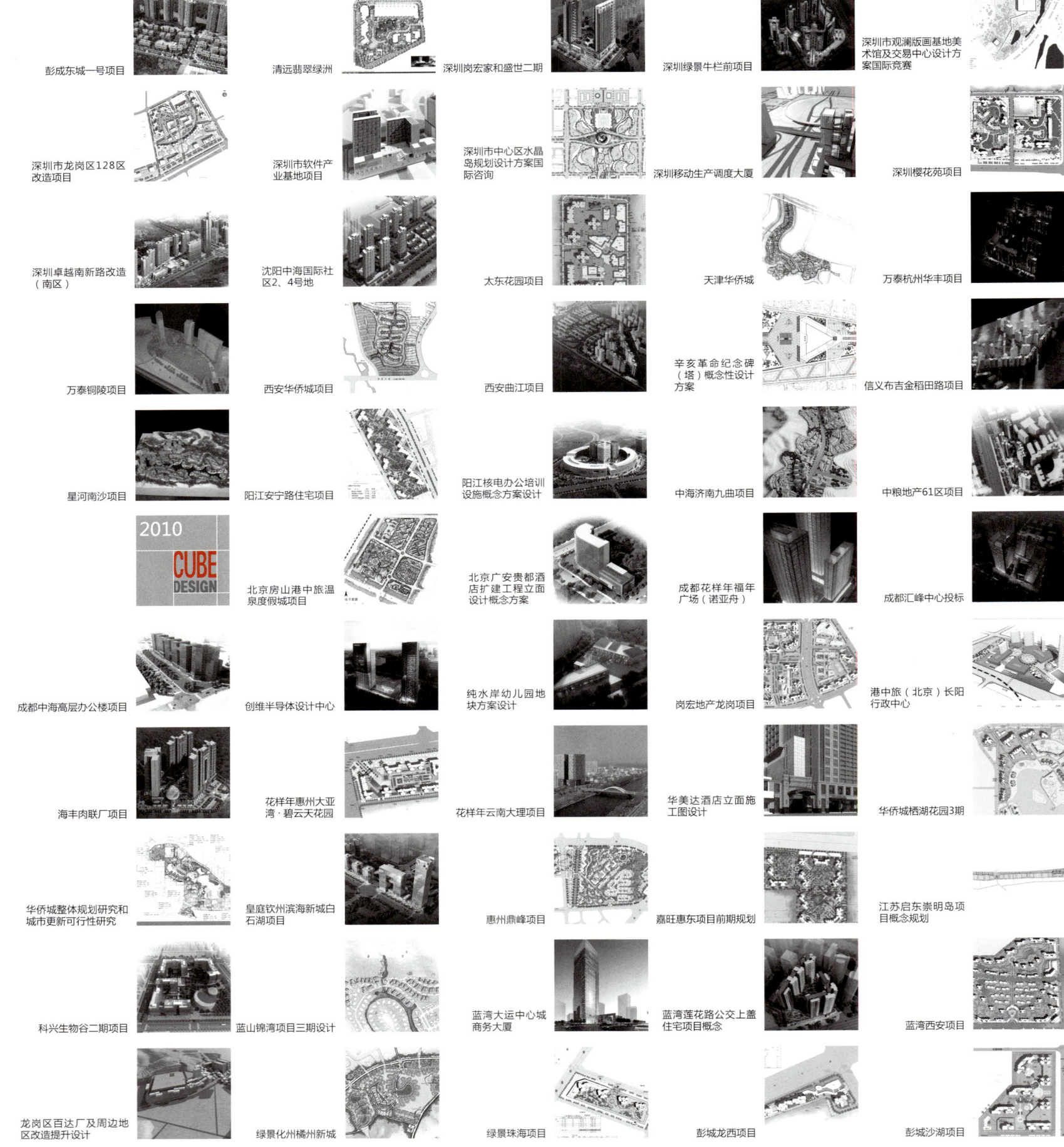

《立方设计》10周年纪念特刊现正发行。在此向深圳市立方建筑设计顾问有限公司的各位领导、各部门员工以及香港理工国际出版社表示衷心的感谢。在编辑过程中，尽管遇到了许多困难，但在大家的谅解与支持下，最终顺利完成。同时，也正因为有各方领导的大力支持与工作人员的努力配合，《立方设计》才得以都圆满成功。最后，也希望大家继续支持深圳市立方建筑设计顾问有限公司。希望立方设计的下一个十年更加精彩灿烂！

香港理工国际出版社　项目编辑组

The 10th anniversary edition of "CUBE Design" is now being issued formally. Hereby sincere thanks to all the staffs and leaders of Shenzhen Cube Architecture Design Office as well as the HKPIP, with the support and understanding of who the book was finally completed although there were so many difficulties while editing. Because of the hardworking, teamwork of all the staffs and the substantial support of the leaders, the "CUBE Design" could achieve the complete success. Last but not the least, wish you could support the Shenzhen Cube Architecture Design Office as well as before and a more bright next decade to the Shenzhen Cube Architecture Design Office !

Project Editing Group of HKPIP

图书在版编目(CIP)数据

立方设计/深圳市立方建筑设计顾问有限公司，香港理工国际出版社主编. — 北京：中国建筑工业出版社，2011.6
 ISBN 978-7-112-13086-3

 I. ①立… II. ①深…②香… III. ①建筑设计—研究 IV.①TU2

中国版本图书馆CIP数据核字（2011）第052973号

策　　划：香港理工国际出版社
责任编辑：徐　纺　徐明怡
编　　辑：高雅雯　唐丹丹
美术指导：易　帅　Krlly
编　　委：邱慧康　范纯青　向大庆　彭光曦　何光明　黄镇辉
　　　　　贺李渊　梁怡攀　易　帅　高雅雯

立方设计
深圳市立方建筑设计顾问有限公司　　主编
香港理工国际出版社
*
中国建筑工业出版社出版、发行（北京西郊百万庄）
各地新华书店、建筑书店经销
利丰雅高印刷（深圳）有限公司　制版
利丰雅高印刷（深圳）有限公司　印刷
*
开本：889×1194毫米　1/12　印张：57　字数：1150千字
2011年6月第一版　2011年6月第一次印刷
定价：638.00元（上、下册）
ISBN 978-7-112-13086-3
　　　（20497）

版权所有 翻印必究
如有印装质量问题，可寄本社退换
　（邮政编码　100037）

CUBE DESIGN

立方设计（上册）

深圳市立方建筑设计顾问有限公司 / 香港理工国际出版社 主编

中国建筑工业出版社

Stride
day by day

......

关于本书

这是立方自2001年成立以来的第一本作品集,它记载着一个本土新锐设计团队长达十年的实践轨迹,以及建筑师职业责任的自我觉醒过程。

书中收录的作品,有的在投标中中标,有的在竞赛中落选;有的带给我们成就感,也有的留下些许遗憾;书中也许没有鸟巢、CCTV这样的明星建筑或里程碑作品,亦少有实验建筑和前卫宣言般的作品;书中更多的是大众喜闻乐见、日常生活中的平实建筑与场所的展示,更多的是立方对建成环境点滴改善的积累。

有缺憾才完美,坚持可持续才更完美。

背景与愿景

立方成长的十年,正是中国经济高速增长、城市化进程全面展开的十年,涌现了不少的明星建筑、新理念、新技术等。可是,在高速发展的背后我们也看到了诸多的不平衡:当我们走出时尚前卫的新地标建筑时,它周边的环境与建筑却依然陈旧不堪;当我们在为北京奥运的盛会欢呼时,鸟巢粗壮的钢结构已悄悄渗出锈渍;当我们在飞驰的地铁里享受地铁的高效时,交通换乘却大费周章;当我们的同事还在家中回味1989年搬入新居的幸福时,下一秒却收到了拆迁的通知单……

在很多方面我们的确是体现了进步,可我们进步更多的是在思潮、理念上,相对却忽略了实体的建筑与环境;我们的亮点往往出现在局部城市或领域,而在更广泛、全面的层面上则更期待提升。我们的改善往往局限在重要的建筑上,而环境及城市品质水准方面较少;我们的转变更多的是在形式和表面上,而内在功能和资源的系统整合上较少;我们的建设速度加快了,而建筑的寿命却似乎变短了。

因此,我们明确了自己的责任与目标,应以切实改善城市环境、改善大众生活品质为己任,积极参与到广泛的城市建设潮流之中,在主流的设计领域为大众提供尽量多的高品质建筑与环境,在快速建设过程中通过自身的努力去达到"慢"设计的效果。

功能与形式观

功能与形式一直是关于建筑的经典主题。逐步成熟的功能观与形式观是我们展开建筑实践的行动指南。

功能观——寻找关键点

在我们看来,功能不完全等于甲方的任务书,每个建筑的真实功能包括三个层面:

首先,甲方要求。它关乎客户与项目的利益与效益,但我们不能仅仅被它束缚,而需要协助甲方进行系统的梳理;

其次,真正使用者的要求。甲方往往并非是未来真正的使用者,真正使用者的需求是建筑师在与甲方的反复沟通中逐渐明晰的,这是我们极其关注的一方面;

第三,城市要求。建筑从它诞生的那一刻起,就作为公共形象出现。它是环境的一分子,是城市的一部分,影响到每一个利益相关者。因此,我们关注的是建筑建成后是否给环境带来积极且持续的改变。

我们将这三方面的需求有机地结合起来分析,寻找平衡解决的途径。我们坚信,上帝一定有一个最好的安排,一定存在一个最佳切入点能同时满足三方面的利益。这个点就像扣动扳机击发的瞬间,也如高尔夫球击出的一刹那,一切在轻松中迎刃而解。这个关键点就是我们在设计过程中追寻的关键临界点,而导向这点的核心工作,我们依赖两个字:"创新"。

形式观——品质与标准化

功能观解决了建筑设计中的个性,而形式观导向对建筑"生命"的关注——如何让我们不再建造很快就变旧的建筑,如何让建筑保持青春?

品质,很抽象,但很容易被人直接感受到。建筑品质相对于建筑的个性,更具有普世的现实意义。而在我们看来,在当前的时代条件下,标准化是品质实现的关键。在发达国家,为什么相对容易造出高品质的作品,原因在于系统的标准化。建筑相关的不同局部及生产工艺都服从于房屋系统的标准,在标准的平台上每个子项都有更成熟的积累,整体组合获得最佳的品质感。

在当前快速建设的时代,设计周期短,设计量大,参与者参差不齐,想要建出高品质的建筑就得像手工艺人花数倍的精力,倾心打磨。其实我们更需要标准化:在一个项目的设计中,贯彻系统的标准化、模块化概念,结合施工、材料、工艺寻找合适的建成方式。

这样的功能观和形式观伴随我们走过十年来的建筑实践,它们就像是一双筷子,同样长、同样粗、同样重要,而且需要协调配合才能发挥其最佳作用。它也是我们目前对城市、建筑、建造认识的概括,并将随时代主旋律的变奏而与时俱进。

下一个十年

我们正在将服务团队建成一个完整的"设计链",建成一个具有相对统一价值观的开放平台。

一方面,我们健全了从规划、建筑到室内、景园以及工程设计各工种的完整纵向设计链,提供全过程系统的服务和高完成度的作品,并分别独立平等地承担不同项目的设计工作;另一方面,各工种可分别作为公司对外合作与配合的对接环节,使我们可以同外界设计团队形成更开放的工作群机制。

我们坚信这样的模式能保证我们团队持续地发展,信息渠道的开放、竞争力的强大、创新力的旺盛,保证我们能不断地自我升华,从而提供更多的优异作品并总有出乎意料的惊人之举。

Whose
scientific
design nature
and comfort
are……

Preface

About This Book

This is the first collection of works published since Cube was established in 2001. It records a new local design team's development in a decade as well as architects' self-awareness of professional responsibility.

Some of the works recorded in the book were bids won by us, some lost in contest, some brought us a sense of accomplishment, and some left regrets and an improvement space. The book contains neither star building & milestone works like Bird's Nest and CCTV Tower nor experimental architecture & prospective works. It presents more popular and daily-used buildings and sites and focuses more on the accumulation of building improvements.

Regrets lead to perfection and sustainable progress makes buildings more perfect.

Background and Vision

The decade during which Cube grows is a decade in which China's economy grows rapidly and urbanization is in full swing. Lots of star buildings, new ideas, and new technologies have emerged. However, behind the rapid development, we also see an unbalance in many aspects: when we step out of fashionable landmark buildings, we see their surrounding environment and buildings are still dilapidated; when we cheer for the Beijing Olympic Games, the steel structure of Bird's Nest has exposed rust stains; when we enjoy high-efficiency metro services, we have to take great pains in changing means of transportation; when our workmates are still recalling the happy moment of moving into a new house in 1989, they receive a notice of demolition...

We do make much progress in many aspects but our progress is reflected mainly through ideologically and conceptually instead of through buildings and environment. Our highlights always appear in some urban areas or fields and they are expected in a wider and all-round scope. Our improvement is confined to important buildings but environment and urban quality is seldom improved. Our change is embodied more in form and is skin-deep but intrinsic functions and systematic resource integration seldom change. We are building at a faster speed but it seems that our buildings have a very short service life.

Therefore we define clearer responsibilities and goals. We will shoulder the burden of improving urban environment and life quality of the public practically, participate actively in urban construction, provide more high-quality buildings and environments to the public in the mainstream design field, and make efforts to realize a "slow" design effect in the process of rapid construction.

Function and Form Philosophies

Function and form are always classic themes of buildings. Gradually-matured function philosophies and form philosophies are a guide to our construction practices.

Function Philosophy – to find a "key point"

In our opinion, function is not equal to Party A's assignment. Each building's true functions include three aspects:
1) Party A's requirements. Such requirements concern profits and interests of customers and projects. However we should not be limited by them but help Party A straight out its system.
2) Real users' requirements. In many cases, Party A is not a real user in the future. Real user's demands come out gradually through repeated communication between architects and Party A. This is an aspect on which we pay much attention.
3) Requirements of a city. From the moment a building was born, it appears as a public image. It is a component of environment and a part of a city. It has an effect on every stakeholder. Therefore we concern about whether a building could bring active and continuous change to environment after its completion.

We analyze demands in these three aspects as a while to find a balanced solution. We firmly believe that God has made the best arrangement and an optimal point of penetration is in existence to meet demands in three aspects most appropriately. This point looks like the moment at which a trigger is pulled or a golf ball is hit. It solves all problems easily. This key point is the key critical point we try to find in the process of design. We rely on innovation to complete core work leading to this point.

Form Philosophy – Quality and Standardization

Function philosophy endows buildings with personalities while form philosophy trends to focusing on the "life" of buildings – how to complete buildings that will not become outdated soon and how to keep them young?

Quality is abstract but is easily sensed by people directly. In comparison with building personalities, building quality is more significant practically. In our opinion, standardization is a key to realize high quality buildings under current conditions. The reason that it is relatively easier to create high-quality works in developed countries is that there is standardized system in those countries. All parts and technologies in relation to building must follow the standards of housing system. Under a standard platform, all sub-items experience a more matured accumulation and integration process to obtain an optimal sense of quality. We are now in an era of rapid construction characterized by short design period, high design workload, and different design levels of designers. A high-quality building needs lots of efforts. In fact what we need urgently is standardization, i.e. to follow systematic standardized and modularized concepts and take into account construction, materials, and technology in the design of a project so as to find an appropriate construction way.

Such function and form philosophies have accompanied our practices for a decade. They look like a pair of chopsticks with the same size and both are important. Only when they act in concert with each other could we get the most out of them. This is a summation of our understanding of cities, buildings, and construction and it will keep pace with the times with the change of main themes of the times.

Design Chain – Next Decade

We are building our service team into a complete "design chain" and an open platform with relatively unified values.

On the one hand, we build a complete longitudinal design chain covering planning, construction, interior design, landscaping, and engineering design. On the other hand, each type of work may be used respectively as an object for which our company works with others. Thus we build an open workgroup mechanism to cooperate with other design teams.

We firmly believe that such mode can ensure that our team realizes sustainable development and has an open information channel, powerful competitive strength, strong innovation capacity and that we can keep improving to create more masterpieces and go beyond other's expectation.

Passion makes the architects creating miracles from the stones……

序

在上个世纪末，随着中国加入WTO，建筑设计市场逐渐开放，越来越多的境外公司涌入中国参与到轰轰烈烈的城市建设中。与此同时，在深圳有一批有才华有理想的年轻本土建筑师抛开铁饭碗，组成一个个民营设计公司与事务所。十多年来，他们历经风雨，从生存到发展，并逐渐成为设计市场的中坚力量。立方设计则是其中的佼佼者之一。

我和立方的几位创始合伙人相识很早，他们曾是港资公司的精英或是传统大院的骨干。立方成立之后，他们的组合以团结、稳定著称，这在风云变幻的深圳业界也显得非常可贵。我经常在各种竞赛竞标中看到他们的获胜，也不断听到同行对立方建筑作品的好评。如今拿到立方沉甸甸的作品集时仍有不少的惊喜和感动。

我被立方十年来的坚持、执着和愈发清晰的责任意识所感动。商业化培养了他们高度的职业服务意识，却并未磨去他们的棱角，他们依然坚持着"根植于市场而高于市场"的不懈追求，并且变得更加成熟；同时，我也欣喜地看到立方的持续创新能力，在山寨流行的市场环境中，他们坚持对每一个项目、每一块用地积极探寻独辟蹊径的解决方案，不断超越自我；更加难能可贵的是，立方对建筑品质的积极实践，他们不断尝试在现有国情下利用现有施工技术去实现高完成度的建成作品，这在快速发展与建设的年代尤其具有现实意义。

中国社会的高速发展带来了建筑设计行业的空前繁荣，也使我们感受到巨大的责任和挑战；我们必须在最短的时间内完成大量的设计，以满足超速建设的需求，效率之高也使国外同行难以置信；高速增长后的社会将进入平稳发展期，我们快速设计的成果必须为将来的平稳期留下可持续完善的框架，而非供拆毁的建筑垃圾；设计关乎民生，我们必须通过设计去切实改善大众的生活，提升生活品质，推动社会的广泛进步，真正让进步影响到生活更广泛和更基本的层面。因此，我们必须加速行业的职业化步伐，而不是过于明星化和娱乐化的作秀，我们必须在职业意识、技术改进、理念更新、建造品质等方方面面进行坚实的积累，实现更快的自我进化，更好地为我们的国民服务。

希望更多的本土建筑师、本土设计团队能明确我们行业当前的发展方向，向着打造"设计之都"，实现"中国创造"的目标迈进，并祝立方公司的未来更加辉煌！

孟建民

Preface

With China's accession to the WTO at the end of previous century, the architectural design market was opened gradually and more and more overseas companies entered China to participate in the flourishing urban construction. In the meantime, a group of young, talented, and ambitious Shenzhen-based architects abandoned their lifelong secure jobs and founded some private design companies and firms. In the past decade, they underwent hardship, survived and then became the backbone of the design market. Cube Design is a top design company of them.

Long time ago, I had known the founders of Cube Design, who were once elites of Hong Kong-funded companies or the backbone of traditional design institutes. After Cube Design was founded, the team is famous for its solidarity, which is commendable in the constantly-changing design industry in Shenzhen. I often hear of its victory in various competitions and peers' praise to its architectural works. Now I still feel astonished and moved as I get the collection of works of Cube Design.

I am deeply moved by Cube Design's persistence, dedication, and increasingly-clear consciousness of responsibility in the past decade. Commercialization enhanced their professional service consciousness but did not remove their edges and corners. They still stick to the pursuit of "being based on the market but standing higher than the market" and become more mature. I am glad to see Cube Design's continuous innovation capacity. Though copying is popular on the market, they still strive to find a unique solution for every project and every plot of land and try to go beyond themselves. More importantly, they are active to improve architectural quality. They try to use existing construction technology to realize construction works of high degree of completion, which is practically significant in the rapidly-developing age.

The rapidly-developing social development in China gives rise to an unprecedented prosperity of the architectural design industry and also brings a strong sense of responsibility and a great challenge to us. We have to complete a number of designs in a short time in order to meet the demands of ultra-fast construction. Even overseas peers cannot believe such a high efficiency. A stably-developing period will come after a rapidly-growing period. Our fast design outcomes must leave a sustainable and perfect framework, instead of building rubbishes to be demolished, for the forthcoming stably-developing period. Design concerns people's livelihood. Our designs must better people's life, improve their life quality, push on social progress widely, and truly influence more basic aspects of life. Therefore, we must speed up the professionalization step in our industry instead of offering shows for entertainment. We must realize a solid accumulation and faster self-evolution in respect of professional consciousness, technological improvement, idea updating, and construction quality, etc. so as to serve our country and people in a better way.

Hope more local architects and design teams can know exactly current development direction of the industry and go ahead toward the goal of "building a design metropolis" and "invented by China". I wish Cube Design a brighter future.

Project Distrbution

项目分布

Office Building, Office
写字楼、办公

Talents Garden • Shenzhen
人才园 • 深圳
Foshan News Center • Foshan
佛山新闻中心 • 佛山
Changsha Broadcasting and TV Center • Changsha
长沙广电中心 • 长沙
Shenzhen Software Industry Base • Shenzhen
深圳软件产业基地 • 深圳
China Mobile Shenzhen Branch Production Dispatching Building • Shenzhen
深圳移动生产调度大厦 • 深圳
Jintang Building • Tianjin
金唐大厦 • 天津
Liantai Building • Shenzhen
联泰大厦 • 深圳
Taitai Pharmaceutical Office Building • Shenzhen
太太药业办公楼 • 深圳
Huizhou Daily Newspaper Office • Huizhou
惠州日报社 • 惠州
Bao'an District Construction Quality Inspection Building • Shenzhen
宝安区工程质检大楼 • 深圳
Future Plaza • Shenzhen
花样香年广场 • 深圳

Commercial and Urban Complexes
商业及城市综合体

Planning & Design of Foshan Complex • Foshan
佛山综合体规划设计 • 佛山
Lsea Asia International Center • Nanning
利海亚洲国际中心 • 南宁
Central District Crystal Island • Shenzhen
中心区水晶岛 • 深圳
Emperor Star City • Shanghai
英皇明星城 • 上海
Chengdu Future Plaza • Chengdu
成都香年广场 • 成都
Maoye Baizixiang • Chongqing
贸业百子巷 • 重庆
Sheffield Land Plaza • Chengdu
谢菲尔德置地广场 • 成都
City Commercial Plaza • Guigang
城市商业广场 • 贵港

Education & Culture
文化教育

Museum of Contemporary Art and City Planning Exhibition Hall • Shenzhen
现代艺术馆与城市规划展览馆 • 深圳
Hu Yaobang Memorial Hall • Xiangtan
胡耀邦纪念馆 • 湘潭
Art Gallery at the Guanlan Engraving Base • Shenzhen
观澜版画基地美术馆 • 深圳
Guangxi Kettledrum Museum • Nanning
广西铜鼓博物馆 • 南宁
Venice Bridge Museum • Venice (Italy)
威尼斯桥博物馆 • 威尼斯（意大利）
New Campus of Xidian University • Xi'an
西安电子科技大学新校区 • 西安
University Town International Conference Center • Shenzhen (Supplementary Project)
大学城国际会议中心 • 深圳
Ezhou No. 8 Middle School • Ezhou
鄂州第八中学 • 鄂州
Longgang Pingshan Gymnasium • Shenzhen
龙岗坪山体育馆 • 深圳
Guangxi Art Gallery • Nanning
广西美术馆 • 南宁
Beirut Culture and Arts Center • Beirut (Lebanon)
贝鲁特文化艺术中心 • 贝鲁特（黎巴嫩）
Xinhai Revolution Monument (Tower) • Wuhan
辛亥革命纪念碑（塔）• 武汉
Guangxi City Planning & Construction Exhibition Hall • Nanning
广西城市规划建设展览馆 • 南宁
Shenzhen Production, Teaching and Research Base of Huazhong University of Science and Technology • Shenzhen
华中科技大学深圳产学研基地 • 深圳
Foshan Polytechnic College • Foshan
佛山职业技术学院 • 佛山
Shenzhen Bay Middle School • Shenzhen
深圳湾中学 • 深圳

Hotels
酒店

Egyptian Hotel • Cairo (Egypt)
埃及酒店 • 开罗（埃及）
Luneng Sanya Hotel • Sanya
鲁能三亚酒店 • 三亚
Guanling Hotel, Beihai, Guangxi • Beihai
广西北海冠岭酒店 • 北海
Dameisha Hotel (Liantai Club) • Shenzhen
大梅沙酒店（联泰会所）• 深圳
Dongjiu Building • Yixing
东氿大厦 • 宜兴

Urban Design and Landscape Planning
城市设计与规划

Huaqiangbei Three-dimensional Street and City Design • Shenzhen
华强北立体街道城市设计 • 深圳
Overseas Chinese Town LOFT • Shenzhen
华侨城LOFT • 深圳
Eastern Square of Nanning Conference & Exhibition Center • Nanning
南宁会展中心东广场 • 南宁
Design for the Region around the Sci-Tech Park in New Guangming District • Shenzhen
深圳光明新区科技公园周边地区城市设计 • 深圳

Modern Fashion
现代时尚

Sunny Shangeli Villa Garden • Beihai
阳光香格里别墅花园 • 北海
Tian'an Golf Sea View Garden • Shenzhen
天安高尔夫海景花园 • 深圳
Liantai Dameisha Villa • Shenzhen
联泰大梅沙别墅 • 深圳
Dongguan 0769 • Dongguan
东莞0769 • 东莞
Futong in Bao'an 96 District (Phase IV, 2006) • Shenzhen
宝安96区富通4期 • 深圳
Wei Zhen Fu • Shenzhen
唯珍府 • 深圳
Overseas Chinese Town Pure Waterfront • Chengdu
华侨城纯水岸 • 成都
Overseas Chinese Town Portofino Pure Waterfront (Phase VII) • Shenzhen
华侨城波托菲诺纯水岸七期 • 深圳
Cheng Han Man Cheng (Phase IV) • Shenzhen
承翰慢城四期 • 深圳
Bi Hai Yun Tian • Shenzhen
碧海云天 • 深圳
Nice Homestead • Wuhan
美好家园 • 武汉
Renheng Apartment • Lanzhou
仁恒住宅 • 兰州
Junyue River Paradise • Changsha
君悦香邸 • 长沙
Hua Hao Yuan Garden • Shenzhen
华浩源景园 • 深圳
Xiang Fu Jia Cheng • Changsha
湖南湘府嘉城 • 长沙
Vanke City Garden • Wuhan
万科城市花园 • 武汉
Jia Ri Lan Wan • Qinhuangdao
假日蓝湾 • 秦皇岛
Liantai Dameisha Apartment • Shenzhen
联泰大梅沙公寓 • 深圳
Overseas Chinese Town Singles Apartment • Shenzhen
华侨城单身公寓 • 深圳
Fantasia Blooming Town • Chengdu
花样年花样城 • 成都
Ao Yun Garden • Shanghai
澳韵花园 • 上海
Tao Hua Ling • Yichang
桃花岭 • 宜昌
80 Hou Jie • Shenzhen
80后街 • 深圳
Vanke Jin Yu Lan Wan • Zhuhai
万科金域蓝湾 • 珠海

Traditional and Classic
传统经典

Excellence Repulse Bay • Shenzhen
卓越浅水湾 • 深圳
Horoy Xi Yuan • Shenzhen
鸿荣源熙园 • 深圳
Zhi Di Chun Feng Ju • Shenzhen
置地春风居 • 深圳
Horoy Park Land • Shenzhen
鸿荣源公园大地 • 深圳
Hua Sheng Ling Yu • Shenzhen
华盛领域 • 深圳
China Overseas Property -Long Gang Olympic New Town • Shenzhen
中海龙岗奥体新城 • 深圳
Nan'ao Kai Xuan Bay Garden • Shenzhen
南澳凯旋湾 • 深圳
Shi Dai Tian Jiao • Wuhan
时代天娇 • 武汉
Grace Royal Apartment • Changshu
华府世家 • 常熟
China Overseas Property -Banyan Coast • Chengdu
中海翠屏湾 • 成都
China Overseas Property - International Community • Suzhou
中海国际社区 • 苏州
China Overseas Property - Xu Jiang Project • Suzhou
中海胥江项目 • 苏州

Regional Style
地域风情

Portofino Swan Castle • Shenzhen
波托菲诺天鹅堡 • 深圳
Zhong Cheng Li Jing Xiang Shan • Changsha
中城丽景香山 • 长沙
Li Jing Shan Zhuang • Shenzhen
荔景山庄 • 深圳
Xiayang Villa District • Sanya
下洋别墅区 • 三亚
Qipan Mountain Villa District • Shenyang
棋盘山别墅区 • 沈阳
Bi Lin Wan • Shanghai
碧林湾 • 上海
Du Shi Yi Jia • Shanghai
都市宜家 • 上海
Sunshine Palm Garden (Phase III) • Shenzhen
阳光棕榈院三期 • 深圳
Tonghe - Story of South-bank • Hangzhou
通和 - 南岸故事 • 杭州
Lian Tai Mangrove Bay • Shenzhen
联泰红树湾 • 深圳

写字楼、办公 Office Building, Office

- 人才园·深圳 Talents Garden · Shenzhen ...5
- 佛山新闻中心·佛山 Foshan News Center · Foshan ...13
- 长沙广电中心·长沙 Changsha Broadcasting and TV Center · Changsha ...27
- 深圳软件产业基地·深圳 Shenzhen Software Industry Base · Shenzhen ...35
- 深圳移动生产调度大厦·深圳 China Mobile Shenzhen Branch Production Dispatching Building · Shenzhen ...43
- 核电办公培训设施·阳江 Nuclear Power Office Clerk Training Complex · Yangjiang ...49
- 金唐大厦·天津 Jintang Building · Tianjin ...57
- 联泰大厦·深圳 Liantai Building · Shenzhen ...65
- 太太药业办公楼·深圳 Taitai Pharmaceutical Office Building · Shenzhen ...71
- 惠州日报社·惠州 Huizhou Daily Newspaper Office · Huizhou ...77
- 宝安区工程质检大楼·深圳 Baoan District Construction Quality Inspection Building · Shenzhen ...85

Contents 目录

立方设计（上册）·公共建筑

花样香年广场·深圳 Future Plaza·Shenzhen…91

商业及城市综合体 Commercial and Urban Complexes

佛山综合体规划设计·佛山 Planning & Design of Foshan Complex·Foshan…103
利海亚洲国际中心·南宁 Lsea Asia International Center·Nanning…113
中心区水晶岛·深圳 Central District Crystal Island·Shenzhen…119
英皇明星城·上海 Emperor Star City·Shanghai…129
成都香年广场·成都 Chengdu Future Plaza·Chengdu…135
贸业百子巷·重庆 Maoye Baizixiang·Chongqing…145
谢菲尔德置地广场·成都 Sheffield Land Plaza·Chengdu…149
城市商业广场·贵港 City Commercial Plaza·Guigang…157

教育文化 Education & Culture

现代艺术馆与城市规划展览馆·深圳 Museum of Contemporary Art and City Planning Exhibition Hall·Shenzhen…167

胡耀邦纪念馆·湘潭 Hu Yaobang Memorial Hall·Xiangtan...175
观澜版画基地美术馆·深圳 Art Gallery at the Guanlan Engraving Base·Shenzhen...183
广西铜鼓博物馆·南宁 Guangxi Kettledrum Museum·Nanning...189
威尼斯桥博物馆·威尼斯（意大利） Venice Bridge Museum·Venice (Italy)...199
西安电子科技大学新校区·西安 New Campus of Xidian University·Xi'an...205
大学城国际会议中心·深圳 University Town International Conference Center·Shenzhen (Supplementary Project)...219
鄂州第八中学·鄂州 Ezhou No. 8 Middle School·Ezhou...229
龙岗坪山体育馆·深圳 Longgang Pingshan Gymnasium·Shenzhen...237
广西美术馆·南宁 Guangxi Art Gallery·Nanning...247
贝鲁特文化艺术中心·贝鲁特（黎巴嫩） Beirut Culture and Arts Center·Beirut (Lebanon)...255
辛亥革命纪念碑（塔）·武汉 Xinhai Revolution Monument (Tower)·Wuhan...261
广西城市规划建设展览馆·南宁 Guangxi City Planning & Construction Exhibition Hall·Nanning...269
华中科技大学深圳产学基地·深圳 Shenzhen Production, Teaching and Research Base of Huazhong University of Science and Technology·Shenzhen...275

佛山职业技术学院·佛山　Foshan Polytechnic College·Foshan…281
深圳湾中学·深圳　Shenzhen Bay Middle School·Shenzhen…289

酒店 Hotels

埃及酒店·开罗（埃及）　Egyptian Hotel·Cairo (Egypt)…297
鲁能三亚酒店·三亚　Luneng Sanya Hotel·Sanya…305
广西北海冠岭酒店·北海　Guanling Hotel, Beihai, Guangxi·Beihai…313
大梅沙酒店（联泰会所）·深圳　Dameisha Hotel (Liantai Club)·Shenzhen…319
东氿大厦·宜兴　Dongjiu Building·Yixing…325

城市设计与景观规划 Urban Design and Landscape Planning

华强北立体街道城市设计·深圳　Huaqiangbei Three-dimensional Street and City Design·Shenzhen…335
华侨城LOFT·深圳　Overseas Chinese Town LOFT·Shenzhen…341
南宁会展中心东广场·南宁　Eastern Square of Nanning Conference & Exhibition Center·Nanning…347
深圳光明新区科技公园周边地区城市设计·深圳　Design for the Region around the Sci-Tech Park in New Guangming District·Shenzhen…353

现代时尚 Modern Fashion

阳光香格里别墅花园·北海　Sunny Shangeli Villa Garden·Beihai...366

天安高尔夫海景花园·深圳　Tian'an Golf Sea View Garden·Shenzhen...376

联泰大梅沙别墅·深圳　Liantai Dameisha Villa·Shenzhen...386

东莞0769·东莞　Dongguan 0769·Dongguan...394

宝安96区富通4期·深圳　Futong in Bao'an 96 District (Phase IV, 2006)·Shenzhen...404

唯珍府·深圳　Wei Zhen Fu·Shenzhen...412

华侨城纯水岸·成都　Overseas Chinese Town Pure Waterfron·Chengdu...422

华侨城波托菲诺纯水岸七期·深圳　Overseas Chinese Town Portofino Pure Waterfront (Phase VII)·Shenzhen...434

承翰慢城四期·深圳　Cheng Han Man Cheng (Phase IV)·Shenzhen...444

碧海云天·深圳　Bi Hai Yun Tian·Shenzhen...450

美好家园·武汉　Nice Homestead·Wuhan...456

仁恒住宅·兰州　Renheng Apartment·Lanzhou...462

君悦香邸·长沙　Junyue River Paradise·Changsha...468

Contents 目录

立方设计（下册）·创新住宅

华浩源景园·深圳　Hua Hao Yuan - Jing Yuan·Shenzhen...474
湖南湘府嘉城·长沙　Xiang Fu Jia Cheng·Changsha...478
万科城市花园·武汉　Vanke City Garden·Wuhan...484
假日蓝湾·秦皇岛　Jia Ri Lan Wan·Qinhuangdao...490
联泰大梅沙公寓·深圳　Liantai Dameisha Apartment·Shenzhen...496
华侨城单身公寓·深圳　Overseas Chinese Town Singles Apartment·Shenzhen...500
花样年花样城·成都　Fantasia Blooming Town·Chengdu...504
澳韵花园·上海　Ao Yun Garden·Shanghai...510
桃花岭·宜昌　Tao Hua Ling·Yichang...514
80后街·深圳　80Hou Jie·Shenzhen...518
万科金域蓝湾·珠海　Vanke Jin Yu Lan Wan·Zhuhai...524

传统经典 Traditional and Classic

卓越浅水湾·深圳　Excellence Repulse Bay·Shenzhe...530
鸿荣源熙园·深圳　Horoy Xi Yuan·Shenzhen...536

置地春风居·深圳 Zhi Di Chun Feng Ju·Shenzhen...542
鸿荣源公园大地·深圳 Horoy Park Land·Shenzhen...548
华盛领域·深圳 Hua Sheng Ling Yu·Shenzhen...558
中海龙岗奥体新城·深圳 China Overseas Property -Long Gang Olympic New Town·Shenzhen...560
嘉旺阅山华府·深圳 Jiawang Mountain Mansion·Shenzhen...568
南澳凯旋湾·深圳 Nan'ao Kai Xuan Bay Garden·Shenzhen...574
时代天骄·武汉 Shi Dai Tian Jiao·Wuhan...580
华府世家·常熟 Grace Royal Apartment·Changshu...584
中海翠屏湾·成都 China Overseas Property -Banyan Coast·Chengdu...590
中海国际社区·苏州 China Overseas Property - International Community·Suzhou...594
中海胥江项目·苏州 China Overseas Property - Xu Jiang Project·Suzhou...598

地域风情 Regional Styles

波托菲诺天鹅堡·深圳 Portofino Swan Castle·Shenzhen...604
中城丽景香山·长沙 Zhong Cheng Li Jing Xiang Shan·Changsha...612
荔景山庄·深圳 Li Jing Shan Zhuang·Shenzhen...618
下洋别墅区·三亚 Xiayang Villa District·Sanya...624
棋盘山别墅区·沈阳 Qipan Mountain Villa District·Shenyang...630
碧林湾·上海 Bi Lin Wan·Shanghai...634
都市宜家·上海 Du Shi Yi Jia·Shanghai...638
阳光棕榈院三期·深圳 Sunshine Palm Garden (Phase III)·Shenzhen...644
通和 - 南岸故事·杭州 Tonghe - Story of South-bank·Hangzhou...646
联泰红树湾·深圳 Lian Tai Mangrove Bay·Shenzhen...650

Office Building, Office

Office Building, Office

写字楼、办公

Through international competition and domestic bidding, we have designed lots of office buildings, such as Shenzhen Future Plaza, Shenzhen Talents Garden, etc. A good office building not only provides a good office environment to people but also takes into account the fact that people spend one third of a day in the office to exchange with other, including human-human communication, human-thing exchange, and human-machine communication, etc. In our works, a human-based, natural, relaxed, and convenient living environment becomes a necessary supplement to a high-efficiency mechanized office space.

我们先后通过国际竞赛和国内投标实践了大量的办公建筑,如深圳香年广场、深圳人才园等一批办公建筑方案设计。好的办公建筑不仅仅要提供给人一个好的办公环境,更多地是要考虑人们将在这里度过全天时间的三分之一,并且一活动的主要生活方式是交流,包括人与人的交往、人与物的交换、人与机器的交流等等。在我们的作品中,人本、自然、放松、方便的生活环境成了高效率机械化办公空间的必要补充。

This building is a main living hall and also a window of Shenzhen that shows the city's particular features and image to people who stay Shenzhen for business and living.

该建筑是一个主要的生活大厅,以及代表深圳向国内外来深创业的生活者展示城市独特魅力和形象的窗口。

Talents Garden · Shenzhen

Shenzhen | Project Location
2008 | Project Date
83,000m² | Project Scale
plan design, interior design, and landscape design | Design Phase
not started | Project Status

人才园·深圳

深圳市 | 项目地点
2008年 | 设计时间
8.3万m² | 项目规模
方案设计、室内设计、景观设计 | 设计阶段
未建 | 项目现状

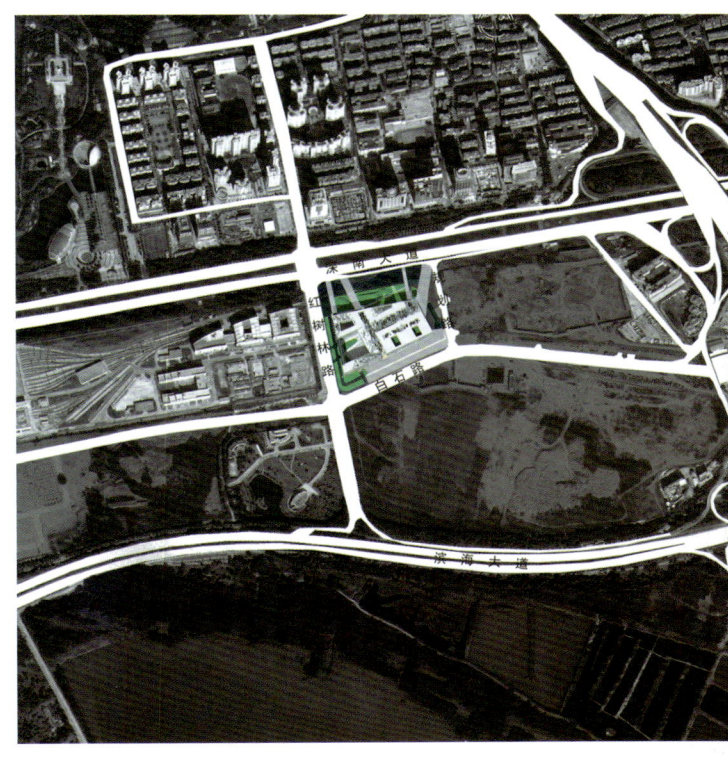

区位分析

深圳市人才园位于深圳市福田区竹子林片区，距离深圳市中心区只有十分钟车程。基地东侧隔规划路毗邻福田交通枢纽大厦，南临白石路，西侧隔红树林路毗邻地铁车辆段，北临深南大道，周边均属办公区。南面为红树林自然生态保护区，基地内可远眺深圳湾景色；西北面为优美的园博园区，咫尺可达。

Location Analysis

Located in the Zhuzilin Area in Futian District, the Shenzhen Talents Garden neighbors a planned road (adjacent to Futian Traffic Hub Building) to the east, Baishi Road to the south, Hongshulin Road (adjacent to Metro Depot) to the west, and Shennan Avenue to the north. Its surrounding is an office area. In its south is Hongshulin Nature Reserve and Shenzhen Bay and in its northwest is Shenzhen International Garden and Flower Expo Park, where the environment is very beautiful and people are easy to reach.

总体布局

基于地块的条件和功能及环保的需要，建筑选择了南北向布局，将主要的公共服务功能布置在北侧及东侧，面朝深南路开口，便于与城市公共交通连接。办公部分则布置在西侧，呈"回"字形布局，内有庭院。场地西南角为人社局前广场。通过对高差的设计，建筑公共与办公部分被巧妙分隔，避免了相互干扰。

General Layout

Based on the plot's conditions and function and environmental protection needs, the building is laid out from south to north. Main public service functions are set in the north and east, facing the entrance of Shennan Road, where the access to urban public traffic facilities is easy. With courtyards, office sections are set in the west, looking like a "回" shape. In the southwest corner of the site is a square in front of the Human Resources and Social Security Bureau. Through a design of height difference, the building's pubic section is separated from its office section. Thus no disturbance will occur between them.

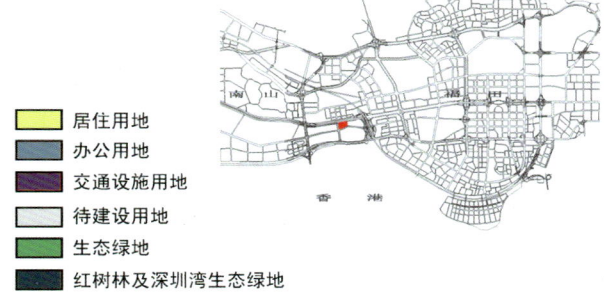

- 居住用地
- 办公用地
- 交通设施用地
- 待建设用地
- 生态绿地
- 红树林及深圳湾生态绿地

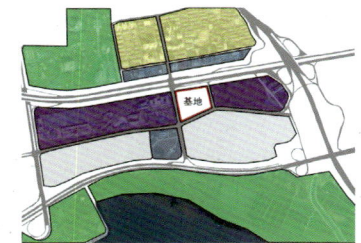

人才街区

我们希望人才街区可以是一个城市综合体,一个人才"mall"。设计巧妙地利用了这一建筑与外部空间交融的灰空间对来访者进行有效的组织、接待、引导和分流,使其成为城市的一部分,以突显其城市性、开放性、亲和性的特征,寓意人才园拥揽八方人才之志。

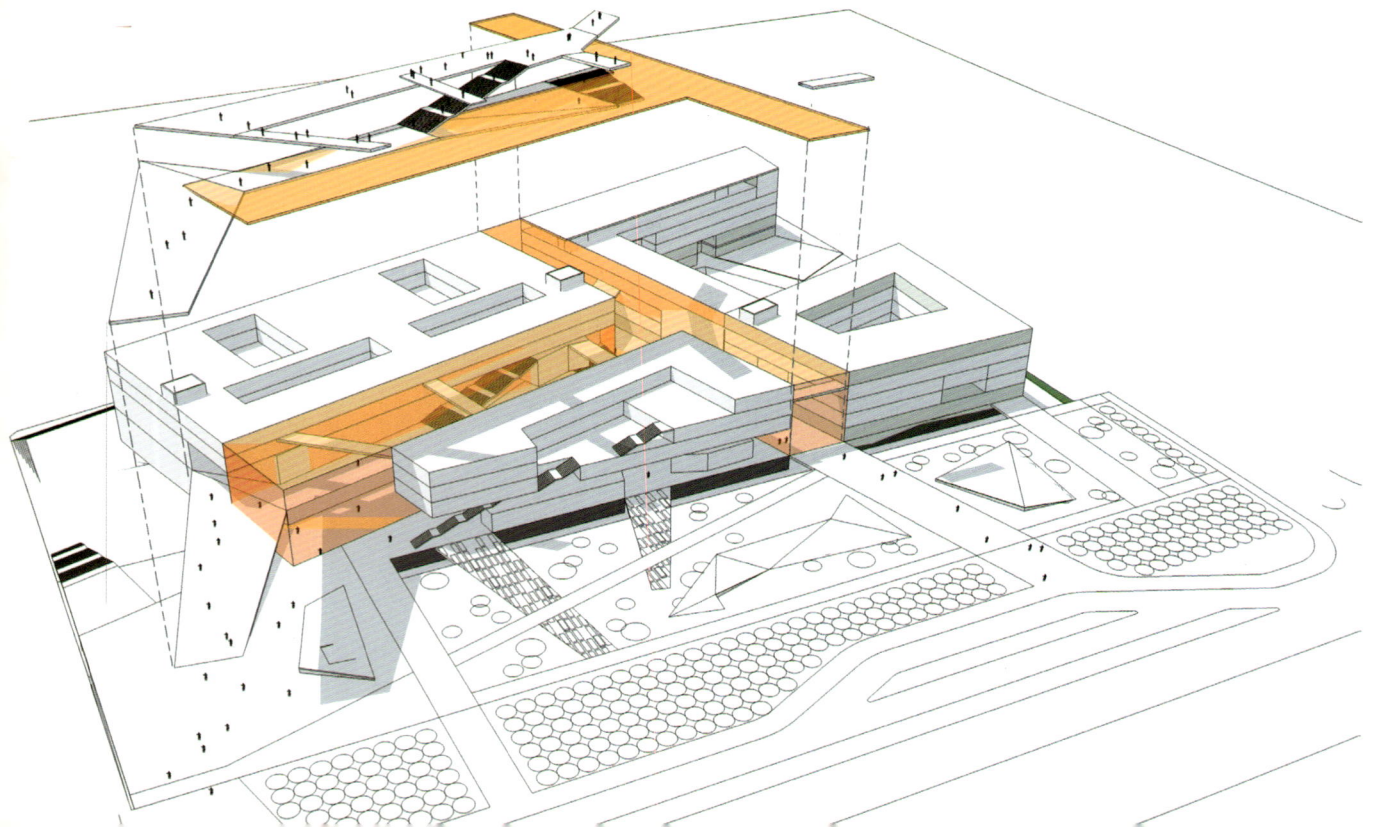

Talents Block

We hope that our talents block is an urban building complex and a talents "mall". The design makes use of the grey space between the building and external space to organize, receive, guide, and split visitors effectively and makes it become a part of the city, showing its urban, open, affinity features and implying an ambition of owning talents from all directions.

高集约式的立体分区

在功能布局上，我们提出集约化用地的思路。设计采用平面与垂直相结合的分区方式，在容积率最高的公共综合服务大厅部分结合基地北高南低的地形特点，利用垂直分区的方式将其中招聘大厅、公共服务、考试中心三个大型功能空间叠加在一个建筑体量当中，以节地带来巨大的经济效能。

Highly-intensive Three-dimensional Division

We bring forward a train of thought of intensive land use in respect of function layout. In the design, planes are combined with vertical spaces. For the public service hall with the highest plot ratio, three large functional spaces in the public space, i.e. employment hall, public service, and exam center, are put in one building by means of vertical division according to the base's landform feature (north side is higher than south side). As a result, great economic benefits can be realized on a limited parcel of land.

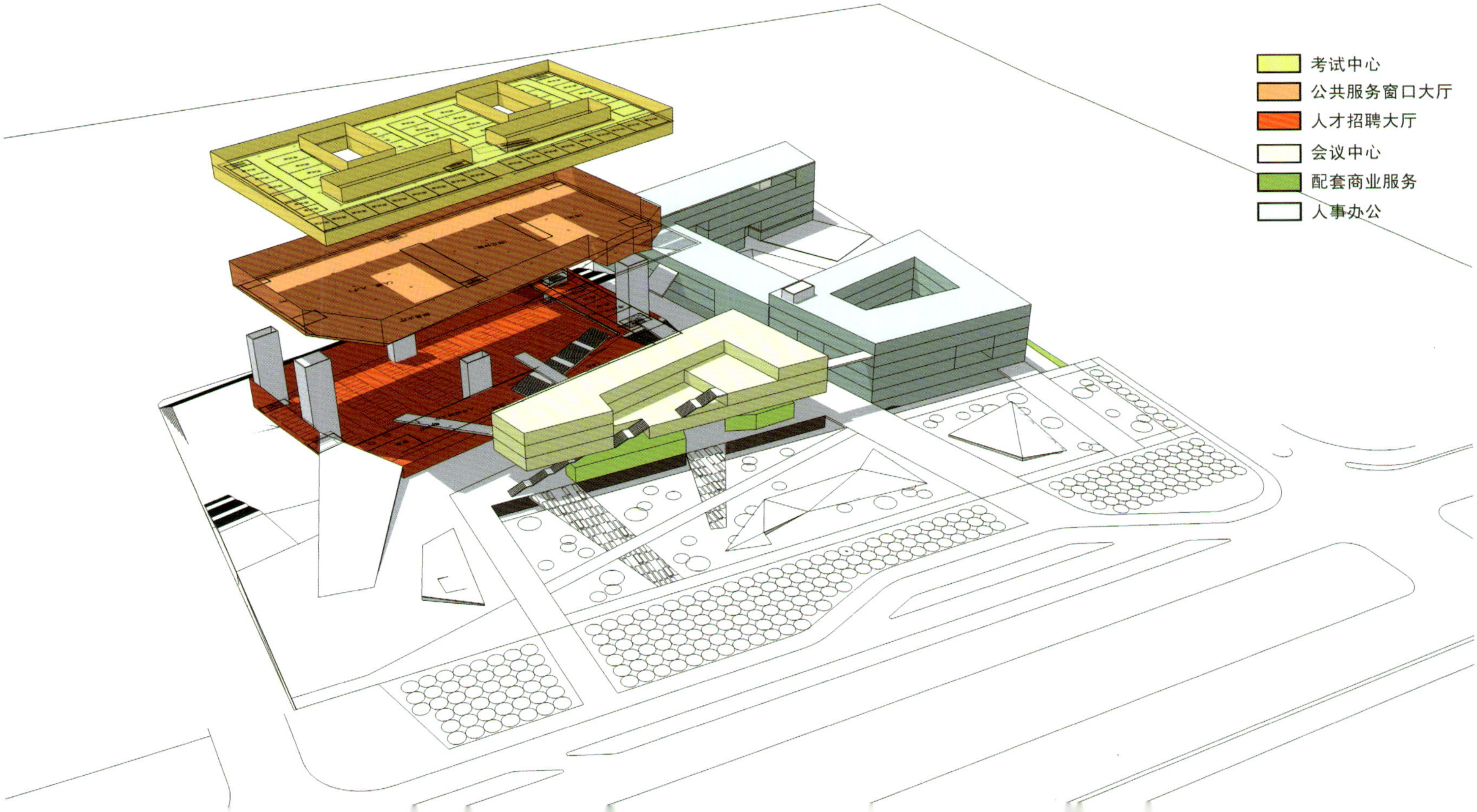

考试中心
公共服务窗口大厅
人才招聘大厅
会议中心
配套商业服务
人事办公

人流立体交通组织方式

设计引入了机场式的人流立体交通组织方式，在多处设置了自动扶梯、电梯、楼梯等垂直交通体，使人流更便捷、快速地到达各个空间。
巨型折坡及其空中连廊系统：巨型折坡及其空中连廊系统穿插于建筑各体量之中，将建筑群各功能空间有机地联系在一起，满足人们交流沟通的需求，并有利于实现建筑功能一体化的目标。

Airport-style Pedestrian 3D Traffic Treatment

The design introduces an airport-style pedestrian 3D traffic organization means. Vertical traffic facilities, such as escalator, elevator, stair, etc., are set at many places. Thus pedestrians can reach all spaces easily and quickly.
The giant slope and its air corridor system: the giant slope and its air corridor system go into all parts of the building so as to connect all functional spaces of the building complex organically, meet people's communication needs, and realize the goal of integration of architectural functions.

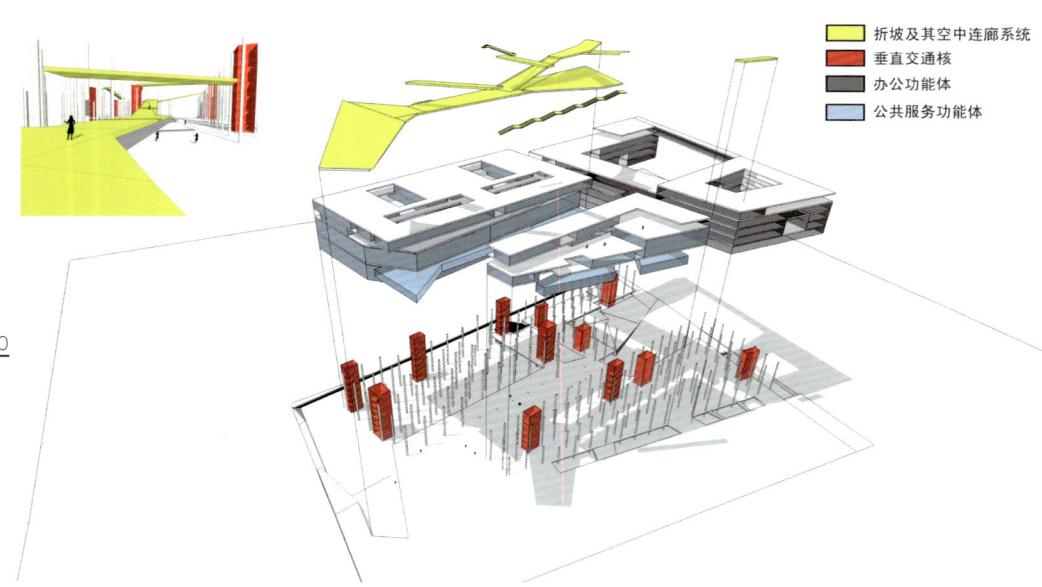

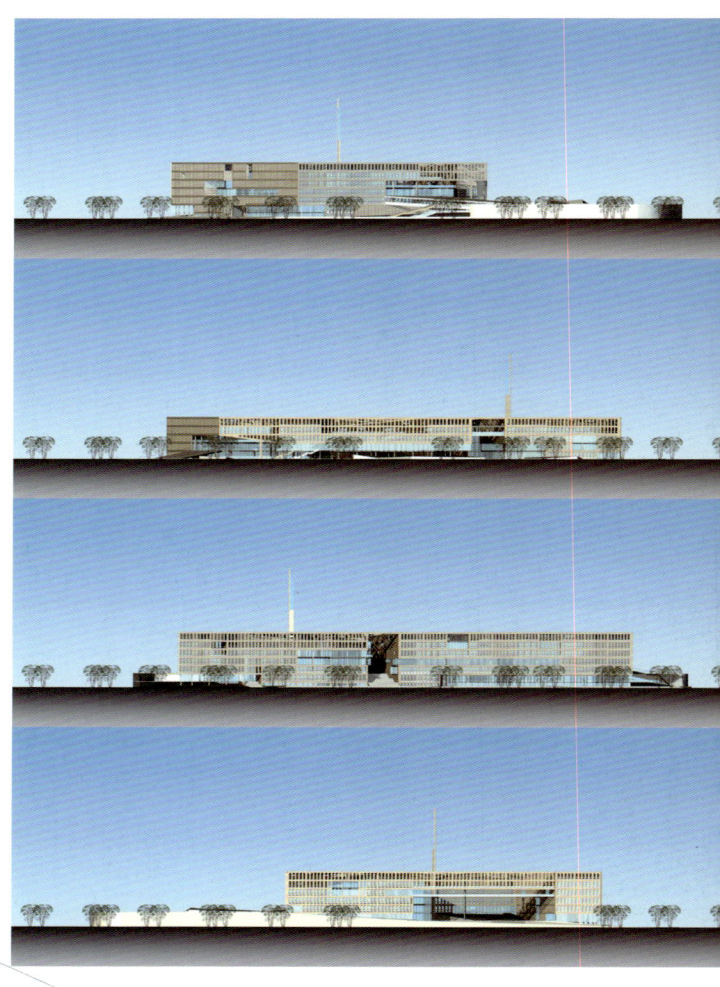

"地毯式"生态建筑

屋面及地面的巨型"地毯式"折坡体系形成一套网状汇水系统，可收集雨水，进行回收再循环利用。雨水在经过直接或简单处理后可用于冲厕、清洁、浇灌庭院、消防和回灌地下水等，将带来巨大的经济效益。"地毯式"的屋顶花园不仅是办公人员活动休息的理想场所，同时也是建筑生态固碳的有效方式。

Carpet-style Eco-building

The giant "carpet-style" slope system on the roof and ground forms a set of net-shaped water collection system, which can collect rain water and make it recycled. After being treated directly or simply, rain water may be used for toilet, cleaning, courtyard irrigation, fire fighting, and refilling of underground water. In this way, tremendous economic benefits can be created. On the one hand, the carpet-style roof garden is an ideal place for office clerks to rest. On the other hand, it is also an effective way of ecological architectural carbon fixation.

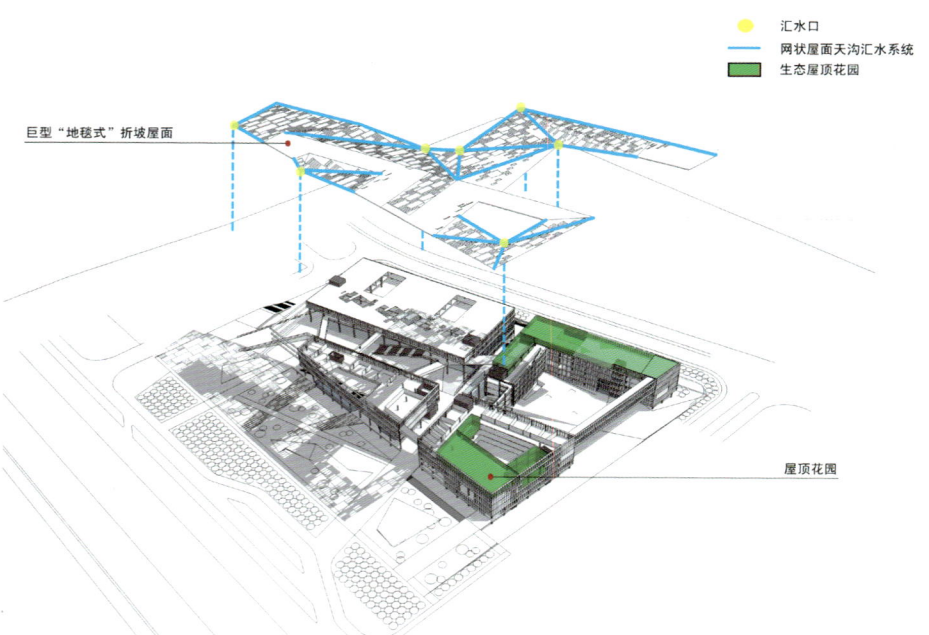

"零台阶"式多层公共空间

设计利用巨型折坡体系，模拟丘陵的起伏，提高空间形态的连续性，让大量人流以轻松、自然、安全、高效的方式完成不同高度间的转换，从而实现整个基地范围多方位多层次空间的"零台阶"衔接。

"No-step" Multi-layer Public Space

A giant slope system is used to simulate up-and-down hills so as to enhance the continuity of spatial form, make lots of people to see different spaces easily, naturally, safely, and efficiently, and make the base's multi-layer spaces in various directions connect each other without step.

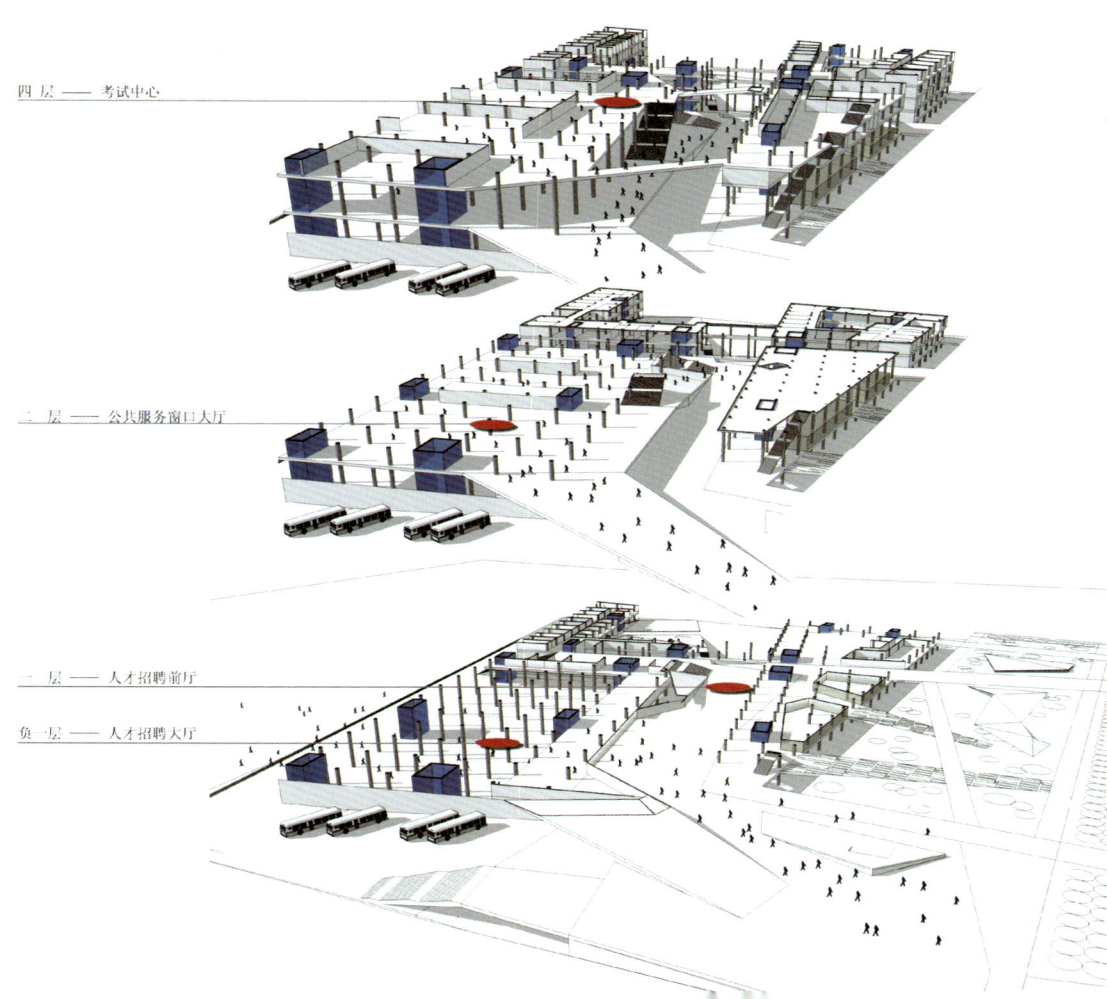

四层 —— 考试中心

三层 —— 公共服务窗口大厅

二层 —— 人才招聘前厅

负一层 —— 人才招聘大厅

This building is an achievement by creating good integration of functions and forms and compatibility between single person and the society. It is the building of the new times.

该建筑是功能和形式、社会和人共同作用的结果,是时代在创造建筑。

Foshan News Center • Foshan

Foshan | Project Location
2003 | Project Date
88,000m² | Project Scale
conceptual design, preliminary design | Design Phase
construction completed | Project Status
Japan Design | Design Partner

佛山新闻中心·佛山

中国,广东省,佛山市 | 项目地点
2003年 | 设计时间
8.8万m² | 项目规模
方案设计 初步设计 | 设计阶段
已建 | 项目现状
日本设计 | 合作单位

项目总览

本项目采用清晰方格网为结构脉络的半围合群体建筑布局形式，充分地表现了新闻中心功能的复合性及其外部空间的开放性特征，与自然形成和谐共生的关系。新闻中心高度密集，快速地收集、过滤、传递、反馈、处理和发布信息，像一台城市主板上的信息中央处理器，具有鲜明的时代同构特质。

30m标高的可调节遮阳通透构架天幕使建筑群与周边山丘整合，"人工化的自然"产生了内与外、虚与实的边缘美。简洁的整体形象突出了新闻中心的标志性。主体办公区与服务区之间立体穿插的天桥，穿越山丘的甬道，空中飘浮的参观道与天幕成为新闻中心这个极具时代感的中央处理器高效性的象征。

Project Overview

A semi-enclosed building complex layout is used in this project, with a distinct square grid as the structural basis. It fully demonstrates the function of the News Center and the openness of its exterior space, forming a harmoniously relationship with nature. The News Center collects, filtrates, transmits, respond, handles, and releases information intensively and quickly, just like an information "CPU" on a city motherboard, with distinct characteristic of the period.

The 30m-high semi-transparent canopy with adjustable structure integrates the building complex with surrounding hills. Such "man-made" nature presents a dreamlike beauty integrating the "inside & outside" and the "fiction & reality". Its simple but distinct appearance turns the News Center into a landmark. The overpass between the main office and the service area, the passageway crossing the hill, the floating sightseeing path and the canopy together symbolizes high efficiency of the News Center, and the modern representation of the "CPU".

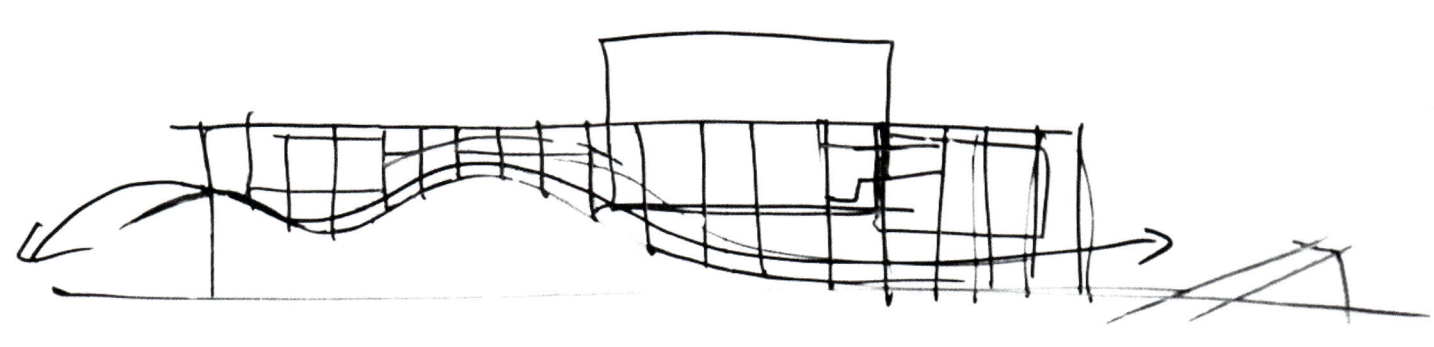

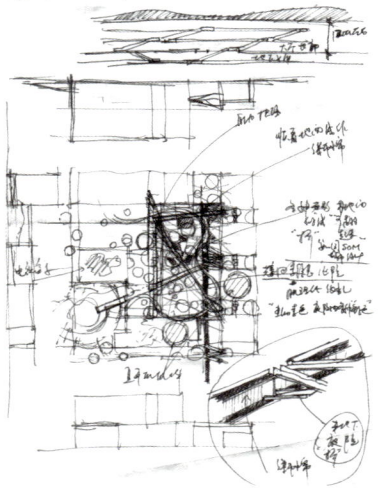
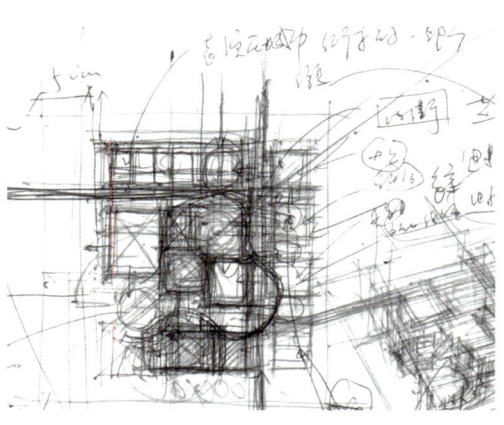

功能布局

理性、高效、逻辑的功能组合是作为社会信息中央处理器的新闻中心的必要条件。以下沉的信息庭院和服务街为交集，功能分为两大区：西边为新闻中心的主体功能区；东边是公共服务功能区。

主体功能区

山丘上由四大功能块组成：演播大厅、电视台、电台、报社。山丘下主要有电视台、电台、报社各自的外向性功能房间以及电视台的主要演播区，电视台、广播电台、报社共享的以及向信息庭院开放的营业大厅。

1,500m²的演播大厅：这个"大电视盒子"相对独立地布置于主体功能区的中部。

电视台和广播电台：因其运作特点相似，电视台和广播电台并置在用地北侧，却又各自独立，是建筑群中最高的体量。

报社布置在南部，由于报纸发行的独特性，在南向邻街设计了依"山"就势的外置广告街，缓缓抬升的坡道串联起一个个广告工作室。从一层至二层动态地展示了报纸广告的外向功能特质，在南面城市街景中形成一道多媒体亮线。

Functional Layout

Rational, efficient, and logic combination of functions is a prerequisite of the News Center to serve as a social information CPU. With the information courtyard and service street as the boundary, the News Center is divided into two functional areas, main functional area in the west of the center and public service area in the east of the center.

Main Functional Areas

On the hill there are four function blocks, performance hall, TV station, broadcasting station, and newspaper office. At the foot of the hill are functional rooms of the TV station, broadcasting station and newspaper office, main performance area of the TV station, and a business hall which is shared by the TV station, broadcasting station, and newspaper office and open to the information courtyard.

1,500m² performance hall: this big "TV box" is set separately in the middle of the main functional area.

TV station and broadcasting station: because the TV station and broadcasting station have similar operating features, they are set in parallel respectively in the north side. They are the tallest structures in the building complex.

The newspaper office is located in the south. To cater to the unique demand of newspaper distribution, an outdoor advertising street section on the neighboring south-facing street is designed, which leans against the hill. A slowly-rising ramp connects all advertising studios. The dynamic display of the newspaper advertisings from the first floor to the second floor features a multimedia highlight among the southern city street views.

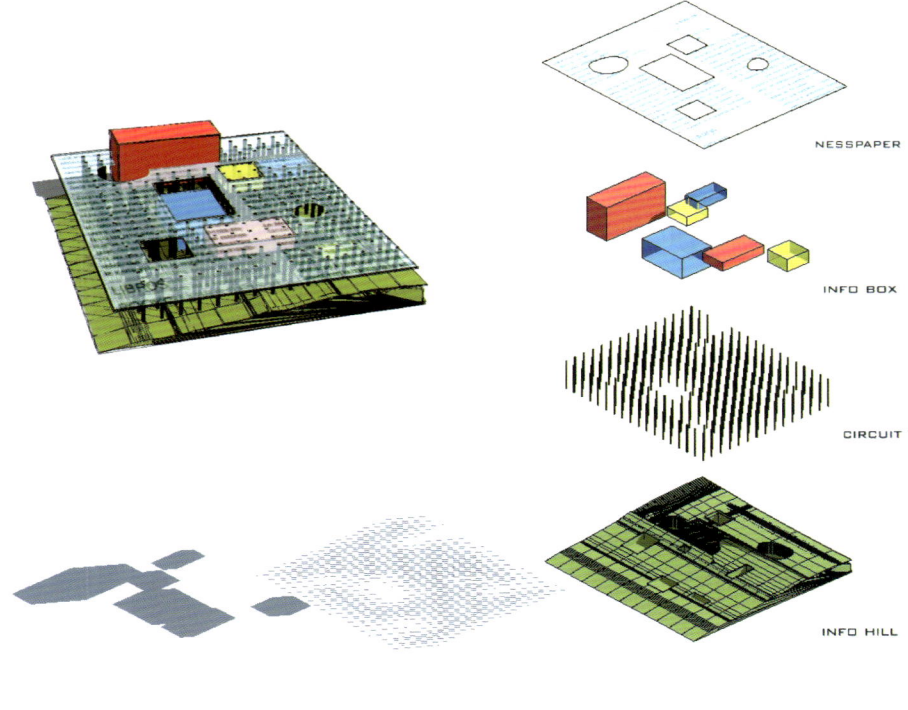
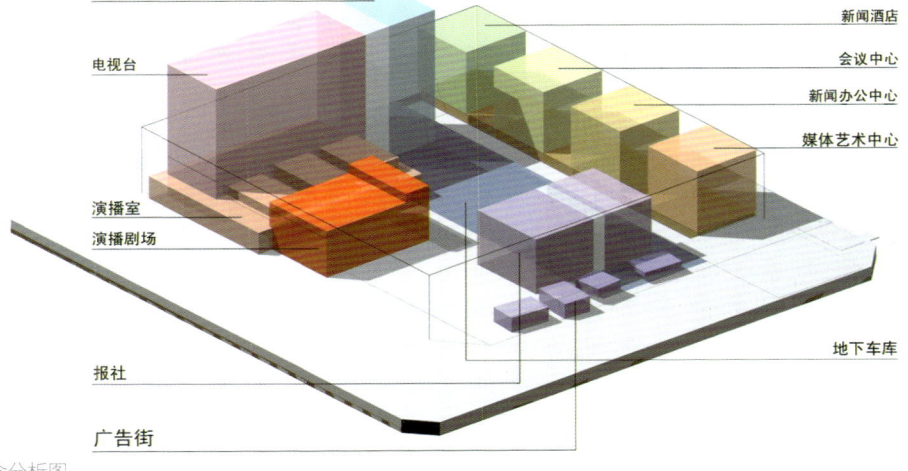

概念分析图

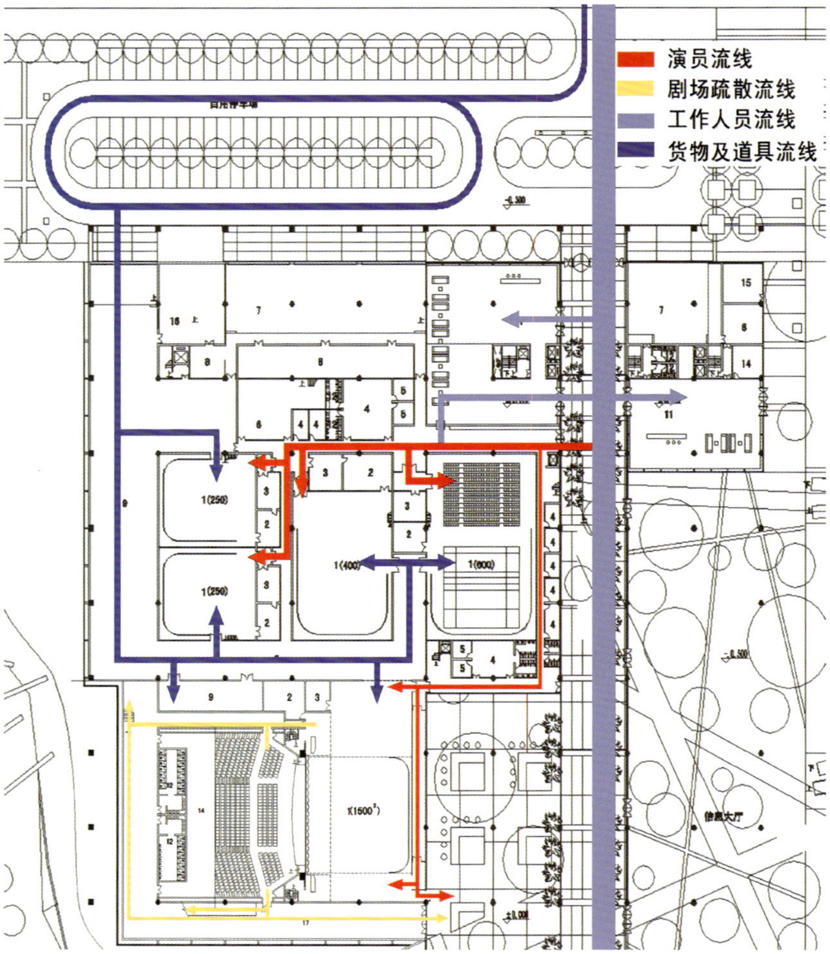

演员流线
剧场疏散流线
工作人员流线
货物及道具流线

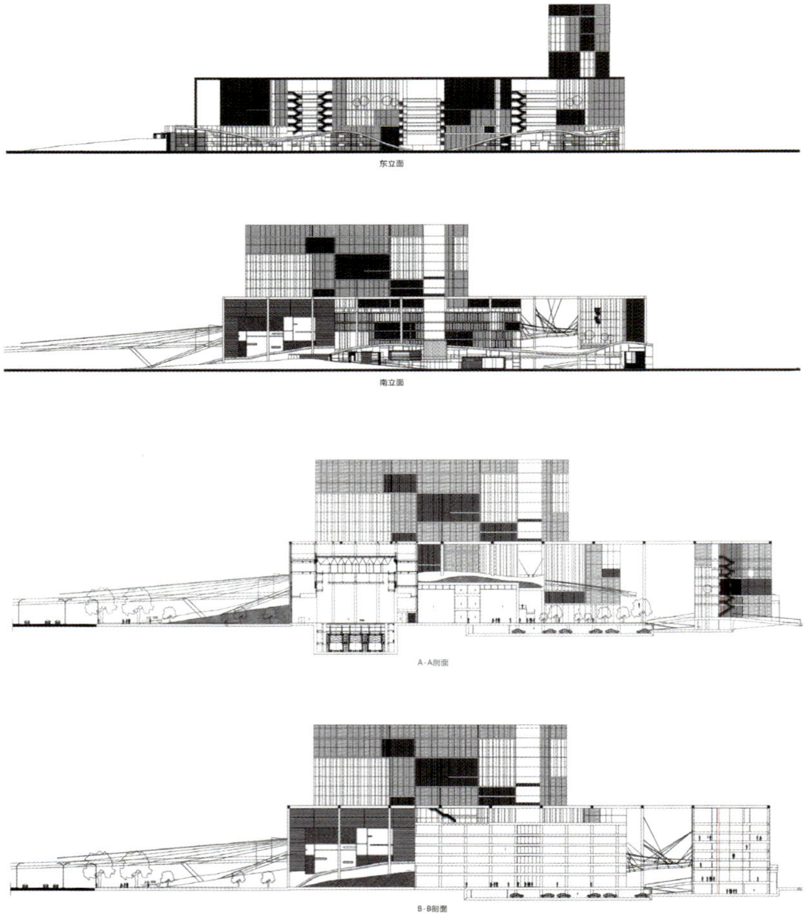

东立面

南立面

A-A剖面

B-B剖面

建筑内部流线

立面图

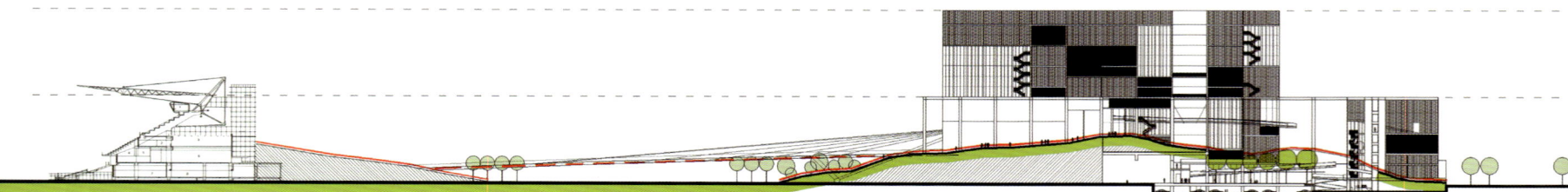

节能与环保设计

生态山丘

采光要求较低的功能房间覆土成丘，"无屋顶"建筑节能、环保、隔热、降噪，为市民提供更多"绿色屋顶"式的公共活动空间。市民在缓缓的山坡上，可以俯瞰体育中心广场的活动，在空间取向和视觉上与体育中心的联系更为紧密。随着山丘高度的上升，它的开放性自然逐渐减弱，实现由体育中心无序开放的公共广场向内部功能区公共交往空间的自然过渡。

Energy Saving and Environment Friendly Design

Ecological Hill

Functional rooms, requiring little sunlight, spread over the hill. "Roofless" buildings are characterized by energy saving, environmental friendliness, heat insulation, and noise reduction. The "green roof" type public activity spaces are offered to citizens. Staying on the slope, citizens can overlook the Sports Center Plaza. Thus they feel closer to the Sports Center visually and spatially. From the foot to the top of the hill, its openness to the public becomes weaker gradually, realizing a natural transition from openness of the Sports Center to the purpose-bearing communication space of the internal functional areas.

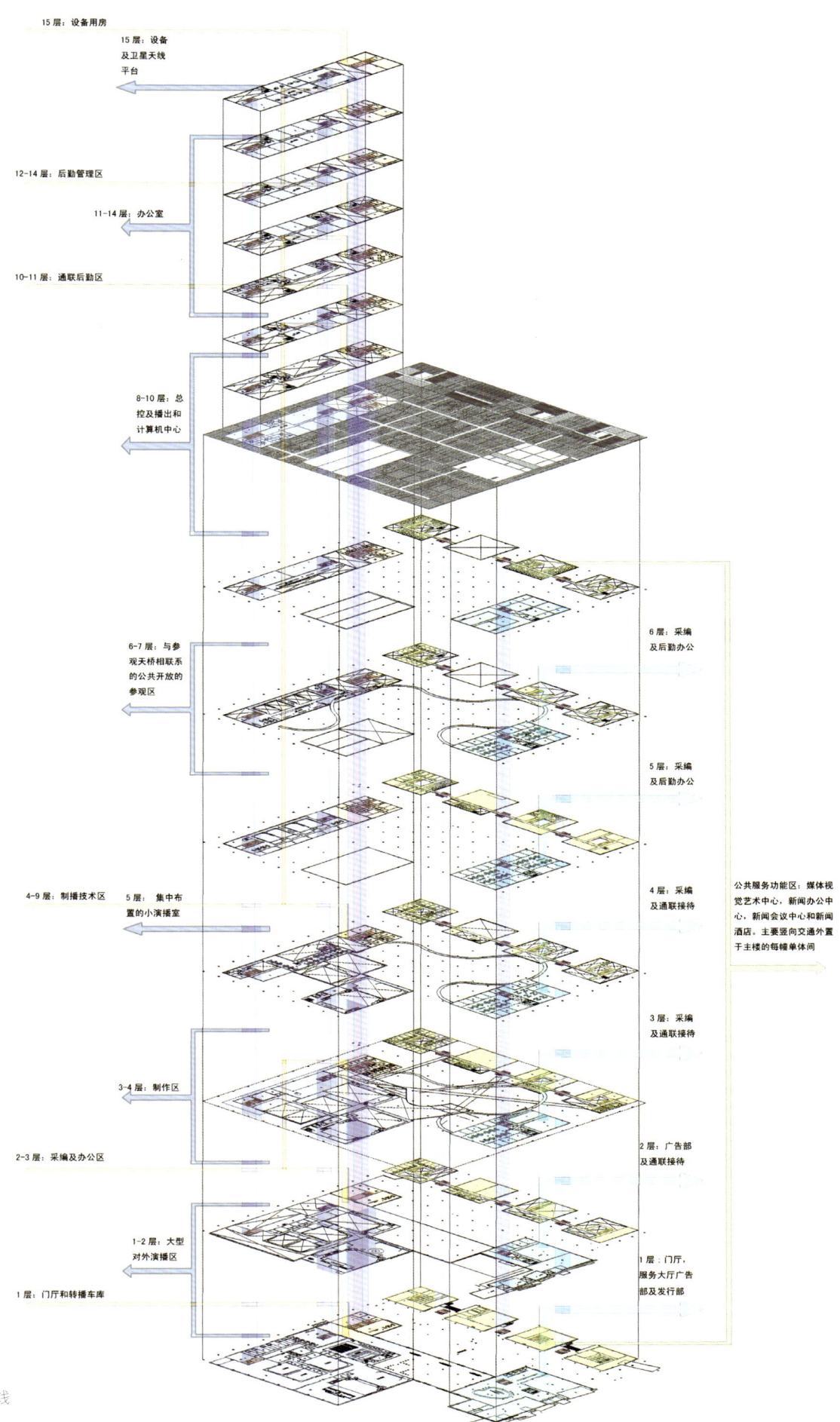

建筑内部流线

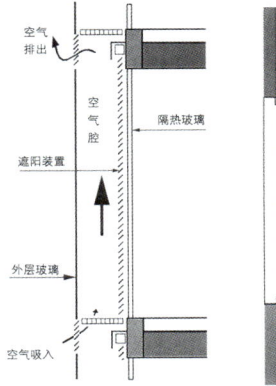
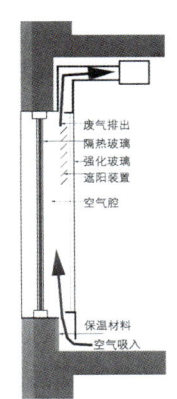
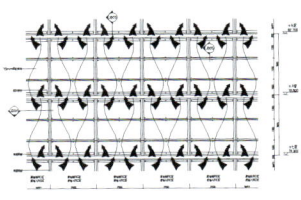

双层幕墙剖面(1)　双层幕墙剖面(2)　双层幕墙通风原理示意图

幕墙

通风幕墙利用"烟囱效应"与"温室效应"的原理,夏天打开上下百叶,通过中间空气层的流动带走温室热量,减少室内空调的能源消耗;冬天关闭百叶,使中间空气层成为封闭温室,减少室内供暖能源的消耗。通风幕墙的外层玻璃选用无色透明玻璃或低反射玻璃,可最大限度地减少玻璃反射带来的光污染。双层过滤阳光,避免直射,无眩光困扰。隔声性能可达55dB,使在室内生活和工作的人们拥有一个清静的环境。另外,无论天气好坏,无需开窗即可将自然空气传至室内,有效降低高层建筑单纯依靠暖通设备机械通风带来的弊病。经统计,双层幕墙比单层幕墙制冷时节能40%~60%,采暖时节能40%~50%。

Curtain Wall

The ventilating curtain wall is built according to the principle of "stack effect" and "greenhouse effect". In summer, upper and lower shutters are opened. Thus the middle air flow takes away indoor heat and reduces indoor air-conditioner's energy consumption. In winter, shutters are closed. Thus the middle layer of air becomes a closed greenhouse to reduce indoor heat supply energy consumption. Outside the curtain wall is covered by colorless transparent glass or low-reflecting glass, which reduces the light pollution caused by glass reflection as much as possible. The double glazing filters sunshine, avoids direct exposure to the sun, and prevents glaring. Its sound insulation performance is 55dB. Thus people staying indoors can enjoy a quiet environment. Furthermore, whether good weather or bad weather, the natural air can go indoors on condition that windows are closed. Therefore, the problems caused by heating and ventilating machinery in high-risk buildings are solved effectively. According to statistics, a double-layer curtain wall can save 40-60% of energy used for refrigeration and 40-50% of energy used for heat supply compared with a single-layer curtain wall.

Information is a particularity of modern society. As an information processor, the media center should response to and process information in an intensive way.

信息已成为现代社会的特质,媒体中心作为信息的"处理器"应集中体现对信息的响应。

Changsha Broadcasting and TV Center · Changsha

Changsha | Project Location
2003 | Project Date
71,000m² | Project Scale
conceptual design | Design Phase
construction not started | Project Status

长沙广电中心 · 长沙

中国，湖南省，长沙市 | 项目地点
2003年 | 设计时间
7.1万m² | 项目规模
方案设计 | 设计阶段
未建 | 项目现状

项目总览

广电中心不只是一个功能的堆砌体,媒体的社会性和开放性决定了它应该是一个可以让人接触的建筑,是市民的城市中心。在这里,市民可以看到广阔的天空、绿色的草地和山丘,还可以听到风从建筑上轻抚过的声音。

Project Overview

The Broadcasting and TV Center is not only a functional building but also an accessible building to the public because of the sociality and openness of media. It is the center of the city in the citizens' mind. Here citizens can see a broad sky, grassland and green hills and listen to the sound of breeze as it passes by the building.

广电中心，在第一时间掌握区域乃至全球的最新资讯。信息在这里汇集，并发生化学变化以折射出更加深刻、广泛的社会意义，成为反映社会需求和城市自身魅力的荧幕。我们的建筑就是要提炼这种同类中的差异性，因为这种差异是地域自身独有的，也是最根本的。将其提取出来并放大，用以彰显城市魅力的所在。

At the Broadcasting and TV Center, the latest regional or even global information can be grasped at the very first time. Information gathers here and then "chemical changes" take place to reflect more profound and wider social significance, become a movie screen reflecting social demands and the city's charm. Our building is to extract difference of the same kind because such difference is unique and fundamental regionally. When such difference is extracted out and magnified, it can show the city's charm.

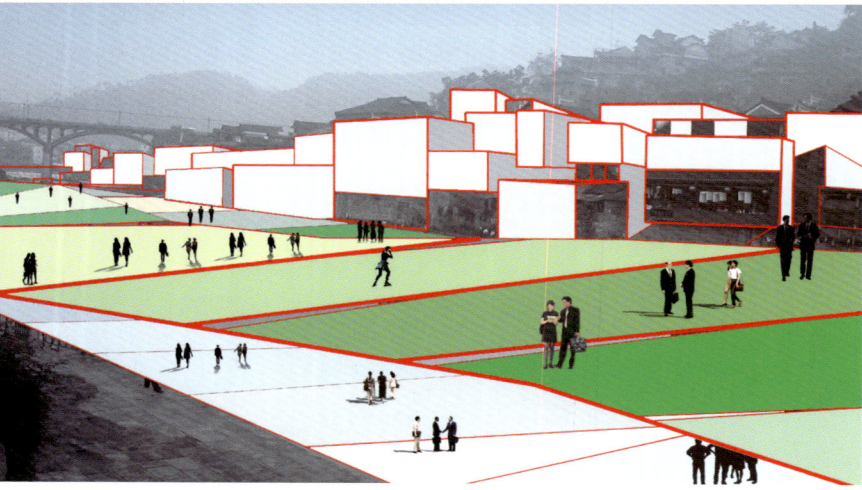

广播电视中心的运作是复杂的。它如同一个都市，具有多样性，充满了无限可能。日常的中心运作、商务、行政、庆典活动甚至是休闲无事的漫游，都需要一个能与之匹配的容器去包容这一切，而非强硬的归纳和操控。

The Broadcasting and TV Center is operated in a complex way. Like a metropolis, it has diversified functions and promotes infinite imaginations. Routine managements, businesses, administrations, celebrations or even wandering in the center requires a suitable building to accommodate all these.

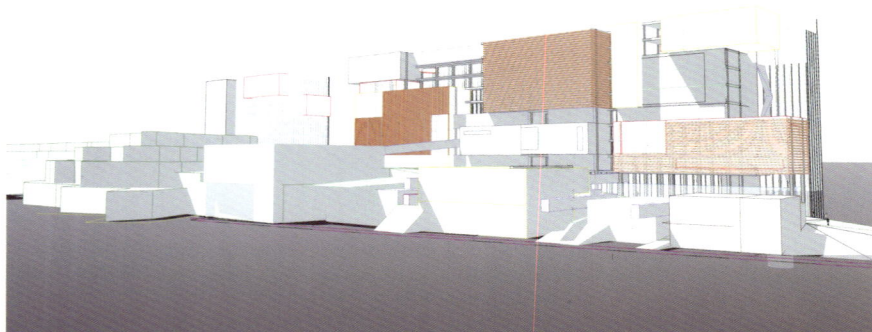

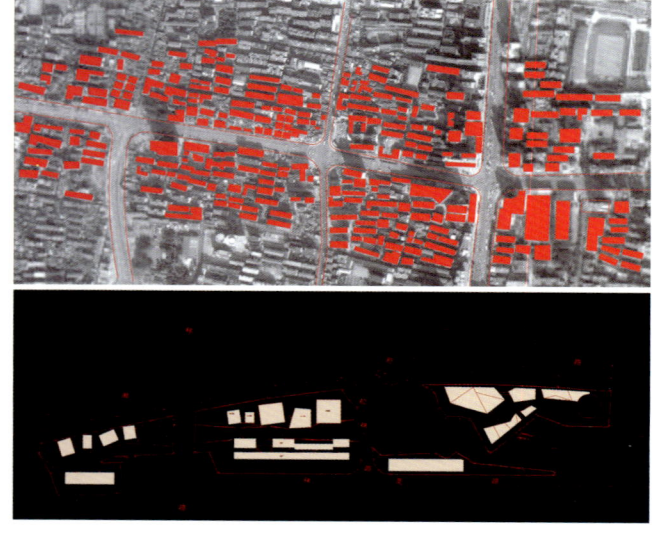
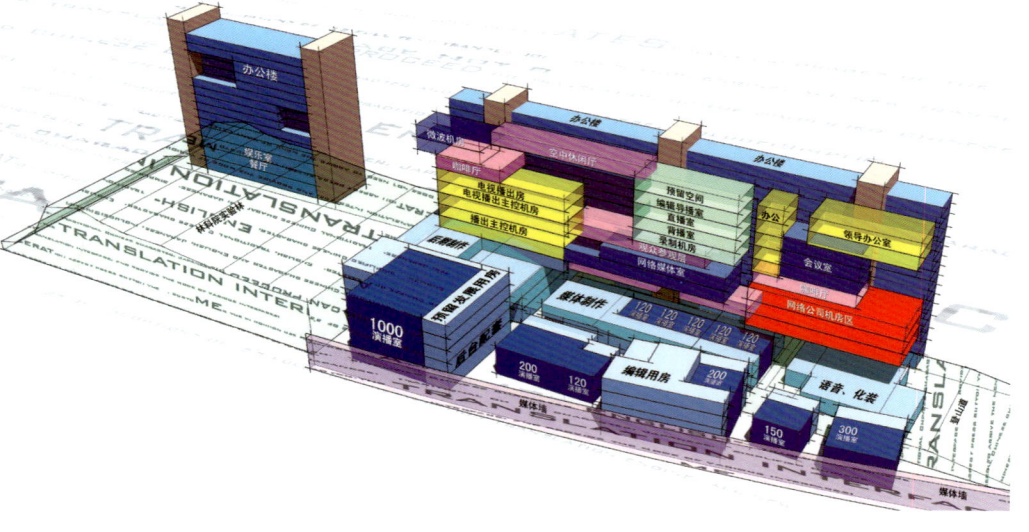

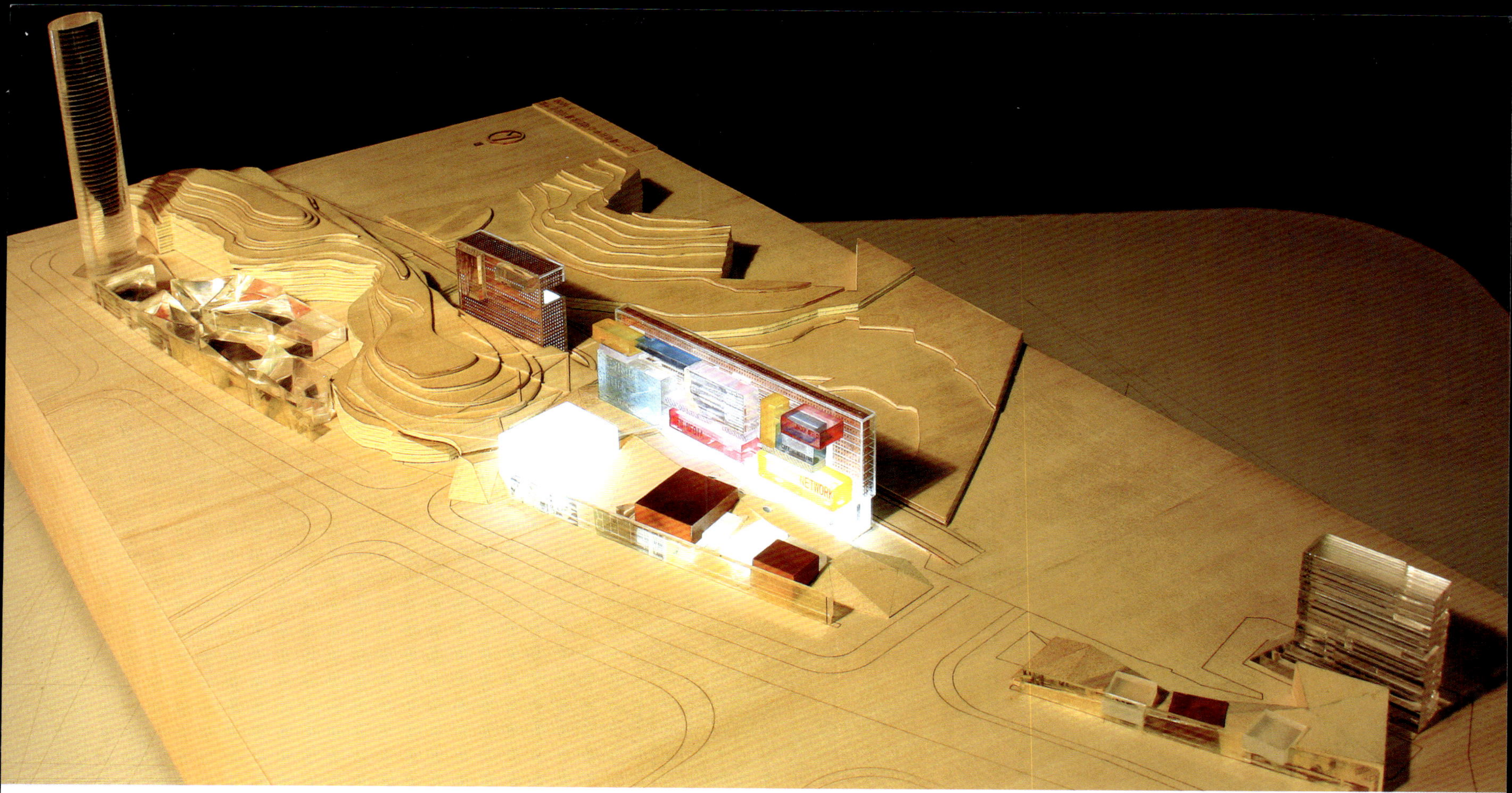

交通空间分析图

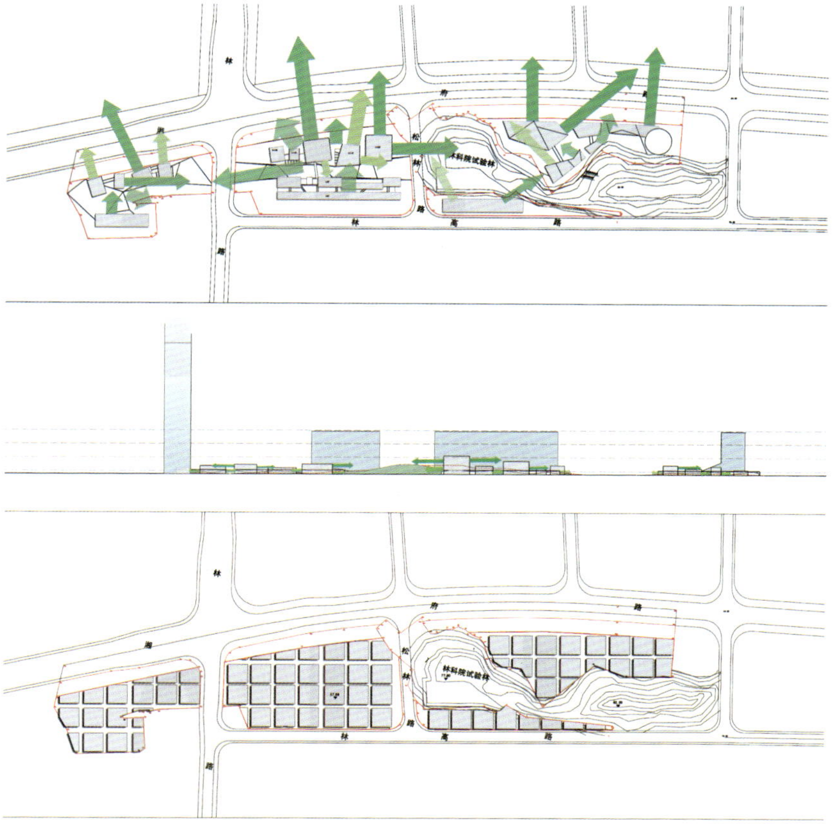

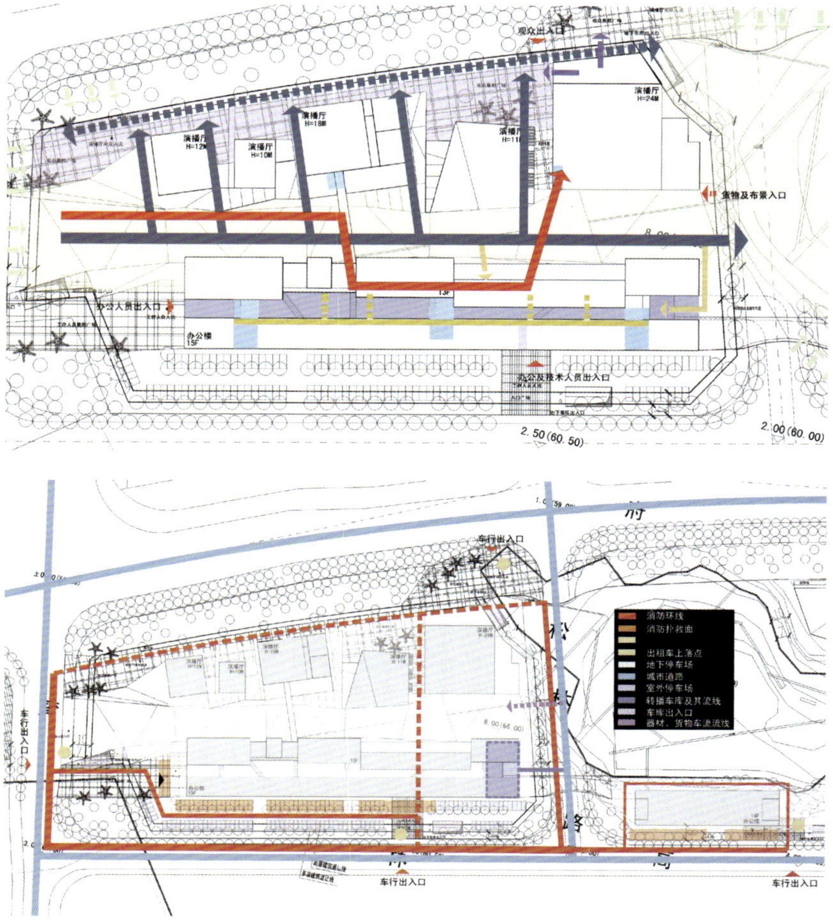

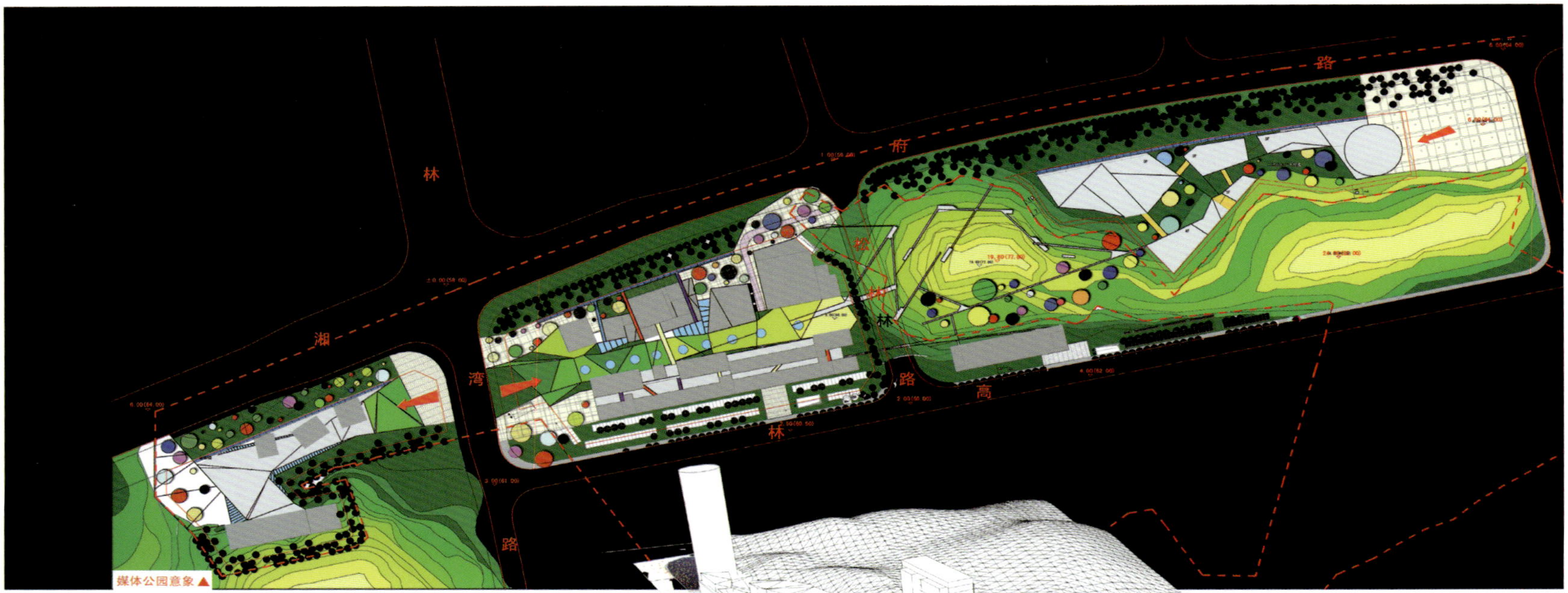

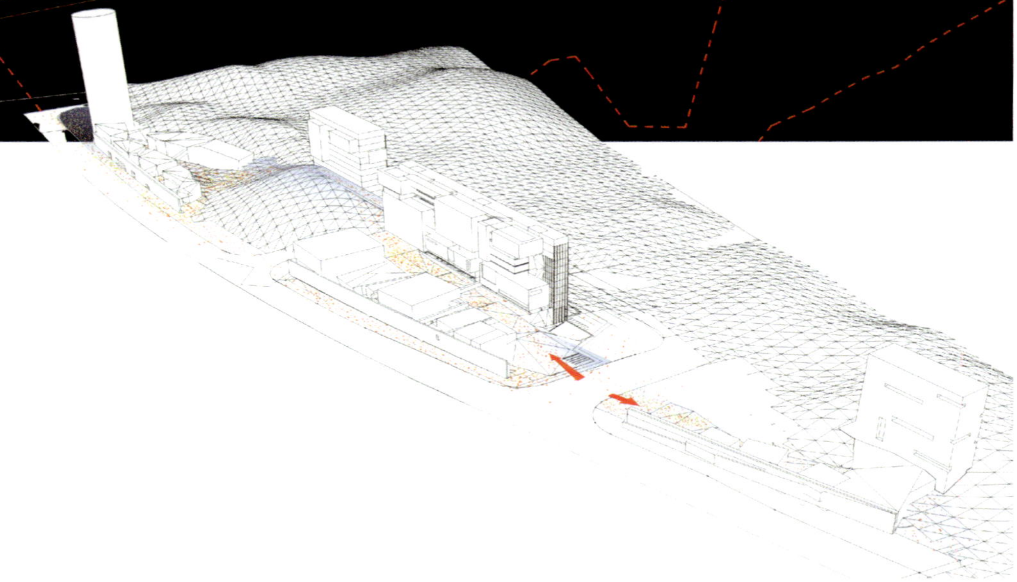

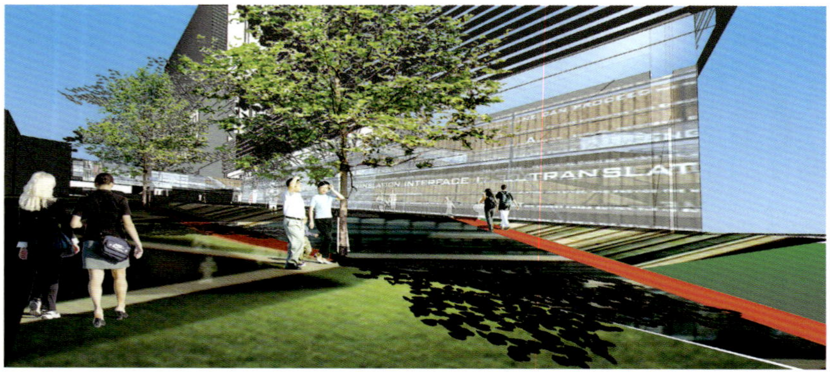

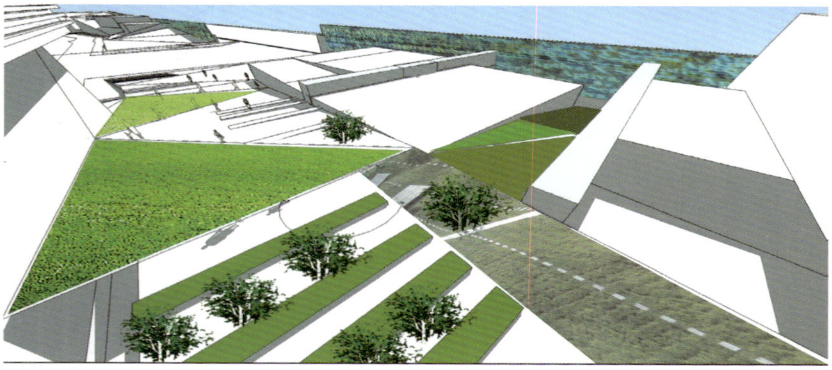

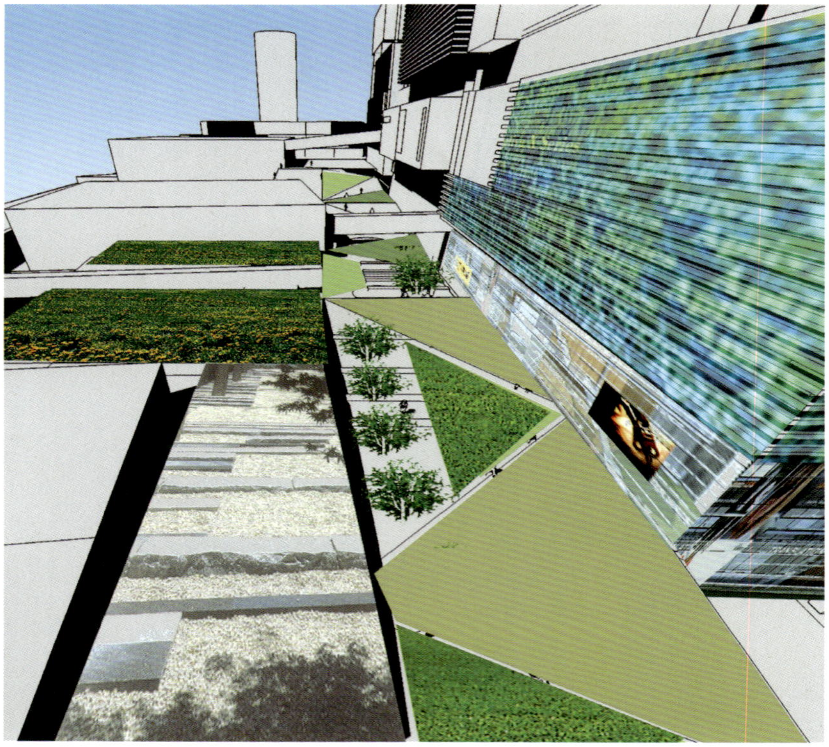

Here an innovative life pattern is created on the basis of the life and working styles of this group of professionals.

这里创造一种创新的生活模式,为这类专业的人群生活和工作的方式而设。

Shenzhen Software Industry Base • Shenzhen

Nanshan District, Shenzhen | Project Location
2009 | Project Date
138,000m² | Project Scale
conceptual design, preliminary design, construction drawing | Design Phase
proposed | Project Status

深圳软件产业基地·深圳

中国,广东省,深圳市,南山区 | 项目地点
2009年 | 设计时间
13.8万m² | 项目规模
方案设计 初步设计 施工图设计 | 设计阶段
在建 | 项目现状
深圳机械院建筑设计有限公司 | 合作单位
南沙原创工作室

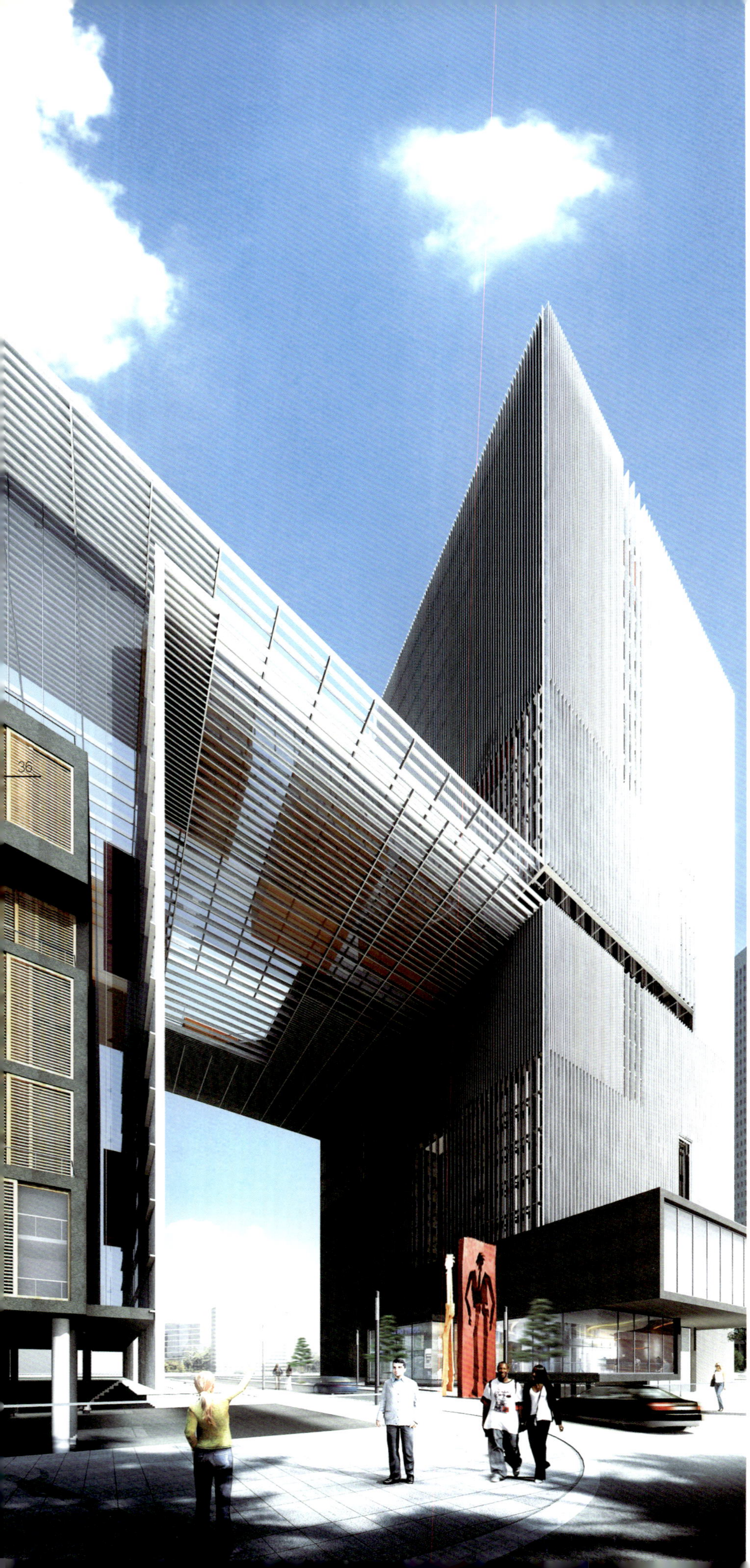

项目总览

本工程位于深圳市高新科技园，即高新南十路和高新南九路的交点，西临白石路。它由办公楼和两座公寓组成，办公楼与人才公寓高度为100m，中间塔楼专家及人才公寓高度为50m。共设3层地下室，首层及二层设有餐厅及会所，地下区域分别为车库与各专业的设备用房。建筑面积约为10.6万m^2，地下室面积约为3.2万m^2。

项目旨在城市的空白记忆区书写一种新的生活方式和对称的生活态度。通过对城市生活气息肌理的提取，打造一处生态立体的居住体系。

建筑不是简单的堆积，它需解读不同兴趣爱好人群的居住习惯，以促进交流。多样性的居住单元能满足不同使用者的居住需求，以此形成一种“里”的质感。公共空间体系串联于其中，并披附一层生态绿色、节能环保的低造价格栅体系，形成一种“外”的态度。

Project Overview

Located in Shenzhen Hi-tech Industrial Park, the project site is near the intersection between Gaoxin Nanshi Road and Gaoxin Nanjiu Road, facing Baishi Road in the west. The project includes an office building and two apartment buildings. The office building and employee residence are both 100 meters in height. The other apartment building for experts and talents, is 50m in height. There are three basement floors. The first and second floors are for restaurants and clubs. On the basement floors, parking lot and rooms for professional equipment are arranged. The total building area of the project is about 106,000 square meters, including 32,000 square meters of basement area.

This project is designed to introduce a new and balanced life style. By adapting to the atmosphere of urban life, it creates an ecological and three-dimensional living system.

Construction is not simple pileup of materials. The living habits of people with different interests and hobbies should be comprehended to promote the communication between them through the design of buildings. The diversified apartments form an "inner" quality that can satisfy different demands of the households. A public space system creating an "external" atmosphere is inserted among the apartments and divided by an ecological, energy-saving, and environmental friendly grid system at low-cost.

主要技术经济指标

第二标段总指标			
总用地面积		25319.19	平方米
总建筑面积		138000	平方米
	地上建筑面积	106000	平方米
	地下建筑面积	32000	平方米
容积率		4.19	
绿地率		18.5%	
地下室建筑面积		32000	平方米
停车数量		900	辆

01-01地块	用地面积		4960.78	平方米
	总建筑面积		47620	平方米
	其中	专家公寓	32800	平方米
		管理办公	6000	平方米
		金融机构	1248	平方米
		中介机构	2496	平方米
		商业配套	4070	平方米
		服务用房	1000	平方米
		通信网络机房	300	平方米
	容积率		9.6	

01-02地块	用地面积		4317	平方米
	总建筑面积		22446	平方米
	其中	专家公寓	5800	平方米
		人才公寓	16646	平方米
	容积率		5.2	平方米
01-03地块	用地面积		4453	平方米
	总建筑面积		36068	平方米
	其中	人才公寓	34268	平方米
		餐饮配套	1800	平方米
	容积率		9.6	

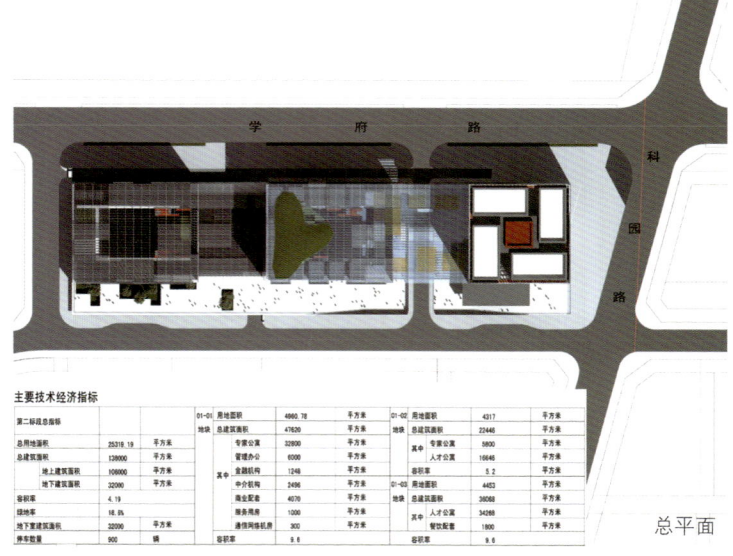

总平面

功能重组、集约、重叠、多维
21世纪，我们呼吁新的城市开发模式，倡导集群化、综合化开发，将一个大规模的综合体项目做到立体化、整体化；我们倡导24小时城市理念，每个综合体都能实现居住、工作、游玩、休憩、学习、创造等多项功能。在这里，我们可以步行上班，可以在家门口购物、娱乐、欣赏高端艺术品展览，可以穿行于各建筑之间，再也没有拥挤车流的干扰，城市不再处处充斥着汽车尾气，完善的地下轨道交通系统也成为未来主要的交通方式。

Functional reorganizing, intensive, superposing, multi-dimensional
In 21st century, we call for new city development modes and advocate centralized and comprehensive development to make a large-size building complex three-dimensional and integral. We advocate a 24-hour city idea and each building complex has multiple functions, including living, working, amusement, rest, study, and creation, etc. Here we can go to work by foot, go shopping, seek entertainment, and appreciate high-end artwork exhibitions at the doorway. We can also walk among buildings without the disturbance of vehicles. Vehicle exhaust no longer exists everywhere in the city. Perfect underground railway traffic system will become a major transportation way in the future.

区域比较

香港科学园

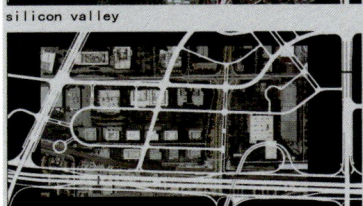
silicon valley

中关村西村

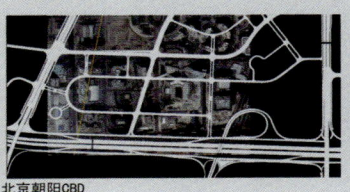
北京朝阳CBD

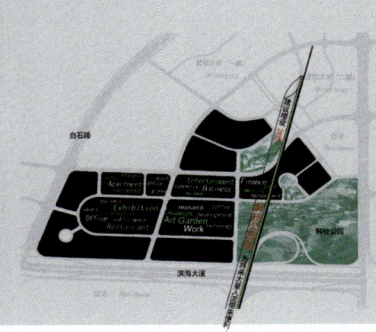

PRIVATE	私密的	Commerce	商业
Business	商务	Walkway	空中联系步行路
office	办公	Exhibition	展示
Research	研发	Entertainment	娱乐
Management	管理		
Apartment	公寓		
Residence	住宅		
PUBLIC	公共的		
Square	广场		
Green	绿地		
Repast	餐饮		

南向界面
端庄，雅丽

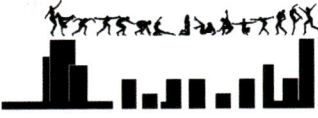
北向界面
俊美，活力

城市空虚 Urban voids

一个没有使用，且在城市居民的集体认知中几乎没有印记的地区。它也是可以利用的地方，充满期待且很强的城市记忆。

A unused area which gives no deep impression to urban residents may also be utilized and bring people expectations and unforgettable memories.

New York City
纽约

美国纽约的BLOCK街区，理性严谨的规划是西方现代大都市的代表。
New York, USA is a representative of western modern metropolis due to its rational and rigorous block planning.

Barcelona Spain
巴塞罗那

巴塞罗那的兰方格形路网，充满了欧洲城市的浪漫情调。
The radioactive road network in Barcelona is full of romance of European cities.

Dafen Village, Shenzhen
深圳大芬村

大芬村，随着城市发展形成的自然肌理体现着城市的记忆。
Dafen Village's natural mechanism, which comes into being with the development of the city, shows memories of the city.

生活方式的移植

交流休憩
可呼吸

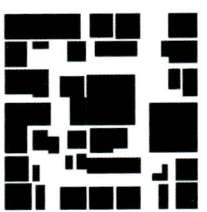
记忆的移植

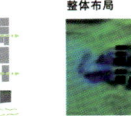
整体布局

"冷巷"生态空间
分散布局

趣味空间组合

生态立体居住

典型地段复制

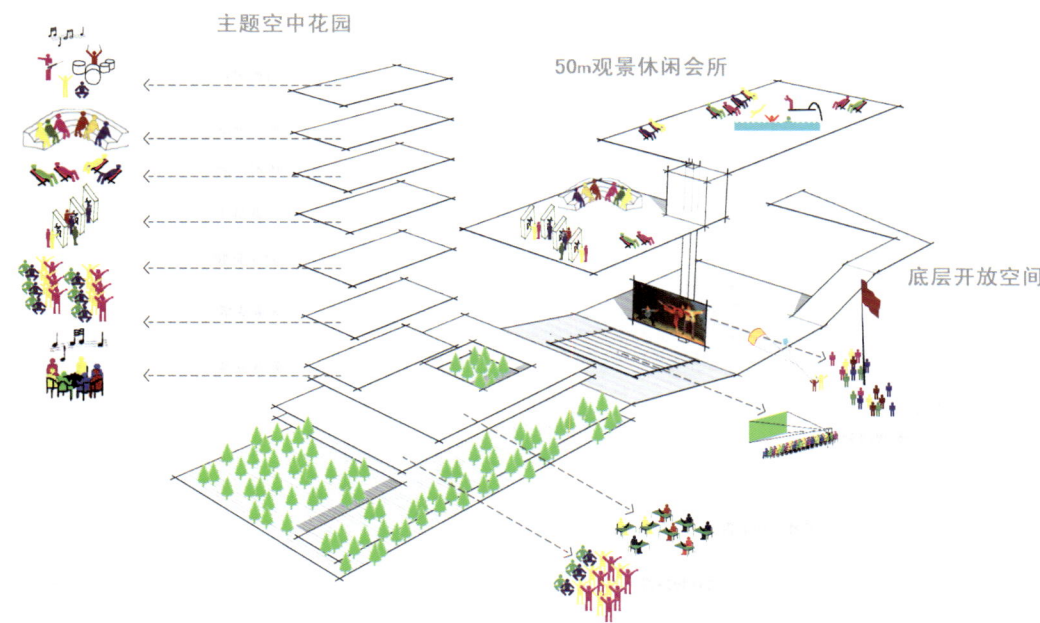

20世纪,我们追求城市功能分区。

居住、商业、行政、CBD、工业等各区独立明确。城市人的进程不断加速,我们的城市不断扩张,明确的城市分区模式也暴露出越来越多的问题:办公区白天繁华,夜晚宛若空城;居住区距离办公区越来越远;CBD成为没有联系和沟通的建筑物聚集区;上下班高峰,城市交通的负荷愈来愈大;汽车尾气污染愈加严重……

In 20th century, we are in pursuit of city functional division.

Residential, business, administration, CBD, and industrial areas are planned separately and definitely. With the acceleration of urbanization, our city expands consecutively. A definite city division mode also exposes more and more problems: office areas are busy in the daytime but look like an empty town at night; residential areas are far away from office areas; CBDs become a building-concentrated area with no contact and communication; the city's traffic burden becomes higher and higher at rush hours; and vehicle exhaust pollution becomes more and more serious ...

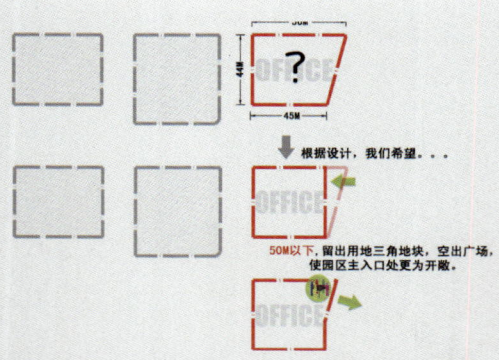

50m以上,建筑挑出,平行红线,呼应用地,在园区主入口处极具标志性,建筑内空出开放空间。

Architectural protrusion of more than 50m and parallel red lines respond to the land. The main entrance of the park is symbolic and there is a large open space inside the building.

办公分析

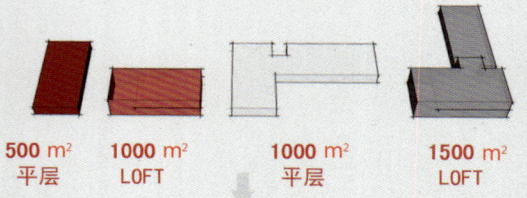

考虑办公空间能灵活适应将来不同规模的未知企业入驻,我们设计出4种类型的办公BOX。

In consideration of office space's adaptability to different sizes of unknown enterprises, we have designed four types of office box.

公共空间体系串联于其中,同时被具有生态绿色、节能环保的低造价格栅体系附着,以形成一种于外的态度。建筑的形态是对功能的直接图解。

A public space system is set among them and, at the same time, it is attached by low-cost ecological, green, energy-saving, and environmentally-friendly grilles so as to form an outward attitude. Architectural form is a direct illustration of functions.

以4层为一个单元体,4种办公BOX叠加组合。保证每个办公单元均有朝东、南的景观视野,组成2种单元体。

A unit consists of four floors and four types of office box are superposed. Thus each office unit has an eastward and southward field of vision, making up two types of unit.

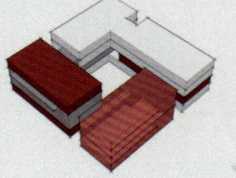

2种单元体交错叠加上升,组成整栋建筑。

Two types of unit superpose each other and ascend to form a whole building.

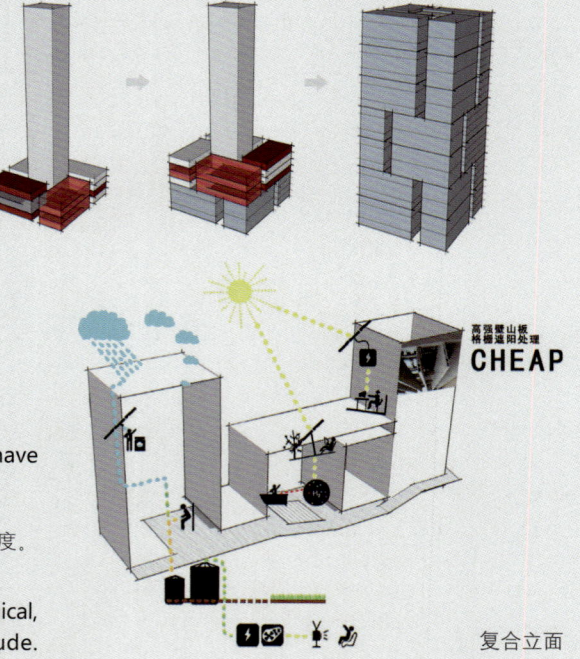

复合立面

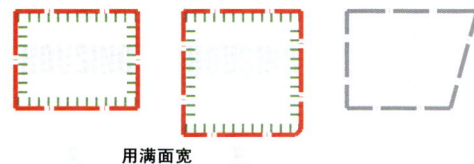

用满面宽

标准层可做38间 A standard floor can offer 38 rooms.
标准层可做42间 A standard floor can offer 42 rooms.
总共可做1254间 There are 1,254 rooms in all.
总共可做672间 672 rooms are available.

全部考虑做单人间时，总共可以做1926间，满足1926人居住；
If all rooms are designed as single rooms, 1,926 single rooms are available to accommodate 1,926 persons.

全部考虑做双人间时，总共可以做1926间，满足3852人居住。
If all rooms are designed as double rooms, 1,926 double rooms are available to accommodate 3,852 persons.

根据设计，我们希望……
According to the design, we hope ...

我们可得出：根据任务书要求解决2,400人居住，所以，用地内无法全部做单人间，且不适合全部做双人间。
We may draw a conclusion:
According to the assignment, we need to accommodate 2,400 persons. Therefore we can design all rooms on the land as single rooms or double rooms respectively.

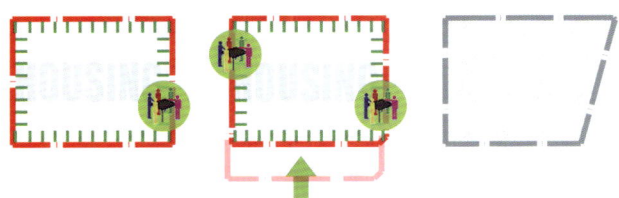

用地保持统一的界面，留出一些供休憩交流的公共空间。
The land should show a unified interface and offers a public space where people can have a rest or communicate with others.

对于人才公寓，我们设计了3种各有特色的户型单元。
With respect to the talent apartment, we have designed threekinds of characteristic house.

更为重要的，我们提倡集落化的居住工作生活体系，这就需要更加强调人性化的设计，设计出个性化的产品以满足年轻人个性化的选择……
More important, we advocate a residence, working, and living system. Thus we need to emphasize humanistic design and design personalized products to meet personalized demands of young persons ...

通过对填海区域的解读，在城市空白记忆区书写一种新的生活方式，一种对称的生活态度。通过对城市具有生活气息肌理的提取，打造一座生态立体的居住体系。
A new life style and a symmetrical life attitude are created in the area with no city memory by understanding the sea reclamation area. A three-dimensional ecological living system is built by withdrawing urban textures full of life atmosphere.

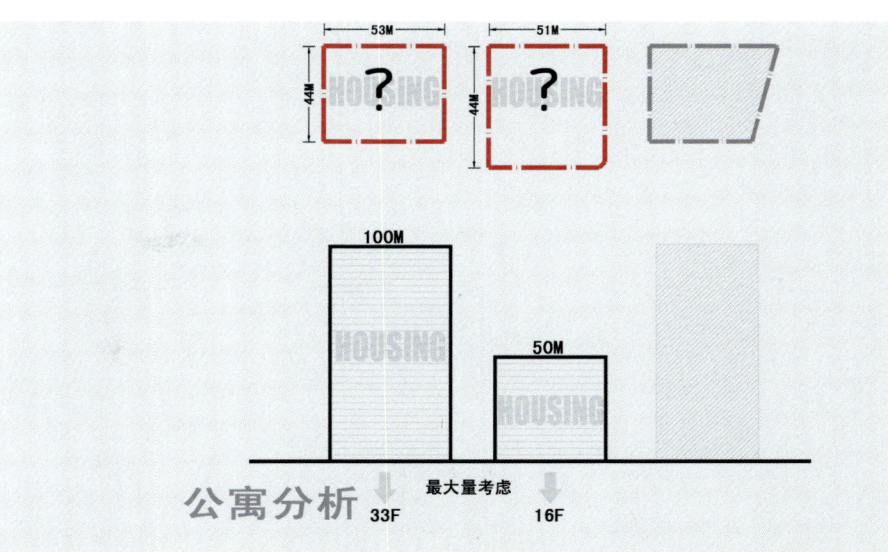

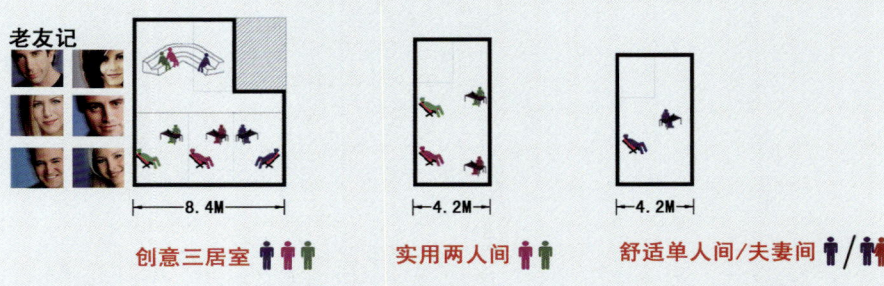

对于专家公寓，我们根据使用性质考虑2种户型单元。
For the expert apartment, we consider two types of house according to the nature of use.

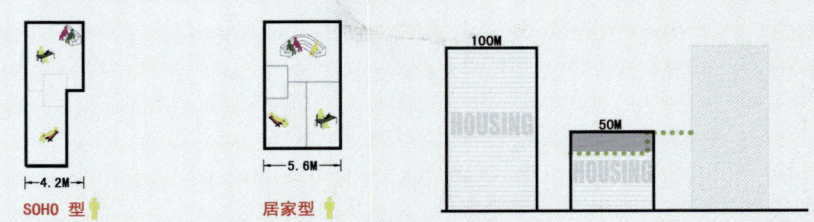

考虑到专家公寓与人才公寓宜分区设置，方便管理。且专家公寓兼有办公性质，故将其置于50m住宅的最上端，竖向上与人才公寓通过空中大堂相分隔，水平向与研发办公楼通过空中连桥联系。
In consideration that the expert apartment and talent apartment are set in different areas for the convenience of management and the expert apartment is also used as a workplace, we design it at the top of 50m residences. It is separated vertically from the talent apartment across the aerial lobby and connects horizontally the R&D building across the aerial bridge.

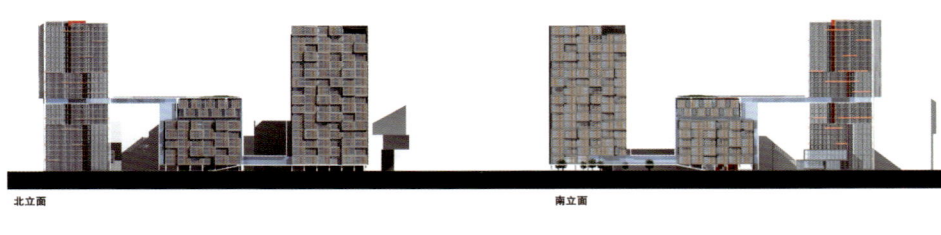

北立面　　　　　　　　　　　南立面

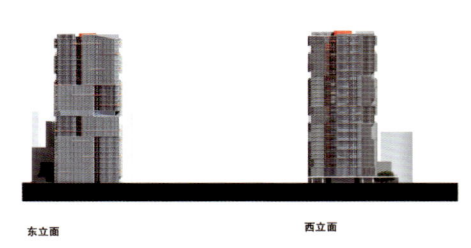
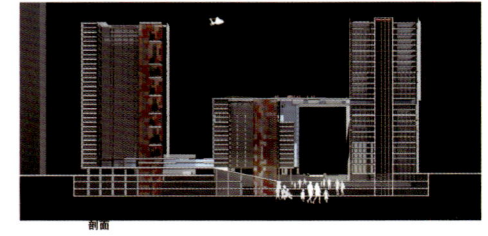

东立面　　　西立面　　　　　剖面

From the ground to the sky, we intend to provide a "future" space and working style at the "corners" of this city and organize three-dimensional vertical urban streets.

我们尝试着从天到地在这个城市的"角落"提供"未来"的空间、工作方式,以及组织具有城市性的立体竖向街道。

China Mobile Shenzhen Branch Production Dispatching Building · Shenzhen

CBD, Shenzhen | Project Location
2009 | Project Date
100,000m² | Project Scale
conceptual design | Design Phase
construction not started | Project Status

深圳移动生产调度大厦·深圳

中国，广东省，深圳市，中心区 | 项目地点
2009年 | 设计时间
10万m² | 项目规模
方案设计 | 设计阶段
未建 | 项目现状

概念设计说明

我们把中国移动提倡互动的理念贯穿于设计之中，认真研究了建筑在城市中的互动功能。我们在本项目建筑的开放区和半开放区引入垂直的城市空间，试图给中心区单一维度的空间现状带来生机。开放的垂直空间，同时也带来了更多建筑单体与城市、群众与建筑之间的互动机会。

分析图

Description of Conceptual Design

We made a study of the building's interaction in the city; as a result, the whole design is consistent with China Mobile's "interaction" philosophy. A vertical space overlooking the city in the open and semi-open area of this building is designed attempting to bring a dynamic force to the single-dimension space in the central area of the building. The open vertical space serves as model that creates an interaction between a single building and the whole city, between the public and buildings.

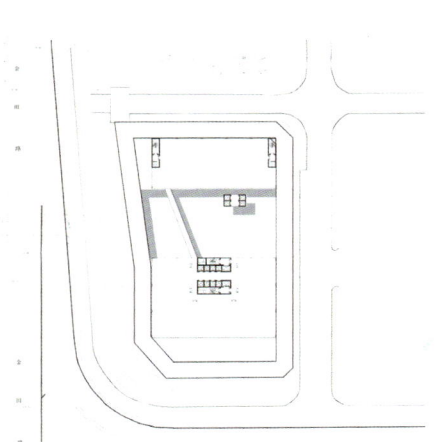
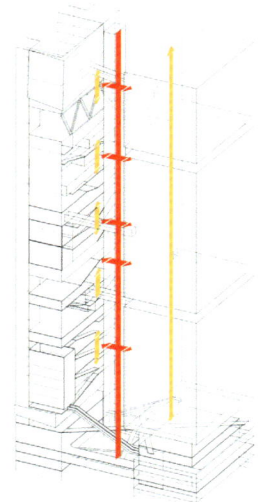

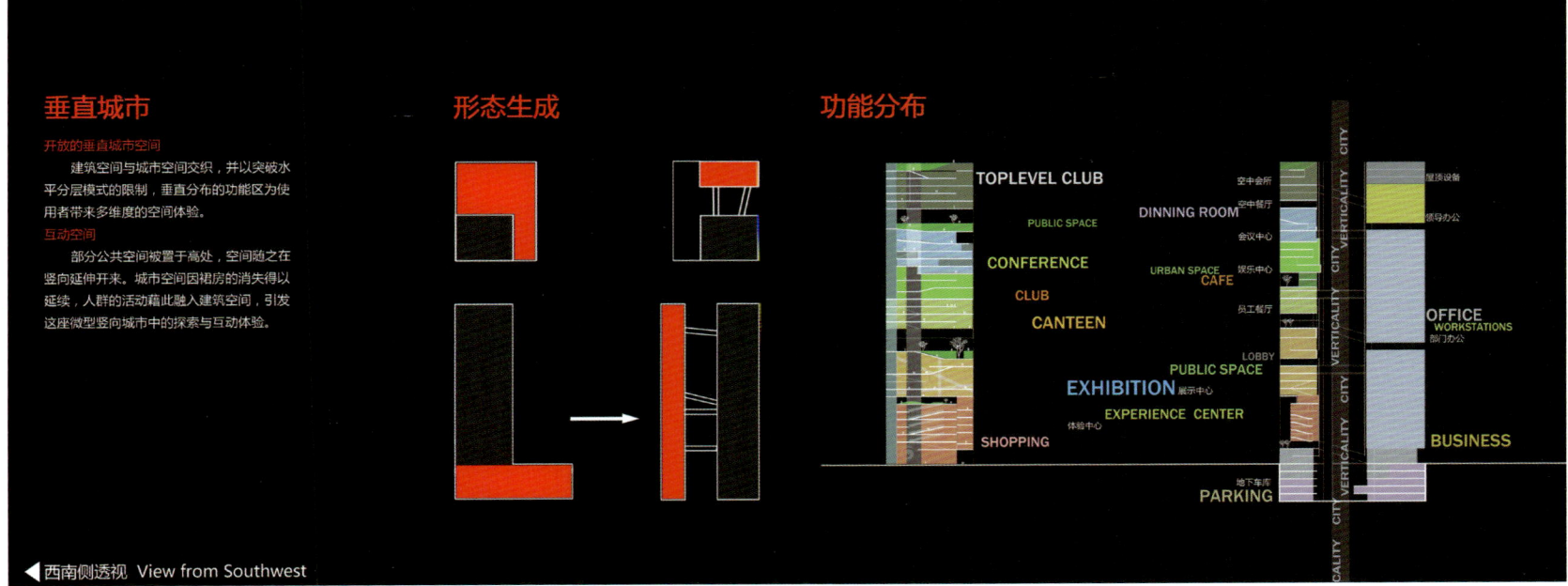

◀ 西南侧透视 View from Southwest

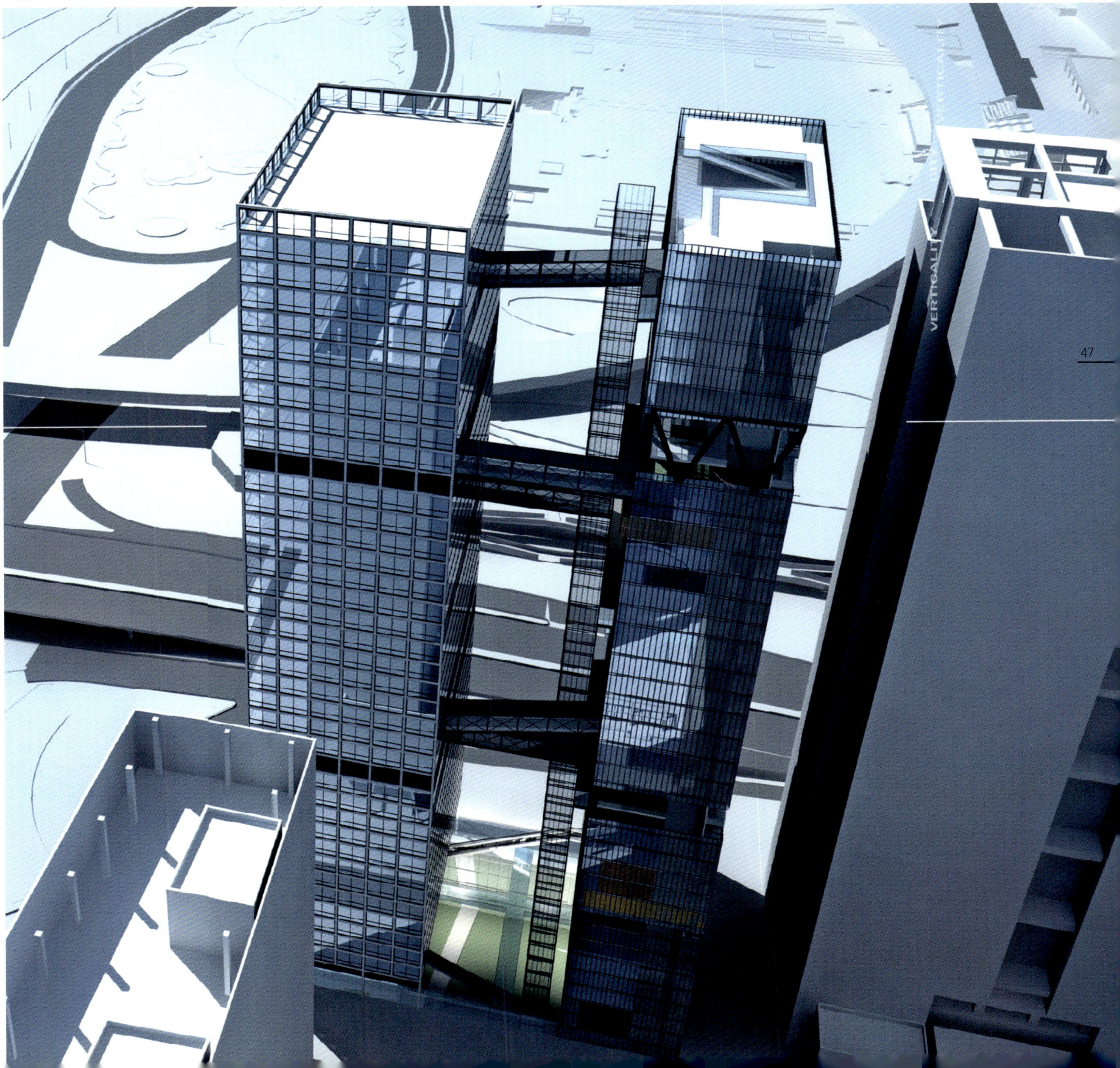

This is not only a training center but a building complex with office, exhibition and living functions. Since it is a carrier of corporate image, we strive to give it a unified image and intrinsic mechanism which stays in perfect harmony with nature.

它不仅仅是培训设施，而是集办公、展示、居住于一体的建筑综合体；作为企业形象的载体，我们力求给予它统一的形象以及同自然和谐共生的内在肌理。

Nuclear Power Office Clerk Training Complex · Yangjiang

Yangjiang, Guangdong | Project Location
2009 | Project Date
55,000m² | Project Scale
competition | Design Phase
construction not started | Project Status

核电办公培训设施·阳江

中国，广东省，阳江市 | 项目地点
2009年 | 设计时间
5.5万m² | 项目规模
方案投标 | 设计阶段
未建 | 项目现状

项目总览

该项目位于阳江东北郊唯一的森林公园处，北倚石塔山，南临共青湖水库。包括办公楼、培训中心、综合楼、公寓、食堂等建筑。总建筑面积约217,532m²，分三期建设，一期工程包括办公楼(6层)、培训中心(6层)。

Project Overview

Located in the only forest park in northeast suburb of Yangjiang City, this complex backs on to Shita Mountain in the north and neighbors Qinghu Reservoir in the south. It includes such building as the office building, training center, multi-functional building, apartment, and dinning hall, etc. With a total building area of about 217,532 square meters, this complex is built in three phases. The phase-I project includes construction of the office building (6 floors) and the training center (6 floors).

构思与概念分析

阶段模型1　　　　阶段模型1　　　　规划模型　　　　一期模型

思考，我们需要一座怎样的建筑去代表企业形象……

项目分析

项目的主要功能是培训与办公，要求在空间划分上对两种功能进行组织，但在景观的营造上有迥然的区别。设计利用水平分区，做到动静分区明确，管理方便。同时，利用开放的屋顶空间将两者联系起来，形成一个交流的场所。

由于考虑到分期开发和将来拓展的因素，项目一期设置在城市主干道的交汇处，为以后的开发形成良好的延伸面。

模型1

模型2

Project Analysis

This complex is used mainly for training and office work. The two functions are required to be organized with a spatial division. Their landscaping should be quite different. The design makes use of horizontal division to realize clear division between activity areas and quiet areas for convenient management. The open roof space links both areas to form a place for communication.

In consideration of such factors as phase-by-phase development and future expansion, the phase-I project is near the intersection between the main roads of the city. Thus easy expansion is available for future development.

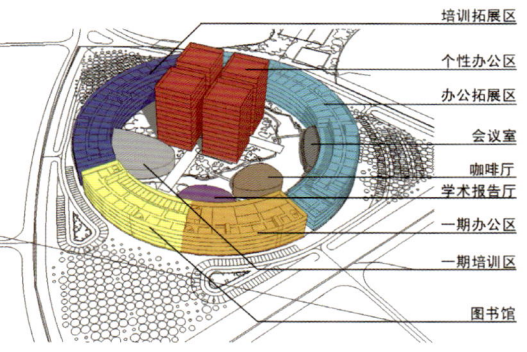

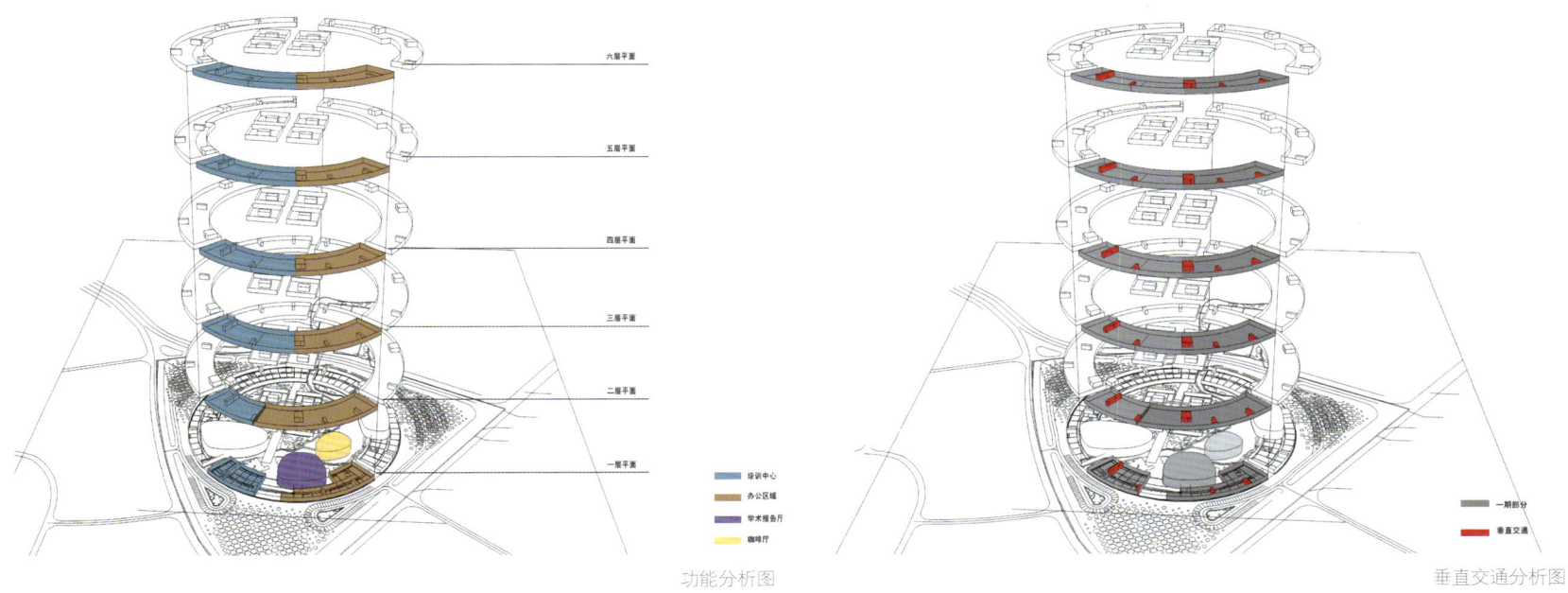

功能分析图

垂直交通分析图

立面图

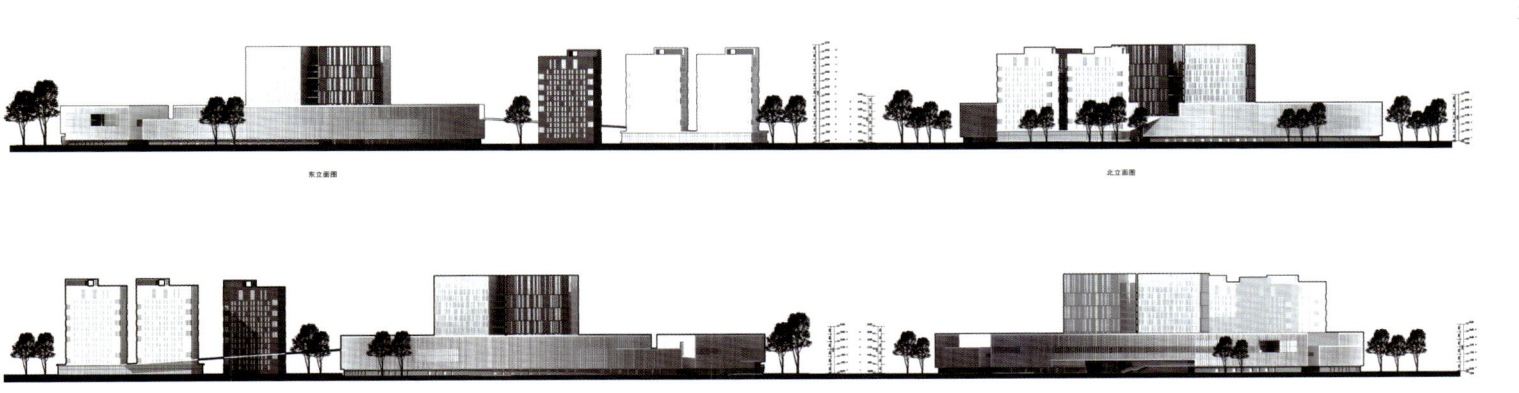

办公楼层平面

办公培训楼功能多样，多元结合在一起形成城市综合体。它是城市的，社会的，是人与自然的。我们在方案中提供了绿色空间的共享，让不同层次进行不同活动的人都可以享受得到平面功能的组合。

With a variety of functions, office training buildings are built together to form a urban building complex, which is of urban, social, human, and natural styles. Our design enables residents to share a green space and people at different levels who engage in different activities to access to a combination of plane functions.

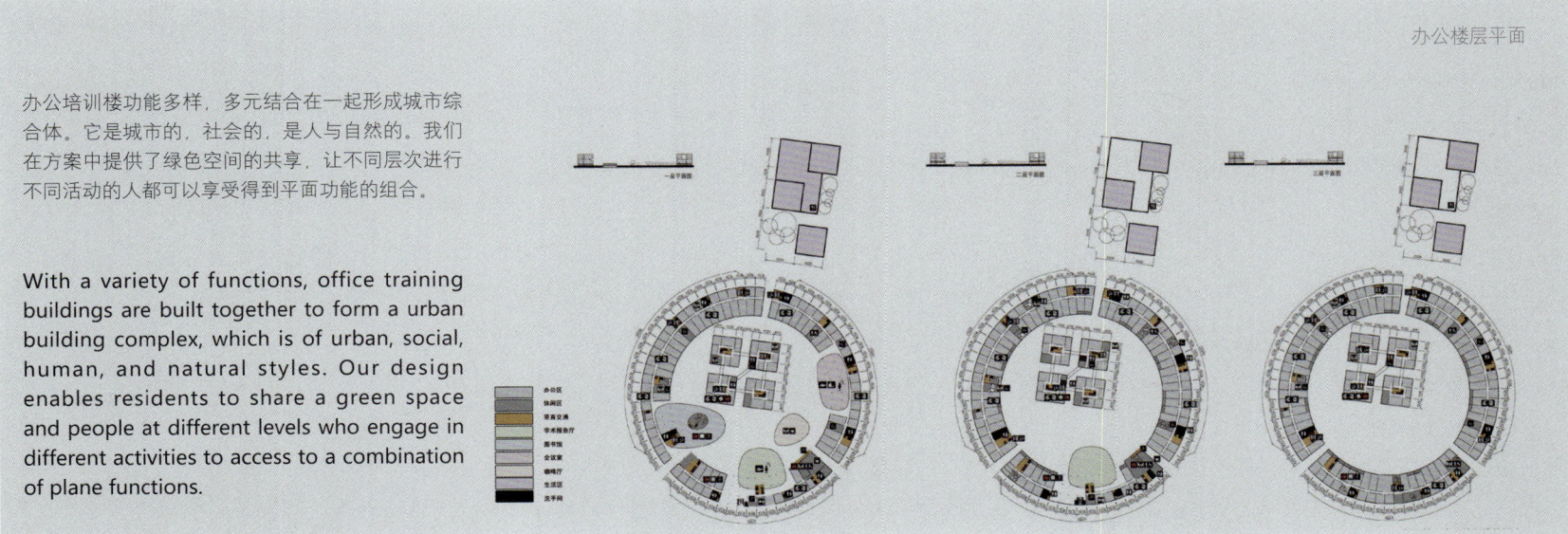

办公培训楼的不同功能单元包括培训教室、实验室、休闲区、办公区、个性化办公区、办公拓展区、培训拓展区和生活服务区及与其相应的配套设施。它们各自相对独立地占据了部分空间，我们通过一个共享的室外中庭把不同的功能单元组合在一起，使它们之间相互有便捷舒适的通道，让它们融合在一起，组成一个完整的综合体。

培训办公区，我们通过底层的架空花园和共享的中庭，使建筑成为一个漂浮在绿色上的盒子。

个性办公区，我们通过大量的空中花园把各个办公区独立开来，使建筑成为一个淹没在绿色中的盒子。

With a variety of functions, office training buildings are built together to form a urban building complex, which is of urban, social, human, and natural styles. Our design enables residents to share a green space and people at different levels who engage in different activities to access to a combination of plane functions.

Training office area: we design an overhead garden and sharable atrium on the ground floor. Thus the building looks like a box floating on a green space.

Personalized office area: we design lots of aerial gardens to separate office areas from each other. Thus the building looks like a box covered by green.

透视1

透视3

The building strives to express the features of and the friendship between two cities as well as the particularity of the project site.

建筑力求表达其所承载的两个城市的特点和友谊,同时也能体现出建设地点的独特性。

Jintang Building · Tianjin

Tianjin | Project Location
2007 | Project Date
125,000m² | Project Scale
conceptual design | Design Phase
construction in progress | Project Status
Shenzhen Machinery Institute Architectural Design Co., Ltd. | Design Partner

金唐大厦·天津

中国，天津市 | 项目地点
2007年 | 设计时间
12.5万m² | 项目规模
方案设计 | 设计阶段
在建 | 项目现状
机械部（深圳）设计研究院 | 合作单位

项目方位图

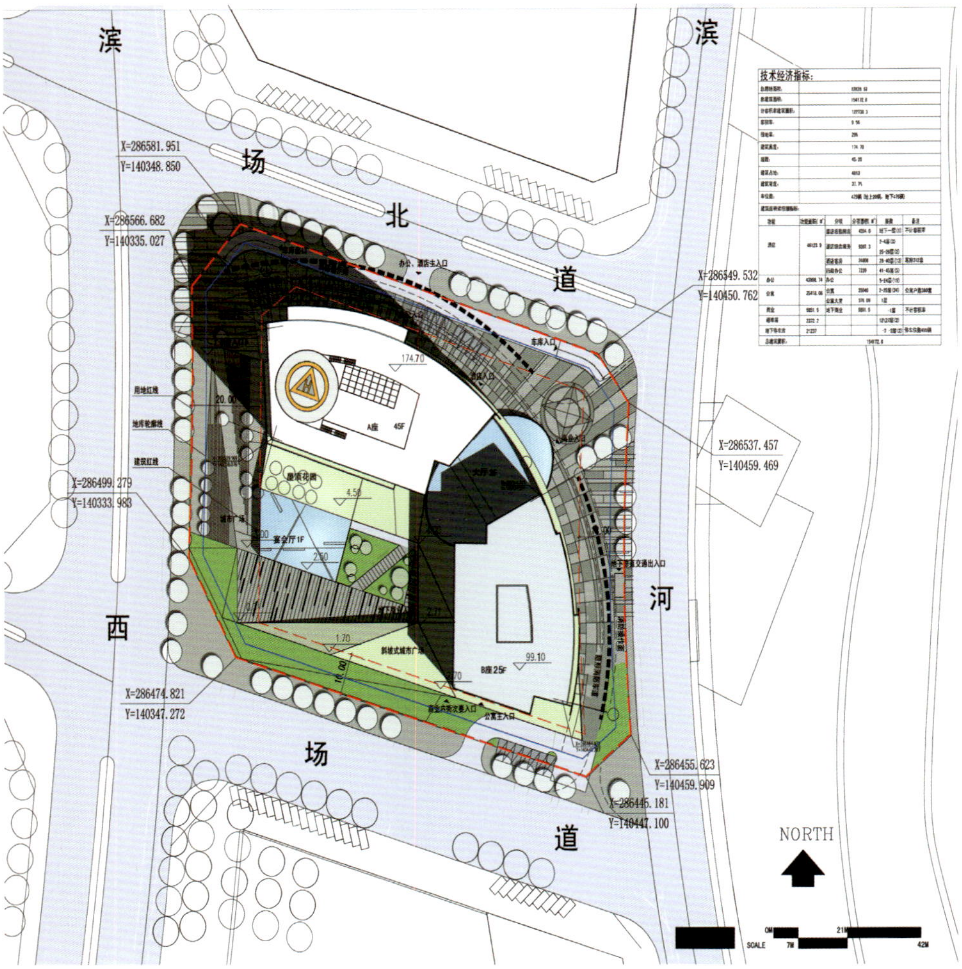

总平面

项目总览

建筑的整体造型犹如巨大的帆船，外立面设计如同波光粼粼的海面。设计中对海洋的比拟，对风帆的引喻，代表着一种乘风破浪、勇往直前的力量。建筑立面表皮从下至上，由实变虚，渐渐融合到天幕之中。弧形的建筑布局，实现了景观与周围环境的最大化利用。该建筑具有多样化的功能，包括了五星级酒店、办公空间以及高级公寓。人们可以通过观光梯抵达位于100m高的空中酒店大堂，欣赏浩瀚的海河景观，也可以步入200m高的行政楼层，享受一览众山小的感觉。该项目中包含地铁以及地下公共车行道的连接，使建筑本身真正融入到这个社区，设计还提供了一个独一无二的半室内半室外垂直走向的空中街市，巧妙地把几个不同的建筑结合成了一个完整、和谐的建筑群体。

Project Overview

The building looks like a giant sailboat and its facade is designed as a shimmering sea. The sea and sail represent the ever advancing force of the wind and the waves. From the bottom to the top, the facades look like that there is a gradual transition from reality to mirage, and finally integrates with the sky. The arc-shape layout allows maximum utilization of landscapes and surroundings. This building has a variety of functions, in which there is a five-star hotel, offices and high-class apartments. People can take the sightseeing elevator to reach the hotel lobby on the floor that is 100 meters high to view the vast ocean and river landscape or go to the administrative floors above 200 meters to enjoy the birds' view. This building is designed with a passage connecting with the subway entrance and underground bus lanes, thus the building has easy access to the neighboring communities. In addition, a unique aerial street partly indoor and partly outdoor is designed to link several towers together to form a complete and harmonious building complex.

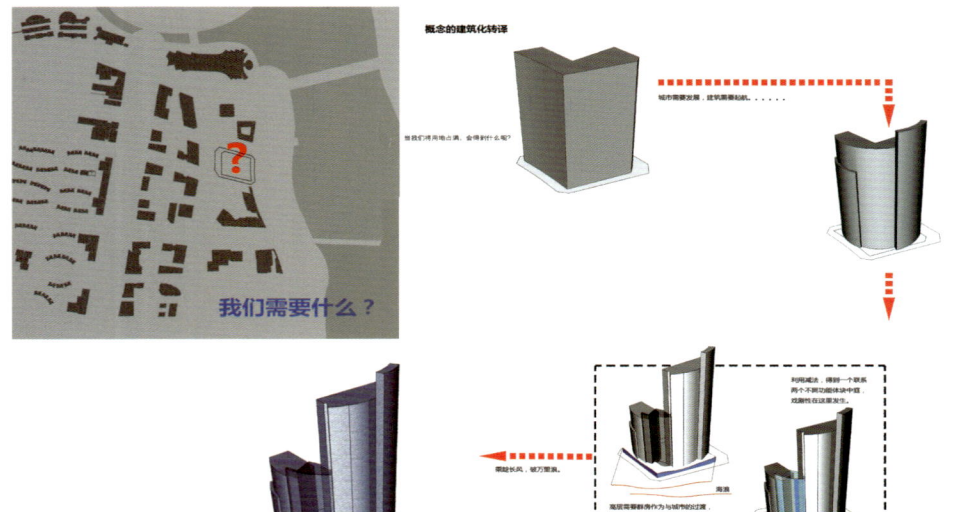

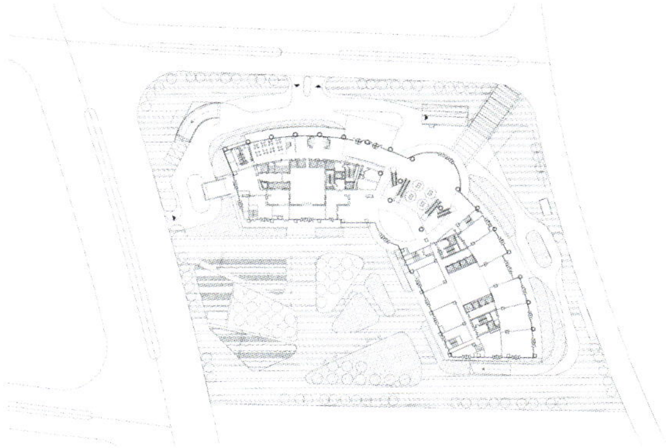

一层平面

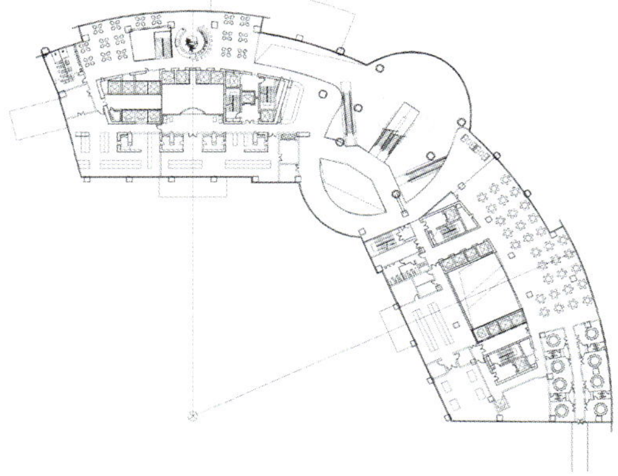

二层平面

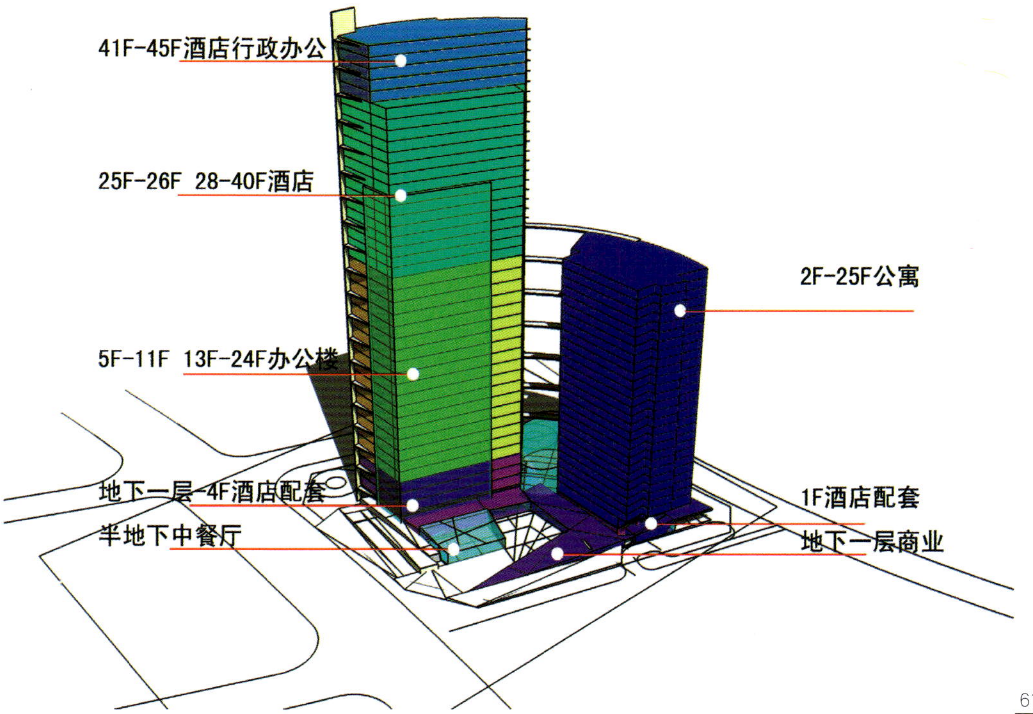

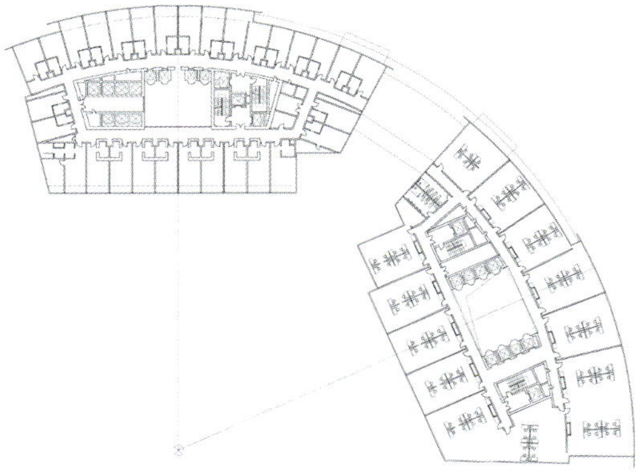

六至十二层平面

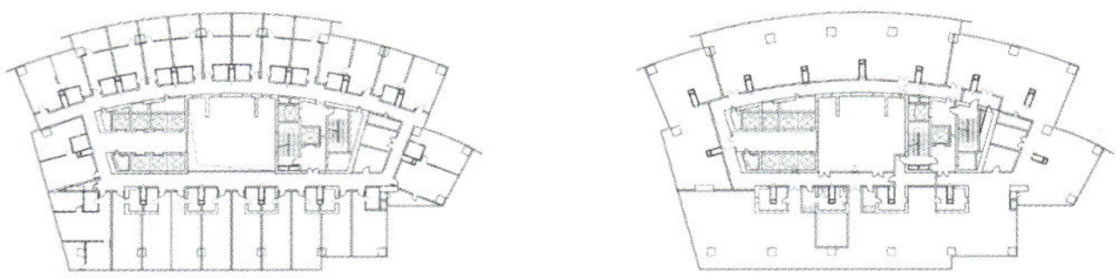

三十二层平面 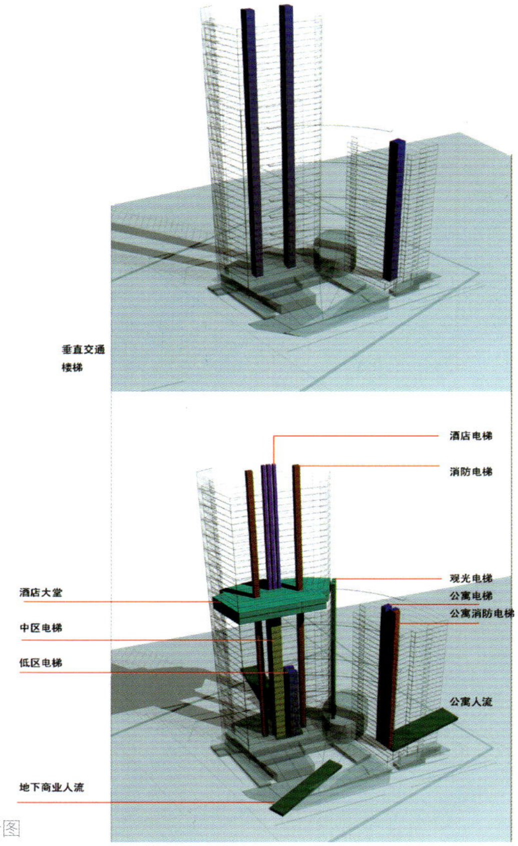　交通空间分析图

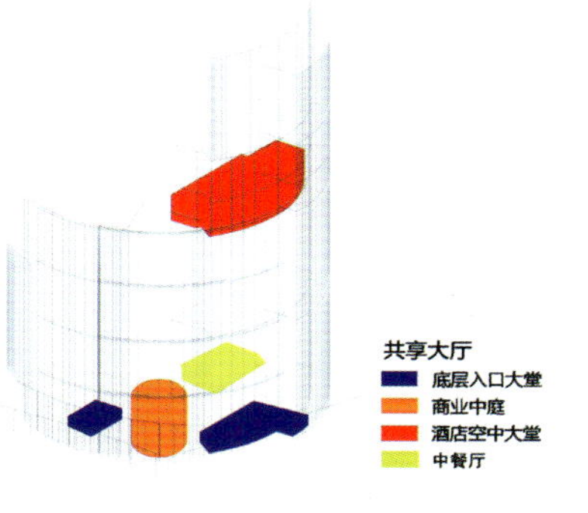

41F-45F 酒店行政办公

25F-26F 28-40F 酒店

2F-25F 公寓

5F-11F 13F-24F 办公楼

地下一层-4F 酒店配套

1F 酒店配套

半地下中餐厅

地下一层商业

流线分析图

垂直交通楼梯

酒店电梯
消防电梯
观光电梯
公寓电梯
公寓消防电梯

酒店大堂
中区电梯
低区电梯

公寓人流
地下商业人流

共享大厅
- 底层入口大堂
- 商业中庭
- 酒店空中大堂
- 中餐厅

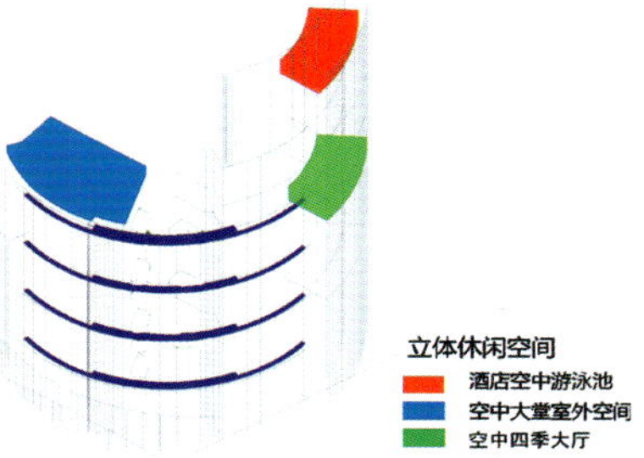

立体休闲空间
- 酒店空中游泳池
- 空中大堂室外空间
- 空中四季大厅

西立面　东立面　南立面　北立面

立面图

立面展开图　　B座侧立面

The former design hopes to offer a building which shows full respect for the neighborhood but as a result it presents a common place architectural form.

原来的设计希望能提供一个对"邻居"有着充分尊重的建筑，结果却是一个普通的建筑形态。

Liantai Building • Shenzhen

Shenzhen | Project Location
2004 | Project Date
25,000m² | Project Scale
conceptual design, preliminary design, construction drawing | Design Phase
construction completed | Project Status
Shenzhen Machinery Institute Architectural Design Co., Ltd. | Design Partner

联泰大厦·深圳

中国，广东省，深圳市 | 项目地点
2004年 | 设计时间
2.5万m² | 项目规模
方案设计 初步设计 施工图设计 | 设计阶段
已建 | 项目现状
机械部(深圳)设计研究院 | 合作单位

项目总览

联泰大厦建设用地为60m见方的正方形，由于周边道路及城市空间的不同，所以它不是一个均质的地块。

通过对用地的分析，我们认为应尽量多地争取南向，用好东向、西北向，最好让每个单位都能看到海。综合起来就是，既要满足城市设计的要求，也要满足开发者、使用者对建筑空间的需求。

Project Overview

The land on which Liantai Building is built is a square with 60 meters sides. But due to the complexity of surrounding roads and space, it is not a plot with homogeneous sides.

Through analysis of the land, we think we should make more rooms face southwards, and create good designs for the eastward and northwestward rooms. It is better to make all apartments face the sea. In a word, we will not only satisfy city planning requirements but also the requirements of the developer and users on this architectural space.

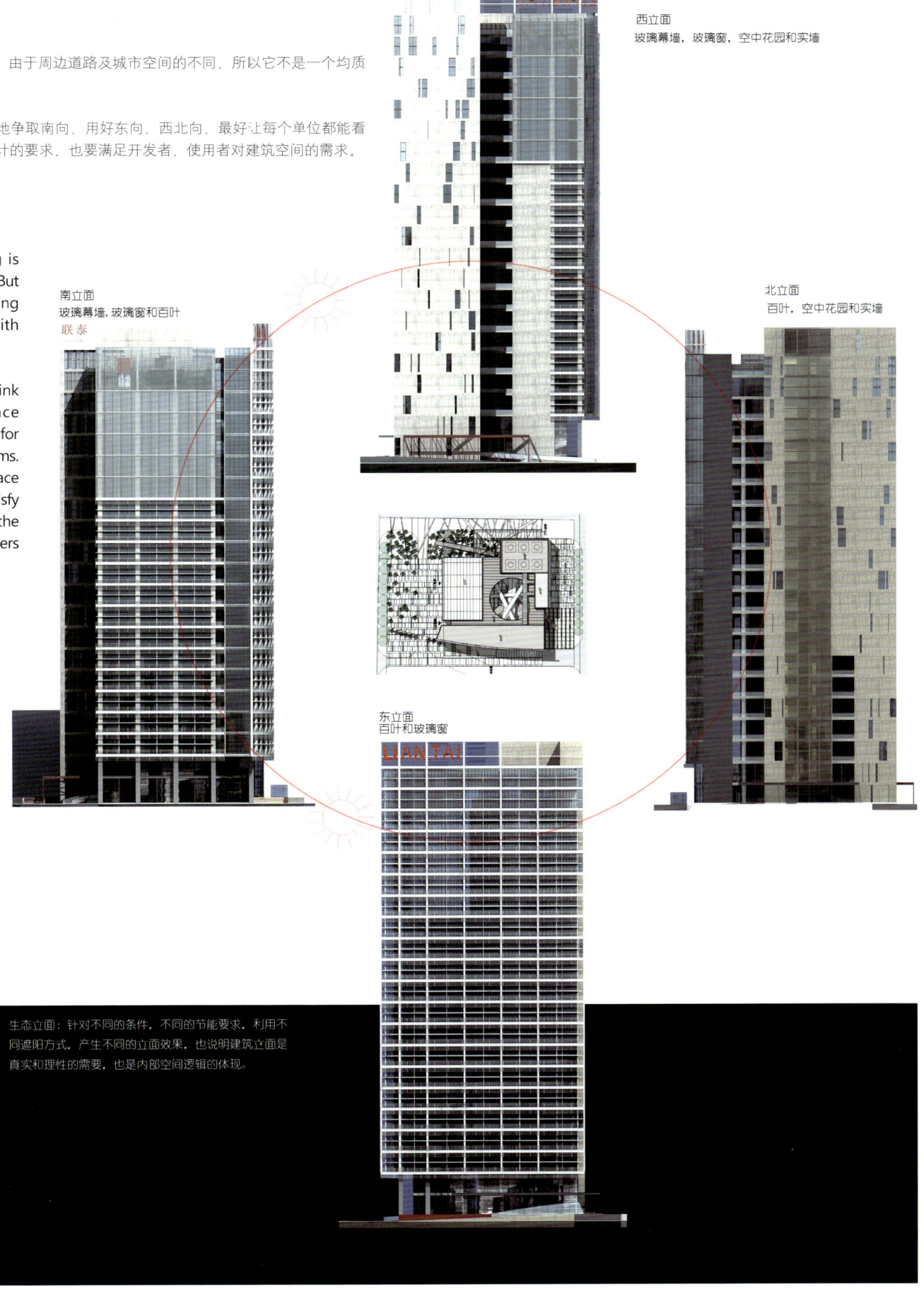

西立面
玻璃幕墙，玻璃窗，空中花园和实墙

南立面
玻璃幕墙，玻璃窗和百叶

北立面
百叶，空中花园和实墙

东立面
百叶和玻璃窗

生态立面：针对不同的条件，不同的节能要求，利用不同遮阳方式，产生不同的立面效果，也说明建筑立面是真实和理性的需要，也是内部空间逻辑的体现。

生态设计立面

设计原则

满足城市设计呼应角色，与周边建筑形成良好的对话，体现城市设计的完整性，以期提供一个联络南北的公共城市空间，使本建筑以及该区域城市获得活力。通过竖向立体中庭的引入，令办公单元享有最佳的建筑物理性，同时也能提供较大的建筑体量，满足开发商建筑体量标志性的要求。

Design Principles

To make urban design live in perfect harmony with neighboring buildings and show the completeness of urban design so as to provide an urban public space connecting south and north and make the building and the region become more dynamic; to introduce a vertical three-dimensional atrium that give the best architectural rationality to offices, and also offer larger architectural structures to meet developer's requirements for architectural structure.

共享中庭造成每个房间都有良好的自然通风

空中绿化

每个房间都拥有东南向景观

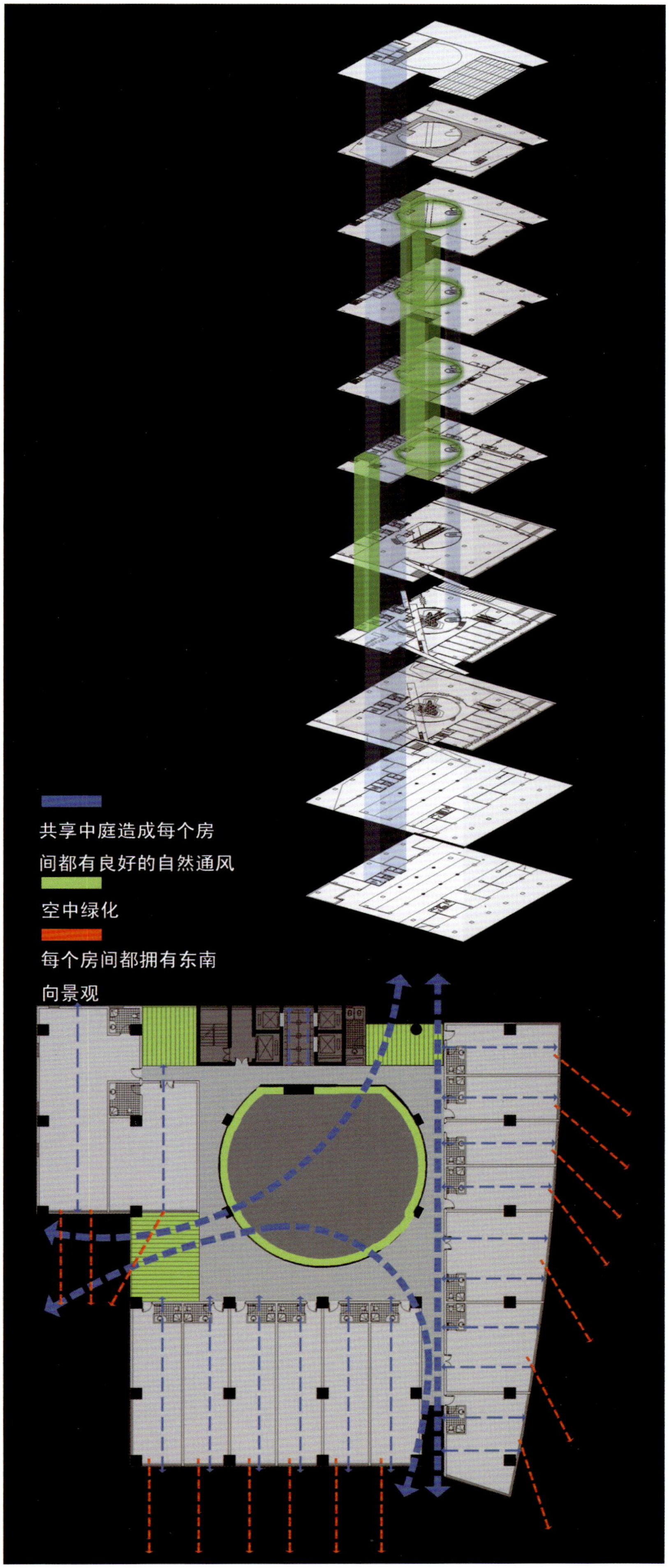

The building stays in perfect harmony with nature. The design featuring sustainable development offers a long service life of the building.

建筑与自然和谐共生,可持续性发展能为建筑提供更长的使用寿命。

Taitai Pharmaceutical Office Building · Shenzhen

Shenzhen | Project Location
2002 | Project Date
11,000m² | Project Scale
conceptual design | Design Phase
construction not started | Project Status

太太药业办公楼·深圳

中国，广东省，深圳市 | 项目地点
2002年 | 设计时间
1.1万m² | 项目规模
方案设计 | 设计阶段
未建 | 项目现状

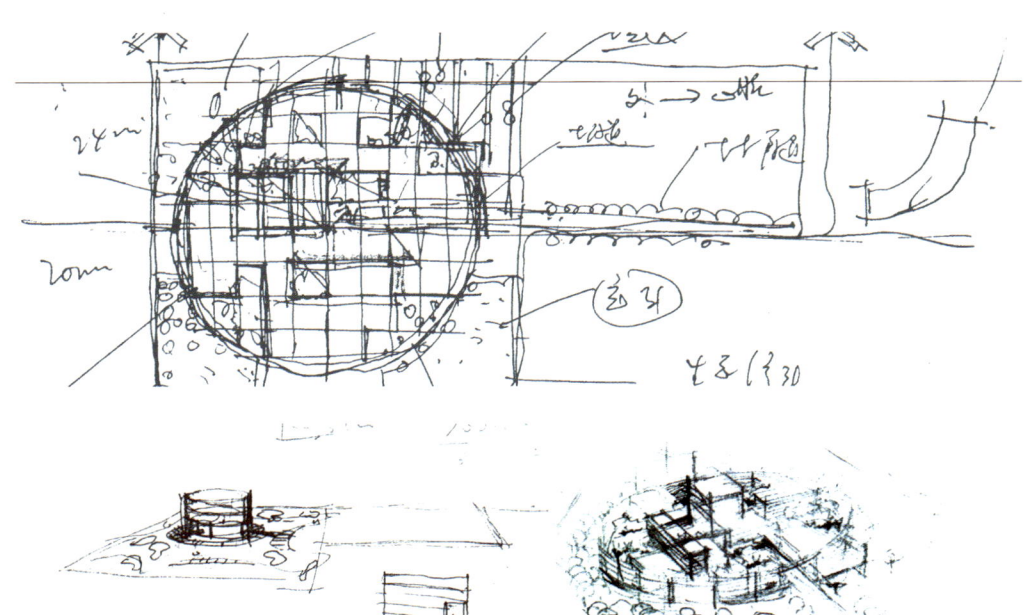

项目总览

建筑不是作为一个简单的以自我为中心的个体凌驾于环境之上的，它的功能空间以群体的形式分散于郁郁葱葱的各部分绿色森林中。此项目不是在建筑外布置绿地，而是在森林中建房子。建筑群总高度控制在24m以下，由不同高度的模块组成，使树木可以高过建筑，大大体会到树荫下办公的生态氛围。同时，24m限高有效节省了投资，减少了土建、消防、人防等投入。

圆形的功能性模糊界面，强化了本项目在科技园区的中心感，又和谐地沟通了厂区及周边其它建筑和区域，使其在聚与散、开与合、交流与内敛中达到一个平衡。

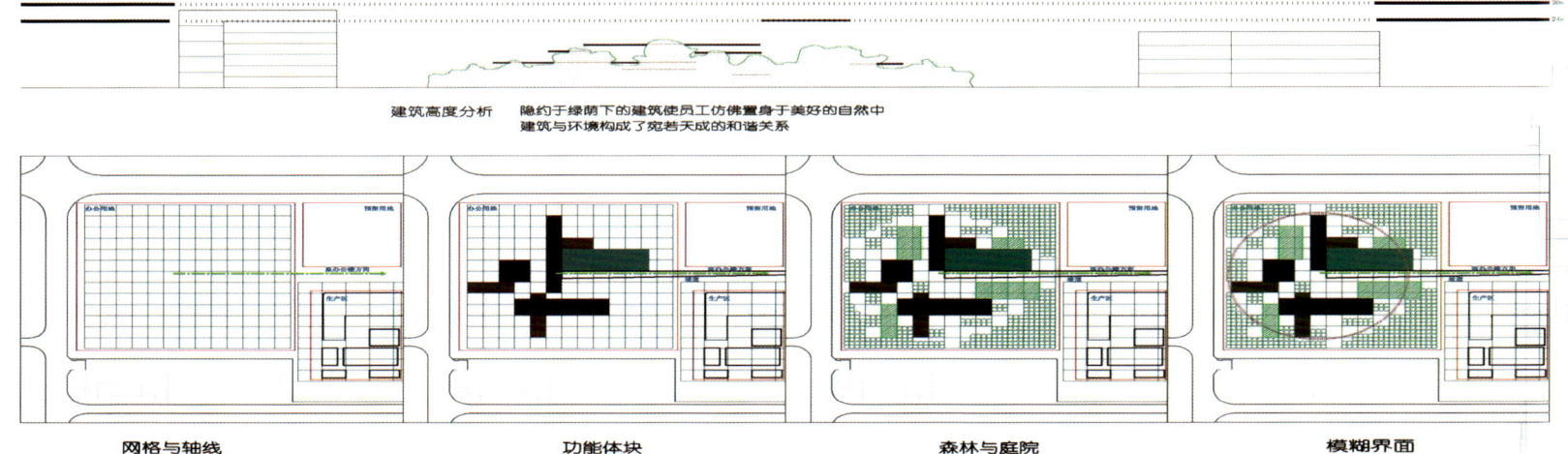

建筑高度分析　隐约于绿荫下的建筑使员工仿佛置身于美好的自然中
建筑与环境构成了宛若天成的和谐关系

| 网格与轴线 | 功能体块 | 森林与庭院 | 模糊界面 |

总平面

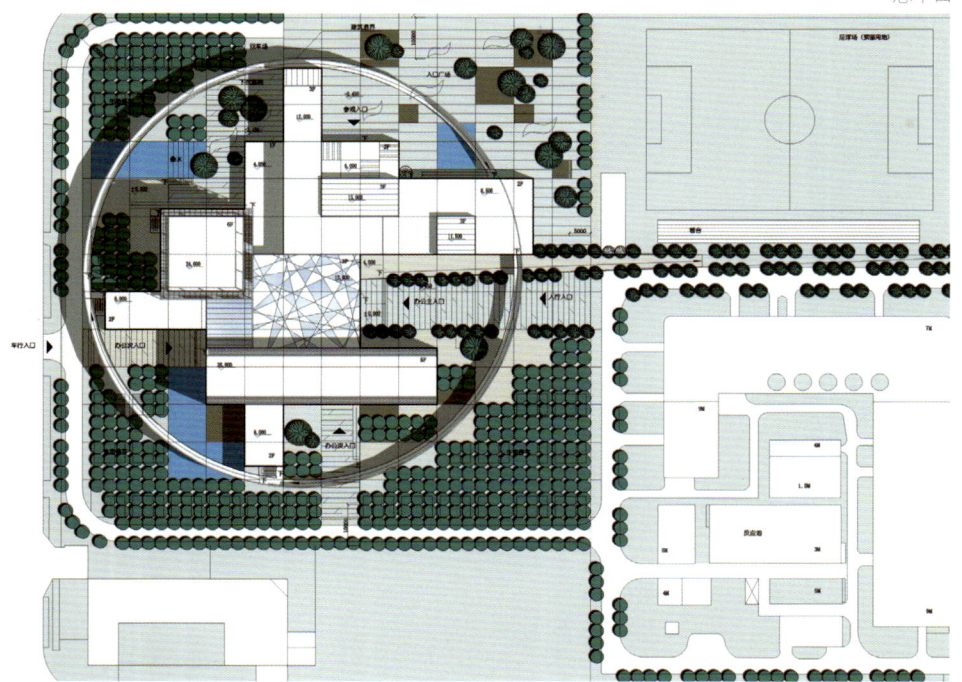

Project Overview

The buildings are not self-centered structures that separate themselves from their environment. The functional modules, although scattered among the woods, form an integrated group. For this project, no green belts is arranged outside the buildings but to build in the woods. The building complex is made up of modules of different heights, but no more than 24 meters. Thus trees are higher than some modules and people feel as if they are working under the trees. Meanwhile, the height limit of 24 meters greatly saves investment because less money is needed for labor, fire protection, and air defense.

The functional modules with a circular facade intensify the central location of this project in the Hi-Tech industrial park and also create perfect harmony with the factory area and its neighboring buildings, realizing a balance between gathering and parting, opening and shutting, expressiveness and undemonstrativeness.

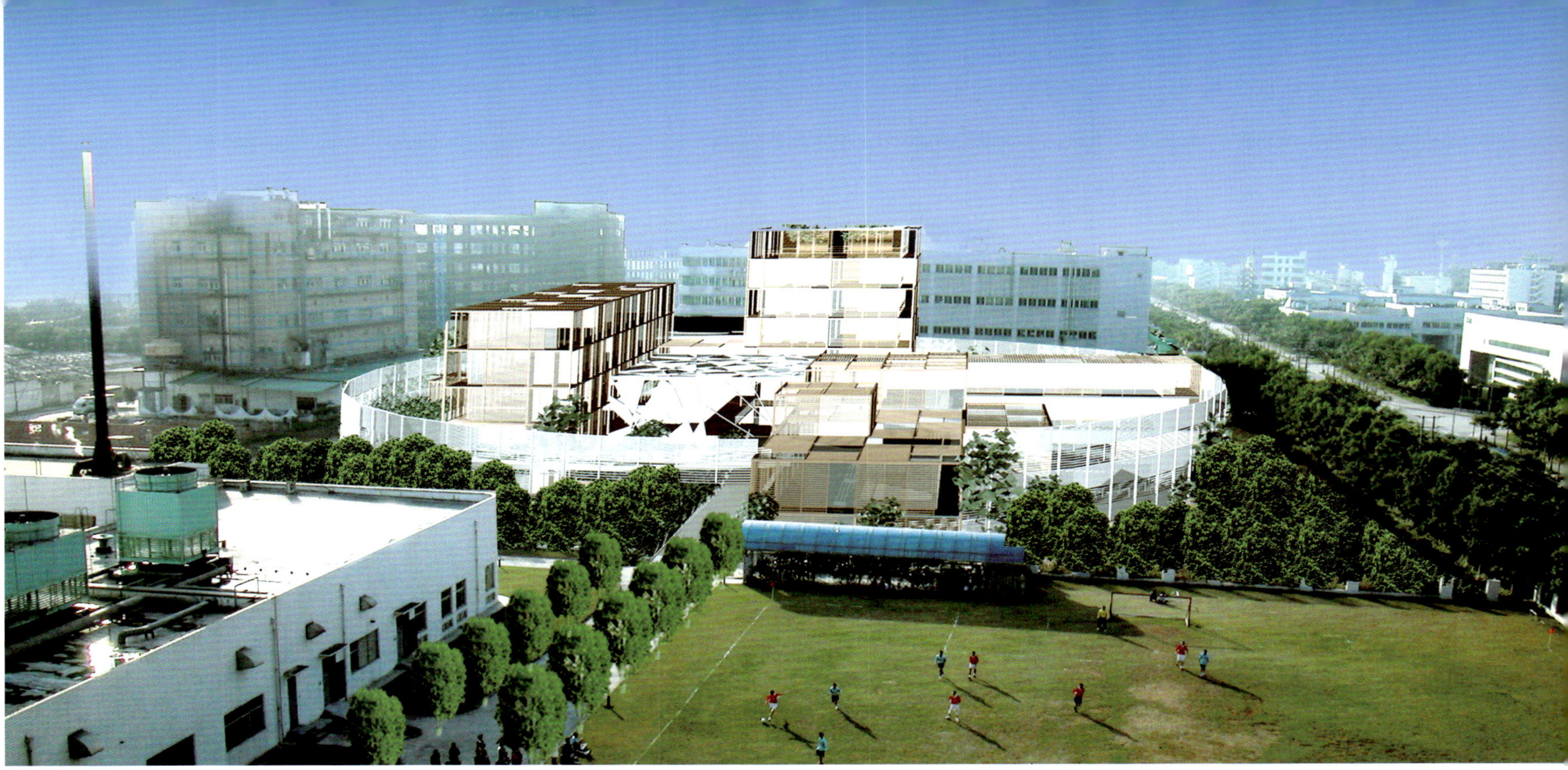

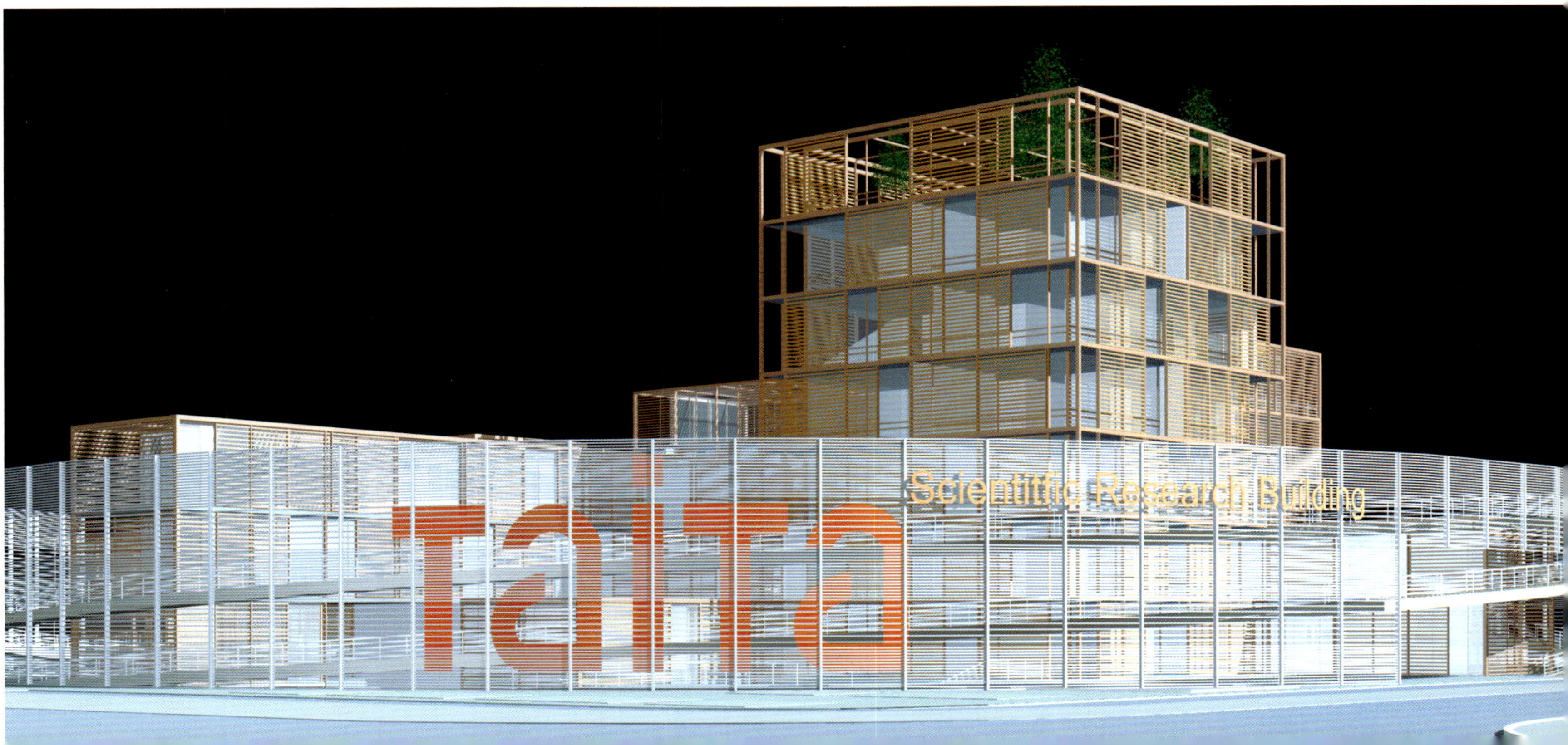

74

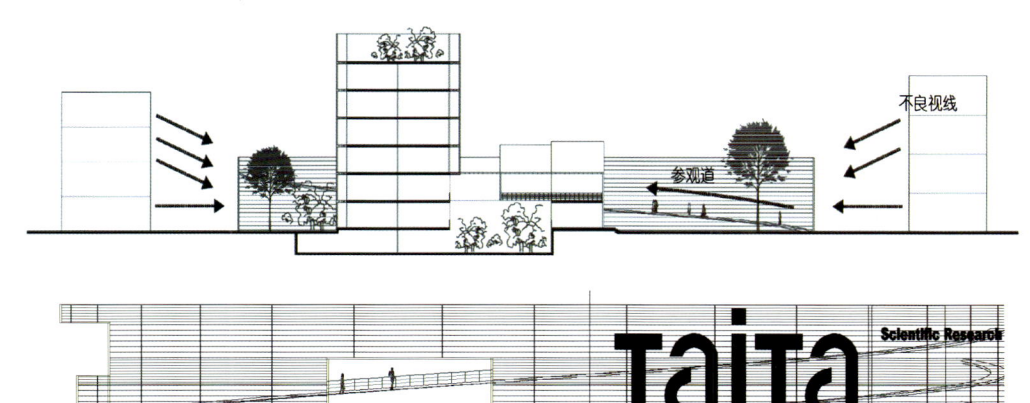

外部空间

由百叶格栅和三段环形盘旋而上的观景坡道组成的环形观景道，将建筑群及其庭院外部空间予以模糊界定，使空间层次更加多样化。坡道徐徐而上，与建筑有机联系。可供客人参观树林中的生态厂区，也可供员工休闲漫步，形成了曲线型的动态趣味空间。

External Space

The spiral-shaped sightseeing passage, installed with blinds and made up of three sections of ascending spiral sightseeing ramps, indistinctly divides the building complex from the exterior courtyard space, showing more diversified spatial levels. The ramps twist up to connect with the building organically. Guests may walk on the ramps to visit the ecological factory area in the woods. Employees may also have a walk on the ramps for leisure, which are of curve-shaped dynamic space.

模块化单元

数字化时代与其理论的共同之处在于基本单元的重组、排列形成多样的世界。从建筑空间、环境、办公单元及外部空间尺度来看，9m×9m×4m的结构单元灵活而实用，并且相同模块的单元，构件尺度统一，较为经济。

Modularized Units

Something in common between the digital age and the attribution theory is that basic units are restructured and rearranged to form a diversified world. From an angle of architectural space, environment, office unit, and external spatial dimension, 9m×9m×4m structural units are flexible and practical. Since units made of same modules have uniform scale, they are cost-effective.

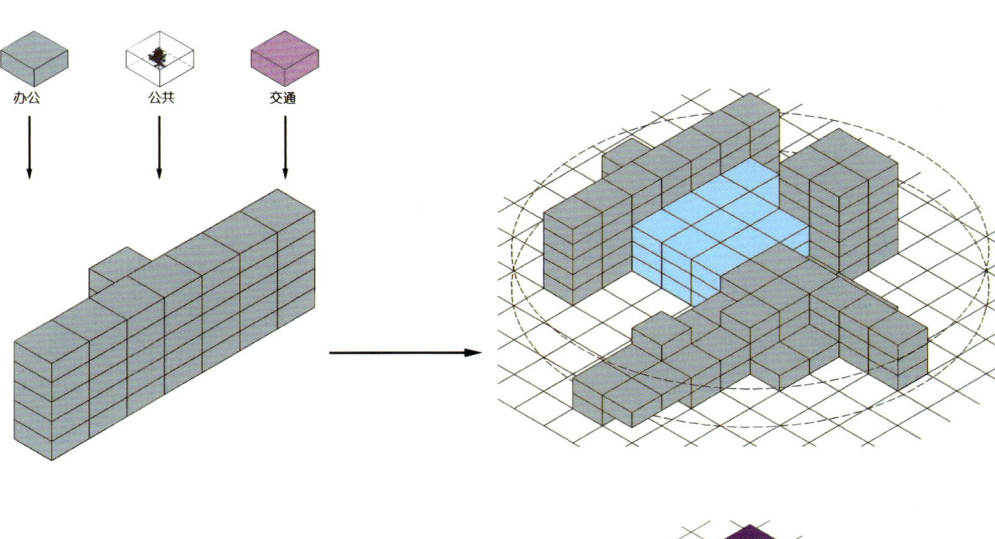

功能模块分析

This is the first part of the building complex planned as a whole. We design it as an integral part of a park and urban landscape.

作为整体规划建筑群的第一部分，我们将其设计成为公园和城市景观的一部分。

Huizhou Daily Newspaper Office · Huizhou

Huizhou | Project Location
2008 | Project Date
113,000m² | Project Scale
conceptual design | Design Phase
construction not started | Project Status

惠州日报社·惠州

中国，广东省，惠州市 | 项目地点
2008年 | 设计时间
11.3万m² | 项目规模
方案设计 | 设计阶段
未建 | 项目现状

场地现状

项目总览

本项目运用现代建筑的手法，糅合中国传统"书院"与江南庭院建筑的结构特征，打造了一幢极富现代气息，又能体现报业媒体主题的区域地标。总体建筑形态采用分布于基地南北两端的半围合"院空间"布局。在一、二期分别围合成各自独立的主题景观庭院。一、二期建筑体量夹合生成中心生态媒体园区，契合中国传统书院空间特征。南北两侧庭院景观向中心媒体园区交辉渗透，结合一条南北向生态绿轴，贯穿整个基地。各单体建筑由庭院空中天桥自由连接，在人们充分享受景观的同时，提升办公机能效率，体现报业大家庭的整体性。

Project Overview

By applying modern construction techniques and integrating structural features of traditional Chinese "academies" with the courtyards south of the Yangtze River, this complex is designed to create a regional landmark showing a strong modern atmosphere and presenting the theme of newspaper media as well. The general architectural layout features a semi-encircled "courtyard space" at the south end and another one at the north end of the complex. An independent themed landscaping courtyard is built respectively in phase-I and phase-II of the project. The phase-I and phase-II buildings encircle a central ecological media zone, complying with spatial designs of traditional Chinese academies. The courtyard landscapes at the south and north ends extend towards the central media zone, forming a north-south ecological axis passing through the entire complex. The individual buildings are connected freely by sky-bridges in the courtyards, thus people can enjoy the beauty of landscapes better; moreover, the office functions and efficiency are improved and thus reflect the integrity of the press community.

总平面

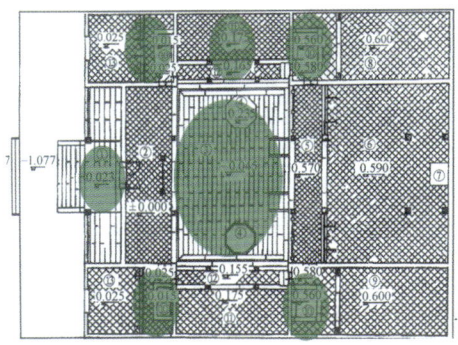
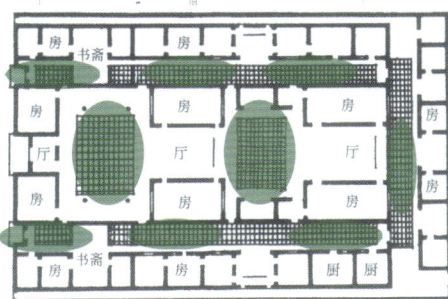

书院　岭南建筑

功能布局

为塑造建筑区域的标识性，本设计将功能地位极其重要的报业大厦及印务中心大楼置于基地最南端，报业大厦与印务中心上下搭叠，一气呵成，尽显雄伟壮观。南北朝向的报业大厦，如一道宽百米的巨大片状体拔地而起，向城市中心区东西横向展开，塑造开放、简洁、动感、力度的现代化报业办公形象，打造高度清晰鲜明的区域地标，同时隐喻报业传媒形象的片状特质。

Layout of Functional Areas

In order to create a landmark in the community, the essential Newspaper Building and Printing Center are designed to be located at the south part of the complex. The Newspaper Building connects the Printing Center at the top and bottom, showing a magnificent and spectacular view. The Newspaper Building with north-south orientation spreads horizontally from east to west in the downtown, just like a tremendous 100m-wide sheet soaring above the horizon. Thus an image of an open, simple, dynamic, strong modern newspaper office and a clear and vivid regional landmark are created. This "sheet" form also implies the newspaper Industry.

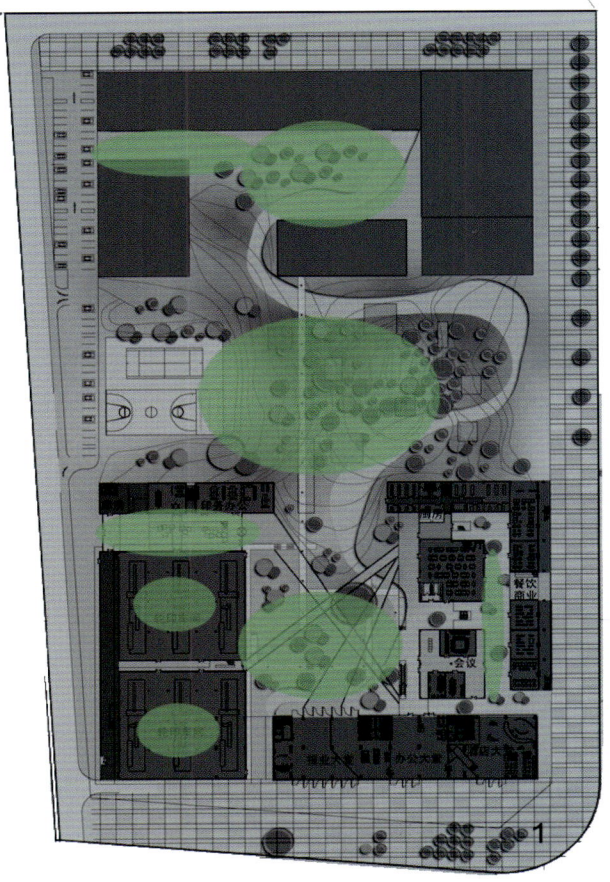

建筑的内在肌理来源于项目原本环境给予的信息，希望项目建成后能唤醒人们对原有环境的回忆。

Intrinsic texture of buildings comes from the information offered by the original project environment. Hope this project can jolt people's memory of former environment after it is completed.

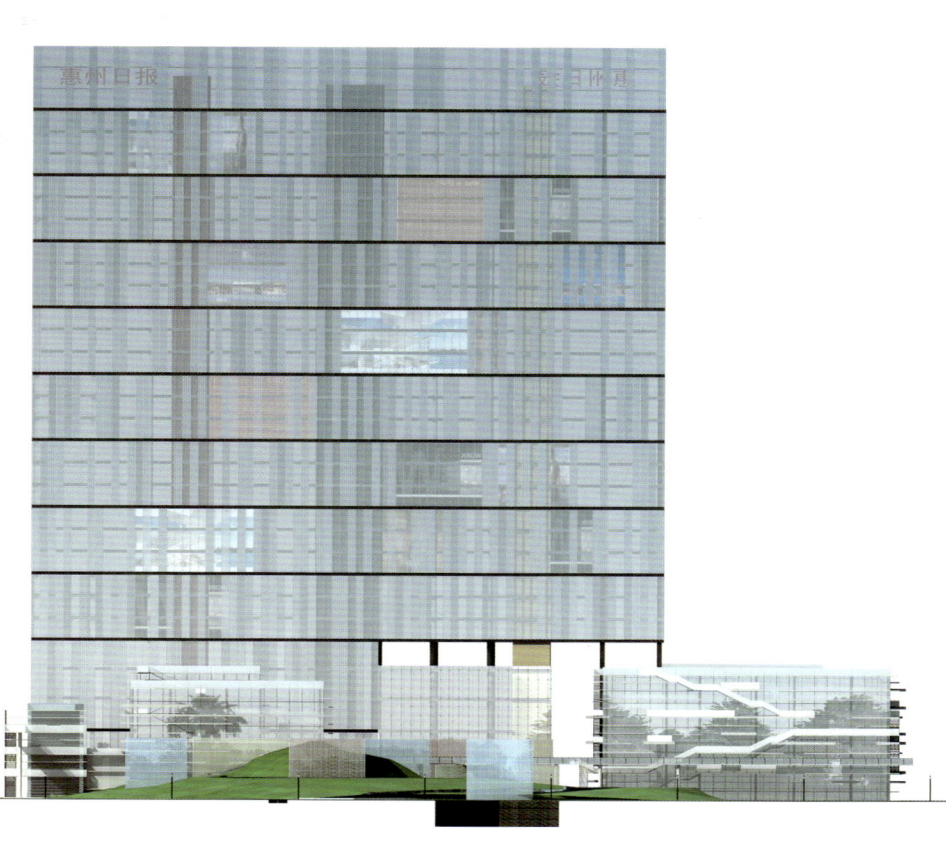

Here we provide an urban public space which is raised up.

在这里，我们提供了一个被举起的城市公共空间。

Bao'an District Construction Quality Inspection Building • Shenzhen

Bao'an District, Shenzhen | Project Location
2003 | Project Date
14,000m² | Project Scale
conceptual design | Design Phase
construction completed | Project Status

宝安区工程质检大楼·深圳

中国，广东省，深圳市，宝安区 | 项目地点
2003年 | 设计时间
1.4万m² | 项目规模
方案设计 | 设计阶段
已建 | 项目现状

本项目旨在打造生态型建筑，构建绿色办公环境。质检站办公楼为城市综合体，功能多样。方案提供了共享的绿色空间，使不同层次的人在进行不同活动时都享受得到绿色带来的舒适。

中庭共享空间是大众交往的场所，完全通透的灰色混凝土构架不仅作为建筑受力体系的一部分，更是建筑表皮呼吸的一部分。外面的阳光、空气、尘埃和来自城市的噪声，通过构架和垂直绿化的过滤后被中庭所接收。

This project aims for constructing an ecological building and creating a green office environment. The quality inspection station building is a complex with a variety of functions. Thanks to a common green space in the building, people can enjoy its comfort when they engage in different activities.

Atrium space is the place for public interaction. The completely porous concrete frames are not only as a part of support structures for the building but also a part of the skin that allows the building to breathe. The atrium receives sunlight, air, dust and noise after being filtrated by the structural frames and its vertical green belt.

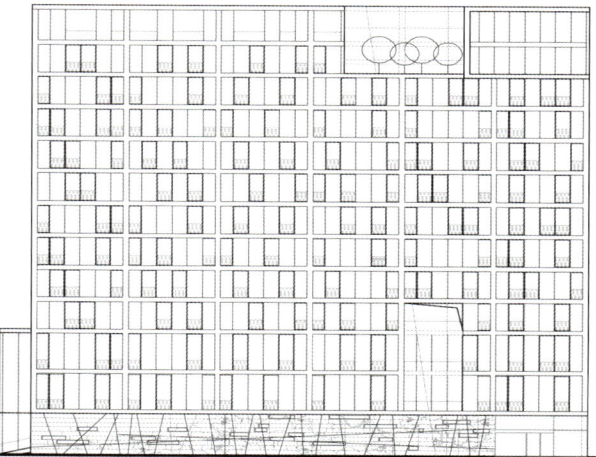

立面图

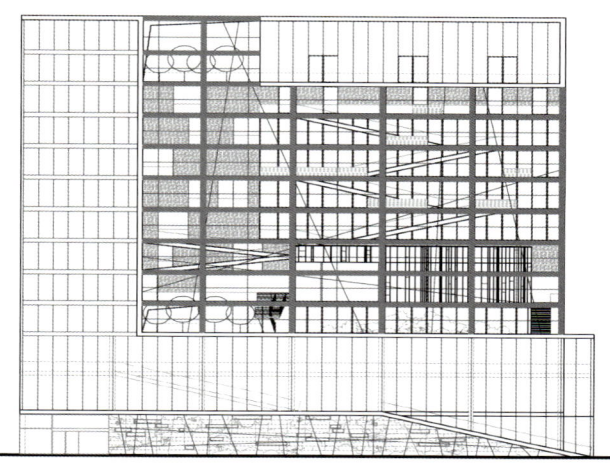

立面图

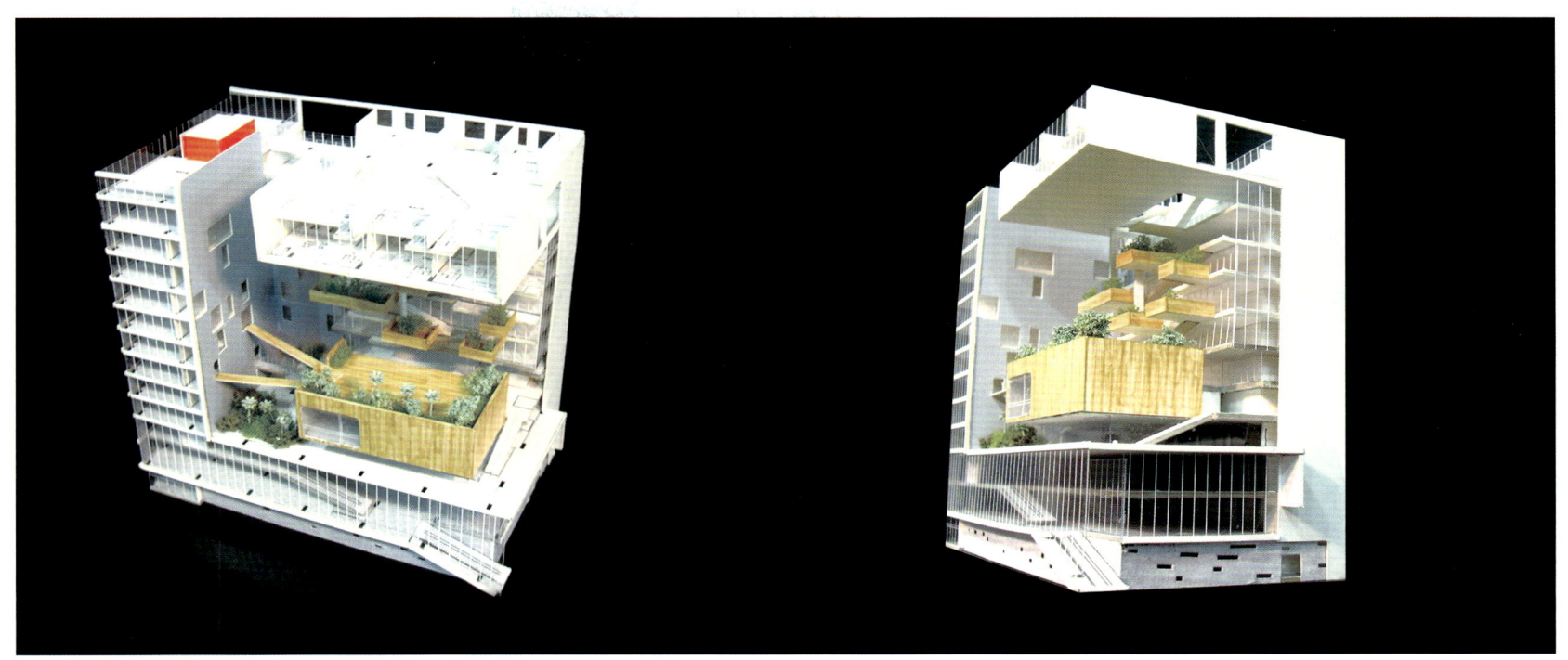

在建筑的底层围绕建筑设计了一圈攀爬的绿色，环抱着整个建筑，使建筑成为漂浮在绿色上的盒子。

This is a building floating on green. A green circle around the building is designed at the ground floor to encircle the whole building. Thus the building looks like a box floating on a green space.

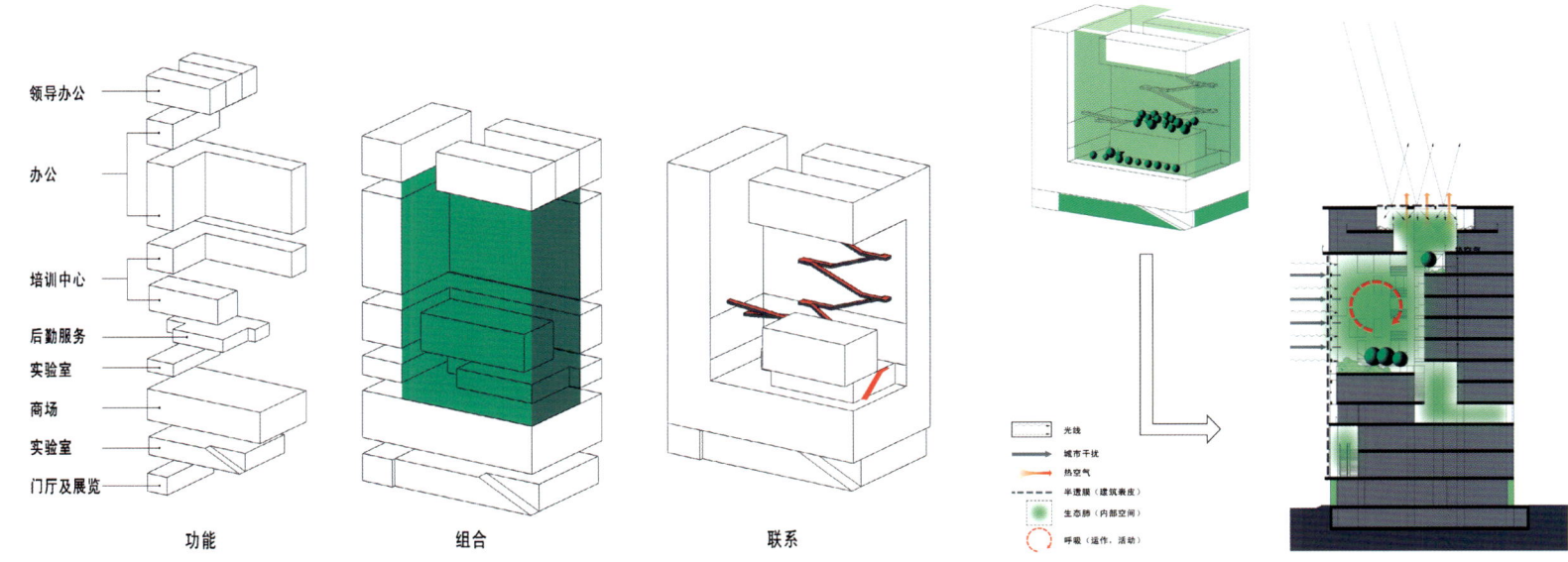

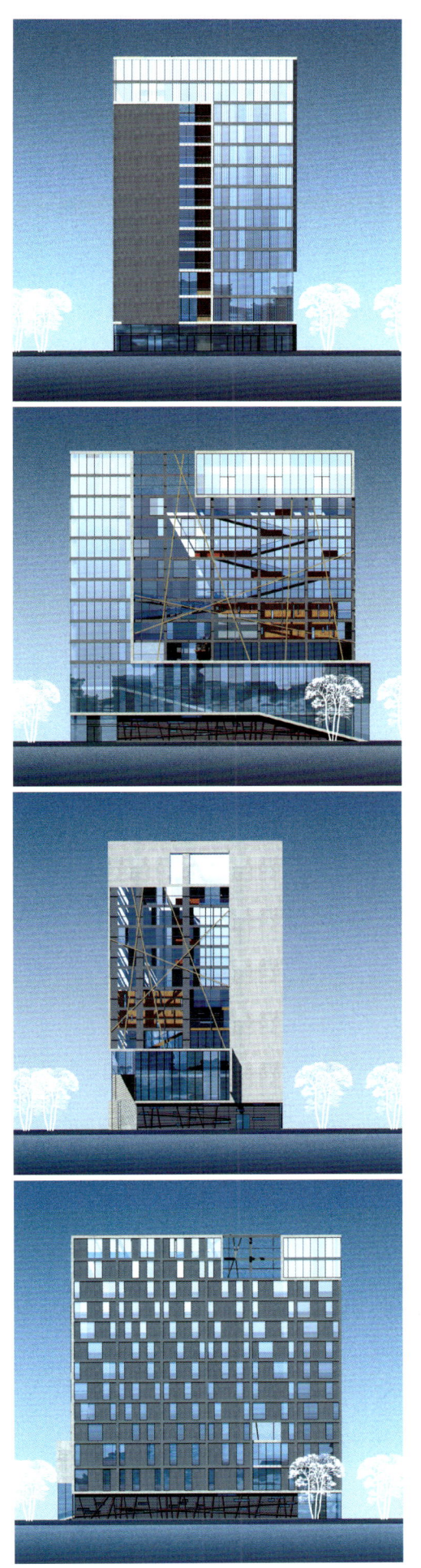

This is a flexible and innovative spiritual space and a place with gathering atmosphere. It has become a region full of designers and art workers of the generation after 1980s.

灵活可变的富有创新的精神空间以及有集聚氛围的场所，这里已经成为80设计艺术工作者扎堆的区域。

Future Plaza • Shenzhen

Qiaoxiang Road, Shenzhen | Project Location
2006 | Project Date
33,000m² | Project Scale
conceptual design, preliminary design, construction drawing | Design Phase
construction completed | Project Status
Shenzhen Machinery Institute Architectural Design Co., Ltd. | Design Partner

花样香年广场·深圳

中国,广东省,深圳市,侨香路 | 项目地点
2006年 | 设计时间
3.3万m² | 项目规模
方案设计 初步设计 施工图设计 | 设计阶段
已建 | 项目现状
机械部(深圳)设计研究院 | 合作单位

设计策划

文化产业基地，现代生态厂房

充分利用南向和东向 纵轴串联各大堂

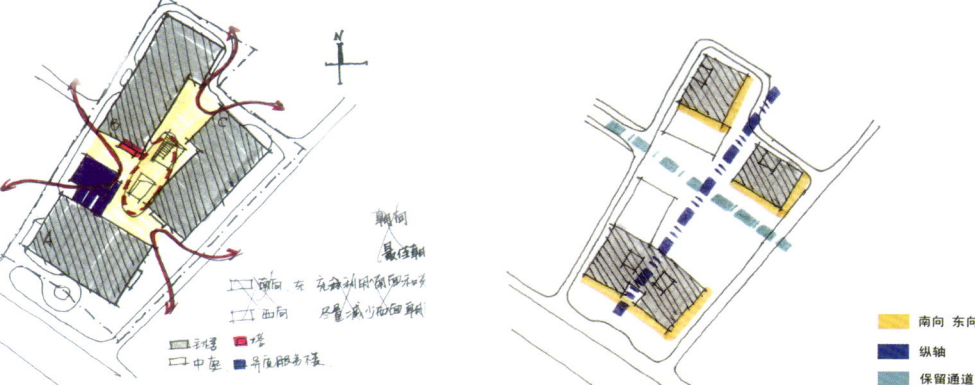

总平面

项目总览

本项目以简约的外在形象，自然通风采光的生态花园，开敞、低密度的工作环境作为设计指导的基本元素。建筑造型上明快、简洁大方，用富有现代感的材料和构图要素营造夺目的现代都市景观。标准的模块化立面单元更是体现了快速发展的工业科技及效率。

Project Overview

In this project, simplistic form, naturally lit and ventilated garden, low density working environment are used as the basic elements of design. The building has a clear, elegant and grand appearance, and uses very contemporary elements of materials and composition to create an eye-catching and modern urban landscape. Furthermore, a standardized modular facade reflects the rapidly-development of industrial technology and efficiency.

功能布局

1、标准层平面
300m²左右的规整矩形为一个单元,以三个这样的单元组合成一个围绕着内天井的平面空间。使交通空间和室外休息空间有开敞的环境和自然的空气,告别单调、黑暗、不通风的走道。

2、工业厂房单元空间
单元层高4.8m,部分空间通高9.6m,不仅可以在平面上布局功能空间,还为使用者在三维立体上提供可能性。在这里,项目为企业的发展扩大留有很大的自由余地,使用者可根据需要创建个性化的工作场所,使空间更具趣味性。

3、公共空间
不同层次的公共空间系统共享天井,成为联系一切公共空间的线索,同时也成为一个容纳各种社会关系的容器,实现自然采光、自然通风、自然绿化。下沉广场、竖向中庭、顶部开放的公共空间,形成连续的竖向序列。

Layout of Functional Areas

1. Standard floor plane
Each unit is a rectangle about 300 square meters. Such three units encircle the atrium. Thus the corridor and outdoor leisure space feature an open environment with natural ventilation, gone are the days of dull, dark and stuffy hallways.

2. Industrial workshop unit space
The unit is 4.8m high and some space can reach 9.6m in height. Functional areas may be arranged on the plane. It is also possible to arrange functional spaces on a three-dimensional basis. A large space is reserved for the future development and expansion of the companies. Users may create personalized workplaces as required, thus making the space more interesting.

3. Public space
The atrium is shared by public space at different floors, which becomes a clue connecting all public spaces and at the same time, a container holding various social relationships. It features natural lighting, natural ventilation, and natural greening. The plaza at lower level, vertical atrium and the open public space at the top form a continuous vertical assembly.

建筑造型

建筑造型明快，简洁大方，用富有现代感的材料和构图要素营造夺目的现代都市景观。标准的模块化立面单元更是体现了快速发展的工业科技及效率。

Architectural Form

The building shows a straightforward and concise form. Modern materials and picture components are used to create eye-catching modern metropolitan landscapes. Standard modularized three-dimensional units are the embodiment of rapidly-developing industrial technology and efficiency.

立面模块化的单元窗墙
600 × 4,800 的单元模块

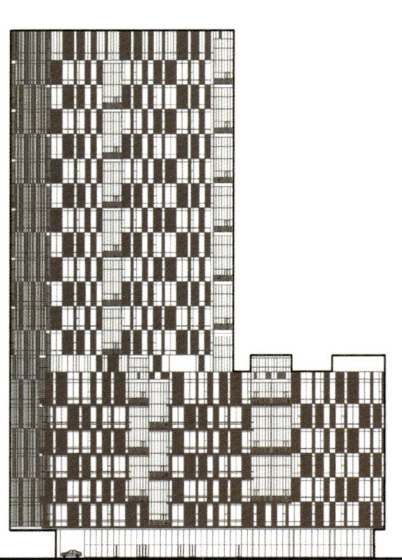

立面图

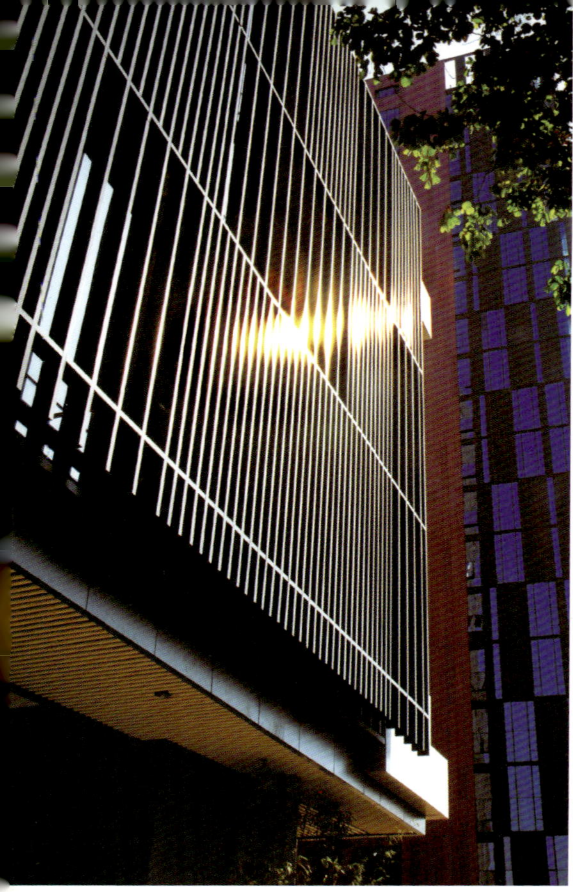

Commercial and Urban Complexes......

Commercial and Urban Complexes

商业及城市综合体

Commercial buildings and multi-functional complex is a focal point of the modern urban life and the "binding agent" of the city. It is called complex and can also be called a "mini-city". Therefore, our view on the design is, on one hand, the new building links the city in terms of space, environment and place and is dissolved into the same, and it also plays the role of "engine" to bring about the vitality to the community; on the other hand, the public openness and multi-functional features of the new building are made prominent; while satisfying the requirements of single commercial activity, more attention is paid to the consequential, potential, rich and humane requirements thus making the brief commercial exchange activity become the prolonged leisure activities; besides, we also design the outdoor area and indoor space on the whole, so as to achieve the architecture of true sense.

商业及多功能综合体是满足现代大众城市生活的集中体现,是城市的"粘合剂"。它被称为综合体,也可以说是"小城市"。因此,我们设计的出发点是:一方面让新的建筑从空间、环境、场所等诸多意义上连接城市的段落,充分地融入进去,并充当一个"引擎"的角色,带给整个社区活力;另一方面,新生的建筑突出其公共开放与多功能的特性,在满足传统单一商业活动的同时,更关注发掘那些伴随的、潜在的、丰富的、更人性的要求,令简短的商业交换活动变成可长时间参与和享用的休闲活动;另外,我们将室外空间环境与室内空间环境同时设计,营造出真正意义上完整的建筑。

The complex becomes a culture harbor, bringing culture, originality and leisure together, and realizing spatial agglomeration with the whole space like a harbor.

建筑化身文化港湾,将文化、创意和休闲集于一身,实现港湾式的空间集聚。

Planning & Design of Foshan Complex • Foshan

Foshan | Project Location
2008 | Project Date
280,000m² | Project Scale
conceptual design | Design Phase
construction not started | Project Status

佛山综合体规划设计·佛山

中国，广东省，佛山市 | 项目地点
2008年 | 设计时间
28万m² | 项目规模
方案设计 | 设计阶段
未建 | 项目现状

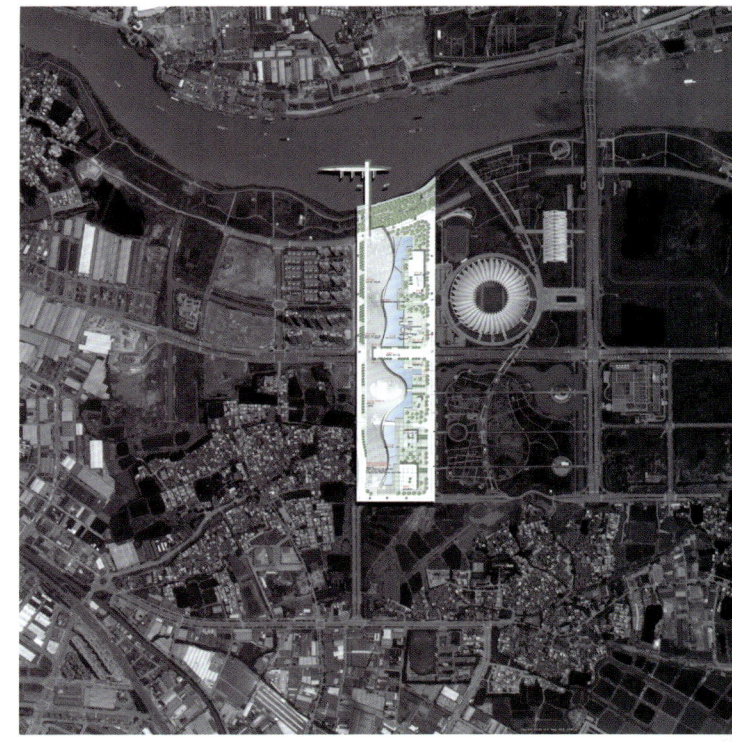

项目总览

未来的东平河将成为东方的塞纳河,而文化综合体就是河上的一个文化码头。综合体将文化、创意和休闲集聚于港湾式的空间,使人能够在这富有活力的建筑空间和令人清新的环境里共生共存。这不仅延用了东平河和世纪莲作为新佛山中心区的主体控制力,而且增强了新中心独一无二的城市形象。

Project Overview

In the future, Dongping River will become "Seine River" in the east while the cultural complex is a "cultural dock" on the river. Culture, originality, and leisure are presented together in the harbor-like space, where people coexist in the dynamic architecture and refreshing environment. The Dongping River and Century Lotus which are major attributes of the new downtown in Foshan are also used as elements in the design of the complex, thus the unique image of the new downtown is enhanced.

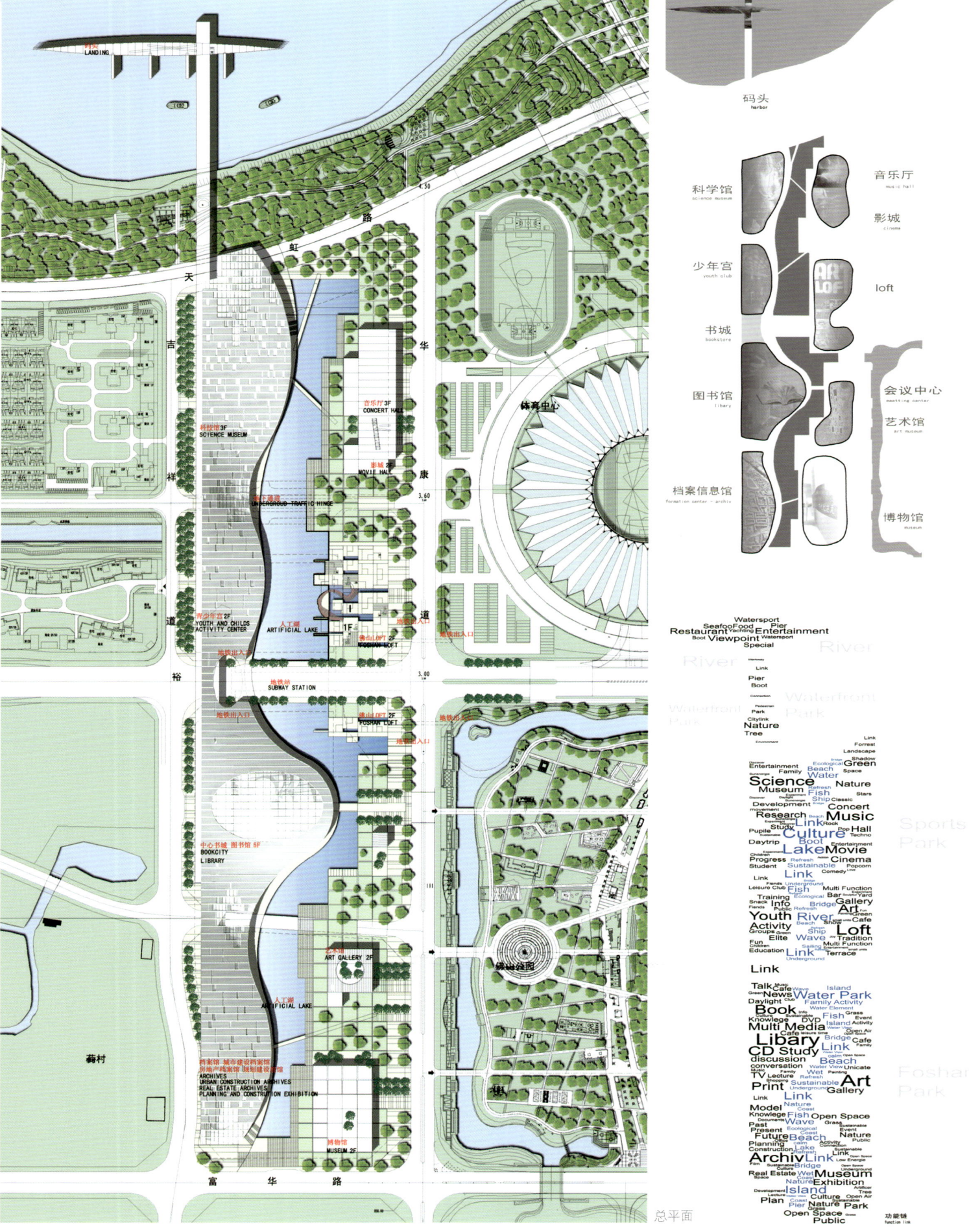

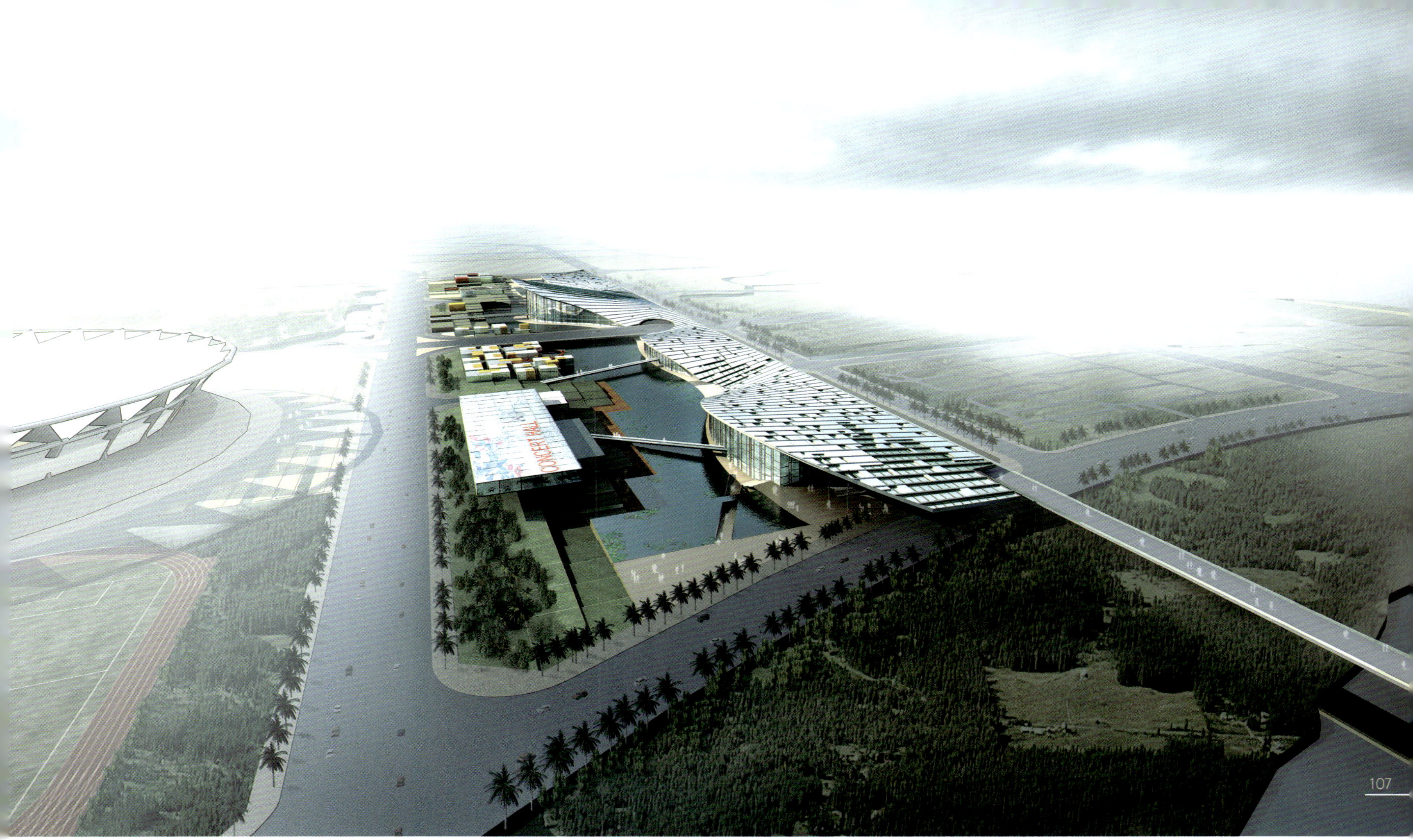

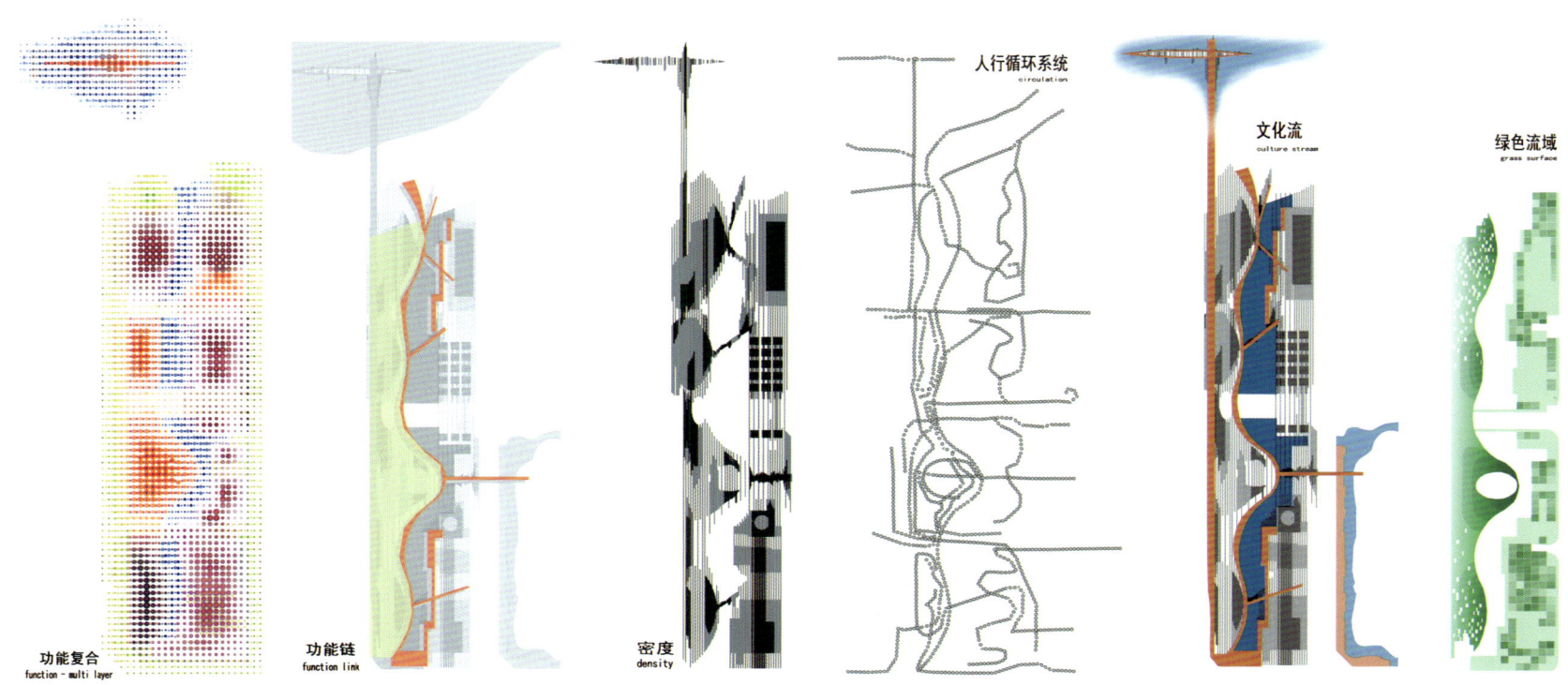

| 功能复合 | 功能链 | 密度 | 人行循环系统 | 文化流 | 绿色流域 |
| function - multi layer | function link | density | circulation | culture stream | grass surface |

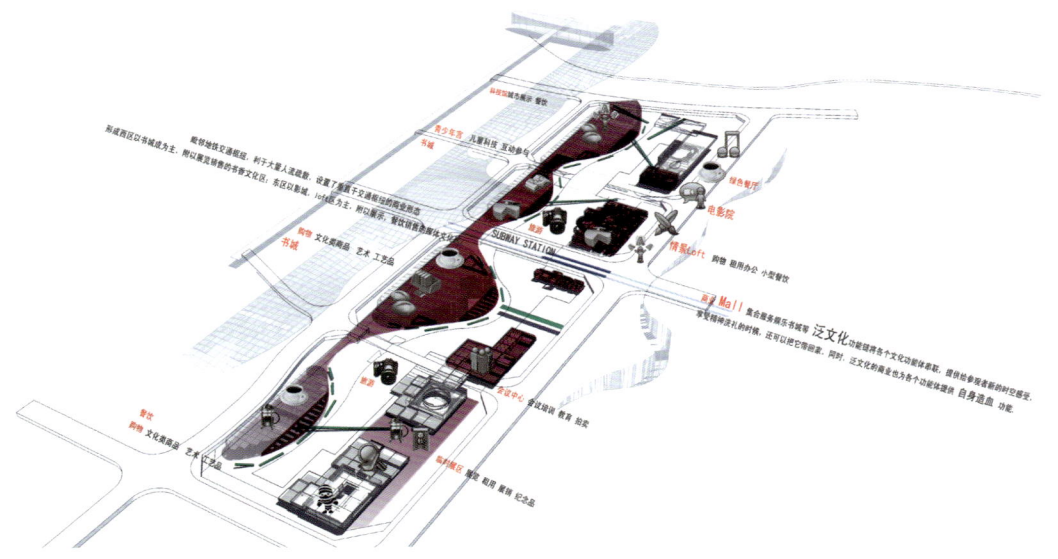

概念与构思

佛山市文化港的建成将实现环境与建筑的一体化。它能够将现有的佛山生态公园、佛山体育公园和佛山滨江公园有机地整合成一个具备完全城市功能、形态丰富、具有地域特色的佛山中央公园，使其作为佛山的城市新地标。

Design Concept and Idea

The Foshan Cultural Complex is designed to realize integration between environment and buildings. It integrates existing Foshan Eco-park, Foshan Sports Park, and Foshan Riverside Park into Foshan Central Park. With complete urban functions, rich shapes, and local characteristics, this integrated park becomes a new landmark of Foshan.

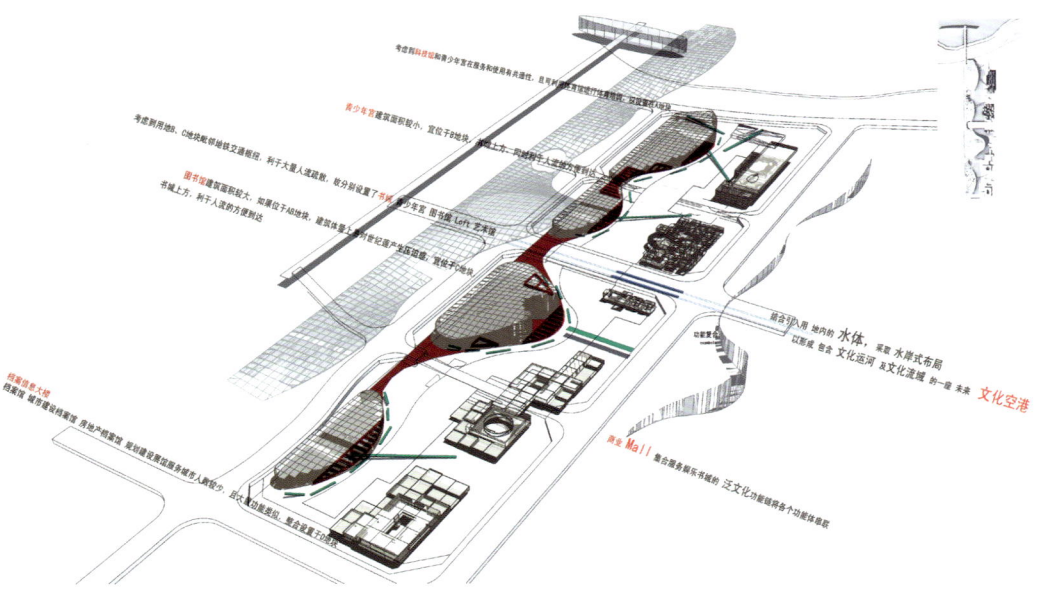

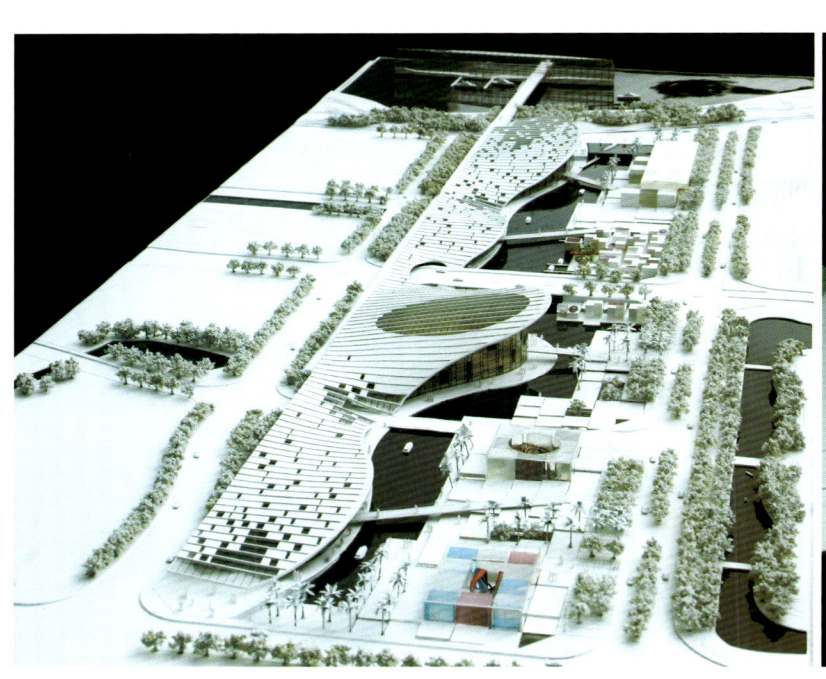

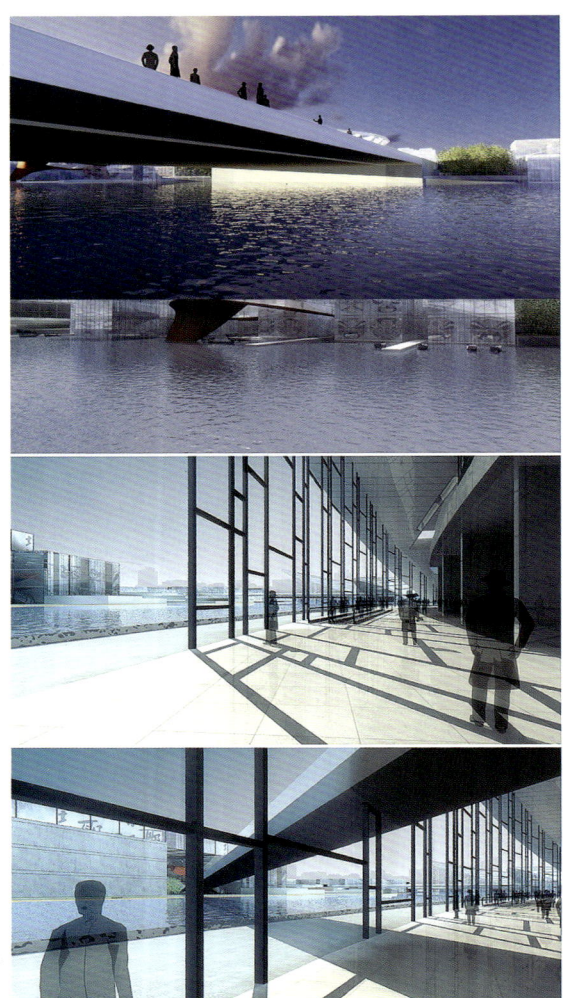

商业功能分析

集合服务、娱乐、书城等泛文化功能链，将各文化功能串连起来，给参观者提供新的时空感受。泛文化的商业也能为各功能体提供自身的造血功能，追求盈利以满足自身的运营。

Analysis of Commercial Functions

The pan-culture function chain, with service, entertainment and bookstore functions, can connect all cultural functions to bring new time-space experiences to visitors. The pan-culture business can also give a "haemopoiesis" function to all businesses in the complex so that they can generate profit and satisfy their own operational needs.

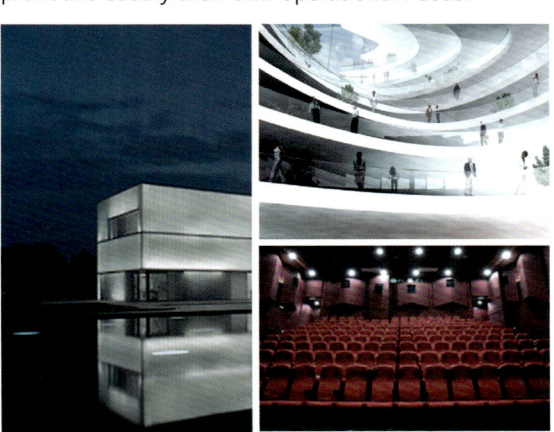

一层平面

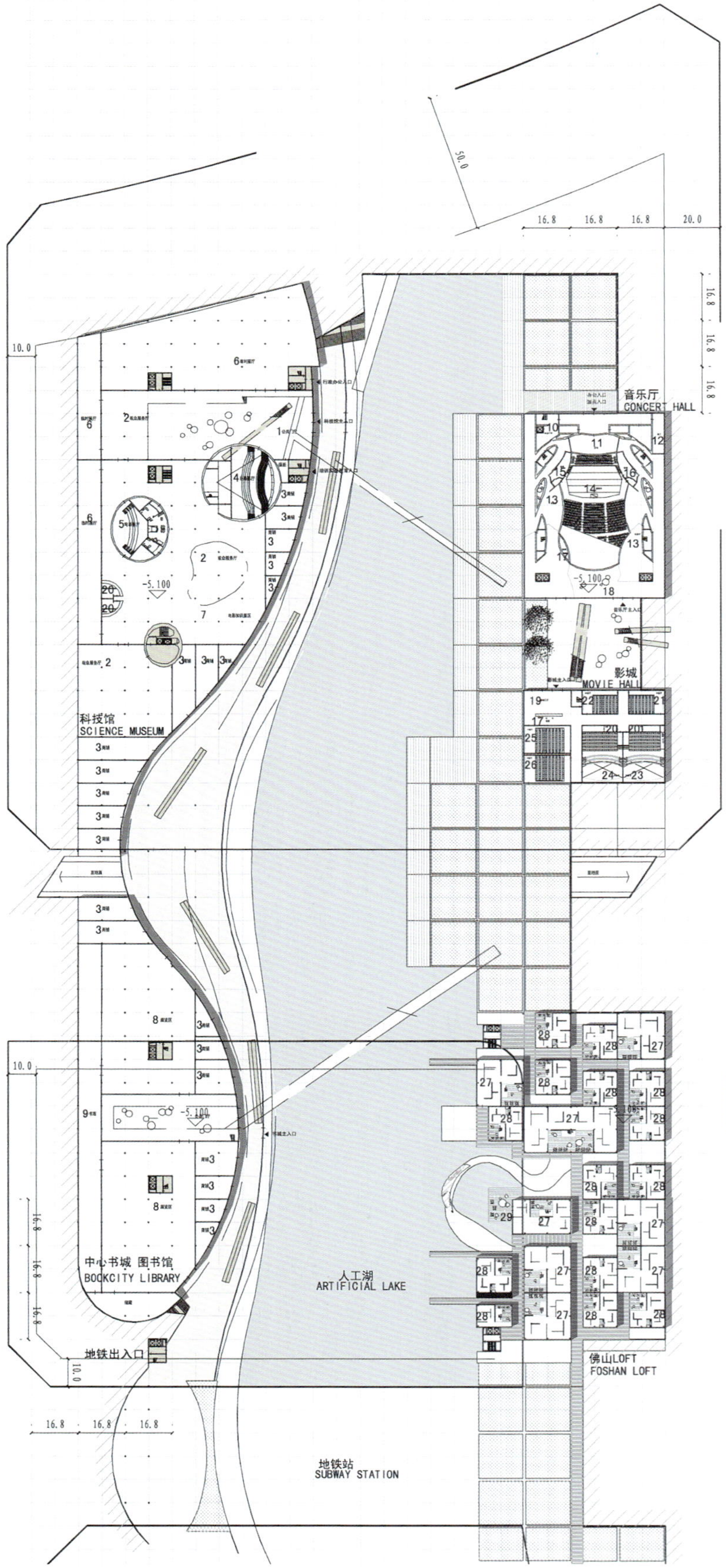
音乐厅
CONCERT HALL
影城
MOVIE HALL
科技馆
SCIENCE MUSEUM
中心书城 图书馆
BOOKCITY LIBRARY
人工湖
ARTIFICIAL LAKE
地铁出入口
地铁站
SUBWAY STATION
佛山LOFT
FOSHAN LOFT
1 公共门厅
2 观众服务厅
3 商铺
4 巨幕影厅
5 穹幕影厅
6 临时展厅
7 电影知识展区
8 阅览区
9 书库
10 办公门厅
11 后场
12 美工
13 休息厅
14 演奏台
15 竖琴
16 钢琴
17 售票
18 音乐厅门厅
19 影城门厅
20 卫生间
21 1#影厅
22 2#影厅
23 3#影厅
24 4#影厅
25 5#影厅
26 6#影厅
27 展厅
28 工作室
29 室外平台

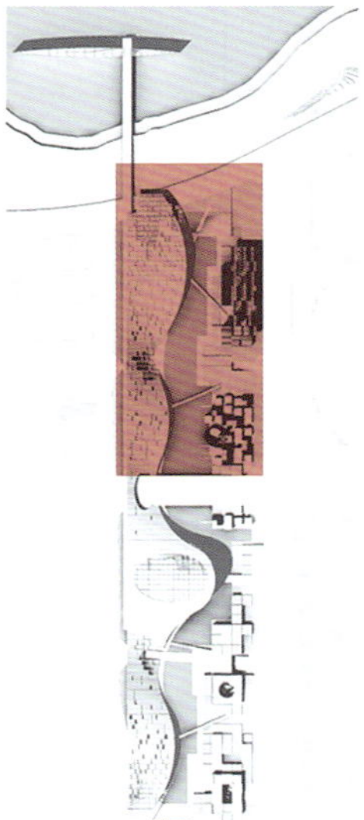

其中项目的营业面积为35,000m²，内部含有大量配套服务设施，结合东侧的影城及毗邻地铁出入口人流集中的Loft区域，以形成整个商业带上的高潮点。

The commercial area of the complex covers 35,000 square meters. With a number of supporting service facilities along with the cinemas to the east and the populous Loft area close to the metro entrance, the complex becomes a "climax" point in the whole commercial zone.

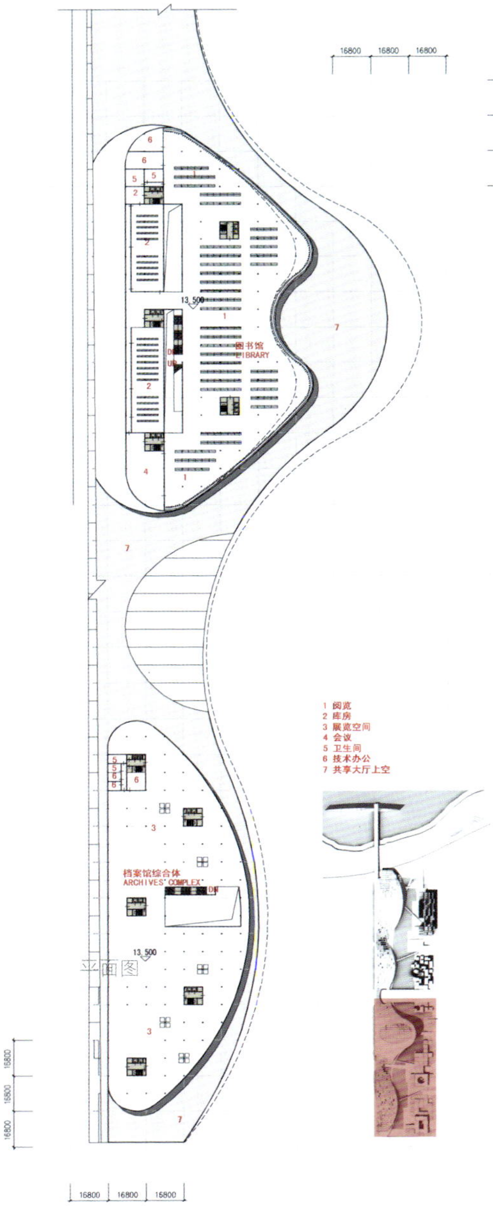

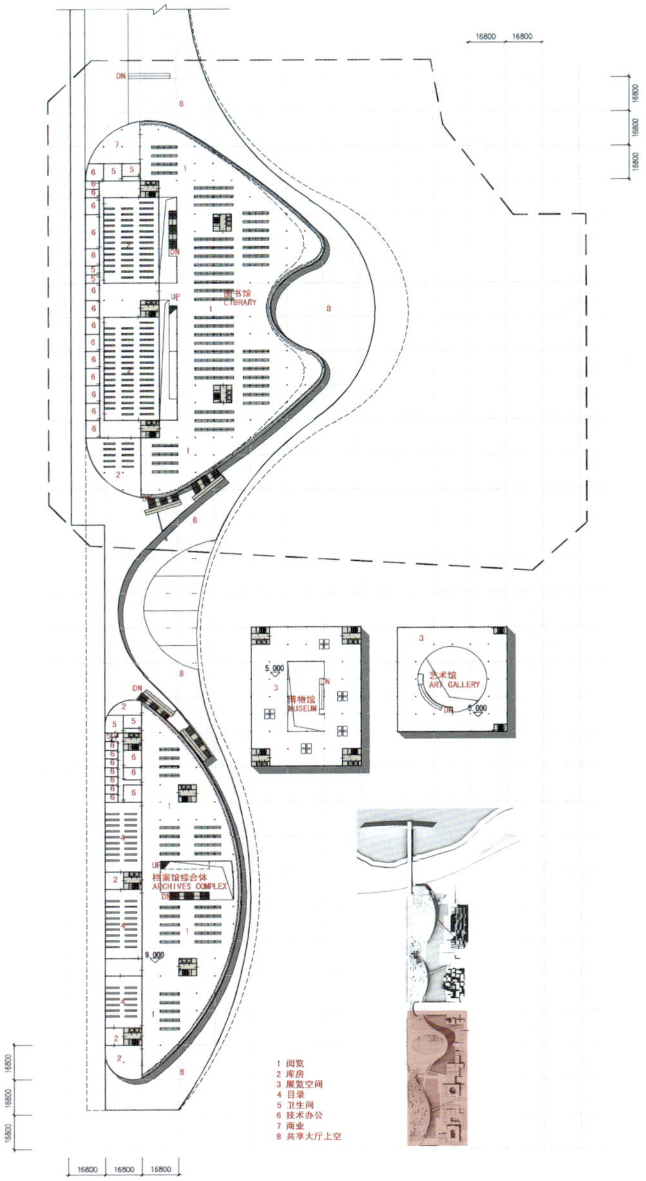

This project covers several blocks. We introduce urban block design ideas into the design of the residential quarters, aiming at creating an urban building complex with living, business, transportation, entertainment and rest functions.

在这一跨越多个街区的项目中,我们在小区的设计中引入城市街区设计的理念,打造以居住为主,集商业、交通、娱乐、休憩为一体的城市综合体。

Lsea Asia International Center • Nanning

New Fengling District, Nanning (In the ASEAN International Business Area) | Project Location
2007 | Project Date
400,000m² | Project Scale
conceptual design | Design Phase
construction in progress | Project Status

利海亚洲国际中心 · 南宁

中国，广西省，南宁市，风岭新区（东盟国际商务区内） | 项目地点
2007年 | 设计时间
40万m² | 项目规模
方案设计 | 设计阶段
在建 | 项目现状

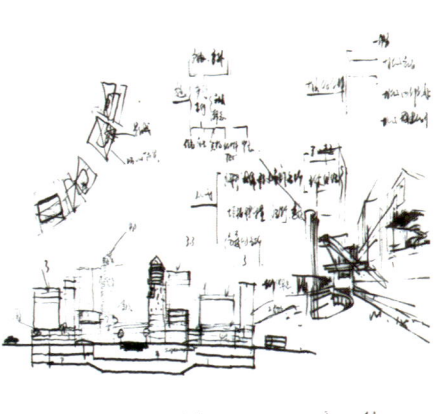
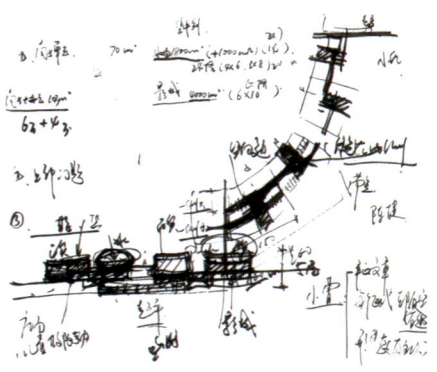

总平面

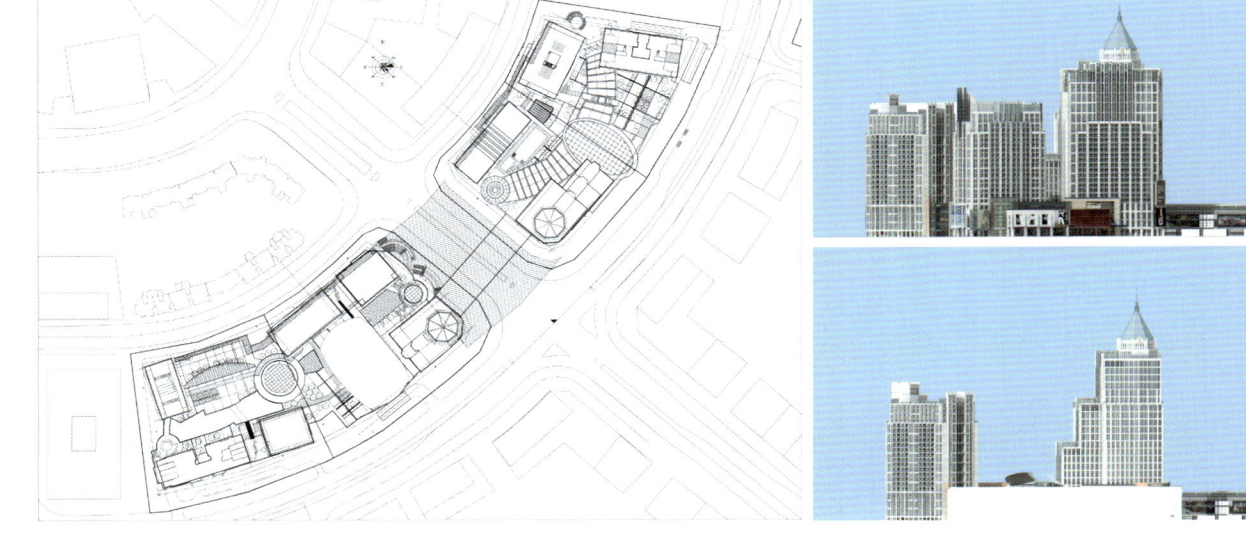

项目概况

利海亚洲国际旨在建立南宁市的"城市之冠",营造浓郁的商业氛围,并紧密联系商业、办公、公寓、住宅,以形成混合发展的概念。

为大众提供丰富的公共空间,营造互动的商业氛围,用前瞻的设计理念演绎现代生活方式,构筑一处安全无扰、舒适宜人的人行商业环境。

Project Overview

This project is aimed to build the "crown" of the city of Nanning, create a strong commercial atmosphere, and follow a development concept of integrating the functions of stores, offices, and apartments.

It is designed to provide large public spaces, create interactive business atmosphere, present modern life styles through perspective design ideas, and build a safe, comfortable and pleasant business environment.

建筑群分A、B两区,裙楼结合各区功能,采用动感的都市化风格,力求集聚该片区的商业氛围。主建筑群采用简洁明快的新古典风格,运用相同的造型逻辑及不同高度连续的街墙组织,形成极富整体凝聚感的城市综合体形象。

The building complex is divided into Area A and B. The skirt building, with various functions, is of dynamic urban style, aiming to build a strong business atmosphere in this region. The main building complex is of simple but bold neoclassical style. It creates an urban complex image with a strong sense of overall coherence through the application of the same modeling logic and continuous street walls of different heights.

立面图

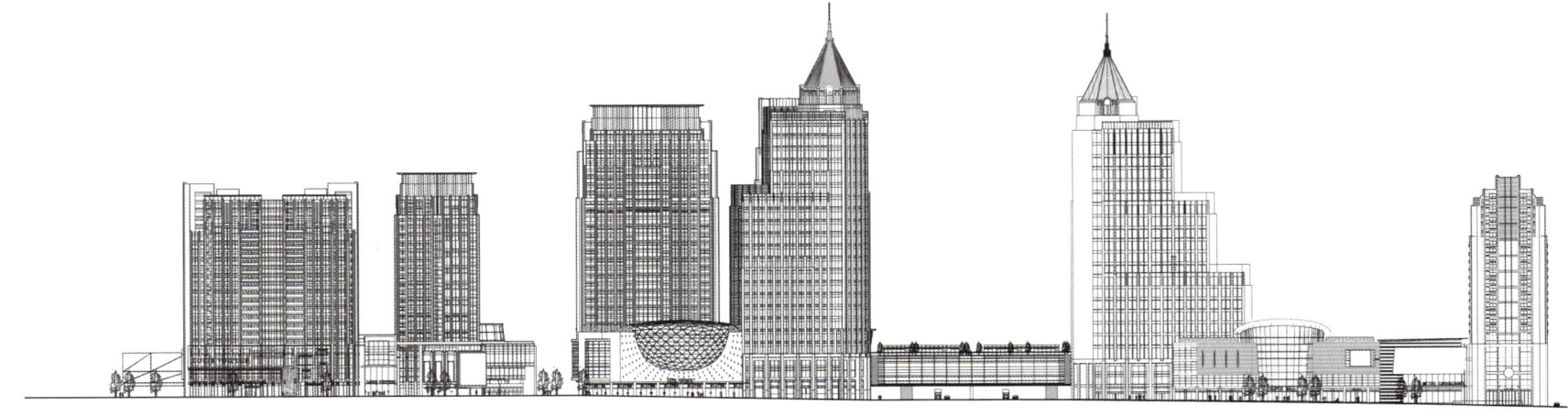

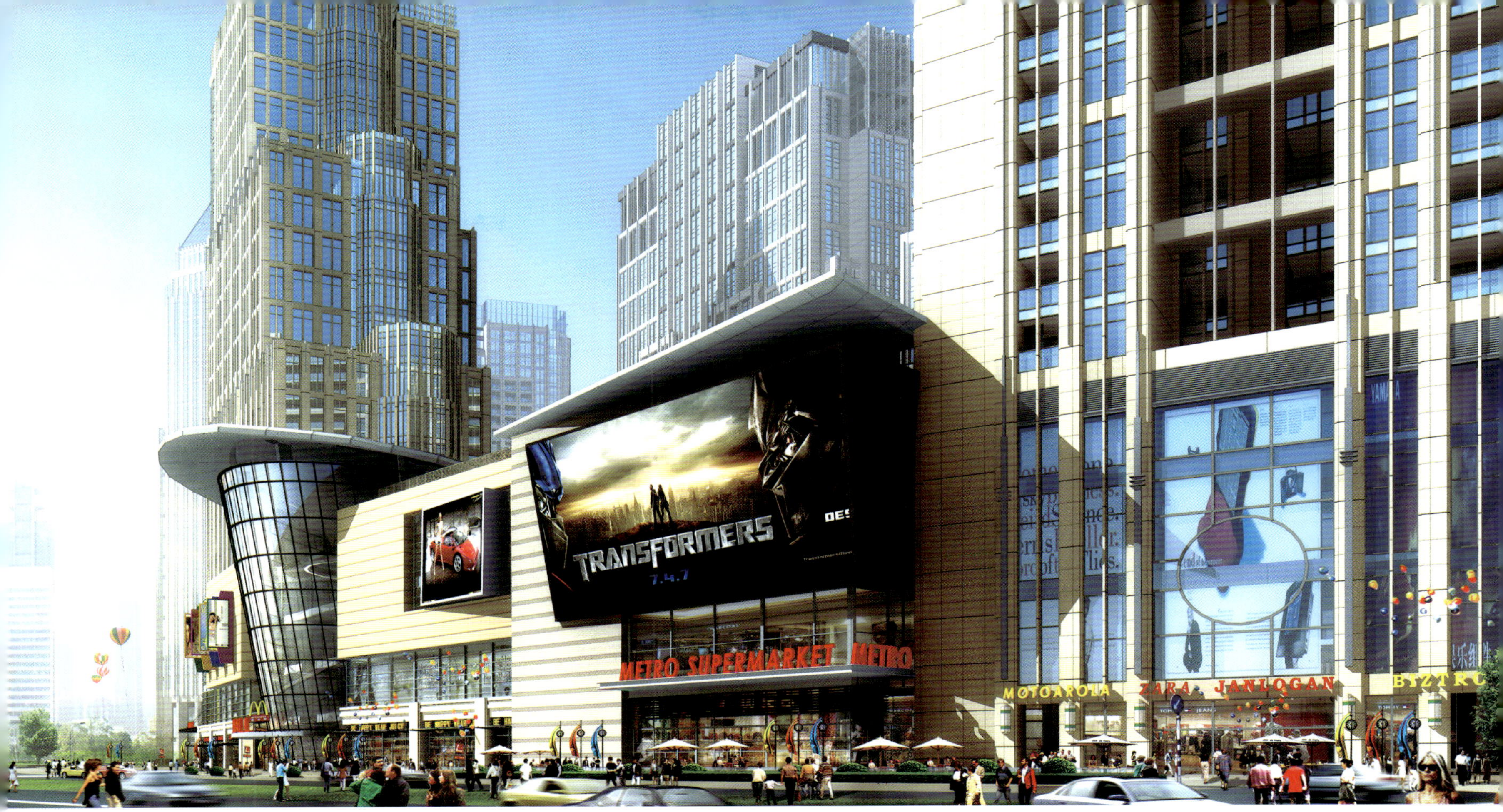
TRANSFORMERS
7.4.7
METRO SUPERMARKET
METRO
ZARA
JANLOGAN

LOUIS VUITTON
Virgin
NO TURNS

Ermenegildo Zegna

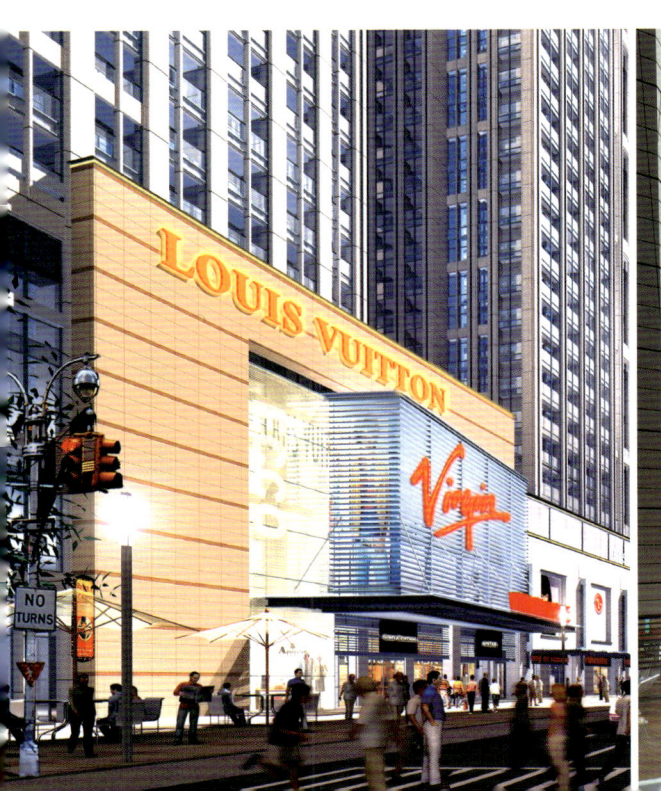

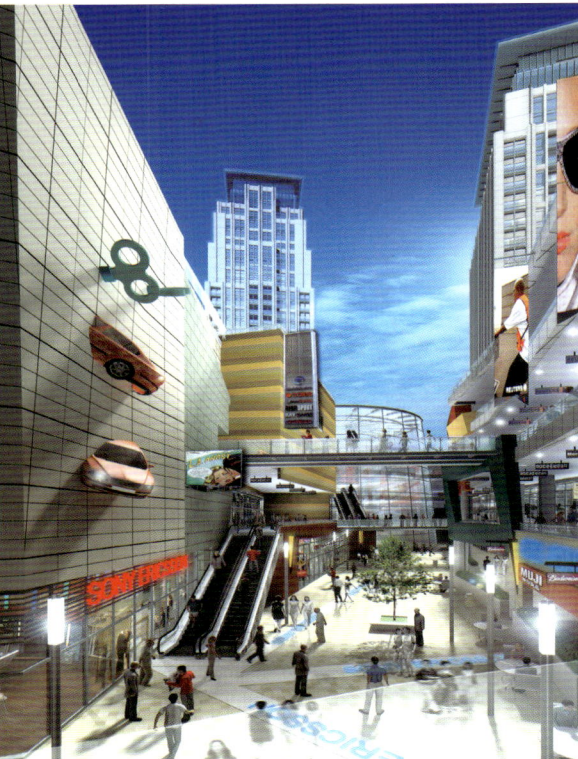

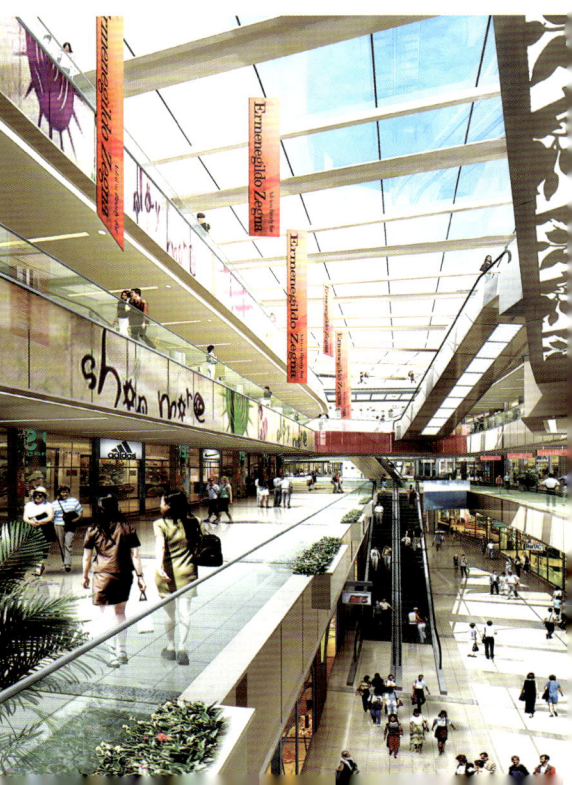

A 30-minute journey and 360-degree slow rotation make you observe this city's achievements through a different dynamic field of vision. In addition, when we need both a central park and a central shopping area, what will we consider?

一段30分钟的奇妙旅程,360°地缓缓旋转,提供另一种动态的视野观看这个城市的成就;另外,当我们既需要中央公园也需要中央购物区的时候,我们该如何思考。

Central District Crystal Island • Shenzhen

Shenzhen | Project Location
2009 | Project Date
64,000m² | Project Scale
conceptual design | Design Phase
construction not started | Project Status

中心区水晶岛·深圳

中国，广东省，深圳市 | 项目地点
2009年 | 设计时间
6.4万m² | 项目规模
方案设计 | 设计阶段
未建 | 项目现状

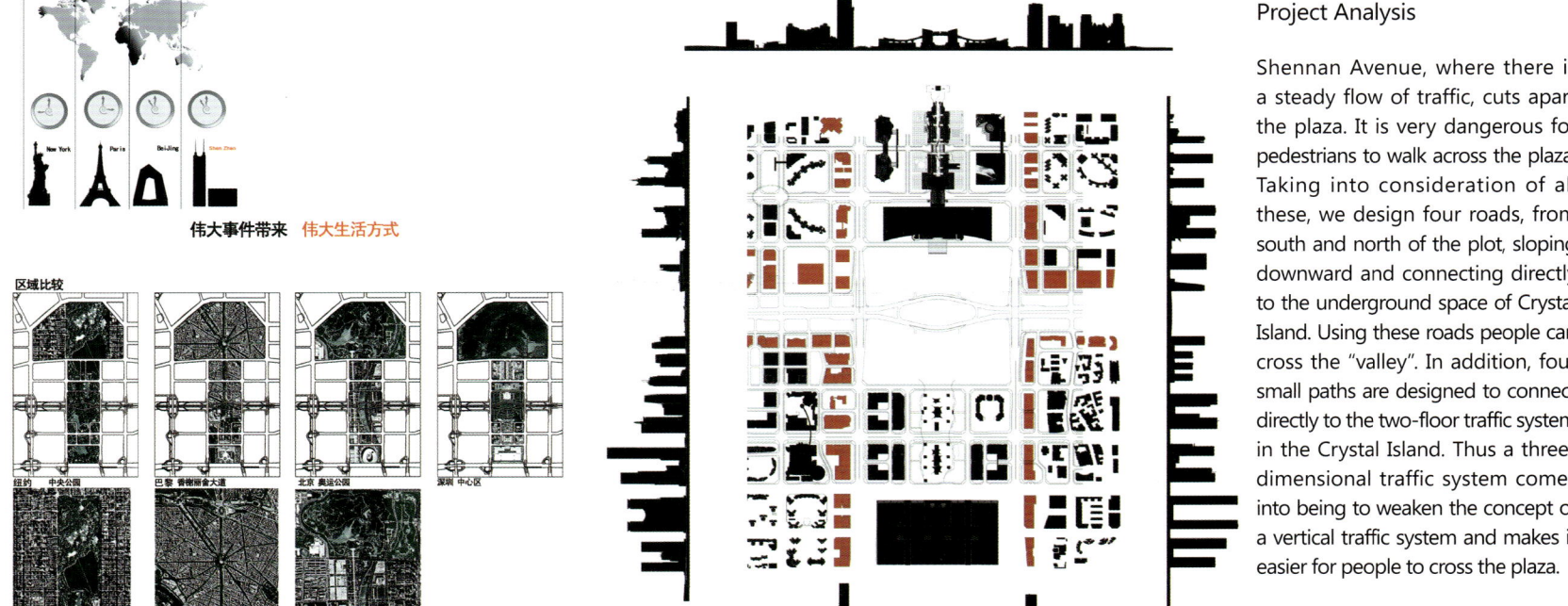

项目分析

车辆川流不息的深南大道将广场切割，行人穿越其中极其危险。针对这样的现状，我们在地块南北引入了四条直通水晶岛地下空间的下沉道路，即我们的"峡谷"穿越路径。另外设有四条小路直接连接水晶岛内二层交通系统，组成一个三维立体的交通体系，弱化了垂直的交通体系，增强广场的穿越性。

Project Analysis

Shennan Avenue, where there is a steady flow of traffic, cuts apart the plaza. It is very dangerous for pedestrians to walk across the plaza. Taking into consideration of all these, we design four roads, from south and north of the plot, sloping downward and connecting directly to the underground space of Crystal Island. Using these roads people can cross the "valley". In addition, four small paths are designed to connect directly to the two-floor traffic system in the Crystal Island. Thus a three-dimensional traffic system comes into being to weaken the concept of a vertical traffic system and makes it easier for people to cross the plaza.

中央公园

绿树就是生态建筑

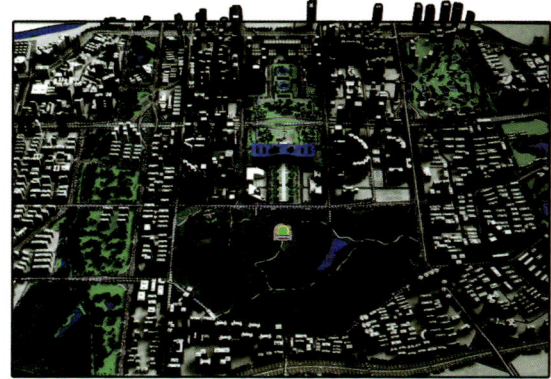

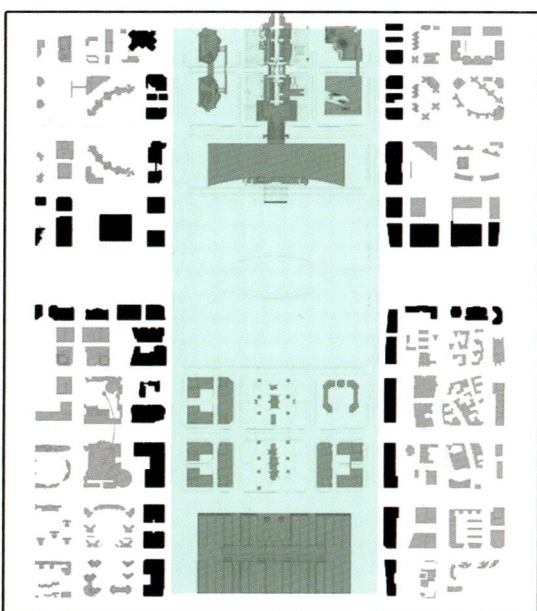

森林飘带 我们在飘带一般舒展柔软的森林环境中休憩、玩耍、健身、交流……

We rest, play, exercise, and exchange ideas in a forest environment which is as stretched and soft as a ribbon ……

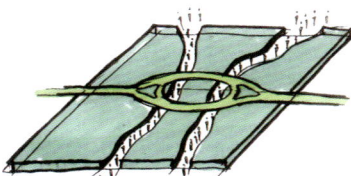

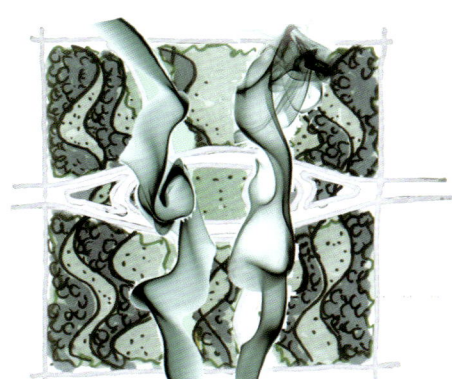

森林峡谷 在峡谷一般的幽径里自由穿行………

The valley winds its way freely along roads …

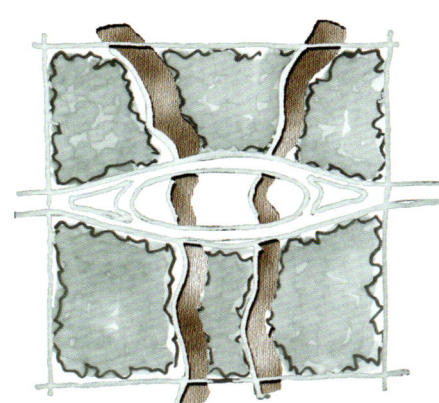

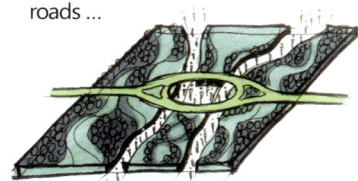

溪流汇聚 市民广场变成充满自然情趣的绿色空间,人们可在此自由呼吸。来自四面八方的人们,追逐着溪流,汇聚于此。

As brooks converge, the plaza changes into a green space with funs of nature, where people can breathe freely. People from all directions, who have a partiality for brooks, get together here.

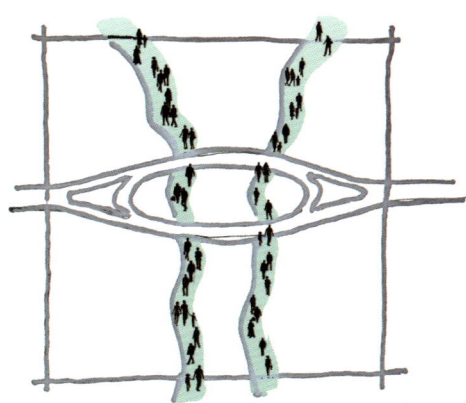

危险!

车辆川流不息

穿越深南大道的行人

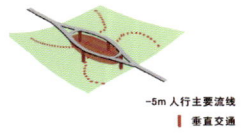

-5m 人行主要流线
垂直交通

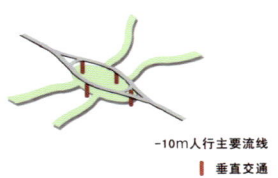

-10m 人行主要流线
垂直交通

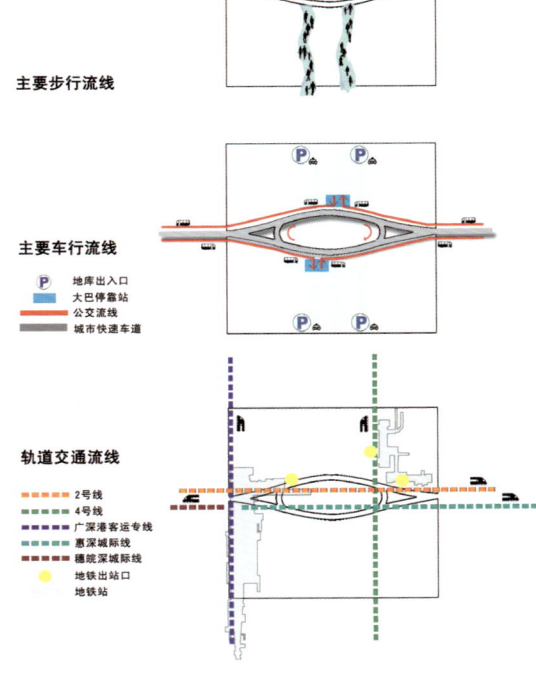

主要步行流线

主要车行流线
- 地库出入口
- 大巴停靠站
- 公交流线
- 城市快速车道

轨道交通流线
- 2号线
- 4号线
- 广深港客运专线
- 惠深城际线
- 穗莞深城际线
- 地铁出站口
- 地铁站

用地现状主要存在着人流可达性差、难以穿越的问题。除了方便人流的穿越，也解决了大量轨道交通换乘人流疏散、地下公交系统人流疏散、地下停车场人流疏散的问题。

The land faces a problem of poor accessibility and crossing difficulty. This system also splits the flow of pedestrians and guides them to the rail traffic system, underground transportation system, and underground parking lot.

标志物 **游玩流程**

30分钟一段奇妙的旅程　　A 30-minute miraculous journey
另一种视野观看市民中心　Makes you observe the Citizen Center from another field of vision.
360°缓缓旋转　　　　　　360-degree rotation
一杯咖啡　　　　　　　　A cup of coffee
一份轻松淡然的心情　　　A relaxed and leisurely mood
优美的音乐飘于耳畔　　　Nice music around the ear
各方友人相聚于此　　　　You meet your friends here
带来一份欢愉　　　　　　To enjoy a good time
带走一份美好的回忆　　　And keep a wonderful memory.

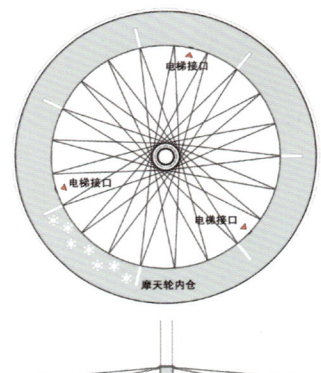

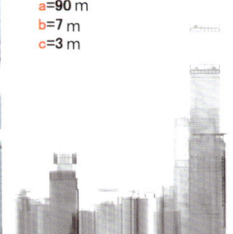

a=90 m
b=7 m
c=3 m

用地功能结构

乔木"森林"

草坡"高尔夫球场"

步行"峡谷"

广场大体由道路、起伏草坡和高大乔木构成，穿越广场的深南大道和两条人行"峡谷景观带"，加上连绵起伏高尔夫球专用的草坡，再叠加上高大的乔木"森林"组合成我们的广场。

The plaza consists of roads, undulating lawns, and tall arbors. Shennan Avenue, which crosses the plaza, two herringbone-shaped valley landscaping zones, undulating lawns used for golf course, and a forest of tall arbors make up the plaza.

整个田地由原市政集会广场、联系中心商业部分的屋顶花园、沟通各方绿色平台的水晶岛中心展厅及四个生态休闲娱乐广场区域构成整个体系。

The whole system is made up of former municipal square, roof gardens of central business buildings, Crystal Island central exhibition hall which connects all green platforms, and four ecological recreation and entertainment areas.

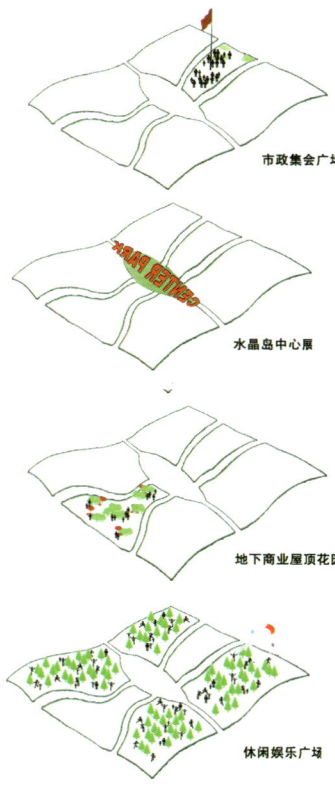

市政集会广场

水晶岛中心展厅

地下商业屋顶花园

休闲娱乐广场

地下空间　**功能布局**

地下空间功能示意
"深圳制造"工业展览馆：
我们希望在这样一个中心的位置能有一个全方位展示深圳的平台，它不是单个的某个建筑，而是一个建筑群，将单个的建筑功能与整体的建筑形态相结合。考虑到地块作为交通节点的特殊性，我们将功能体设计成圆形体量，并以一种相对自由的形态排列，令各方穿越的交通更加灵活多变，同时也营造了丰富的空间形态与周围自然环境很好地融合。在功能安排上，主要设置一些多用途的展示空间，可以举办各种展览，如产品展、作品展。同时也设计了一些满足特殊展示需要的展厅，如汽车展厅和一些附属的报告厅等，满足一种复合的全方位需要。

Underground space functional diagram
"Made in Shenzhen" Industrial Exhibition Hall:
We hope that a platform can appear at this central position to display Shenzhen in all aspects. It will not be a single building but a group of buildings, with single building functions combined with general architectural form. Since this plot is located at a traffic node, we design a circular shape for the project and all structural components are lined up in a relatively-free pattern so that people can cross the plaza flexibly and rich spatial form lives in perfect harmony with surrounding natural environment. With respect to function arrangement, we set some multi-purpose exhibition spaces to hold various exhibitions, such as product or works exhibition. Meanwhile, we also design some exhibition halls to meet special needs, e.g. auto exhibition hall, report hall, etc., which can meet composite all-round needs. .

地下分析图

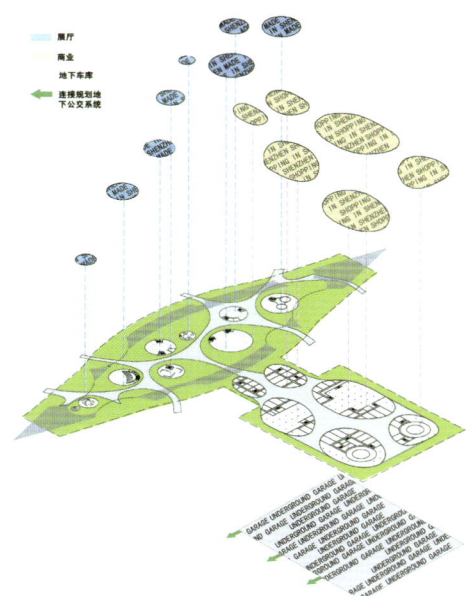

摩天轮结构的方案设计

1. 采用国家和地区的主要规范和规程：建筑结构可靠度设计统一标准（GBJ50068-2001）、建筑结构荷载规范（GB50009-2001）、建筑抗震设计规范（GB50011-2001）、钢结构设计规范（GB50017-2003）、混凝土结构设计规范（GB50010-2002）和建筑地基基础设计规范（GB50007-2002）。

2. 结构体系

本工程属于钢结构体系，整个结构体系包括两个"人"字结构：可上下移动的摩天轮以及中间支撑的巨柱。其中，摩天轮直径约110m，支撑巨柱总高约240m，底部直径约7m。摩天轮由箱型截面钢结构内环梁和外环梁、钢环梁之间的双向预应力钢索（棒）网组成。内环梁通过导轨与中间支撑巨柱连接，中部巨柱采用直径约7m的圆形钢结构。巨柱内部分箱既保证巨柱的刚度和强度，又能满足各类运行设备的独立运行。由于巨柱的总高度较大，为减少鞭梢效应，并增加造型的美感，巨柱分段渐次内收。为满足整体结构的抗风、抗震性能，摩天轮、巨柱两个子结构以及两个子结构之间，全组采用阴尼器结合传感器，利用计算机同步系统，实现了两大子结构的同步协调。

Design of Ferris Wheel Structure

1. Adoption of national and local standards: Unified standard for reliability design of building structures (GBJ50068-2001); Load code for the design of building structures (GB 50009 -2001); Code for seismic design of buildings (GB50011-2001); Code for design of steel structures (GB50017-2003); Code for design of concrete structures (GB50010-2002); Code for design of building foundation (GB50007-2002)

2. Structural system

This project has a steel structure system, consisting of two herringbone-shaped structures, i.e. a Ferris wheel, which moves up and down, and a supporting column in the middle. The Ferris wheel is about 110m in diameter and its supporting column is about 240m in height. Its base is about 7m in diameter. The wheel is made up of internal ring beam & external ring beam of the box-shaped sectional steel structure and the two-way pre-stressing force steel rope (bar) between steel ring beams. The internal ring beam connects the supporting column in the middle via a guiderail. The giant column in the middle is a round steel structure with a diameter of about 7m. The box-shaped structure in the column not only guarantees rigidness and strength of the column, but also ensures that all moving equipment moves independently. Since the giant column is too high, its sections are adducted gradually in order to reduce the whipping effect and make the column look more beautiful. To satisfy the whole structure's wind resisting and anti-seismic performance, damper and sensor are used for and between two sub-structures of the Ferris wheel and its supporting column. A computerized synchronization system is adopted to realize synchronization and coordination of the two sub-structures.

结构体系

在提供城市空间与建筑的同时，还需要提供来这里更多的目的性，我们想到了摩天轮。摩天轮直径约110m，支撑巨柱总高约240m，底部直径约7m。摩天轮由箱型截面钢结构内环梁、外环梁和钢环梁之间的双向预应力钢索（棒）网组成。内环梁通过导轨与中间支撑巨柱连接。中部巨柱采用直径约7m的圆形钢结构。

摩天轮行程最高90m，由5m标高处垂直上升至80m处停住，缓缓旋转下降，整个行程30分钟。

Structural System

In addition to providing traffic functions by the structure, more functions shall be offered to attract people to visit here. So we design a Ferris wheel. The wheel is about 110m in diameter and its supporting column is about 240m in height. Its base is about 7m in diameter. The wheel is made up of an internal steel ring beam and an external steel ring beam with a box section and the two-way pre-stressing force steel rope (bar) between the steel ring beams. The internal ring beam connects the supporting column in the middle via a guide rail. The giant supporting column in the middle is of a round steel structure with a diameter of about 7m.
The maximum height of the wheel is 90m. It ascends vertically from the elevation of 5m to 80m and then descends slowly in a rotary way. The entire rotation takes 30 minutes.

The project has a special requirement for urban culture expression. The building must play the role of serving as a connecting node in the city, and publicizing traditional culture so that it will not be neglected in the future.

项目的特别城市文脉要求,建筑在这里必须起到连接城市、连接历史、展望未来的作用。

Emperor Star City · Shanghai

Shanghai | Project Location
2004 | Project Date
70,700m² | Project Scale
conceptual design, preliminary design, construction drawing | Design Phase
construction in progress | Project Status
Changsha Planning and Design Institute | Design Partner

英皇明星城·上海

中国,上海市 | 项目地点
2004年 | 设计时间
7.07万m² | 项目规模
方案设计 初步设计 施工图设计 | 设计阶段
在建 | 项目现状
长沙规划设计研究院 | 合作单位

灵感来源
Origin Of The Idea

自然形态的山体

购物山

绿色、生态的积极环境与周围高密度、硬质的嘈杂的环境形成对比。
自然、生态的环境加上商业的功能，形成休闲购物的场所。
——"购物山"

本项目用地位于上海浦西极具老上海传统氛围的豫园商圈内，穿过西面沉香阁路即可到达豫园商城的核心区。传统与现代的对比，小尺度与大尺度的协调，主题化步行街与大型商业形态的统一，使本项目别具一格。

设计构思
1．街区式的主题商业空间，以跨时空的"老上海明星街"与"时尚星光大道"营造出动人的商业氛围。
2．以复合的观演式退台中庭空间，强化大型商业中心的共享性城市角色。
3．立体全方位通达的交通动线组织，使本项目与城市保持健康的交流，为项目本身及区域增添活力。

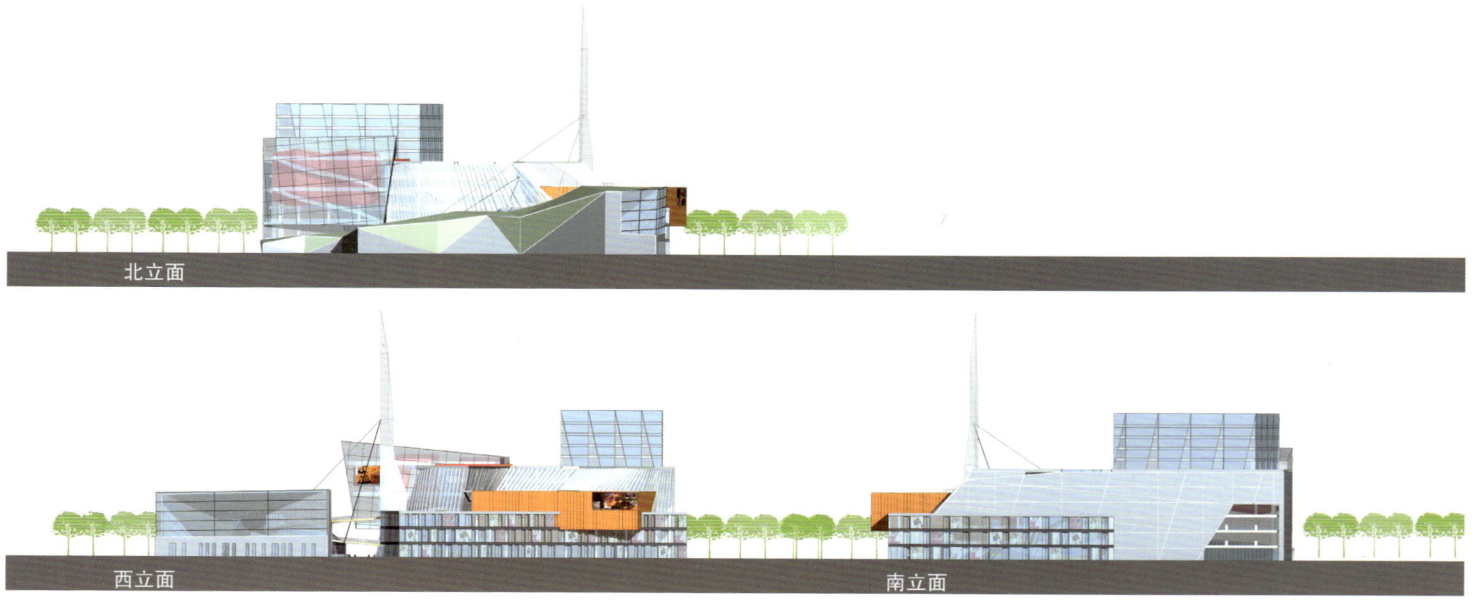

北立面

西立面　　　　　南立面

立面图

The Emperor Star City is located in the Yuyuan Commercial Zone in Puxi of Shanghai, with a strong traditional commercial atmosphere. A visitor coming out of the Emperor Star City - if crossing the Chenxiangge Road in the west, can reach the core area of Yuyuan Tourist Mart. Since the complex presents a comparison between traditional features and modern features, coordination between small and large scale, and integration of a themed pedestrian street and large business facilities - it is of unique style.

Design Concept
1. Street-type themed commercial space: the Old Shanghai Star Street and Popular Star Avenue, with a strong contrast in space-time, create an attractive commercial atmosphere.
2. The composite theatrical-style terrace atrium enhances the role of a large business center as a shareable space.
3. The organization of three-dimensional all-direction traffic lanes makes this complex communicate well with the city. As a result, the complex and the community are invigorated.

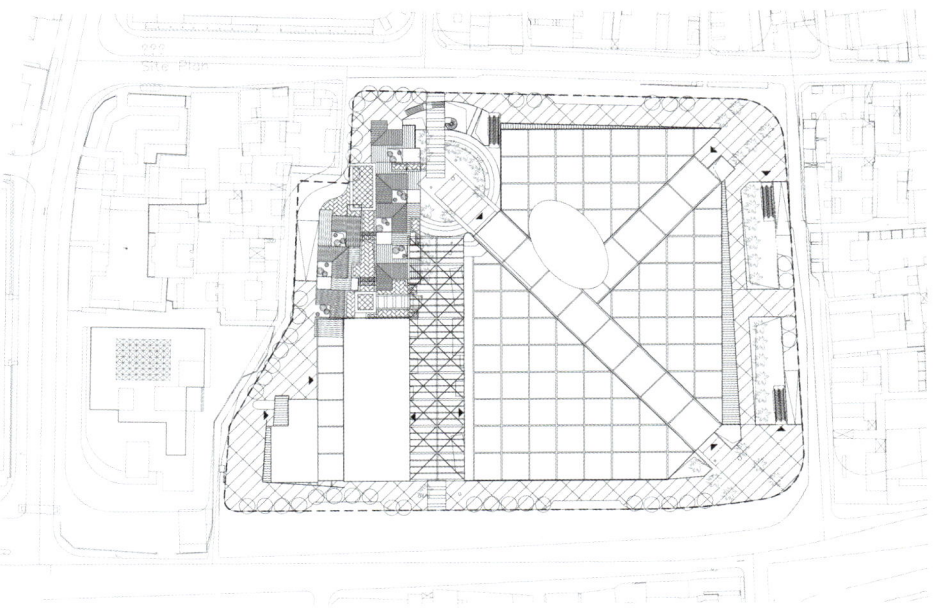

总平面

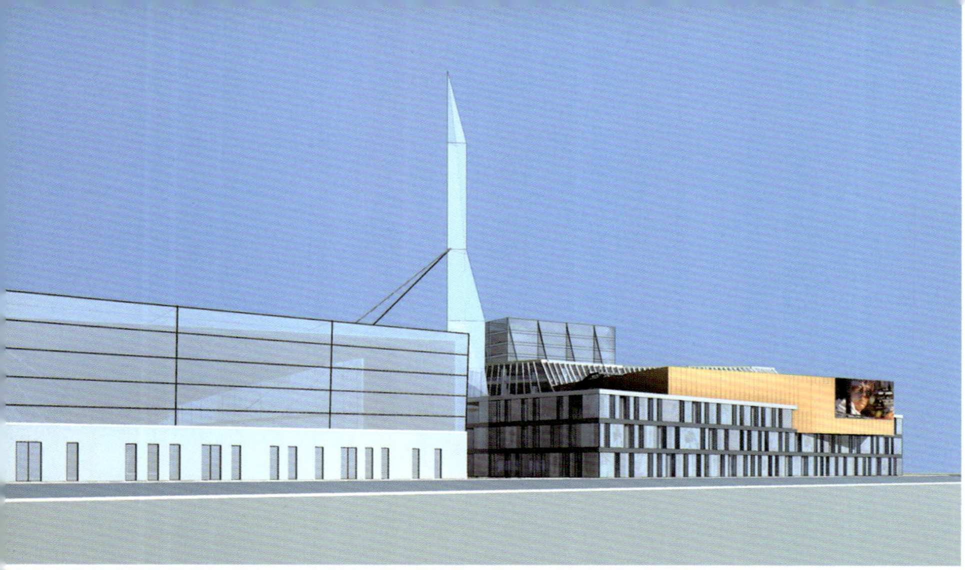

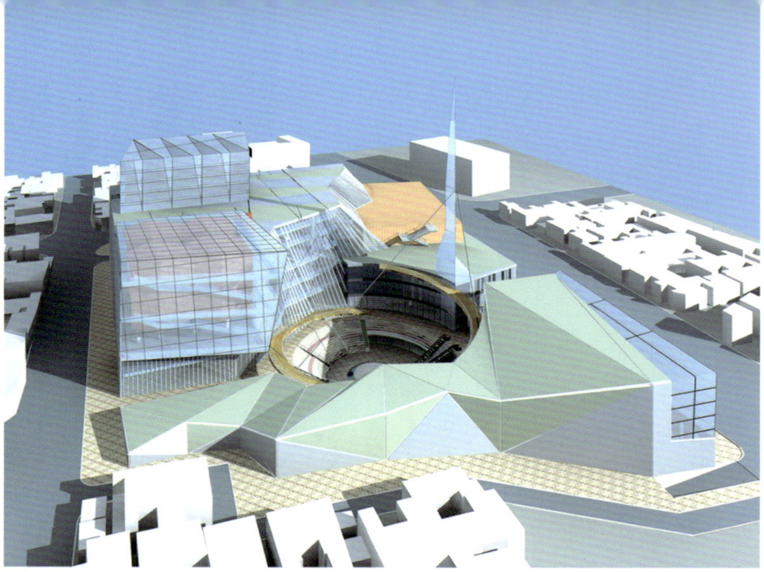

This is a building complex. Therefore a good relationship with the neighborhood must be established and the expression and posture of each single building in the complex must be well designed.

这是一个建筑群,必须解决好同"邻居"的关系,解决好群中每个个体的表情和姿态。

Chengdu Future Plaza • Chengdu

Chengdu | Project Location
2009 | Project Date
200,000m² | Project Scale
conceptual design | Design Phase
construction under planning | Project Status

成都香年广场·成都

中国，四川省，成都市 | 项目地点
2009年 | 设计时间
20万m² | 项目规模
方案设计 | 设计阶段
在建 | 项目现状

项目总览

本建筑小区由两座办公楼和一座公寓组成，其中一座办公楼高度为134m，建筑面积约为90,000m²；另一座高度为100m，建筑面积约为60,000m²；公寓楼高度约为100m，建筑面积约为40,000m²。

Project Overview

This plaza consists of two office buildings and one apartment. One office building, 134m in height, has a building area of about 90,000 square meters. Another one is 100m in height, with a building area of about 60,000 square meters. The apartment is about 100m in height and its building area is about 40,000 square meters.

FUTURE PLAZA
2009.08.25 VERSION ONE

PERSPECTIVE 庭院俯视图

项目规划

最大化地提升住宅与商业的价值，也就是规划上保证客户拥有最大化的好景观朝向；其次，从城市设计角度充分尊重周边的环境，给片区的建设锦上添花。

规划旨在创造一个密集但不拥挤的建筑与城市空间，三栋塔楼呈风车状布置，彼此间留下相对舒适的距离。在场地中设计一条视线通道贯穿整片建筑群，以强调城市空间的开敞性。场地开放式内广场与外部城市空间有三个方向的平层衔接，使建筑更为自然地融入整个高新区。住宅布置在地块的西南角，避免了与周边建筑的对视，拥有良好的景观视野，同时也保证了充足的日照采光。

Project Planning

Maximize the values of the apartments and offices, that is, ensure that the occupants have the best view of landscape and the apartments and offices have the best orientations through planning; achieve great harmony with surroundings from the perspective of urban design so as to add brilliance to the well-organized community.

The planning aims at creating a compact but un-crowded architectural and urban space by arranging the three buildings in a windmill-like pattern, with a relatively comfortable distance between them. A passageway passing through the whole complex with also the function of viewing is designed to emphasize the openness of the urban space. The open internal plaza links external urban spaces in three directions and thus the buildings integrate with the whole hi-tech development zone more naturally. The apartment is located at the southwest corner of the plot, avoiding facing surrounding buildings but letting the occupants have a wide field of vision for landscapes and guaranteeing sufficient daylight.

总平面

常规尺度立面窗扇
window elevation
regul size windows

主体结构-混凝土框架结构
main structure
concrete frame structure

主体结构-混凝土框架结构
main structure
concrete frame structure

公寓立面
Apartment Facade
一使用常规尺度的立面要素以降低结构成本；
一使用常规混凝土结构为建筑的主体结构；
一使用涂料填充色彩。

- Facade components of conventional size are used to reduce cost;
- Conventional concrete structure is used as main structure of the building;
- Paints are used to show more colors.

常规尺度的立面窗扇
window elevation
regular size windows

主体结构-混凝土框架结构
main structure
concrete frame structure

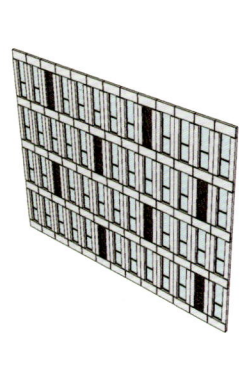 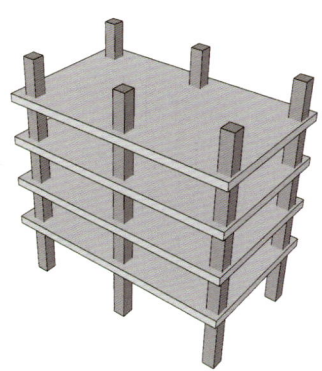 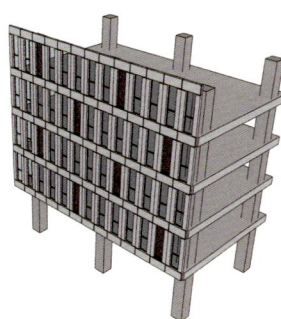

办公立面
Office Facade
一利用常规尺度的立面模数以降低结构成本；
一外立面构架以混凝土为材料；
一以面砖为外饰材料，达到降低成本的效果。

- Facade modulus of conventional size is used to reduce cost;
- External façade framework is made of concrete;
- Face bricks are used for external decoration so as to reduce cost.

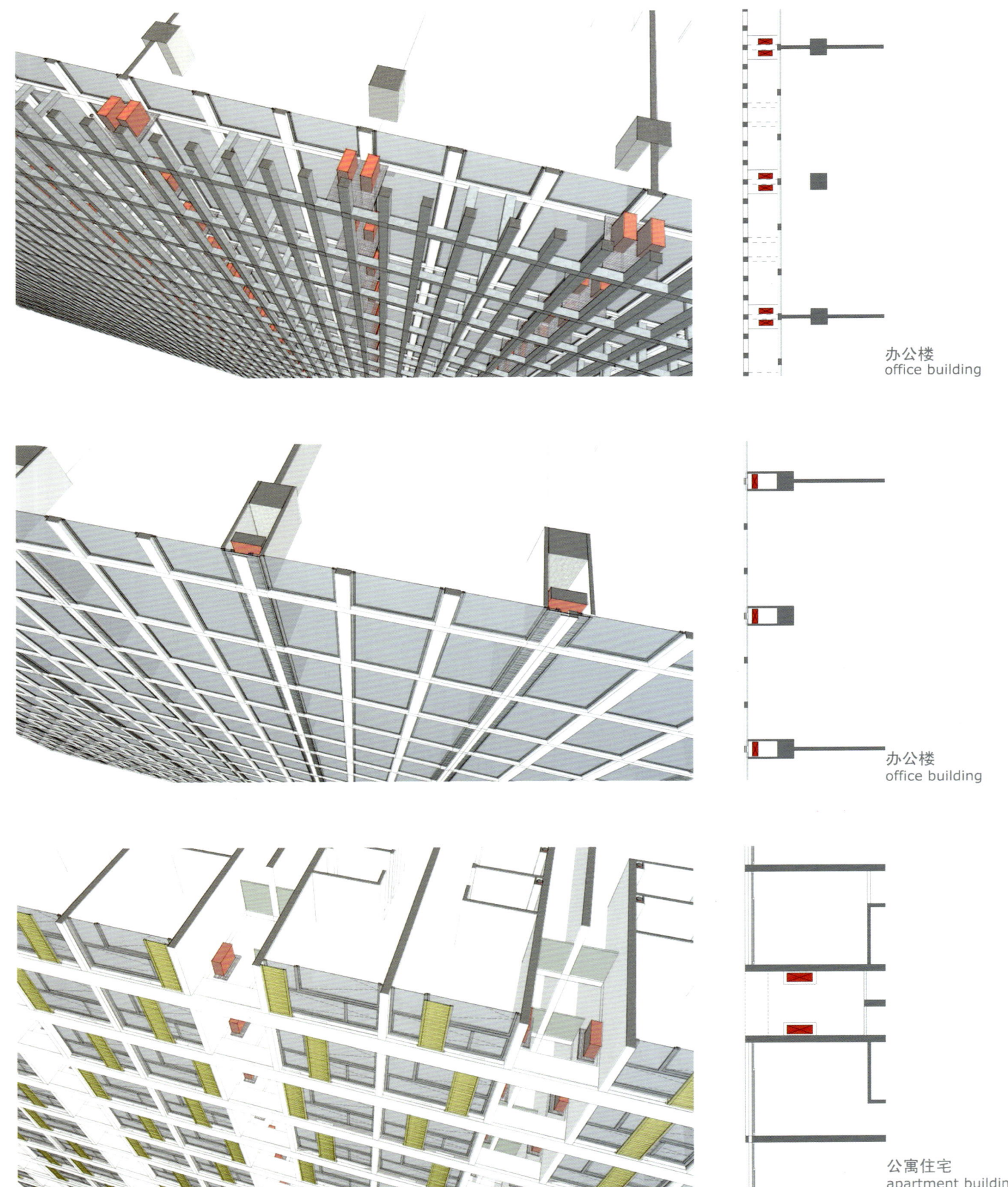

办公楼
office building

办公楼
office building

公寓住宅
apartment building

  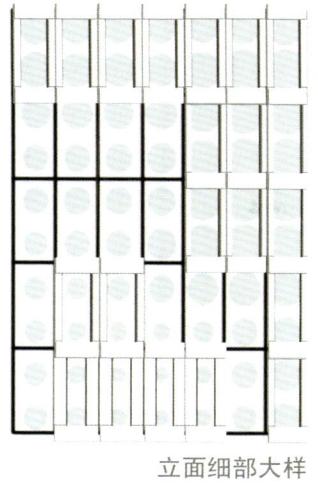

设计原型　　　栅格化　　　立面细部大样

以芙蓉花为胚，设计利用公寓楼立面上的竖向窗间隔墙及横向梁格、空调百叶组成的结构网格，并以其为画布，通过色彩的变化和组合，为城市展现了一副花开芙蓉图，美观且经济，为城市增添了无限的活力及艺术气息。

With hibiscus as embryo, the designer makes fully use of the structural mesh made up of vertical window isolating walls, horizontal beam girds, and air-conditioner shutters on the façade of the apartment building and regards it as a canvas to present a picture of hibiscus in blossom to the city by means of color changes and combinations. The picture is beautiful and economical, bringing an infinite vitality and artistic atmosphere to the city.

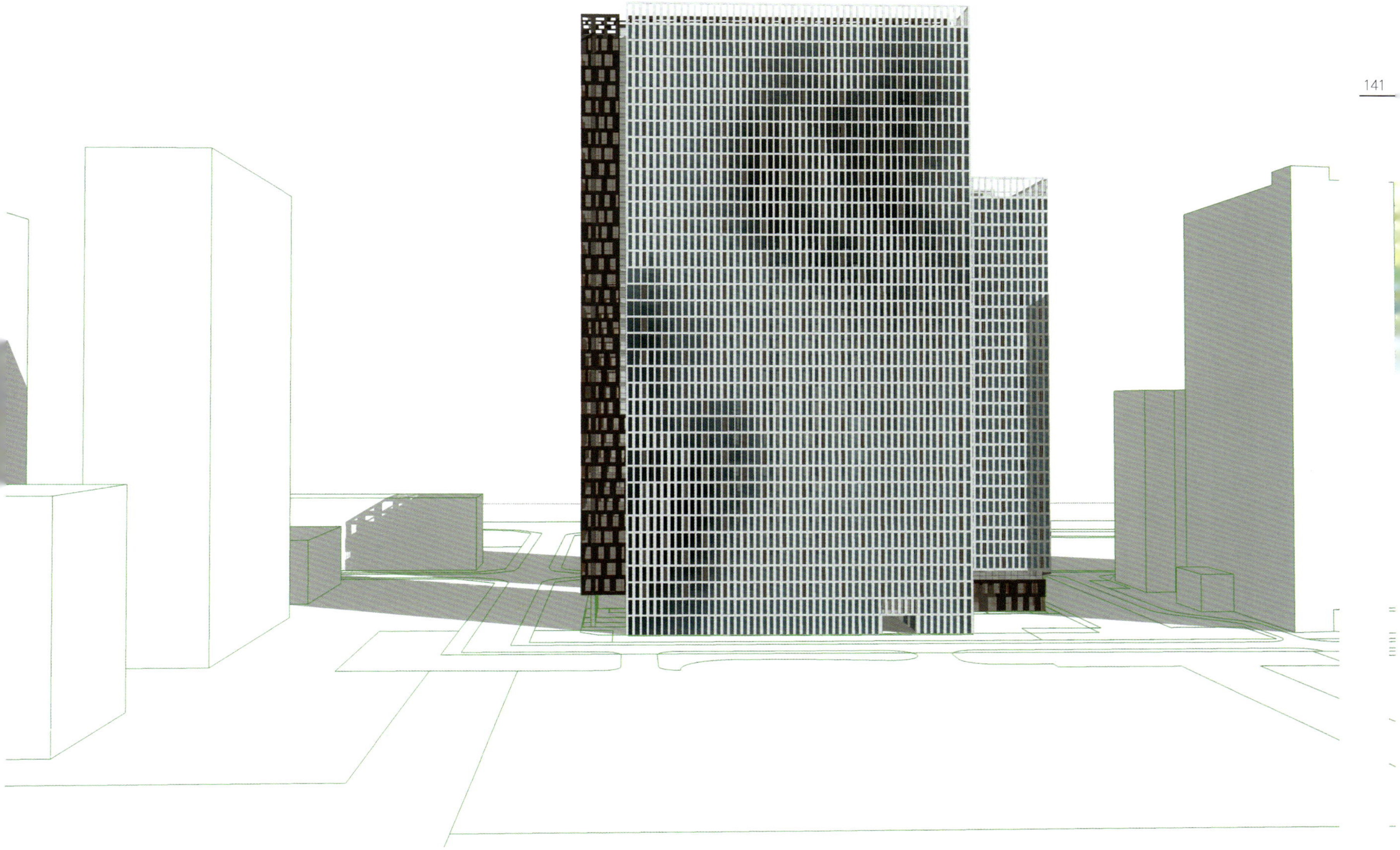

本项目独有的色彩设计将使其与成都周边灰暗的建筑基调形成鲜明的对比。以这些建筑为背景，本项目明快的表皮设计将大大提升其自身在整个成都城市空间中的可辨别度。

Buildings are detailed with color which stands in a contract to the surrounding building and the city of Chengdu. It backed the buildings to create one complex, also creates a unmistakably skin which can be recognized in the urban space of Chengdu.

空调设置

设计中空调储藏方式主要可以分为水平式及垂直式两种。其中，公寓部分采用经典的水平式凸窗下放置，辅以水平百叶遮挡并丰富立面形式；而办公部分则依循立面形式，在垂直方向有韵律地设置空调储藏空间，辅以穿孔挡板遮挡，丰富立面变化而不影响整体立面风格基调。

Air-conditioner setting

Air-conditioners are placed horizontally and vertically in the design. For apartments, air-conditioners are placed under the horizontal bay window, covered by horizontal shutters. Such design enriches the form of façade. For offices, an air-conditioner storage space is set in the vertical direction according to the form of façade, shielded by perforated plates. Such design enriches the façade but does not affect the general style of the façade.

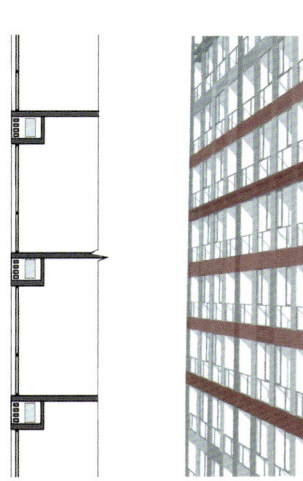

公寓水平式空调储藏剖透视

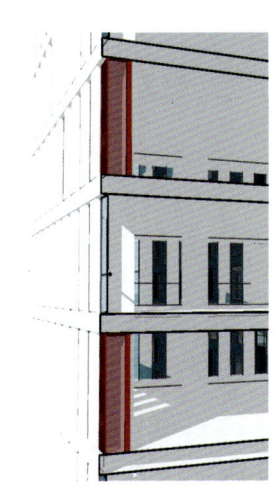

办公垂直式空调储藏剖透视

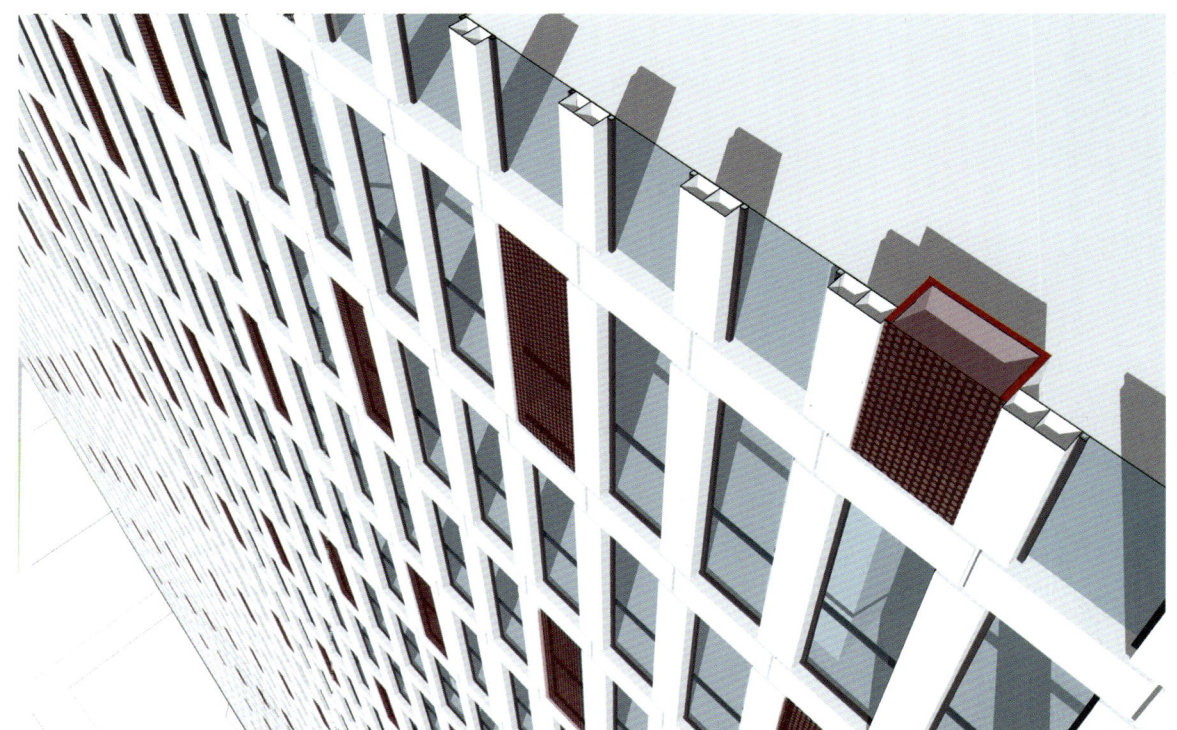

办公垂直式空调储藏剖透视

between the high density of buildings to be built and the landform with great elevation difference. Finally the contradiction is resolved successfully by adopting a design making full use of the elevation difference.

设计的重点在于解决开发强度和巨大的地形差间的矛盾,结果利用地形高差的设计使问题得到了充分解决。

Maoye Baizixiang • Chongqing

Chongqing | Project Location
2004 | Project Date
588,000m² | Prcject Scale
conceptual design | Design Phase
construction not started | Prcject Status

贸业百子巷·重庆

中国，重庆市 | 项目地点
2004年 | 设计时间
58.8万m² | 项目规模
方案设计 | 设计阶段
未建 | 项目现状

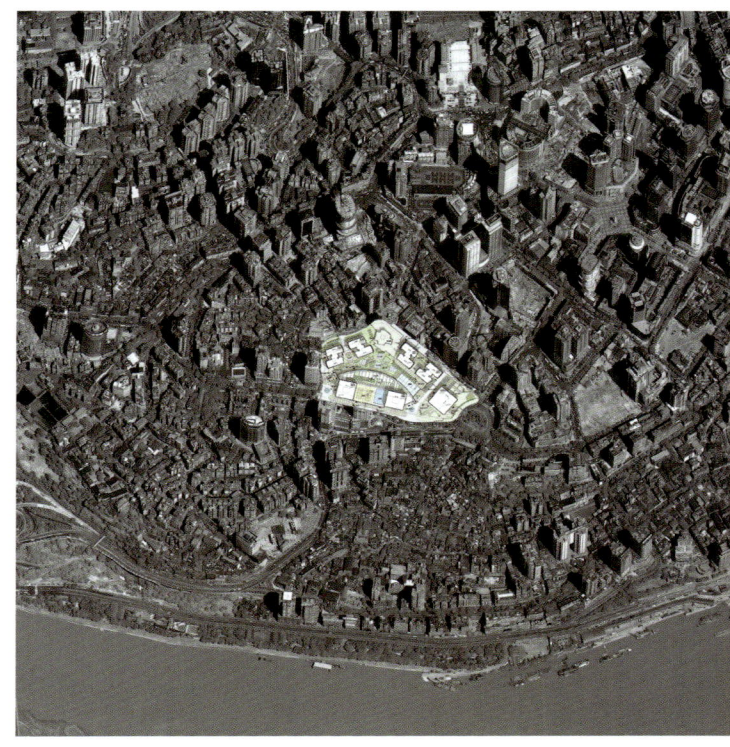

项目总览

本项目位于解放碑商圈"十"字金街南向延伸线端部,城市形象上被定义为渝中区的"城市之冠"。业主拟将其倾力打造成一处集办公、酒店、商业、居住为一体的城市地标性建筑群。

项目采用大围合的布局方式,争取了中心庭园及房屋间距的最大化,同时保证了各功能块的合理分区及校场口、和平路方向的视觉效果。内部商业街的设置很好地延续了城市的解放碑商业圈。立面节点设计精美,尽显重庆"城市之冠"的大气魄。

Project Overview

This complex is located at the end of the south extension line of the "Cross-shaped" Golden Street in the Jiefangbei Commercial Zone. It is defined as the Urban Crown in Yuzhong District. The owner would like to build the project into an urban landmark with integrated functions of office, hotel, business and residence.

The buildings are arranged to encircle a central courtyard, maximizing the area of the central courtyard and the clearance between the buildings and guaranteeing reasonable division of the functional areas and a good view of Jiaochangkou and Heping Road. The internal shopping streets are arranged to integrate well with the Jiefangbei Commercial Zone. The façade nodes are exquisitely designed, exhibiting an imposing spirit of the Urban Crown of the city of Chongqing.

平面图

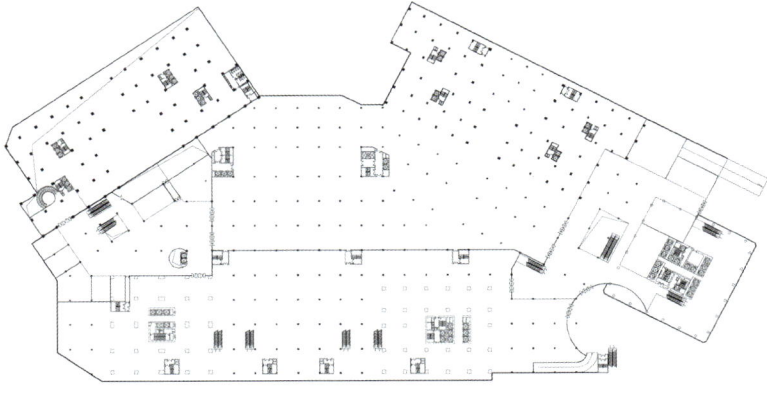

立面图

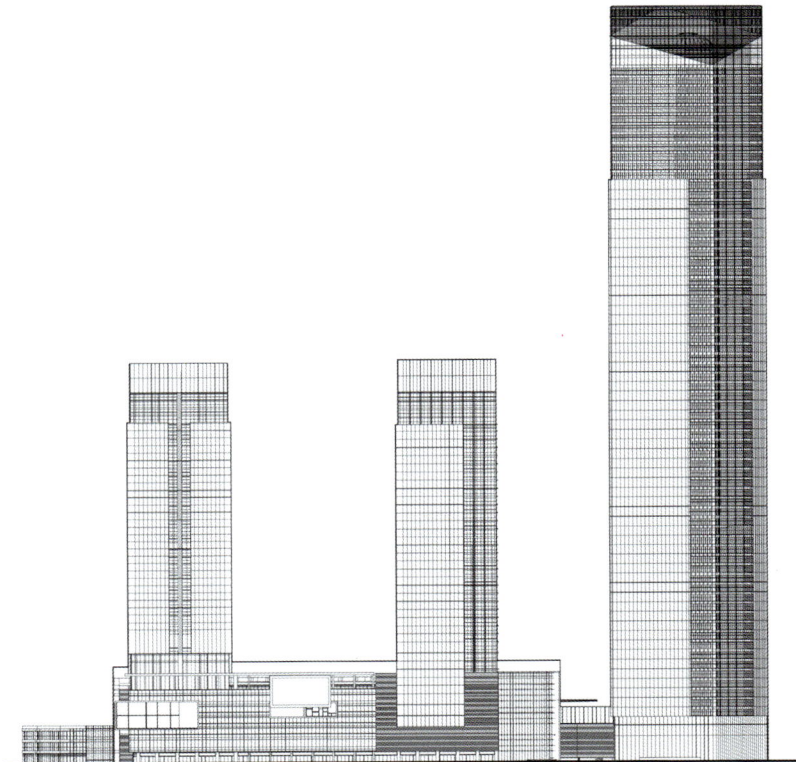

The target is not just erecting independent buildings. The urban space to be created outside the building complex is also important.

目标并非单一的建筑单体，重要的是建筑群所建构的外部城市空间。

Sheffield Land Plaza • Chengdu

Chengdu | Project Location
2009 | Project Date
143,000m² | Project Scale
conceptual design | Design Phase
construction not started | Project Status

谢菲尔德置地广场·成都

中国，四川省，成都市 | 项目地点
2009年 | 设计时间
14.3万m² | 项目规模
方案设计 | 设计阶段
未建 | 项目现状

场地建筑环境分析

布局仔细考量了每栋建筑的视线与采光需求，在避免视线干扰的同时求得了深远的视线以及无阻碍的自然采光。建筑群在南二环沿路方向以较大的退让距离形成商场前广场，写字楼在场地中部围合出为写字楼与商务服务型商业所用的中心广场，使得建筑群以开放的姿态融入城市空间。

Analysis of Architectural Environment

The layout is designed by taking into account the visual link of each building's occupants and the demand for natural lighting, while avoiding interference with line of sight obtained a far-reaching vision and unobstructed natural light. The complex has a relatively long setback distance from Nan Er Huan Road so that a plaza can be built before the emporium. A central plaza, encircled by the office buildings, serves the people in the office buildings and service facilities. Thus the complex integrates with external urban spaces in an open way.

X=15916.953
Y=16214.302
X=15862.986
Y=16256.793
办公入口
X=15690.647
Y=16160.473
X=15793.078
Y=16056.976

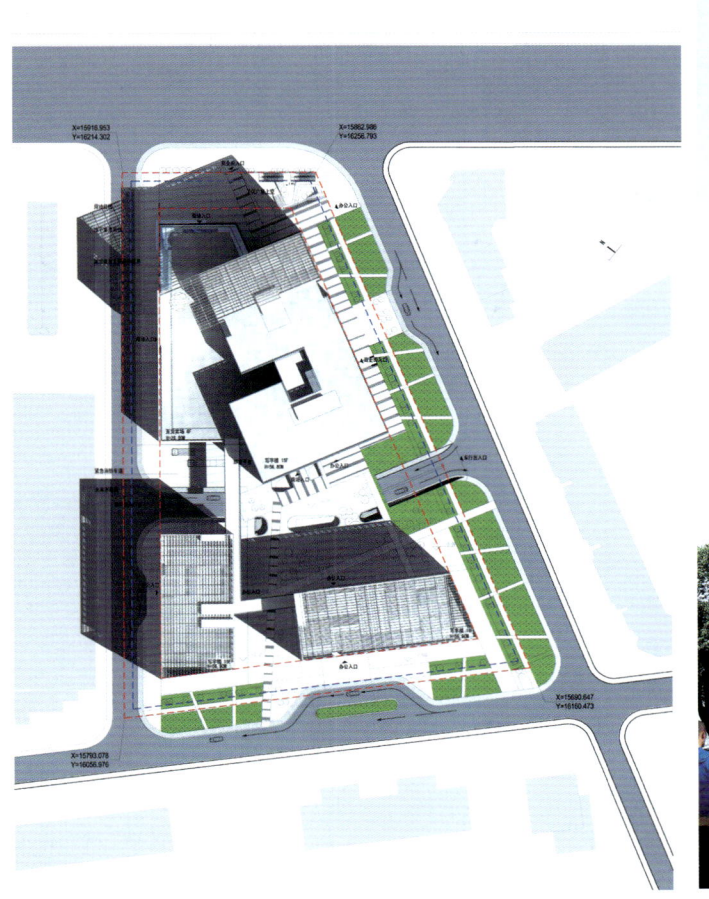

平面图

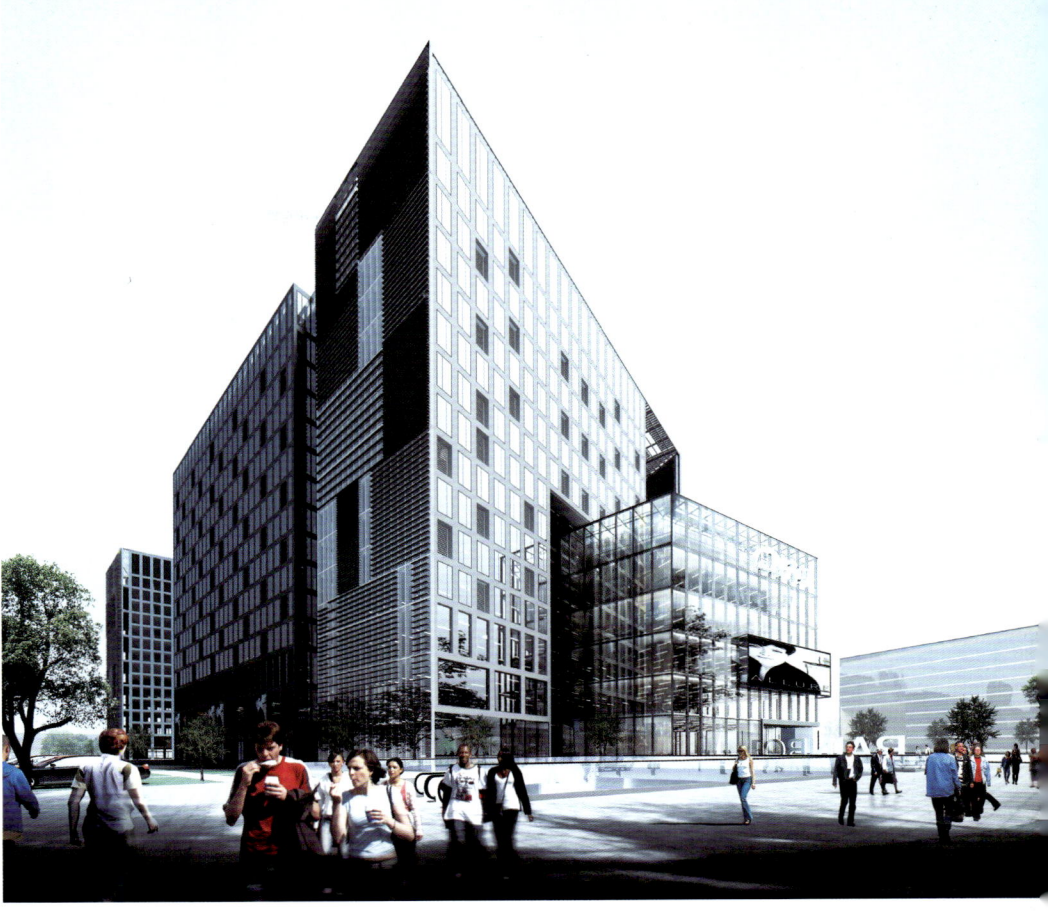

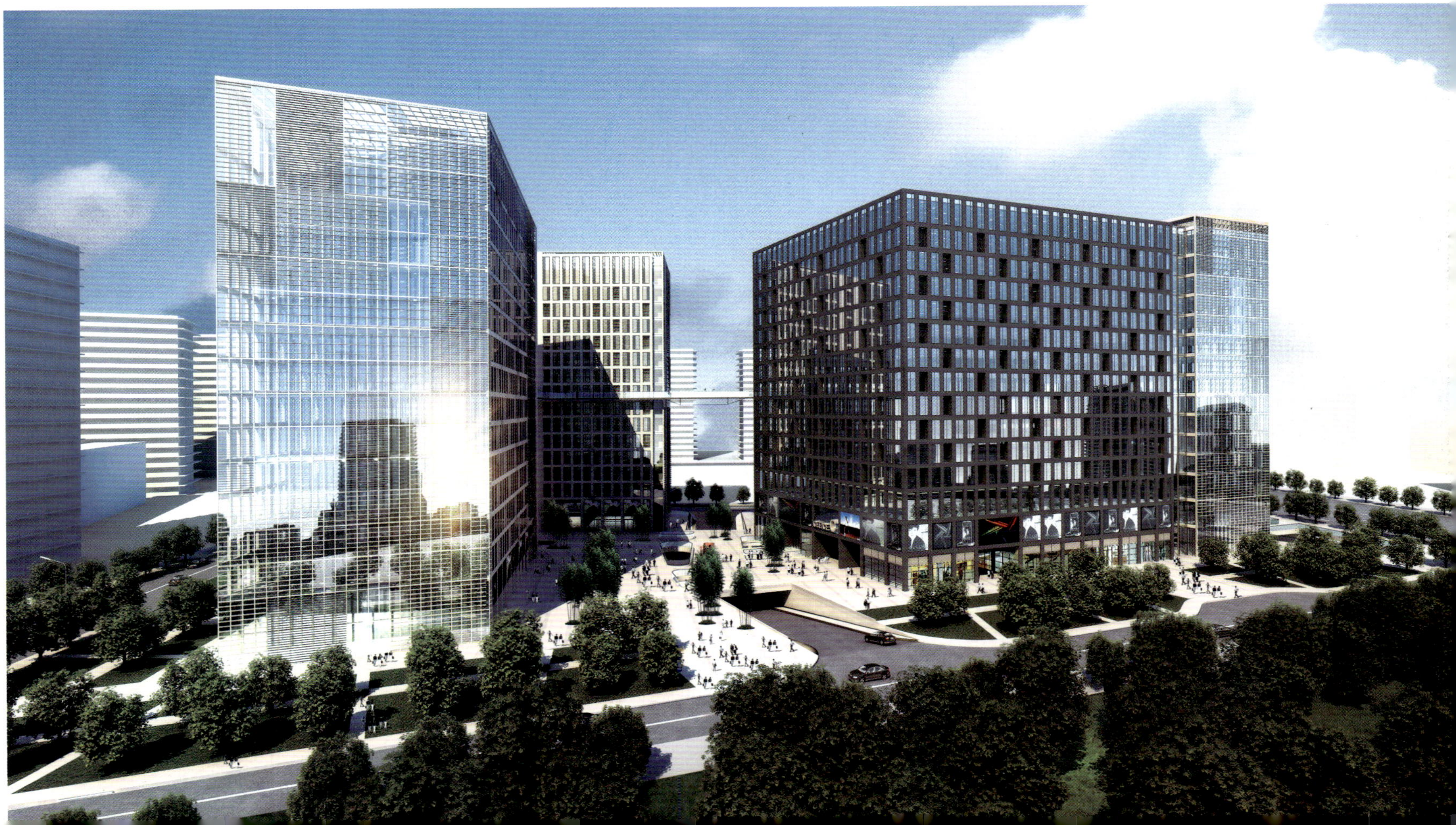

体量研究

地理位置

成都位于中国西南部分,是四川省的省会和副省级城市,也是中国西南部最重要的经济、交通和通信中心之一。

Site Location

Located in southwest China, Chengdu is the capital and a sub-provincial city of Sichuan Province. It is also one of the most important economic, traffic and communication centers in southwest China.

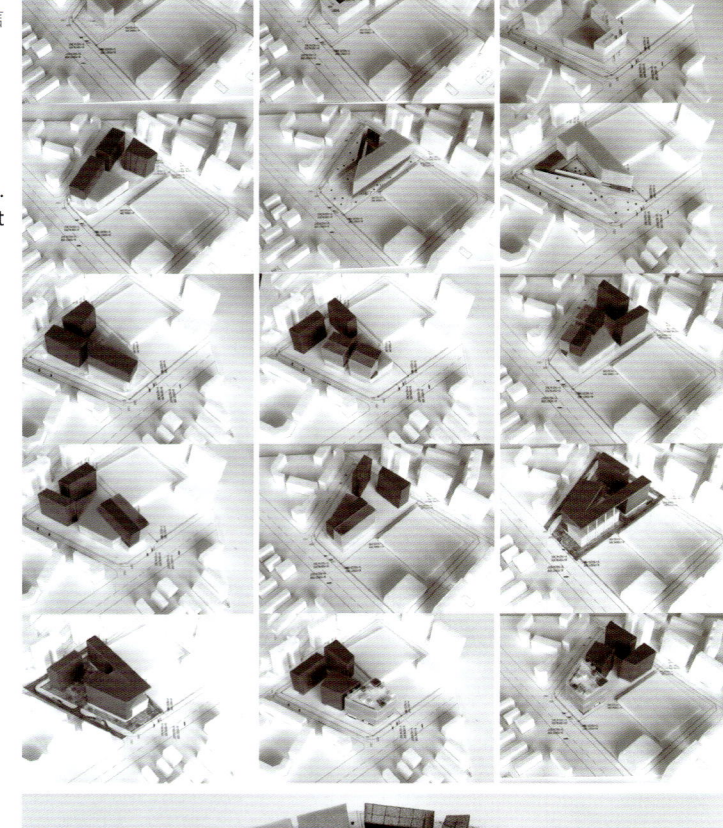

区位分析图

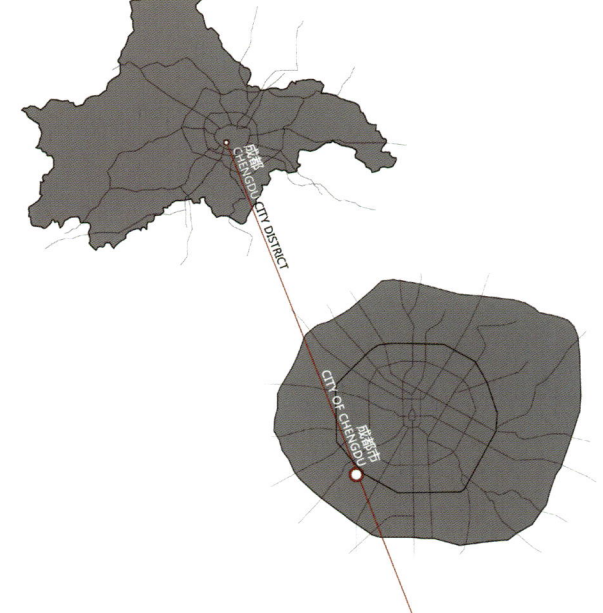

体量研究

通过对任务书的解读,以及对成都市相关规范和形体可能性的研究,辅以理性的比对,最终筛选出符合本项目的目标,并能较好地对项目中产生主要矛盾的方案进行深化设计。

Studies of Size

Through understanding the assignment, studying related standards of Chengdu and shape possibility, and rational comparison, we determine the goal of this project. We are able to carry out in-depth design for main items of the project.

方案解读
BUILDING ANALYSE

方案解读
BUILDING ANALYS

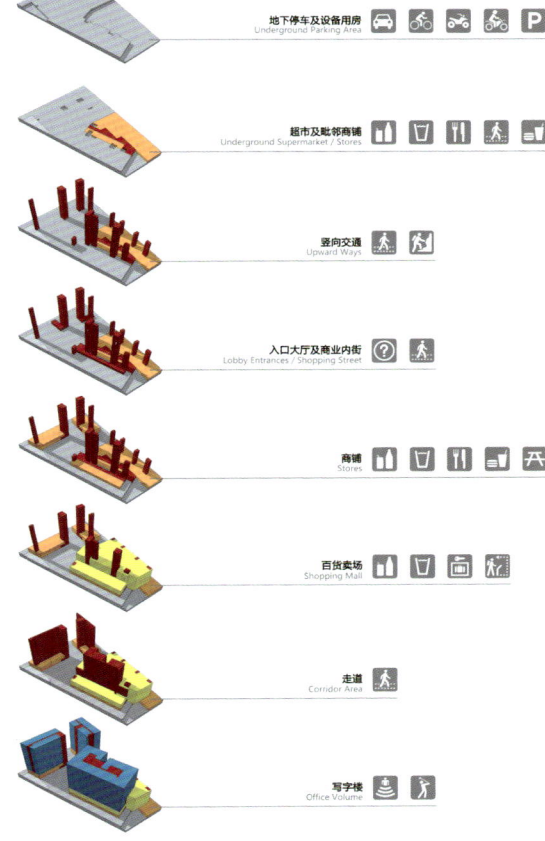

- 地下停车及设备用房 Underground Parking Area
- 超市及毗邻商铺 Underground Supermarket / Stores
- 竖向交通 Upward Ways
- 入口大厅及商业内街 Lobby Entrances / Shopping Street
- 商铺 Stores
- 百货卖场 Shopping Mall
- 走道 Corridor Area
- 写字楼 Office Volume

功能分析图

功能分布图解 Function Diagram

商业店铺 Stores

商业卖场 Shopping Mall

写字楼 Office

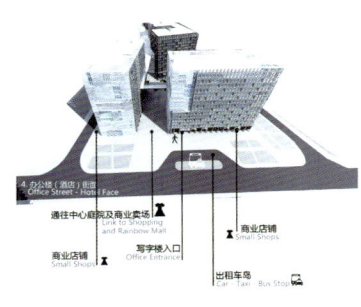

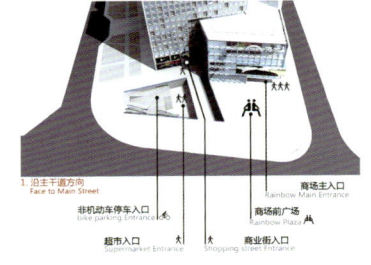

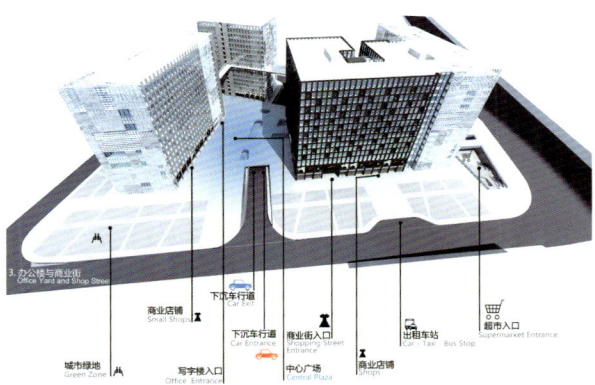
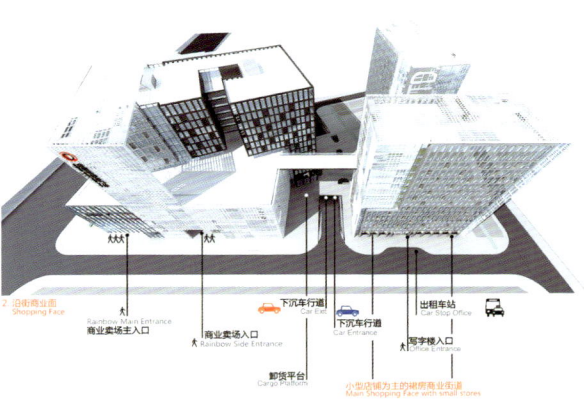

功能布局

竖向空间设计

通过解读任务书发现，写字楼难以避免地会被设置于商业卖场之上，从而导致电梯使用效率低下，商场空间被核心筒打破的问题。设计方案通过局部混合设置写字楼与商业卖场，提高了电梯使用效率，减少了公摊面积，并且利用不同层高获得额外的建筑面积。

Vertical Space Design

After carefully studying the design specification, it is found that the office floors have to be set up above the emporium. As a result, it affects the efficiency of the elevators and the emporium space is broken by the core. The design increases elevator efficiency and reduces shared area by arranging partial offices and the emporium on the same floors, and achieves additional building area by making use of different floor heights.

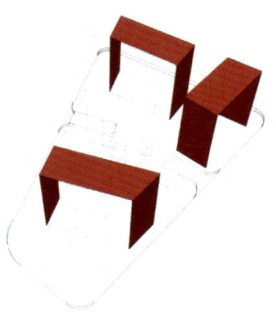 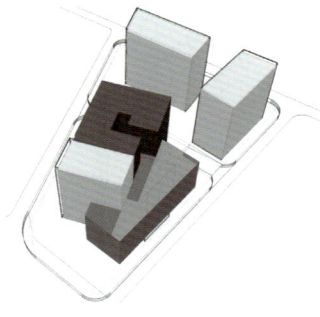 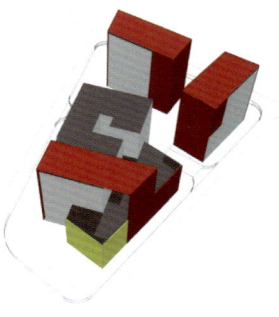

屋顶框架创造出建筑群的整体感。
Frame create a hole building complex.

不同的立面色彩区分出不同的建筑体量
Different colors create different volume.

不同的立面语言界定了不同的建筑功能块
Different facades express the different functions.

建筑立面概念

新时代的办公环境需要高品质的空间、充足的光线和怡人的氛围来维持雇员良好的工作状态，这一切都将在这个未来的商业综合体中得以呈现。

丰富的院落及街道空间伴以混凝土、玻璃与金属构成的表皮凸现出它独特的建筑环境，这三个概念决定了其室内的设计。

Building Facade Concept

A modern office environment calls for a high-quality space, sufficient light, and pleasant atmosphere to facilitate high working efficiency of the employees. All these will be presented in this complex.

Ample courtyard and street spaces plus wall surfaces made up of concrete, glass and metal highlight the superior architectural environment. The interior design is based upon the three concepts mentioned above.

空调机位示意

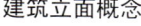
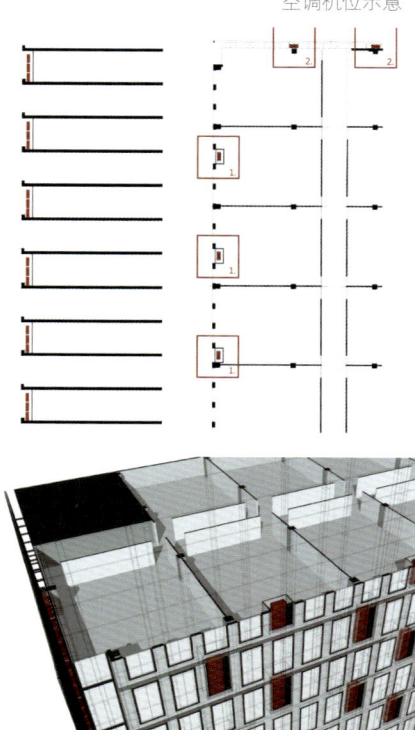

An active landmark image is created on the basis of the principle of integrating the plaza into the city.

以整体的城市观创造积极的具有标志性的"场所"。

City Commercial Plaza • Guigang

Guigang, Guangxi | Project Location
2002 | Project Date
20,000m² | Project Scale
conceptual design, preliminary design, construction drawing | Design Phase
construction completed | Project Status

城市商业广场·贵港

中国,广西省,贵港市 | 项目地点
2002年 | 设计时间
2万m² | 项目规模
方案设计 初步设计 施工图设计 | 设计阶段
已建 | 项目现状

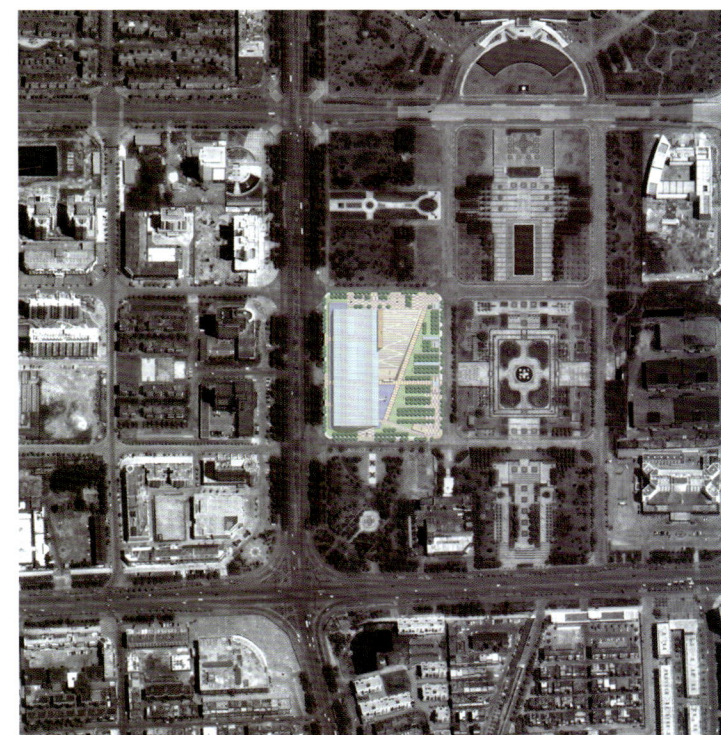

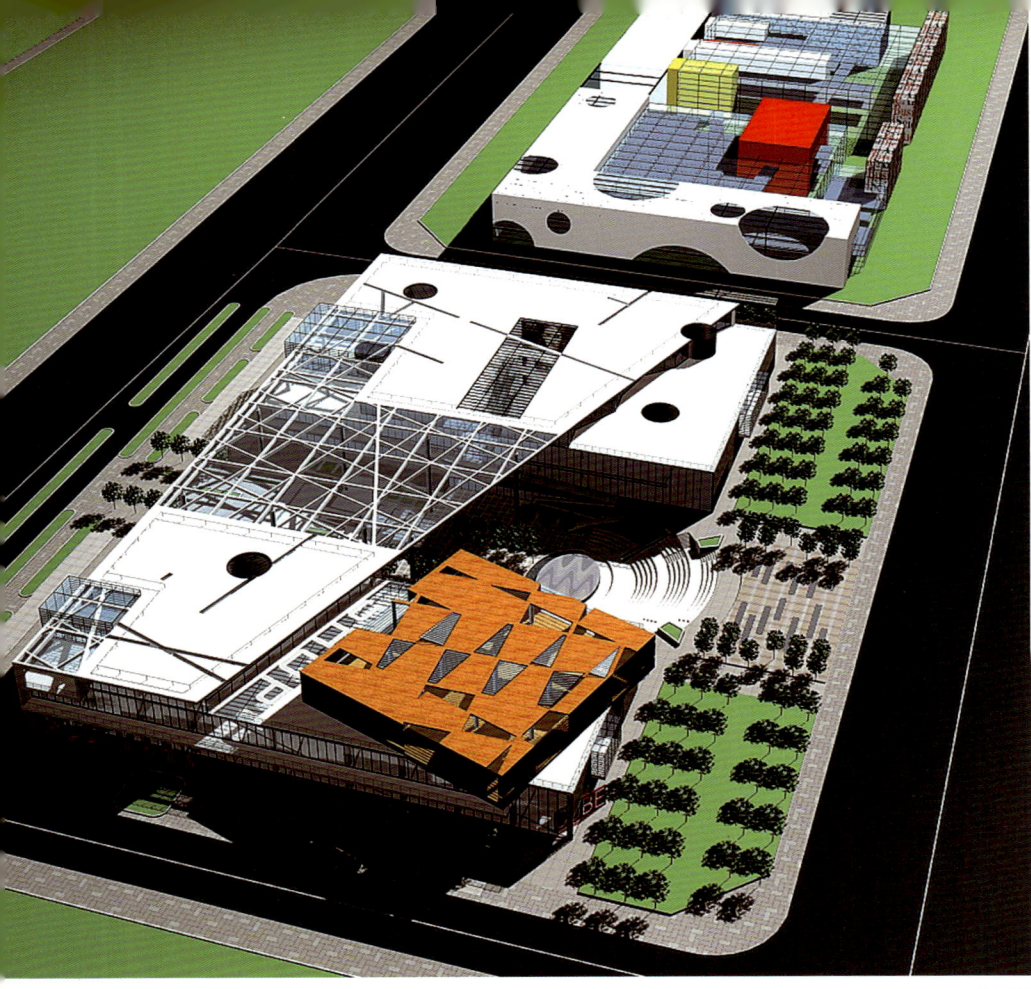

总平面

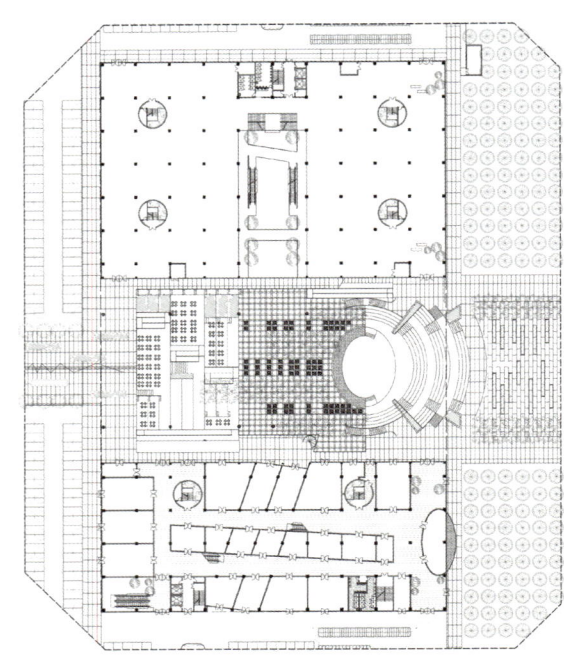

一层平面

项目总览

项目整体规划保留了从中山路看政府广场雕塑的视线通道。主体建筑分为南北两部分,中间自然形成了与政府广场的便捷联系,并成为整个商业广场的主要人行通道和室外景观轴线。同时,阶梯式室外茶座、下沉广场、舞台、露天剧场与政府广场雕塑连为一体。

Project Overview

In the overall planning of this project, the space allowing line of sight is reserved for viewing the statues in the Government Plaza from Zhongshan Road. The main building is divided into southern and northern parts. The passageway in between the two parts connects to the Government Plaza and is a major walkway and the axis of the outdoor landscape in the plaza. The outdoor amphitheater, teahouse, sunken plaza, stage, and open theater integrate perfectly with statutes in the Government Plaza.

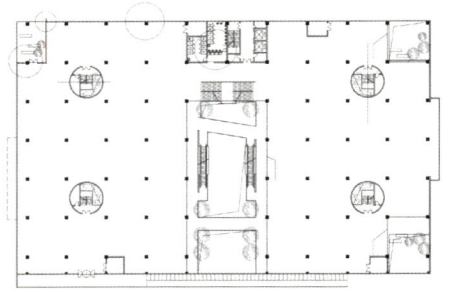

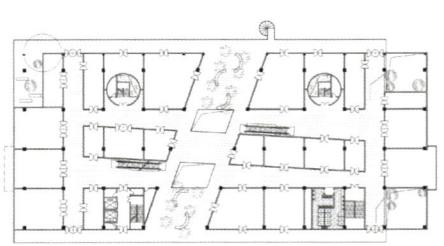

二层平面

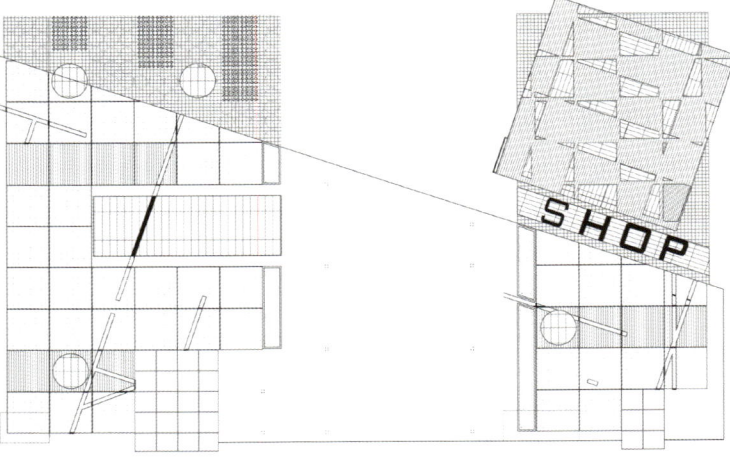

屋顶平面

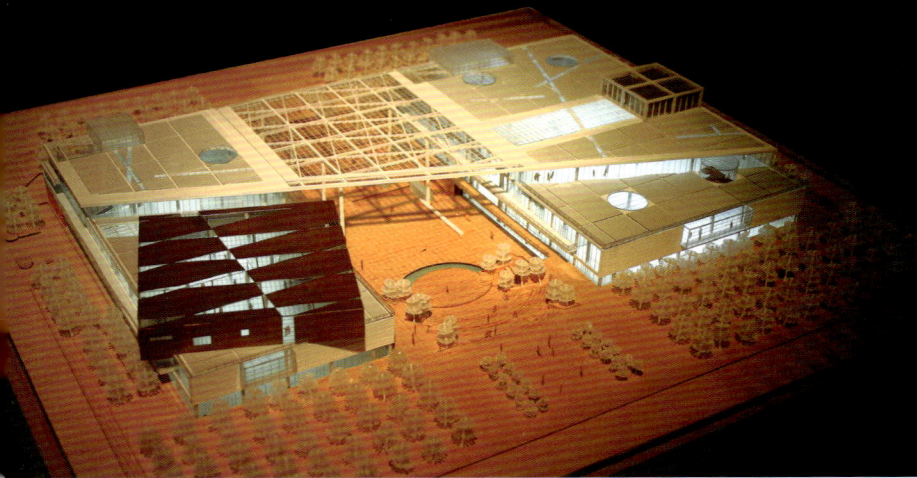

商业广场地下一层，地上三层，每层层高 4.8m。三层建筑体型与行政中心呼应，形成一个面向行政中心的二层屋顶平台。同时三层屋顶架空混凝土梁上覆以玻璃，为阶梯式茶座遮风挡雨，也是政府广场人群避雨的重要场所。

The commercial plaza has one basement floor and three aboveground floors, and the floor height is 4.8m. The shape of the 3-floor building is compatible with the Administration Center, with a roof platform on the second floor facing the Administration Center. The suspending concrete beam on the roof of the third floor is covered with glass, preventing wind and rain from entering the teahouse and provides shelter for people staying in the Government Plaza.

剖面图

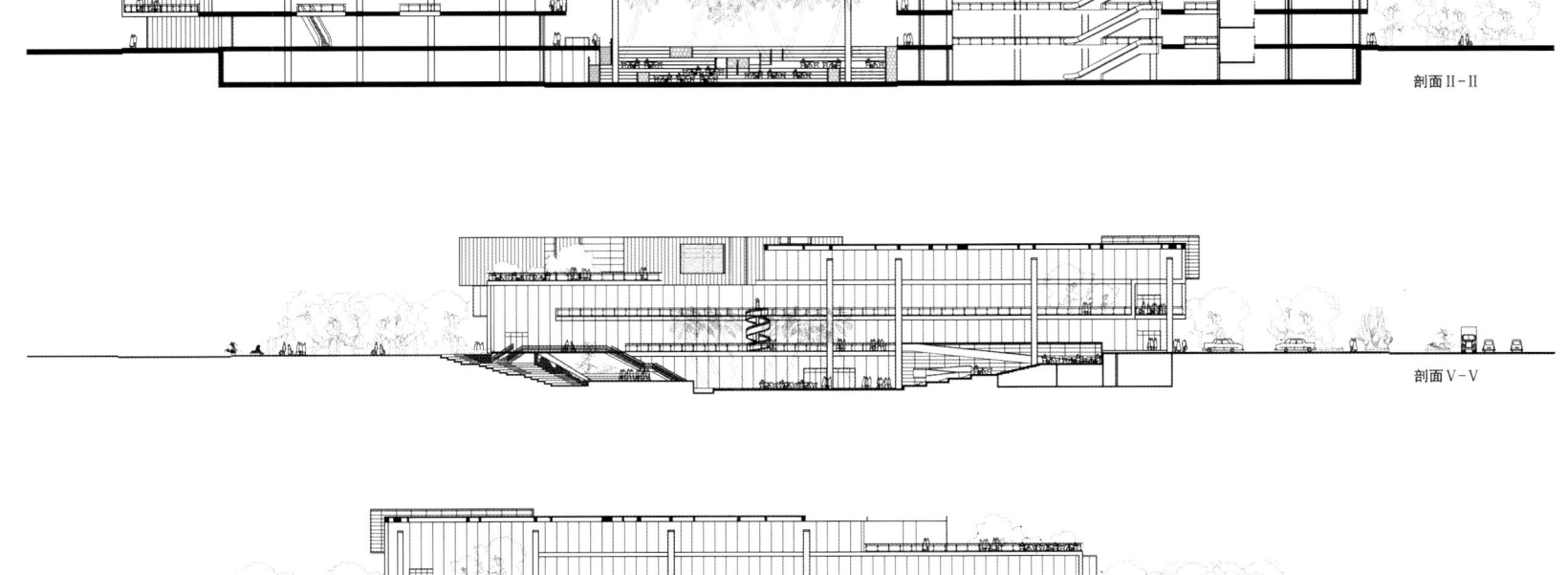

剖面 II-II

剖面 V-V

剖面 I-I

Education & Culture......

Education & Culture

文化教育

A cultural building should present a humanistic spirit. Its architectural image, space, site, and every detail should communicate with people, make people think it over, and enlighten people. We try to realize this result in all memorial halls, exhibition halls designed by us. Buildings are a miniature of human culture. Educational and cultural buildings are a result of this miniature. Our projects reflect a kind of accumulation that focuses on time, past, present, and future, and shows history, culture, and tradition and adapt to life styles, space, and functional demands as well as a consideration and imagination about the future.

文化建筑应具有人文精神。它的建筑形象、空间与场所、甚至每个细节都应该能与人交流，引人思考，给人启迪。我们在每一个纪念馆、展览馆建筑设计中也都朝这个方向去努力。建筑是人类文化的缩影，教育文化建筑则是这一缩影的结晶。在我们的创作项目中，我们贯穿着一种着眼于时间的理念，过去、现在和未来，并体现出历史、文化、传统的沉淀，适应当下的生活方式、空间和功能需求，以及对未来的构思和畅想。

The inspiration derives from labyrinth. A building offers two sets of spaces, one forward one inverse, which are assembled together subtly.

灵感来源于走迷宫。一幢建筑提供两套空间,"正的"和"反的",并且奇妙地组合在一起。

Museum of Contemporary Art and City Planning Exhibition Hall • Shenzhen

Shenzhen | Project Location
2007 | Project Date
80,000m² | Project Scale
conceptual design | Design Phase
construction not started | Project Status

现代艺术馆与城市规划展览馆·深圳

中国，广东省，深圳市 | 项目地点
2007年 | 设计时间
8万m² | 项目规模
方案设计 | 设计阶段
未建 | 项目现状

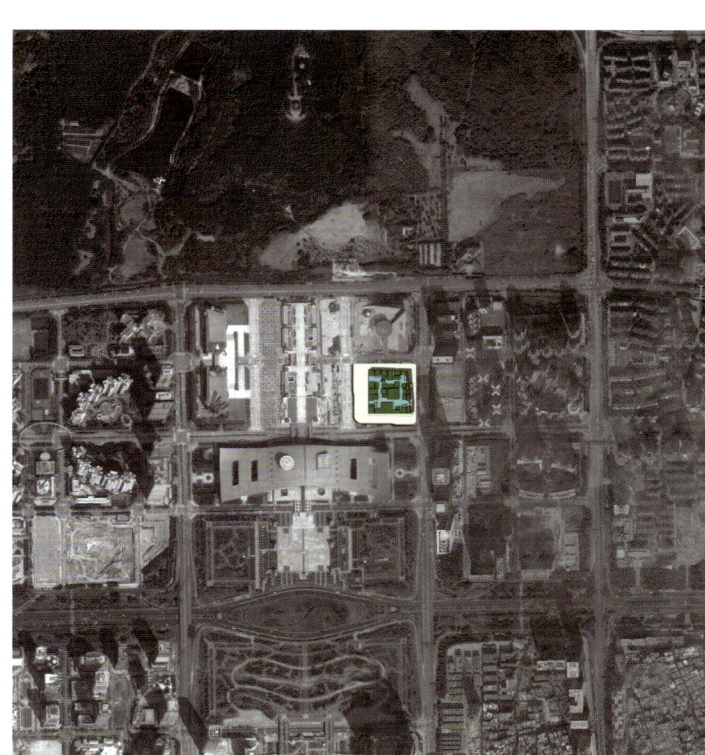

项目总览

深圳市当代艺术馆与城市规划展览馆由1栋3层钢结构建筑物组成。总建筑面积为80,000m²，其中地上面积43,677.50m²，地下面积36,322.50m²。本项目把城市引入到建筑中来，将城市的街巷空间转换为城市规划展览馆，城市建筑内空间转换为当代艺术馆，让实时的"城市活动"成为动态的当代艺术展示。这样，两个展馆就建立起一种互为图底、互为内外的紧密"城市关系"。

另外，通过将建筑物抬起，在两馆和服务配套空间之间为城市提供一个开放和有遮挡的公共空间，这里是另一个户外的展览空间，一个完全属于大众的空间。

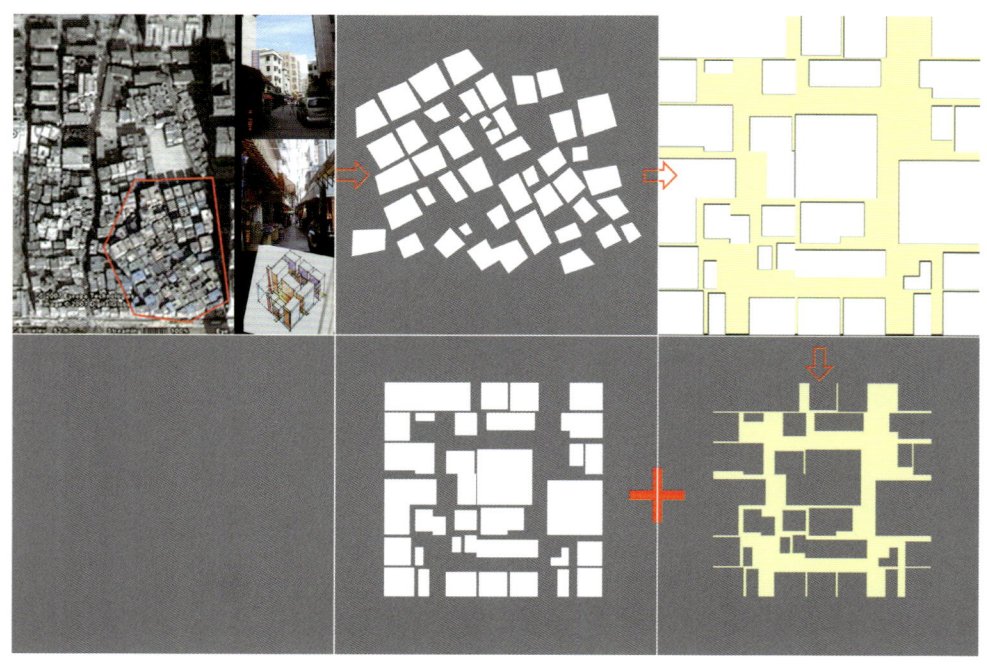
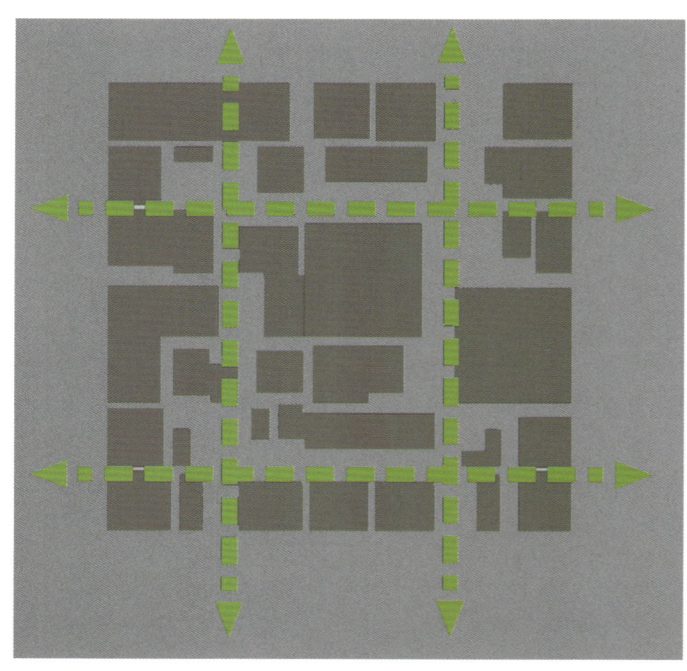

Project Overview

The Shenzhen Museum of Contemporary Art and City Planning Exhibition Hall, a 3-storey steel-structure building, has a total building area of 80,000 square meters, including aboveground area of 43,677.50 square meters and basement area of 36,322.50 square meters. The building is designed by introducing urban elements into the building and converting urban street and lane spaces into city planning exhibition hall and converting indoor spaces of an urban building into a museum of contemporary art. Thus real-time urban activities themselves become dynamic exhibition of contemporary art. In this way, the two exhibition halls establish a close "urban relationship" in which each sets off the other.

In addition, an open and sheltered public space is offered between the two halls and the supporting service space by raising the buildings. It is another outdoor exhibition space, a space wholly for the pubic.

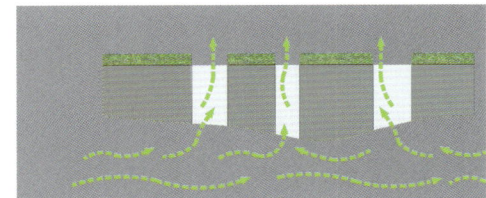

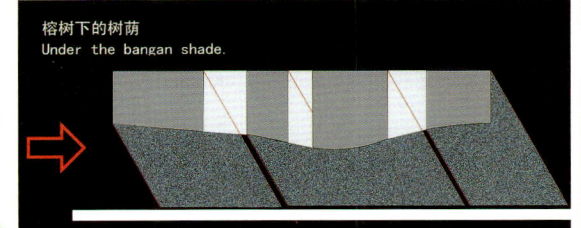

榕树下的树荫
Under the bangan shade.

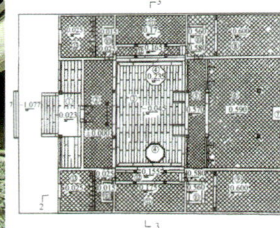

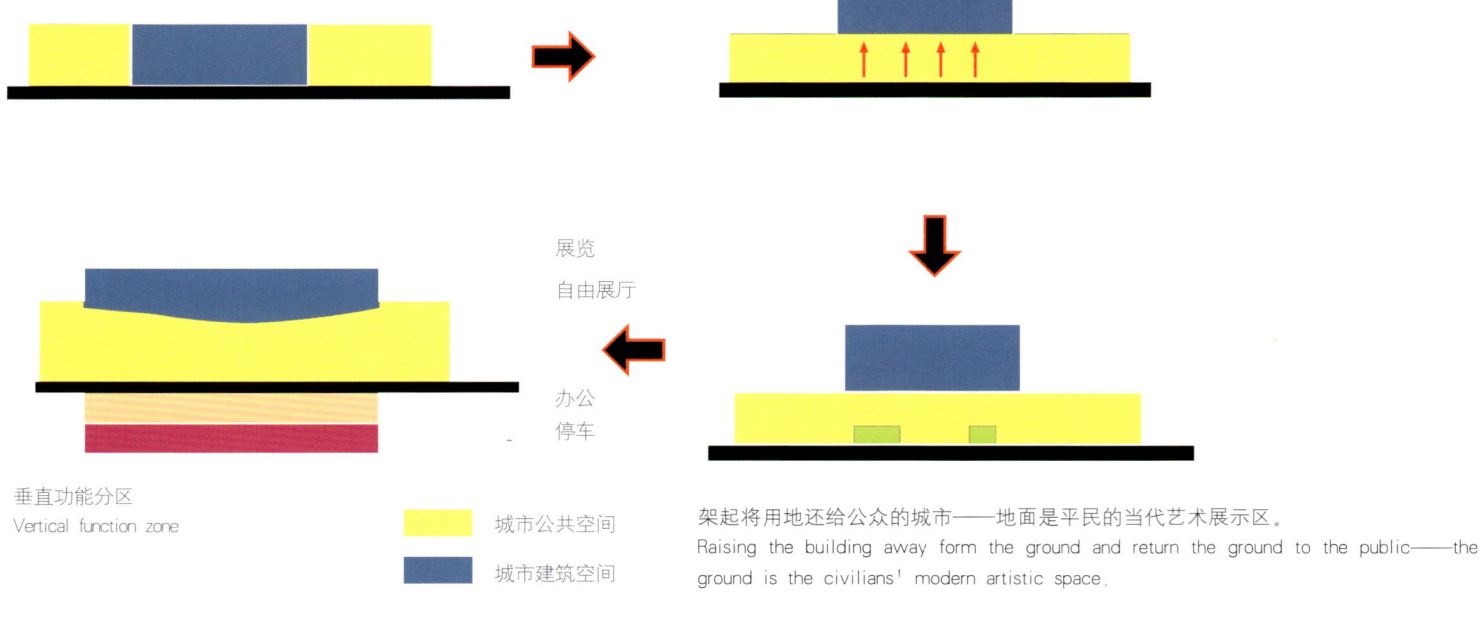

垂直功能分区
Vertical function zone

■ 城市公共空间
■ 城市建筑空间

展览
自由展厅

办公
停车

架起将用地还给公众的城市——地面是平民的当代艺术展示区。
Raising the building away form the ground and return the ground to the public——the ground is the civilians' modern artistic space.

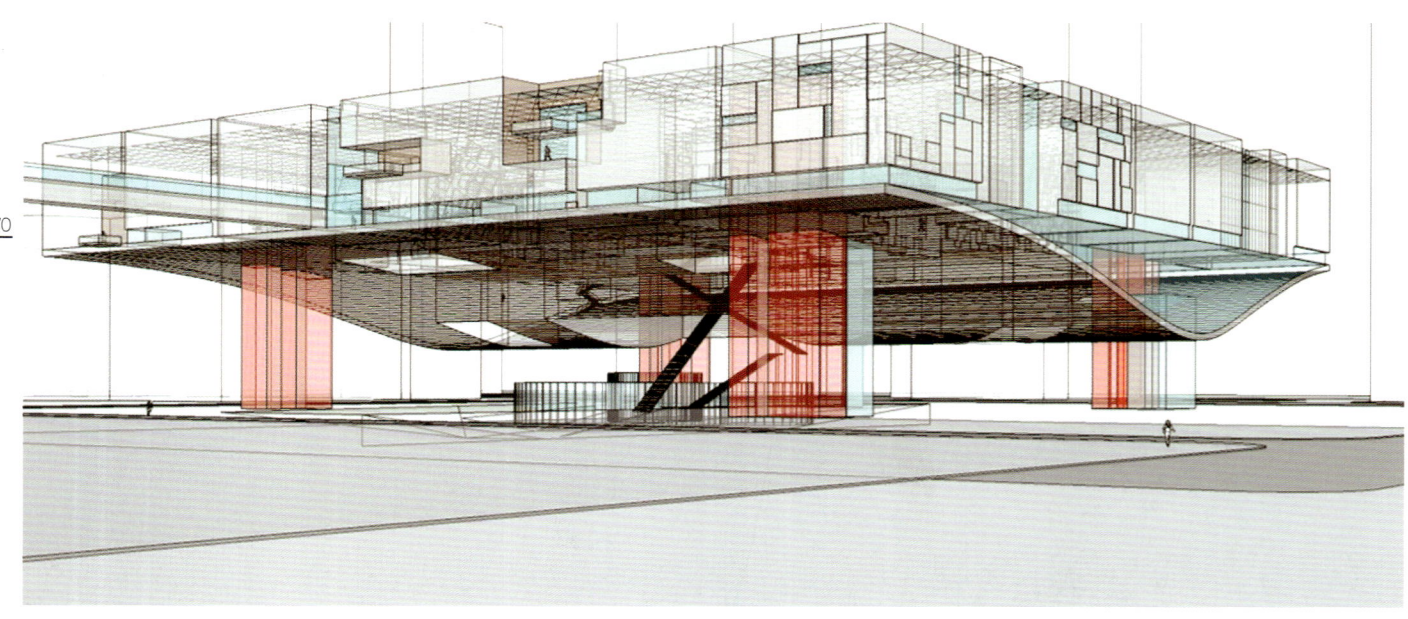

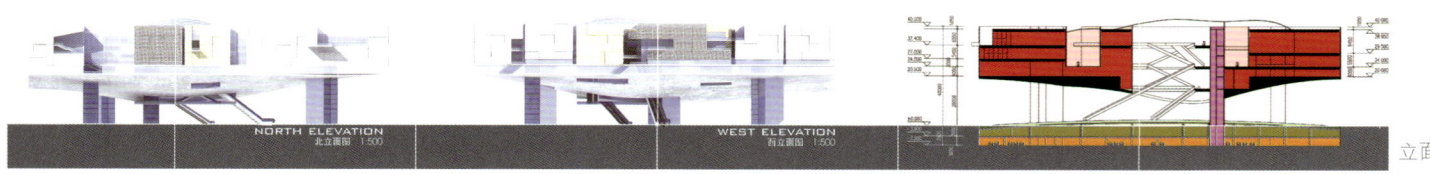

NORTH ELEVATION 北立面图 1:500
WEST ELEVATION 西立面图 1:500
立面图

曲线的生成——curve

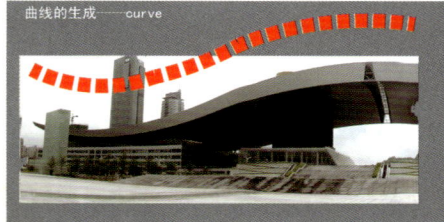
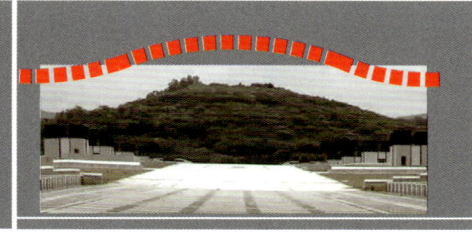
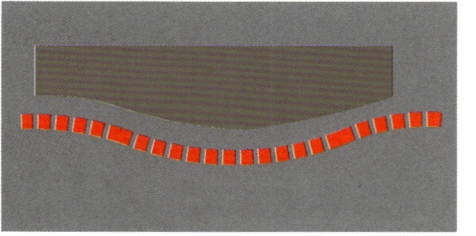

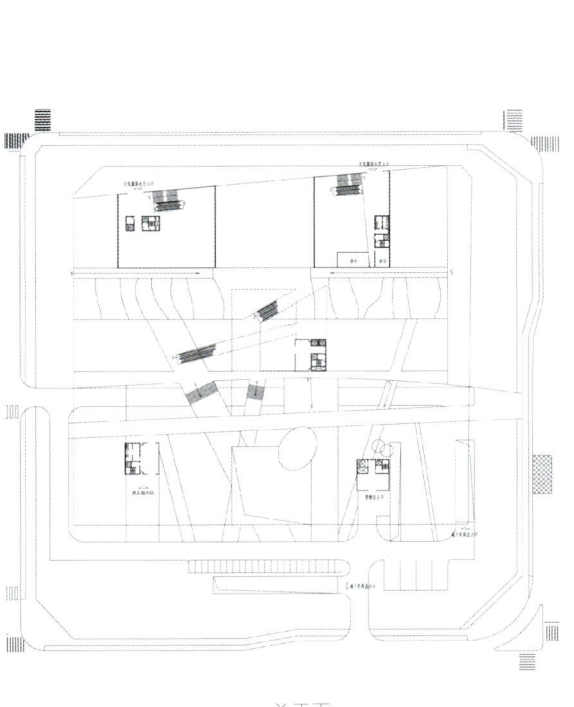

总平面

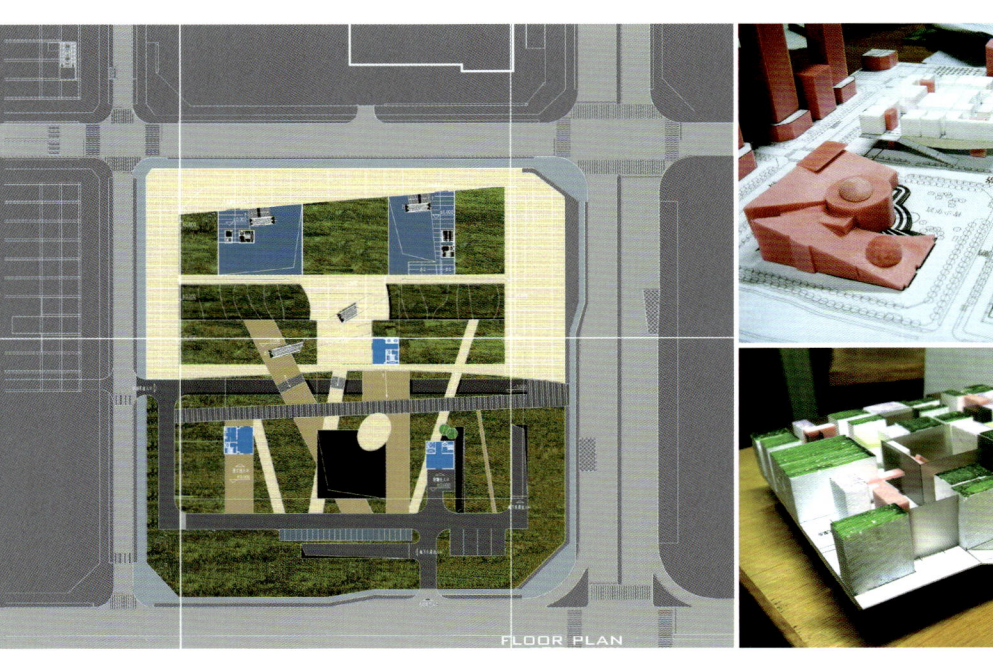

FLOOR PLAN

172

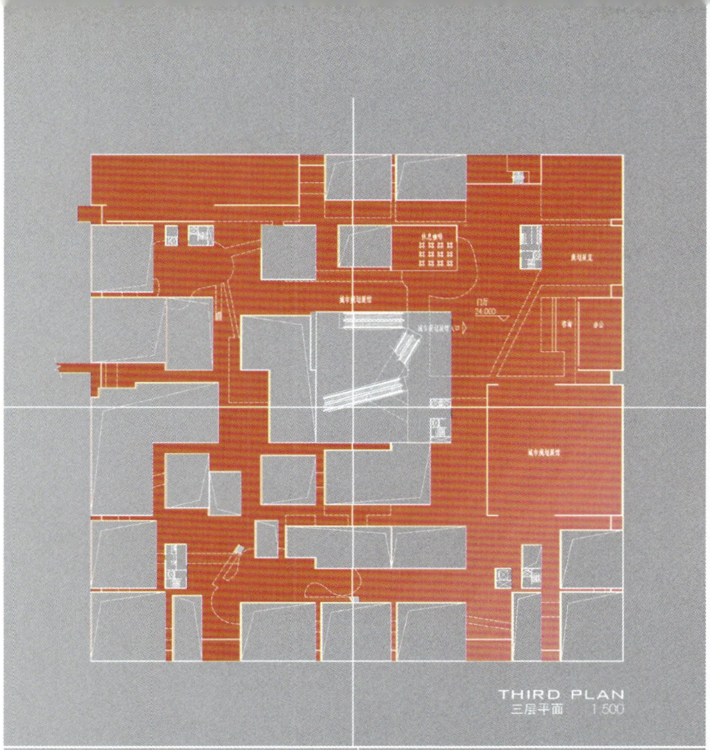

THIRD PLAN
三层平面 1:500

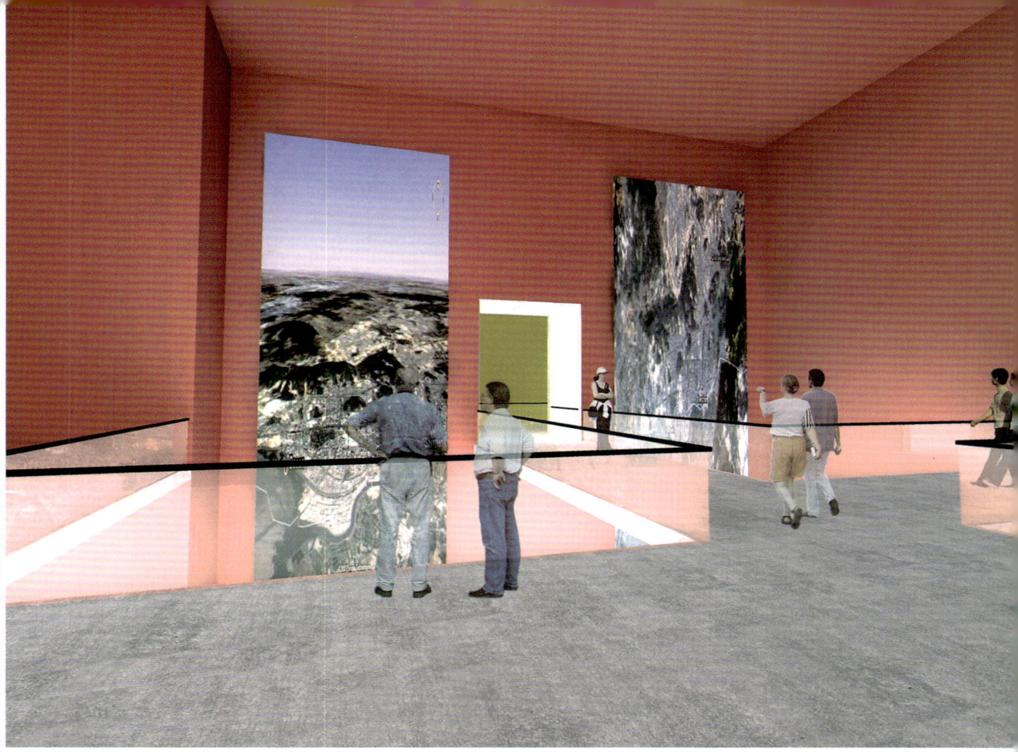

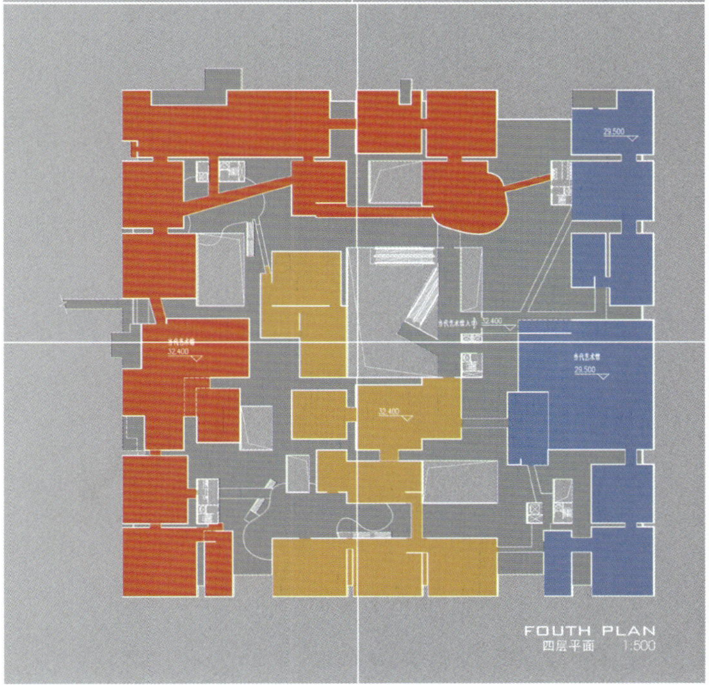

FOUTH PLAN
四层平面 1:500

平面图

0m标高平面

8.40m标高面

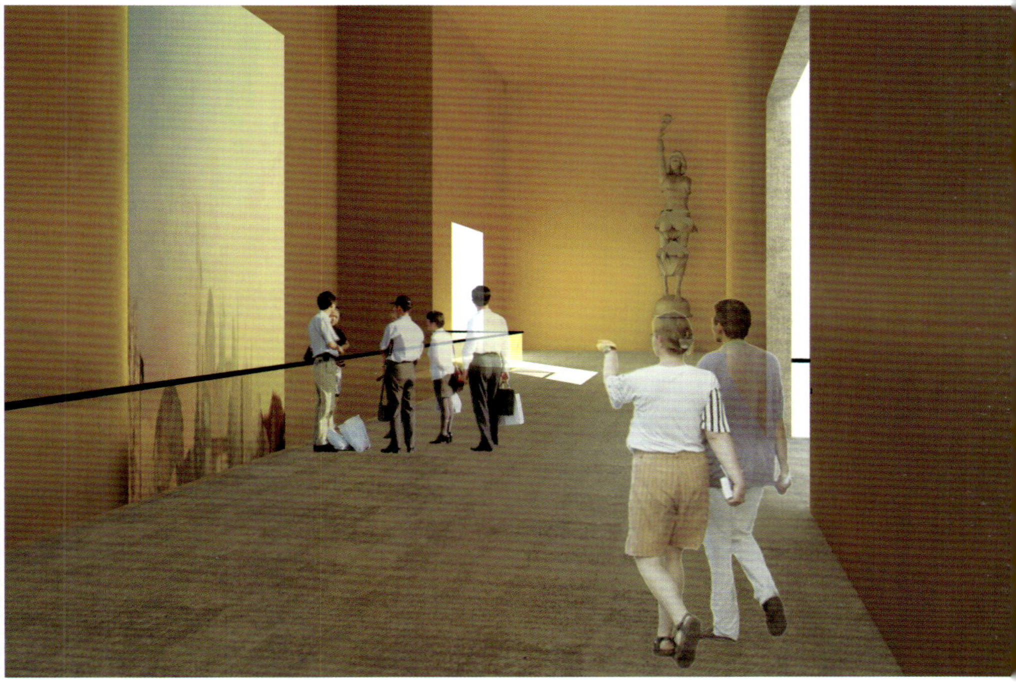

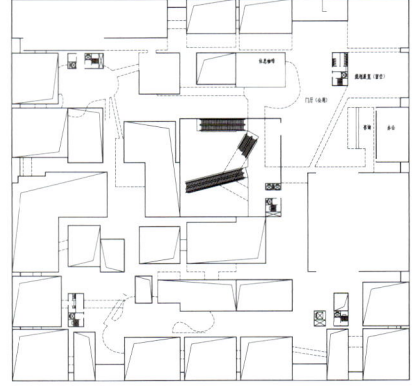

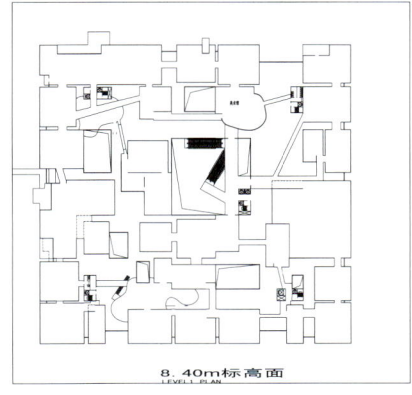

8.4m标高平面

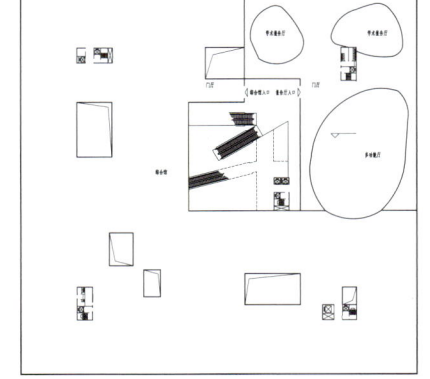

二层平面

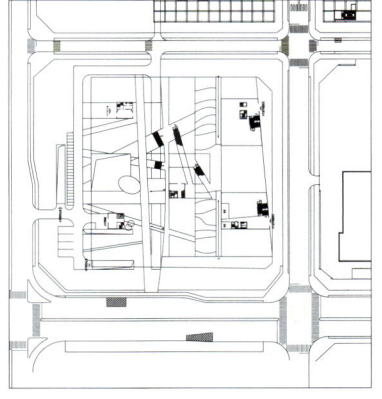

平面模型

The building should express a modest feature, a respect for nature, and a care for mankind. We hope our design makes the building look like that great man.

建筑应该表达出一种足够谦虚的姿态，对于自然的尊重以及对人类的关怀。我们希望设计的建筑看起来就像我们需要纪念和缅怀的那位伟人。

Hu Yaobang Memorial Hall • Xiangtan

Hunan | Project Location
2005 | Project Date
28,000,000m² | Project Scale
conceptual design | Design Phase
construction not started | Project Status

胡耀邦纪念馆·湘潭

中国，湖南省，湘潭市 | 项目地点
2005年 | 设计时间
2800万m² | 项目规模
方案设计 | 设计阶段
未建 | 项目现状

项目总览

本项目方案吸收了当地传统民居的造型因素，建筑所选材料均来源于当地，简朴低调。此外，项目还设计了73步象征胡耀邦73年革命生涯的台阶，一切都如胡耀邦同志一般，亲切而可接近。

Project Overview

The hall is designed to be like a local traditional folk house, simple and unadorned. All building materials are purchased locally. A special design of 73 steps represents Hu Yaobang's 73-year revolutionary career. Everything is as cordial and approachable as Mr. Hu Yaobang.

项目分析

建筑造型和材质

由于展厅与展廊的不同标高，部分建筑底面高于基地。采用吊脚楼建筑形式，正好扩大庭院面积，同时有利于通风和改善小气候，也有利于展品的保存。天井的设计富有湘鄂民居的特色，利于自然采光。

Project Analysis

Architectural Modeling and Materials

Since the exhibition hall's elevation is different from that of the exhibition corridor, a part of foundation bed is higher than the base. The form of hanging house is adopted to enlarge the area of courtyard, good for ventilation, improvement of microclimate, and storage of exhibits. The light-well provides natural lighting and is designed with a style of Hunan and Hubei folk houses.

总体布局

用地周围为山地，山上多茂密的自然植被。纪念馆坐落于丘陵之中，顺应山体走势，尽可能保护山体原貌，并通过桥与山体取得直接便捷的联系。

纪念馆与故居有直接的视线联系。我们用桥、行列树等使纪念馆与故居形成一条视线走廊，加强两者间的联系。

展廊沿山体外立面均为透明玻璃，清山翠林直接映入眼帘。同时，建筑元素取意于周边植被，使整个建筑与环境融为一体。

General Layout

The plot is surrounded by hills with dense natural vegetation. Thememorial hall is located among hills, conforming to the trend ofhill slopes, with the original appearance of the hills preserved asmuch as possible. The hall is connected directly with the hills via abridge.

From the memorial hall, one can see the former residence of HuYaobang. The memorial hall and the residence are connected witha bridge and rows of trees, which forms a corridor guiding the lineof sight of visitors to view the two buildings. Thus the connectionbetween them is enhanced.

The exhibition corridor's walls along the hill are constructed with transparent glass. Thus green hills and woods can be vieweddirectly by visitors in the corridor. The architectural elements implysurrounding vegetation, thus the whole building blends in withthe environment.

东立面

西立面

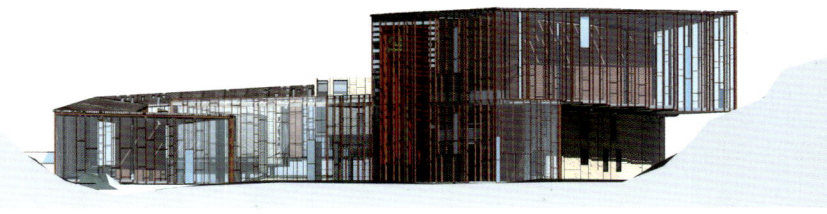

北立面

南立面

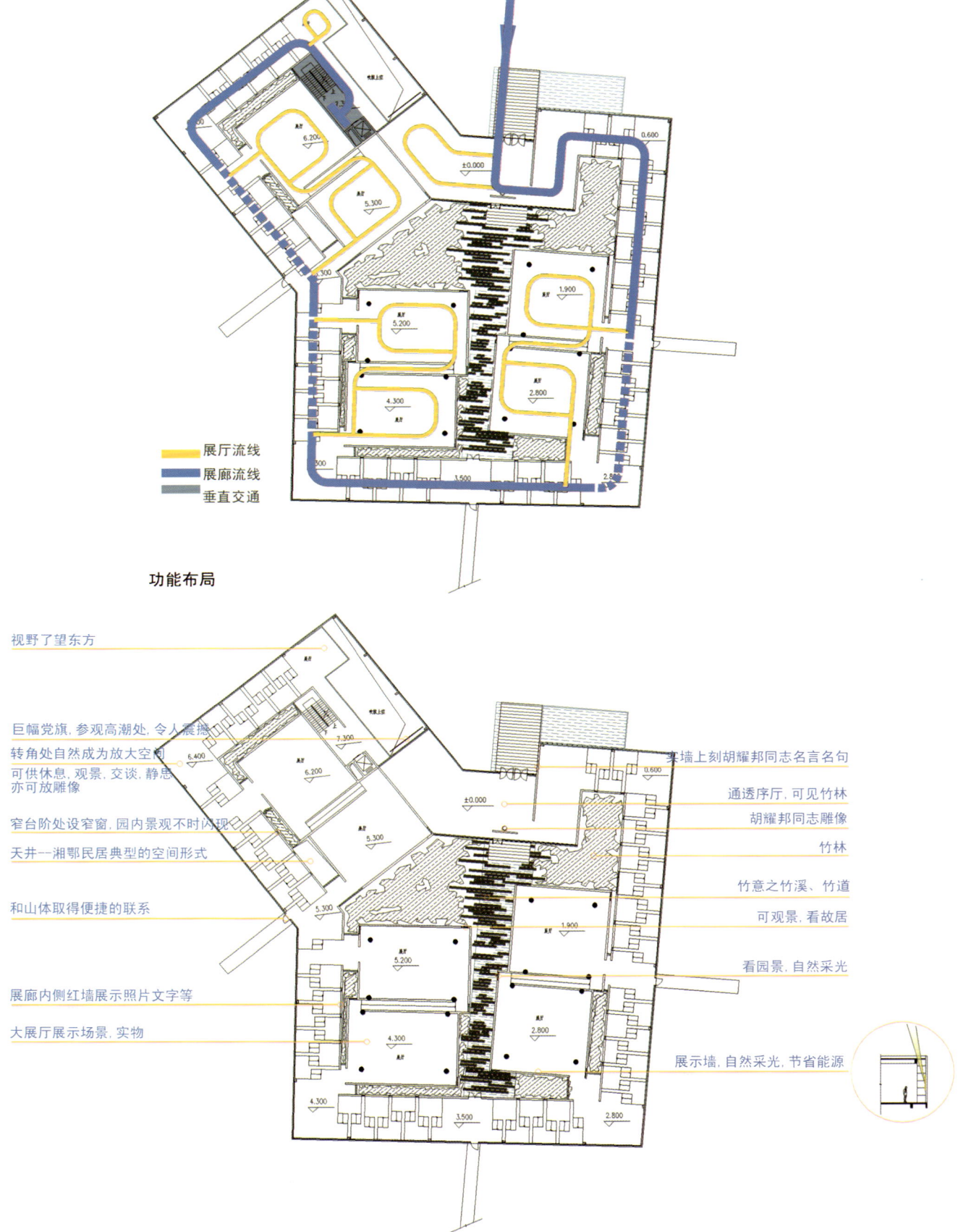

功能布局

功能布局及参观流线

灰色屋顶
红色的砖墙
竹节立面
党旗
透明外立面
73步台阶

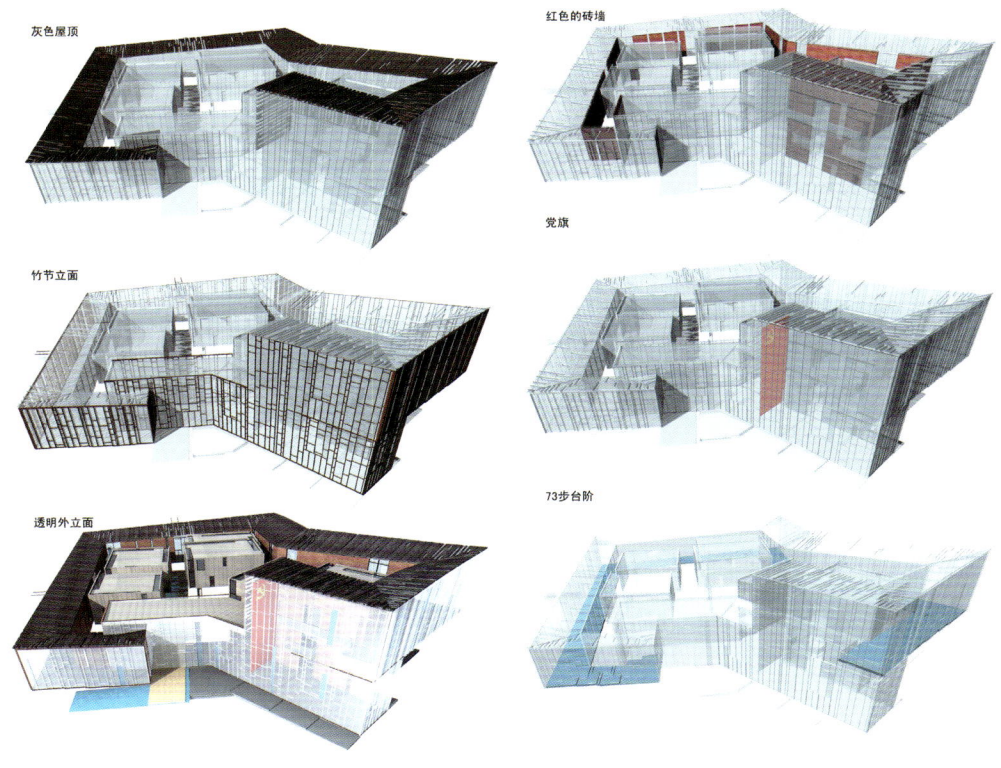

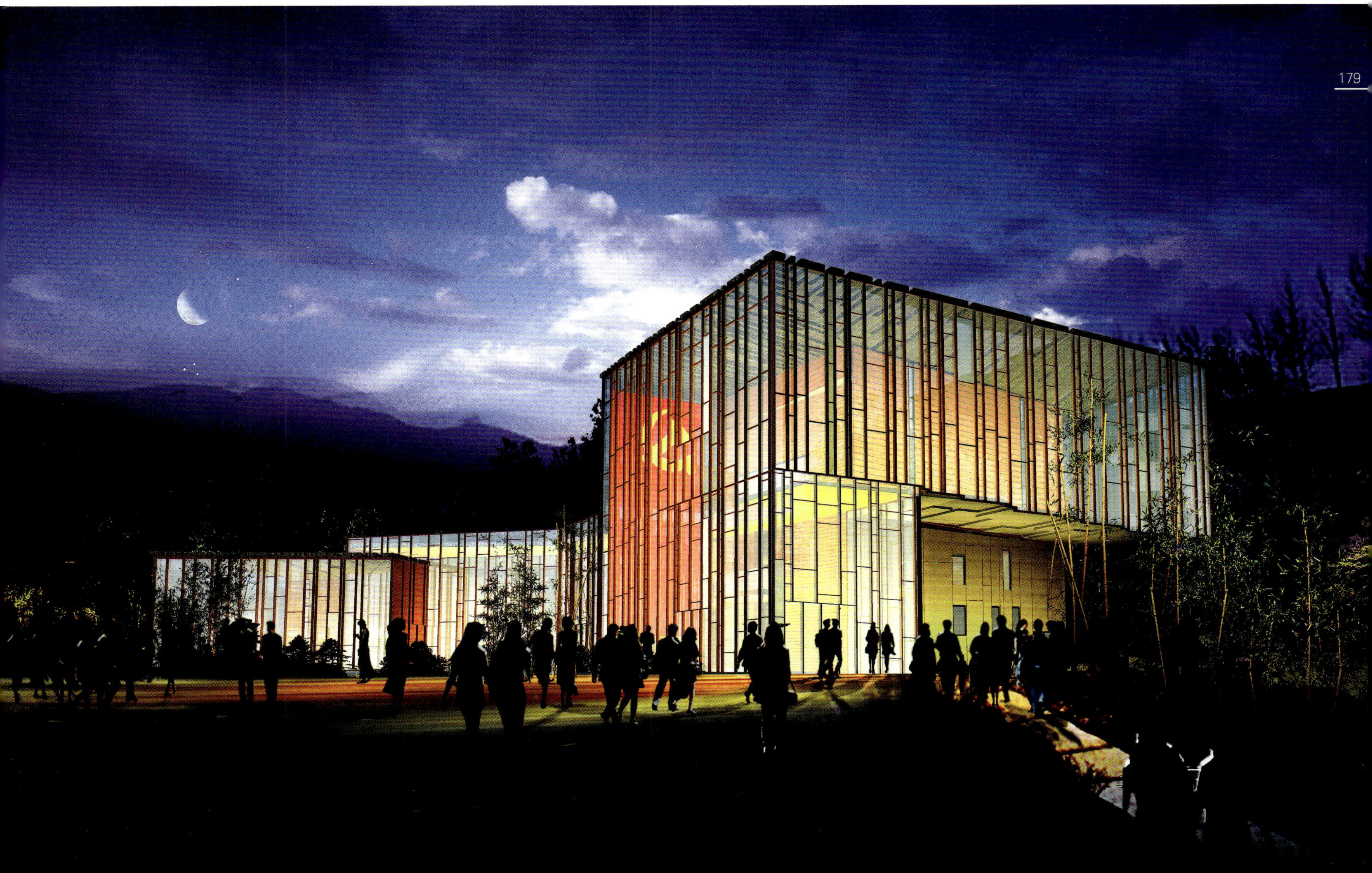

功能布局

经过长长的通道来到广场入口,首先看到的是胡耀邦同志的雕像,沿着台阶走去,一边是记载着历史片断的照片和文字展示,一边是透过玻璃欣赏到的自然景观。沿途还可以参观展廊旁边的展厅,也可以在展廊的休息区休息、静思、交流。走到台阶的尽端,可以看到一面鲜红的党旗挂在东方,阳光透过亮瓦和窗带射入,党旗的鲜红映照整个展厅。

Layout of Functional Areas

There is a long passageway between the hall and the entrance square. First, a visitor can see the statue of Hu Yaobang at the square. On one side of the steps are pictures and description of history on the wall. On the other side is natural beauty behind glass. On the way visitors may all visit the exhibition halls beside the exhibition corridor or have a rest and communicate with others in the rest area. At the end of the steps, visitors can see a red Party flag flying in the east. As sunlight comes from transparent tiles and windows, its redness shines down upon the whole exhibition hall.

The art gallery overhangs in the air, bringing different experience of enjoying beauty. It has interlinked spaces matching the architectural shape, preset functions, and brand new application ways.

悬于空中的美术馆带来不同的寻找美丽的体验、相扣的空间与建筑形态对话、预设的功能和全新的使用方式。

Art Gallery at the Guanlan Engraving Base • Shenzhen

Guanlan, Shenzhen | Project Location
2009 | Project Date
30,500m² | Project Scale
conceptual design | Design Phase
construction not started | Project Status

观澜版画基地美术馆·深圳

中国,广东省,深圳市,观澜 | 项目地点
2009年 | 设计时间
3.05万m² | 项目规模
方案设计 | 设计阶段
未建 | 项目现状

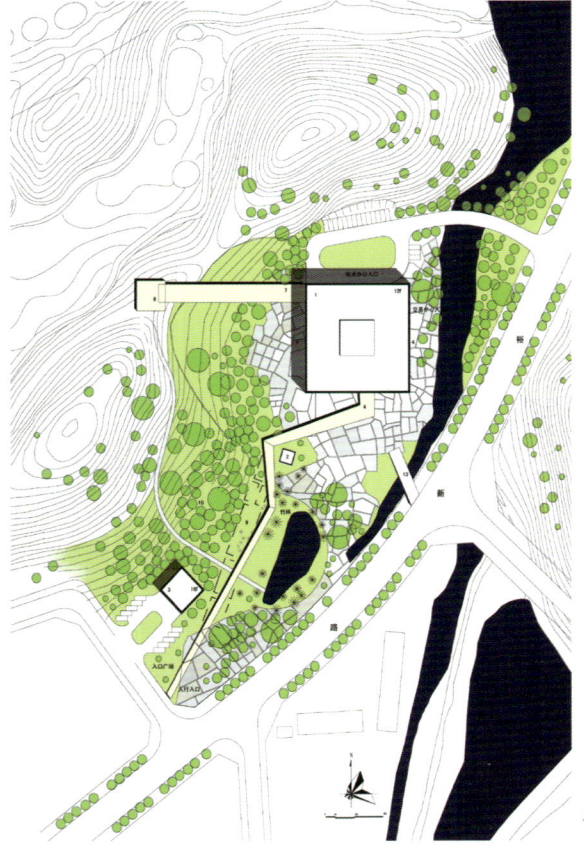

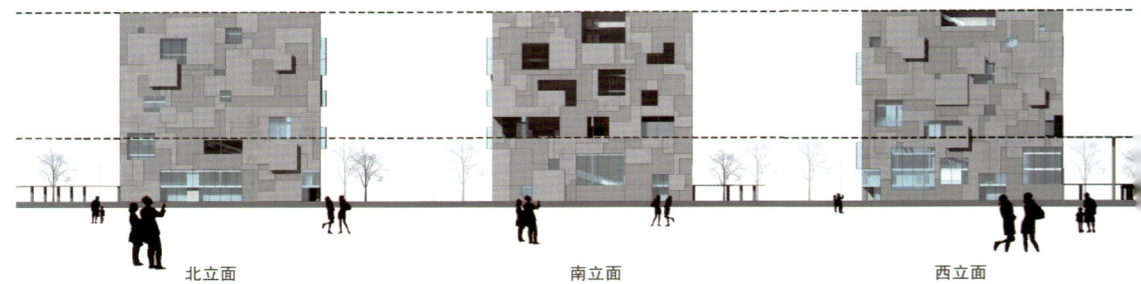

北立面　　　南立面　　　西立面

项目总览

本方案在总体空间功能的规划构思上表现为高度集约。三座大小尺度各异的碉楼，矗立在3万㎡的基地上，它们分别为：54x54x54m的交易中心和美术馆，6.6x6.6x16m由旧碉楼改造而成的信息中心，12x12x32m的停车楼。其中，交易中心和美术馆位于两座山坡之间。

Project Overview

This complex is highly dense in terms of functional spaces. Three towers of different sizes and scales are erected the 30,000m² foundation, that is, the 54x54x54m transaction center and art gallery, 6.6x6.6x16m information center rebuilt from an old castle, and 12x12x32m parking lot building. The transaction center and art gallery are located between two slopes.

面向高尔夫的墙断面示意　　典型展厅外墙断面示意　　马赛克幕墙断面示意

剖面

美术馆和交易中心，由半开放半封闭的表皮包裹，其纹理来自于无限重叠的版画。同时，纹理的丰富也同样表达了对版画特殊设计制作工艺的诠译。

不同方向的立面，对应着不同的剖面，没有完全相同的剖面空间。朝向高尔夫球场西立面的是一个高大开放的吹拔空间，而面向主入口的马赛遮阳板块立面内则多为实墙，为相对封闭的展览空间。

Section

The art gallery and transaction center are covered by a semi-enclosed walls, with textures like infinitely superposed engravings. The rich textures embody the special design and manufacturing technology of engravings.

Facades in different directions correspond to different sections, that is, no two individual spaces have the same section. A grand open patio space faces the western facade of the golf course. At the mosaic shaded facade facing the main entrance, are mostly solid walls, forming a relatively closed exhibition space.

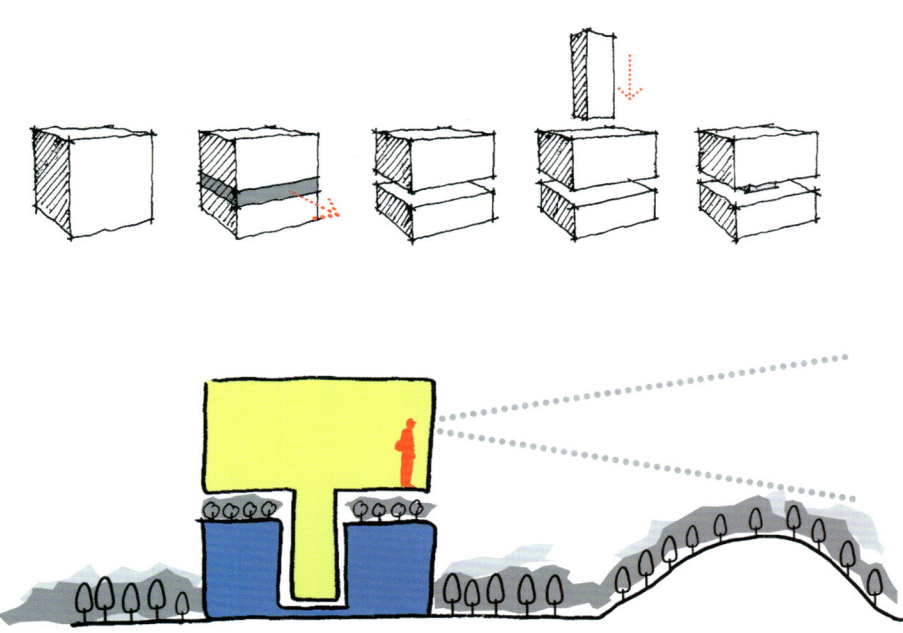

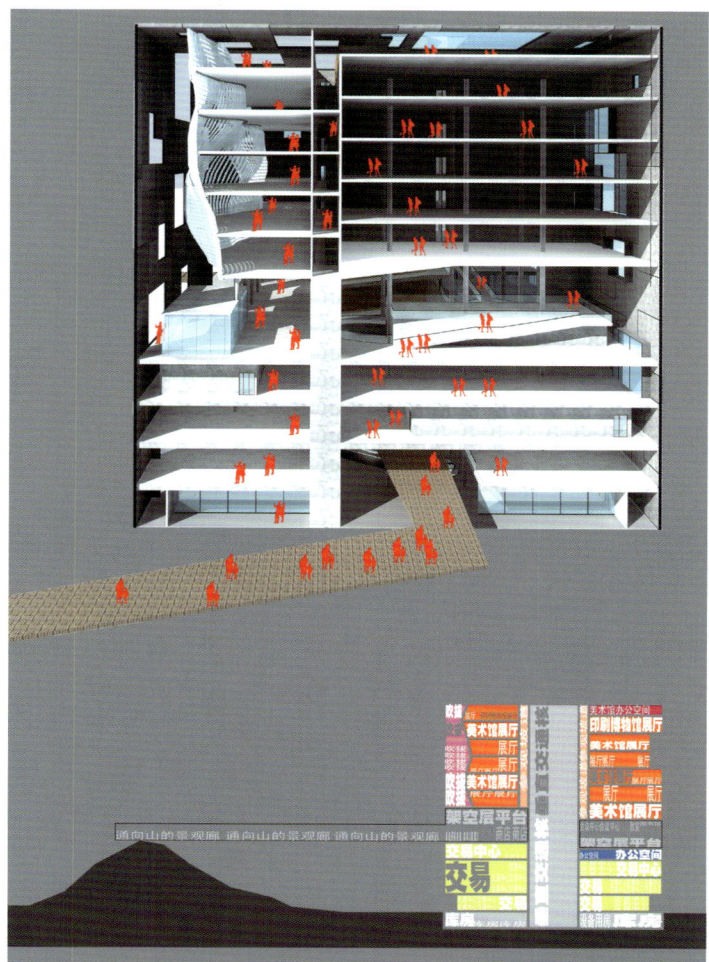

构思分析

空中美术馆

项目位于一个理想的地段，从这里可以一览该地区最好的风景。但是用地和高尔夫球场之间隔着两座小山坡，山高约25~30m，两山腰与用地间高差约16~18m，只要把美术馆举起来，越过山腰就能看到美丽的风景。

一个悬于空中的美术馆

它容纳了所有的展厅，其下方是包围起来的交易大厅空间，建筑中有一组专用的电梯，它直接把人们从美术馆入口送达盒子的顶层。在那里，美丽的风景让人们忘记一切。然而，那只是人们欣赏风景的起点。

参观者自上而下进入各层展厅，欣赏艺术作品。窗外美丽的高尔夫球场也是美术馆的又一展品。

Aerial Art Gallery

The project is located at an ideal section, where the best landscape in this region can be seen. However, there are two hillsides, about 25-30m in height, between the plot and the golf course. The height difference between two hillsides and the plot is about 16-18m. As long as the art gallery is risen up, a person can see beautiful landscapes when he crosses the hillside.

Art Gallery in the Air

It holds all exhibition halls. Under the gallery is a transaction hall. The building has a set of elevator, which sends people directly to the top of the box from entrance of the art gallery. Beautiful landscapes there make people forget everything. However, that place is just the starting point on the way of landscape appreciation.

Visitors enter an exhibition hall from top to bottom to appreciate artworks. The beautiful golf course outside the window is another exhibit in the art gallery.

平面图

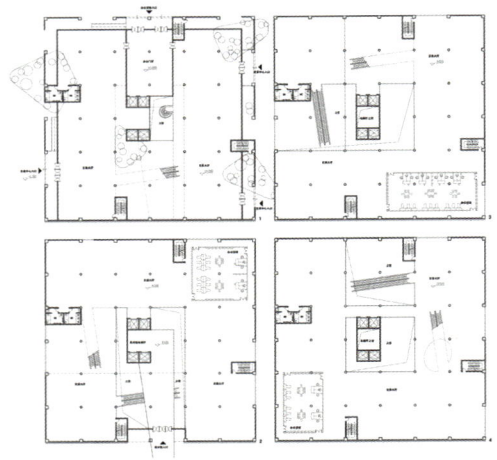

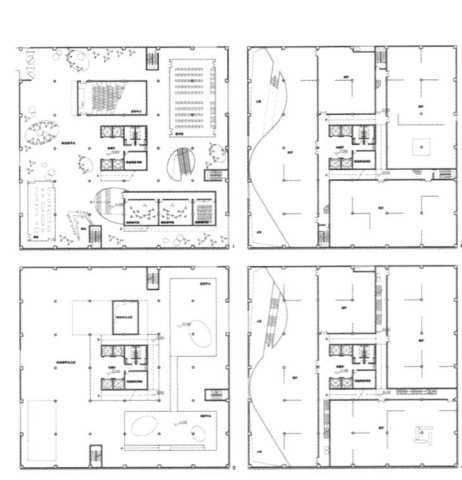

It is the embodiment of traditional culture in a modern way.

它展现出我们对传统文化的当代理解。

Guangxi Kettledrum Museum • Nanning

Nanning | Project Location
2009 | Project Date
30,000m² | Project Scale
conceptual design | Design Phase
construction not started | Project Status

广西铜鼓博物馆·南宁

中国，广西省，南宁市 | 项目地点
2009年 | 设计时间
3万m² | 项目规模
方案设计 | 设计阶段
未建 | 项目现状

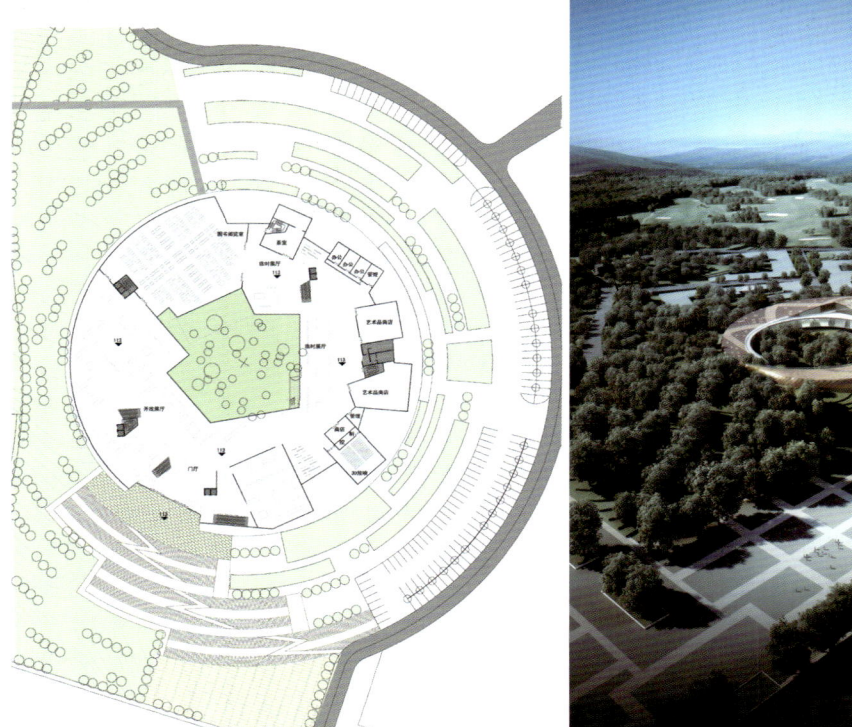

方案一

方案二

方案三

广西当代文化展览 guang xi contemporary culture	苍翠山景 lush hills	悠远水景 green waters	亲水平台 platform	休闲驿馆 relax dak	水路联系 water traffic	山间栈道 trestle	城市干道 roads	广西民俗文化展览 guang xi traditional culture

功能与空间

设计要素

广西壮族人民在特定的地域生活环境、长期的生产实践和社会交往中，形成了独特而富有地域文化特色的民族风俗与聚落形态，这些已成为其民族特征的基本构成要素。

项目设计提炼了"鼓"、"寨"、"道"等要素，作为铜鼓馆设计的出发点。

Design Elements

The Zhuang people in Guangxi have unique and rich folk customs and inhabiting habits characterized by a distinctive local style and flavor, which have been generated during their production and social communication in a particular living environment in a long period. These have become essential elements of their national characteristics.

In the project design, some essential elements, such as drum, stockade village and road, are used as the starting point for design of the Copper Drum Museum.

立面概念

建筑表皮建筑材料采用金属板和镀膜玻璃,屋面采用彩色金属板和镀膜玻璃。同时,不同空间功能根据其需要采用不同折射率的镀膜玻璃,提供柔和的光线,既解决了空间采光问题,也丰富了建筑外立面的变化。

Facade Concept

The building materials include metal plates and coated glass. The roofing is covered with colored metal plates and coated glass. Moreover, for spaces with different functions, coated glass with different refraction indices is used. Such glass allows soft light to coming in, providing day lighting and enriching changes in the exterior facades.

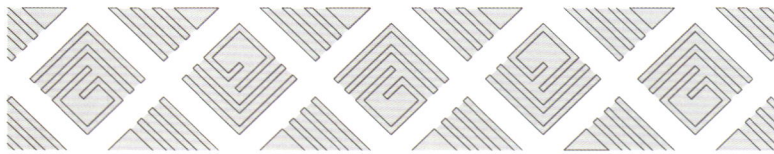

铜鼓馆—表皮

设计以"不挖山，不砍树"为出发点。整个建筑依山就势，充分利用地形地貌，采用"半地下"的形式。占地面积小，降低了对地面的破坏，基本保持了山体原貌。中心庭院为现有的山坡顶部，最大程度地保证了山坡原有植被，既美化环境又能有效防止水土流失。

The design sticks to a principle of "no excavation, no tree-cutting". The whole building is rests on a hill, making full use of the original landform. Since half of the building is located underground, it covers a small area, reducing the damage to ground and maintains original appearance of the hill. The central courtyard situates at the top of an existing slope. Thus original vegetation is protected to the largest extent, the environment is beautified, and water loss and soil erosion is avoided.

立面图

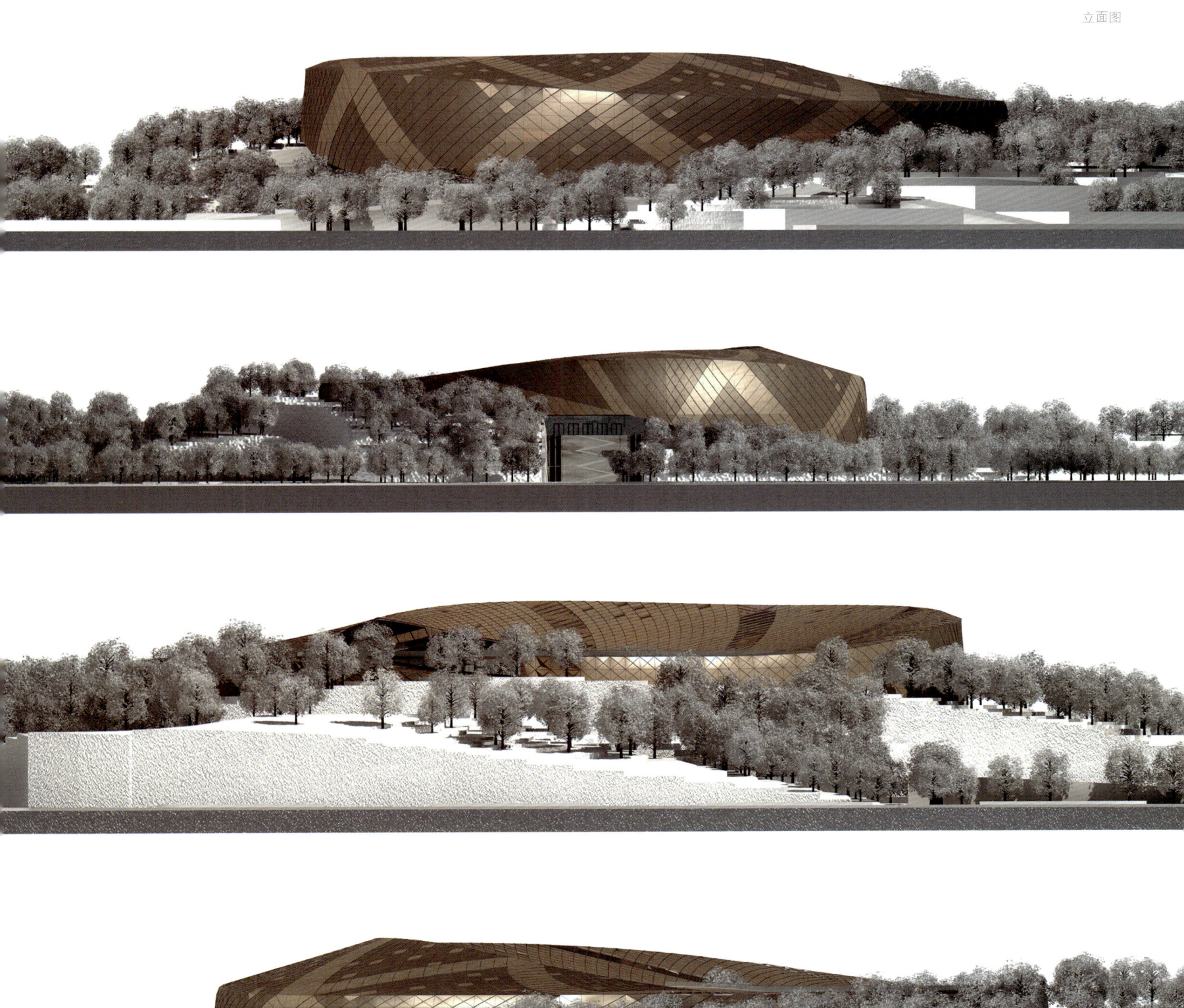

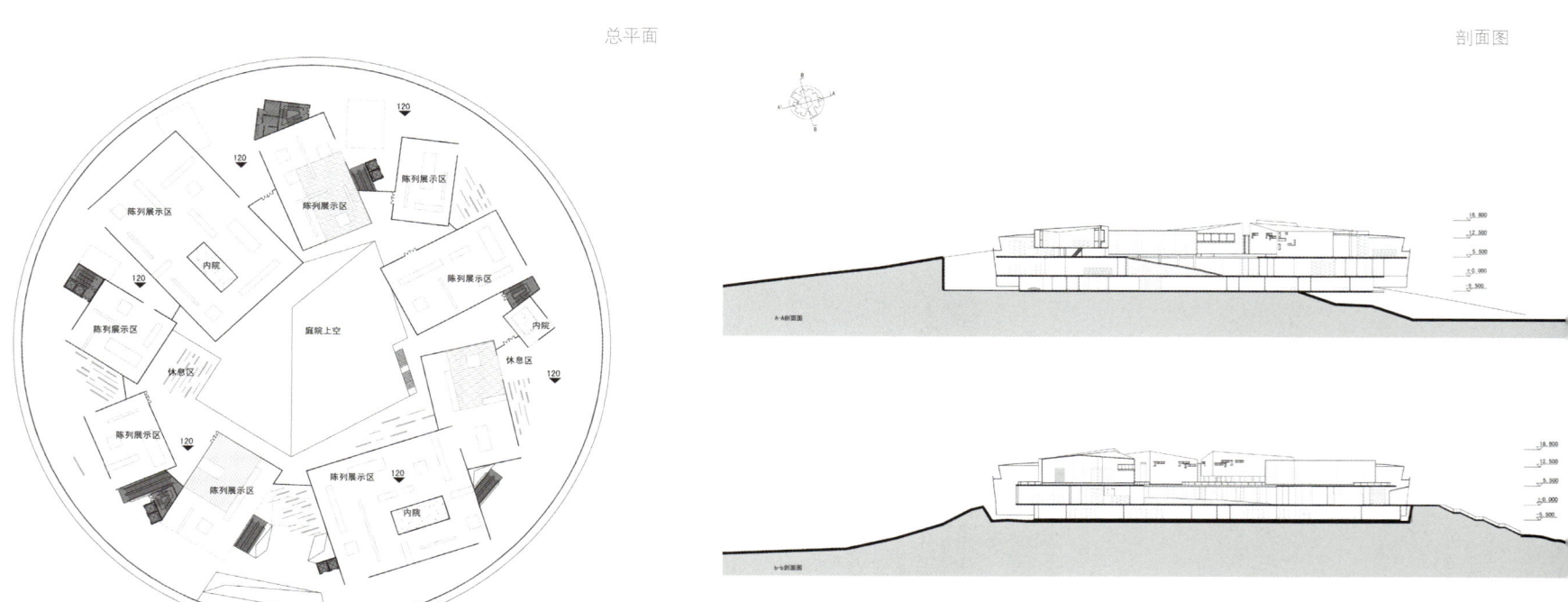

总平面　　　　　　　　　　　　　剖面图

We think the function and space design is the only available one fully meeting the function and space requirements in the design specification but the architectural form is based upon our own understanding of Venice.

我们认为这个设计的功能和空间是任务书功能和空间要求的唯一结果，而形式则出于我们对威尼斯的理解。

Venice Bridge Museum • Venice (Italy)

Venice | Project Location
2006 | Project Date
1,000m² | Project Scale
conceptual design | Design Phase
construction not started | Project Status

威尼斯桥博物馆·威尼斯（意大利）

意大利，威尼斯市 | 项目地点
2006年 | 设计时间
1,000m² | 项目规模
方案设计 | 设计阶段
未建 | 项目现状

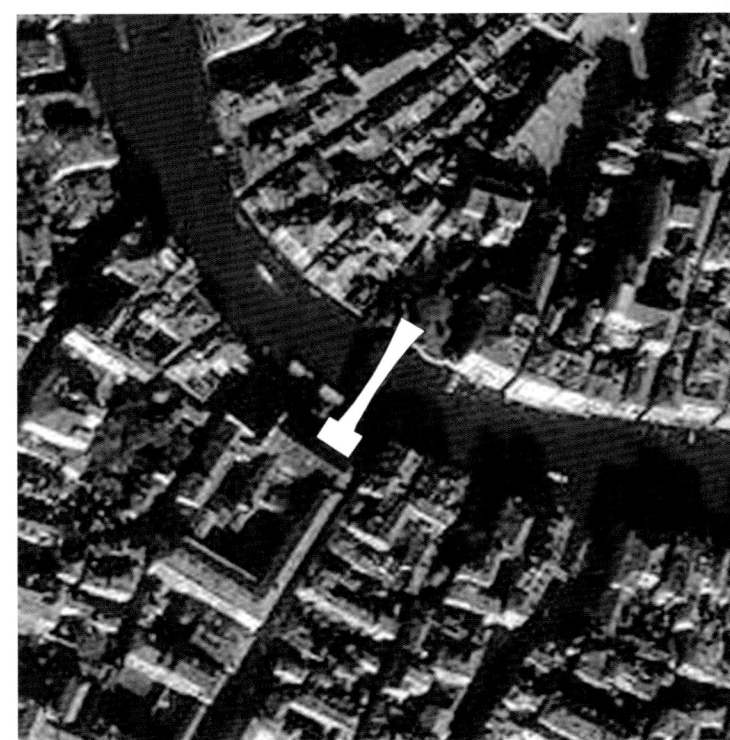

项目总览

这是从亚得里亚海进入威尼斯水城的第一座拱桥。从进入水城的贡多拉上放眼水城、桥、圣马可方场，鳞次栉比的建筑沿着海平面舒缓地延展开来，形成一个包容的姿态，迎接1500年前欧亚大陆贸易的商船，乃至今天世界各地慕名而来的漫漫游人。

在建筑的形体上我们尊重威尼斯城市的历史尺度和文脉的延续。桥的弧线曲度与过去的桥有着共性，但又存在差别。从运河上向新桥望去，原有的视觉线路并没有被阻隔或者改变，甚至透过桥身的玻璃可以微微看到以前不曾见过的对面悠久而美丽的巴洛克建筑，使人同建筑产生新的对话。

Project Overview

This is the first arch bridge on the way from Adriatic Sea to Venice, a city of water. Standing on the gondola at the entrance, you can overlook the city of water, bridge, and St Mark's Square, where rows of buildings spread leisurely along the sea level. The whole city is an inclusive one, welcoming merchant ships for trade in Eurasia 1,500 years ago as well as nowadays' visitors from all over the world attracted by its beauty.

In the design of the architectural shape of the museum, the history and culture of Venice have been taken into account. The arc curvature of the new bridge is similar to but also different from that of the original bridge. As we take a look at the new bridge from the canal, the original view is not blocked or changed. Through the glass on the bridge we even can see the beautiful centuries-old Baroque architecture in the opposite side, which cannot be seen before. This is a kind of new communication with the building.

区位图

构思草图

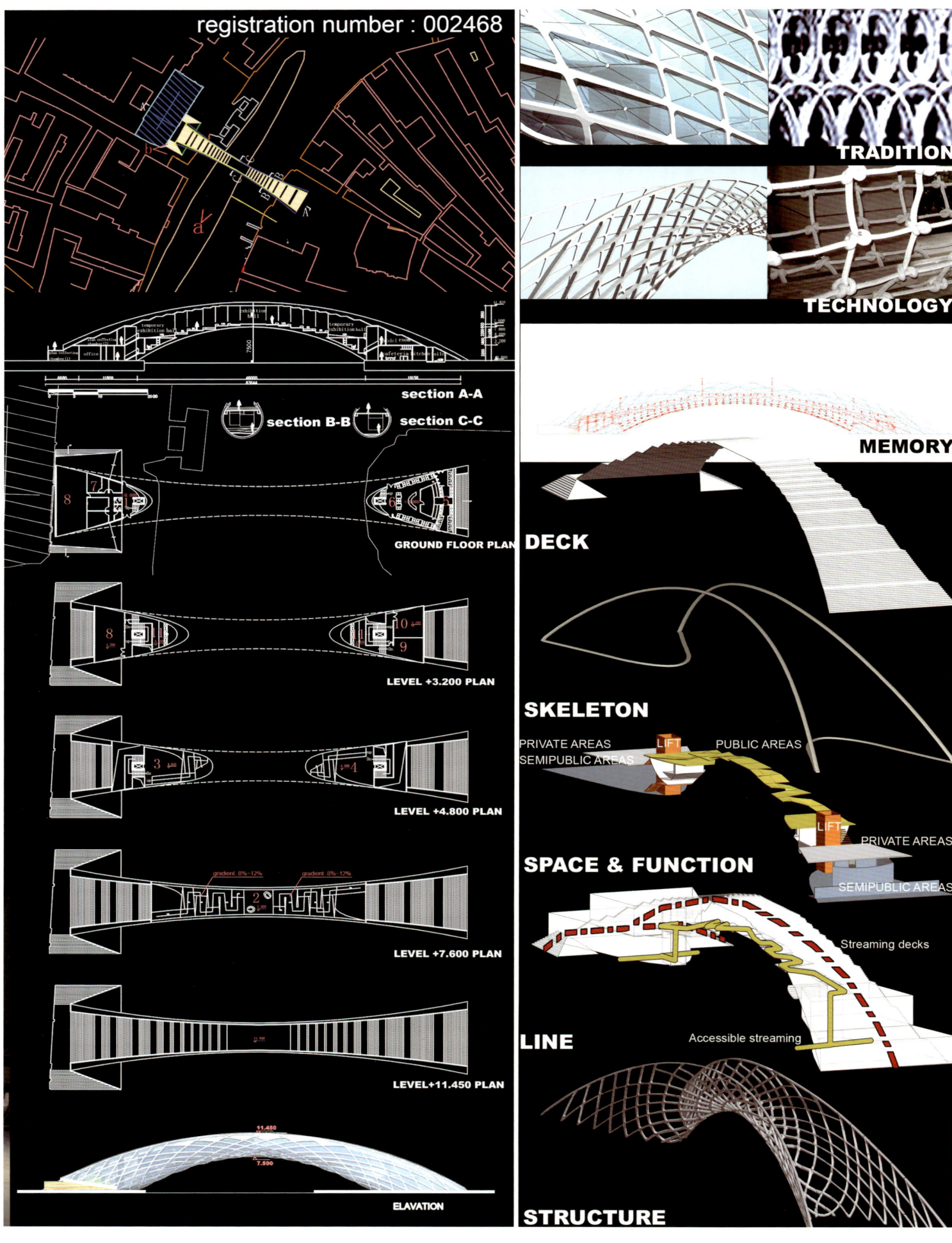

Nowadays many campuses with similar layout and style are built every day. We are always considering how to create the personality and uniqueness for the new campus.

当今时代,每天建成的校园很多,而如何赋予他与其他伙伴不同的个性和差异性则是我们一直思考的要点。

New Campus of Xidian University • Xi'an

Xi'an | Project Location
2004 | Project Date
230,000m² | Project Scale
conceptual design, preliminary design | Design Phase
construction completed | Project Status
Shenzhen Machinery Institute Architectural Design Co., Ltd. | Design Partner

西安电子科技大学新校区·西安

中国，陕西省，西安市 | 项目地点
2004年 | 设计时间
23万m² | 项目规模
方案设计 初步设计 | 设计阶段
已建 | 项目现状
机械部（深圳）设计研究院 | 合作单位

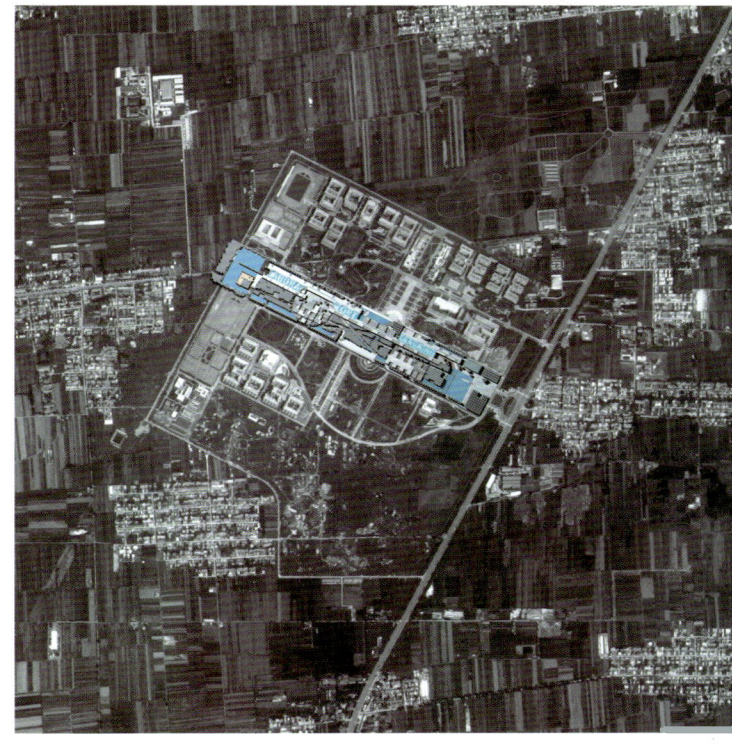

项目总览

西安电子科技大学的巨构不仅是学校的公共教学区，也是学校的核心区。我们选择了一条长约1km，横贯校园东西的学院街将这个区域完美地结合成整体。学院街连接了校园行政中心、中心实验楼、教学楼、阶梯教室、学术会议中心、科技大楼、图书馆、大学生活动中心、校园钟塔等数十栋大小不同的建筑，犹如一个功能齐全的城市。

设计主要采用了18m以内的多层建筑，其建筑美感掩映在校园浓厚的学术氛围与学生朝气蓬勃的活力中，同时也"藏"在了校园的森林之中，是集学校教学研究、办公休息、交流于一体的集约空间。依据不同的建筑功能，设计体现出多样化的空间组织形式和建筑形象。

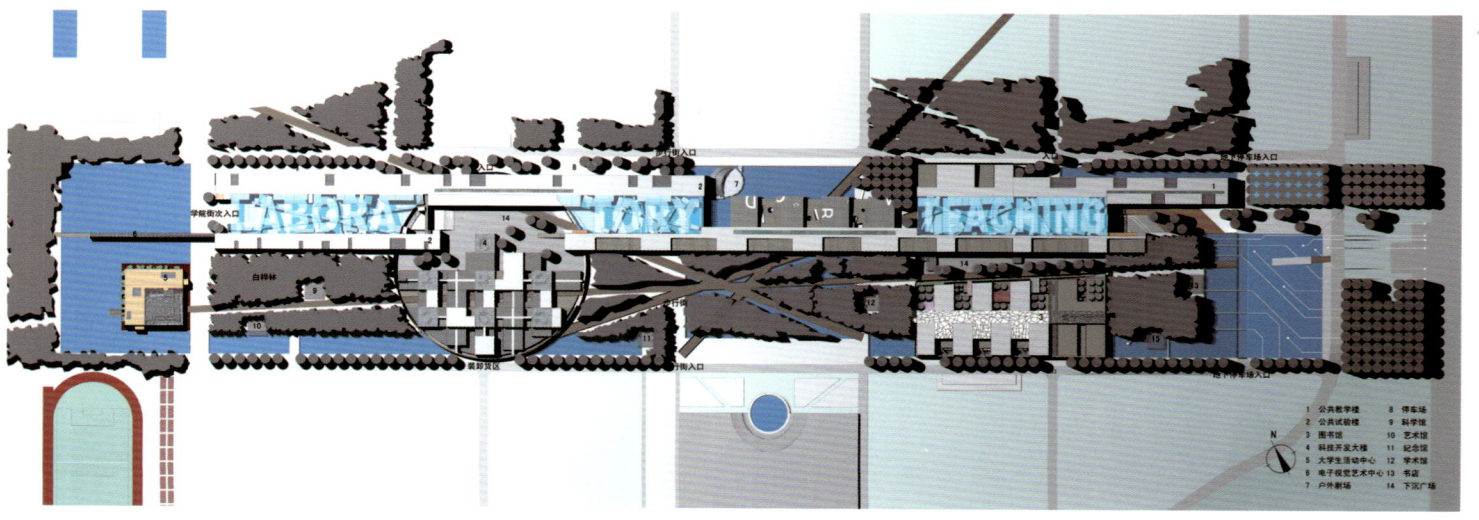

总平面

Project Overview

The giant complex of Xidian University is not only a public teaching area but also a core area of the university. We design an academic street, about 1,000m in length, running from east to west, which connects the whole new campus. The academic street provides access to ten buildings of different sizes, including the administration center, central lab building, teaching building, lecture theater, academic meeting center, sci-tech building, library, student activity center, and campus clock tower, etc., just like a city with all functions.

The buildings are mainly multi-storey buildings with a height no more than 18m. The buildings set off each other, creating a strong academic atmosphere and embodying students' youthful spirit and vigor. Creating an intensive space with such functions as teaching, research, office, rest and communication, the buildings are "hidden" in the campus woods. Diversified space forms and architectural images are designed on the basis of the building functions.

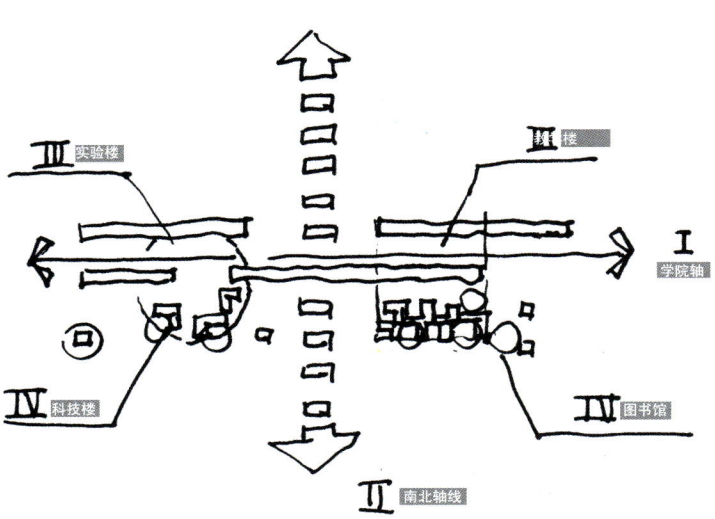

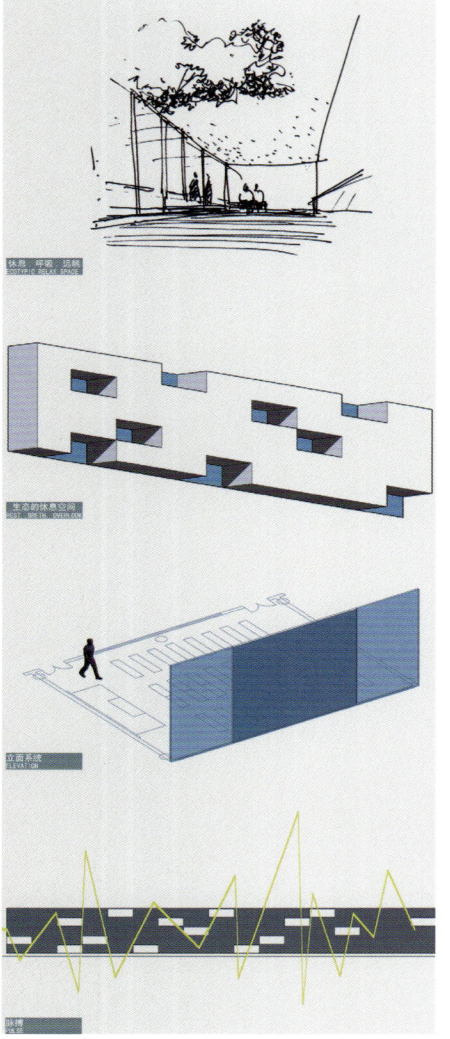
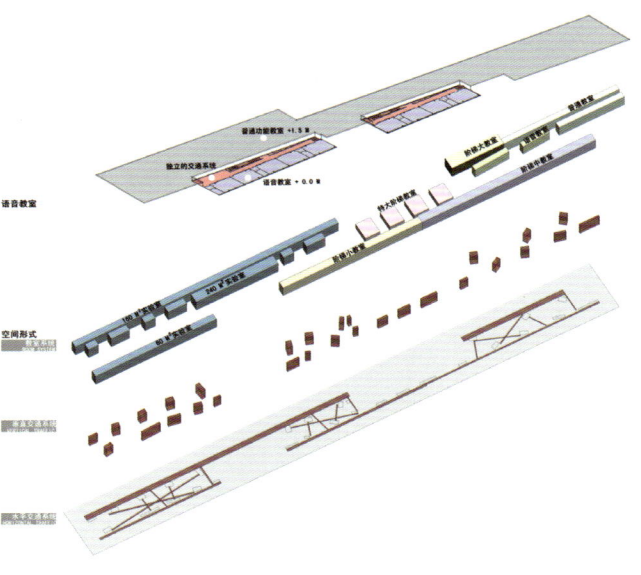
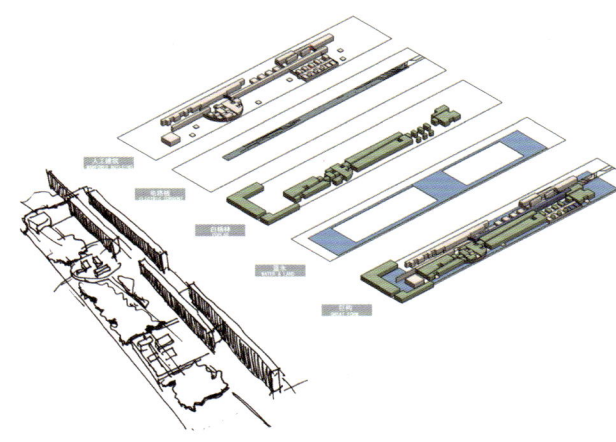

建筑单体——教学楼与实验楼

教学楼、实验楼及现代电子视觉艺术中心组成母体建筑。

教学楼与实验楼

分区：以"一"型为骨架，单向的线性体块布置语音教室、普通教室，大、中、小阶梯教室及特大阶梯教室。空间形式以教室、水平交通、垂直交通三个体系组合而成。网络式的水平交通和外伸岛式的垂直交通使整个空间丰富而有趣，视觉感受步移景异。

生态的设计系统：可供休息的pocket park位于南北向教室相间的室外空间，可作为课间休息、呼吸、眺望的空间，使建筑充分地接近自然，有大有小，有高有低，如同脉搏般的跳跃，充满了韵律感。

教室的立面与教学、学习的行为心理紧密相连。立面视线不同的透视率，使教室自然形成了相对私密和半开放的区域。

相对独立的语音教室，作为教学楼整体的一部分，拥有相对独立的交通核，分区更加明确。

Single building – teaching building and lab building

The building complex consists of a teaching building, lab building, and modern electronic vision art center.

Teaching building and lab building
Division: based on a straight-line framework, audio classroom, common classrooms, large, medium and small-sized terrace classrooms, and super-large terrace classroom are set on the one-way linear block.
The space is made up of three systems, i.e. classroom, horizontal traffic, and vertical traffic systems. Network-style horizontal traffic and protruding island-style vertical traffic make the whole space rich and interesting. Landscape changes after every step.

Ecological design system: located outdoors between south-north classrooms, the Pocket Park may be used for class breaks, breathing fresh air, and overlooking. Thus buildings approach nature as closely as possible. Large and small, tall and low buildings look like a beating pulse, showing a sense of rhythm.

The façade of classroom is closely linked with teaching and learning behaviors and psychology. Since the façade presents different transmittances to lines of vision, each classroom boasts a relatively-private area and a half-opened area naturally.
As a part of the teaching building, the relatively-independent audio classroom has a relatively-independent traffic core as well as more definite division.

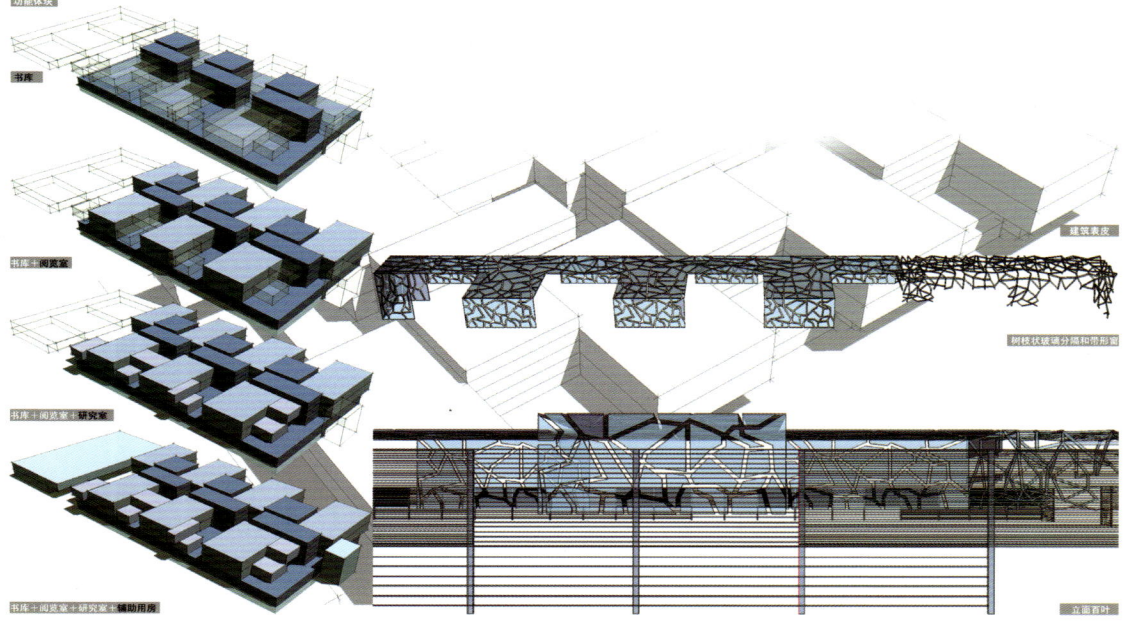

功能布局

该建筑由三大功能块组成,分别是教学区(教学楼、实验楼)、自修研究区(科技楼、图书馆)、休憩和展示区(活动中心、电子视觉艺术展示中心)。三大功能块都采用南北带形的布局,形成高于视觉,让人心情愉悦的强烈印象。外部空间结构简单的说是由两条贯穿东西南北的轴线构成的。内外分别由开放的半私密空间、公共空间、封闭的公共空间、完全公共空间、半私密公共空间和私密的公共空间构成。大小的公共空间以点状、线状、面状的方式予以分布。

Layout of Functional Areas

The giant complex is made up of three major functional blocks, i.e. teaching area (teaching building, lab building), self-study & research area (sci-tech building, library), and rest & exhibition area (activity center, electronic visual art exhibition center). Each of the three functional blocks is arranged from south to north like a belt, creating a strong impression of large scale in sight and making people pleasant. In general, the external spatial structure consists of two axes running from east to west and south to north respectively. For both the indoor and outdoor spaces, they consist of open half-private spaces, public spaces, closed public spaces, open public spaces, half-private public spaces, and private public spaces. Public spaces of different sizes are distributed in shapes of points, lines, and planes.

教学区

教学楼、实验楼及现代电子视觉艺术中心共司组成母体建筑。空间形式以教室、水平交通、垂直交通三个体系组合而成。网络式的水平交通和外伸岛式的垂直交通使整个空间丰富有趣,视觉感受步移景异。

Teaching Area

Main buildings include a teaching building, a lab building and a modern electronic visual art center. The space consists of three systems, i.e. classrooms, horizontal transportation, and vertical transportation. The network-like horizontal transportation and overhanging island-like vertical transportation make the space rich and interesting. Landscape varies greatly as a visitor moves in the space.

215

图书馆——方

A. 设计特点分析
图书馆的设计结合西方图书馆注重外型与中式图书馆注重功用的特点，形成有型实用的院落式生态型空间。

B. 功能分析
从功能分区上看，一条沟通南北、贯穿教学楼的"街厅"是通往教学区的街道，亦是交流、检阅的大厅。街道西面是书库、阅览用房，东面则为学习、办公服务用房，功能分区明确。

Library – square

A. Analysis of design features
The building shape is emphasized in the design of western libraries while functions are emphasized in the design of Chinese libraries. In the design of the library, both the building shape and functions are carefully considered, thus the library has a good shape and useful and ecological courtyard-like space.

B. Analysis of functions
A "street hall", which trends from south to north and runs through the teaching building, is not only a street leading to the teaching area but also a hall for communication and parade. In the west of the street hall are book stacks and reading rooms. In the east are study rooms, offices, and service rooms. The functional areas are well divided in this space.

The eye of wisdom, earth, city, university town ... the church of arts, a holy academic palace ... the stage of stars and college students ...

智慧之眼、人地之眼、城市之眼、大学城之眼……艺术的教堂、学术的圣殿……星的舞台、大学生的舞台……

University Town International Conference Center • Shenzhen (Supplementary Project)

Nanshan, Shenzhen | Project Location
2008 | Project Date
13,235m² | Project Scale
conceptual design | Design Phase
construction not started | Project Status

大学城国际会议中心·深圳

中国,广东省,深圳市,南山区 | 项目地点
2008年 | 设计时间
13,235m² | 项目规模
方案设计 | 设计阶段
未建 | 项目现状

方案构思

1. 形态及表情

   地块同图书馆相邻，会议中心以什么样的形态出现对已建成的图书馆建筑有重要的影响。为了尊重原规划和已建建筑，保证图书馆建筑嵌入到山体的感觉，我们认为将地块南边的山体延续下来，地块上形成坡状的覆土建筑，对地块周边环境和建筑以及整个城区是一个合适的肌理表情。

2. 轴线及视线通道

   人们从城市沿着园区主要道路进入园区，穿过地块可以清晰地看到对面山上的西丽塔，我们认为建筑形体把塔导引入园区应该有意留出这条视线通道。

3. 色彩及标志

   校园现有建筑多采用灰白色调，因此建筑色彩单一。在园区的地理中心，我们认为需要一些色彩给整个园区增添点睛之笔。

Design Ideas

1. Morphology and expression
   The morphology of the conference center would have the important role on the existing library building because of the land adjacent to the library. In order to respect the original planning and existing construction and ensure the library building embedded into the mountain, we consider the land extended downforward in south and terrains formed on the slope of the building that could be appropriate expression of surrounding, architecture and the entire city.
2. The axis and view channel
   People get into the campus by walking along the main road, and then would see the Xi Li Tower on the opposite hill clearly when they go through the block. And we believe that the architectural form should intend to leave this view channel when it introduced the tower into the campus.
3. Colors and sign
   Existing buildings on campus almost use off-white tone and the building has onefold color. In the geographical center of the park, we think it needs to add some color to the whole campus as a perfect finale.

222

设计原则及目标：我们希望，会议中心是一个自然生态的"开放场所"，是大学生学习生活的舞台，是学术的神殿，是"大学城之眼"……

Design principle and goal: we hope that the conference center is a natural and ecological "open place", a stage for college students' study and life, an academic shrine, and the eye of the university town...

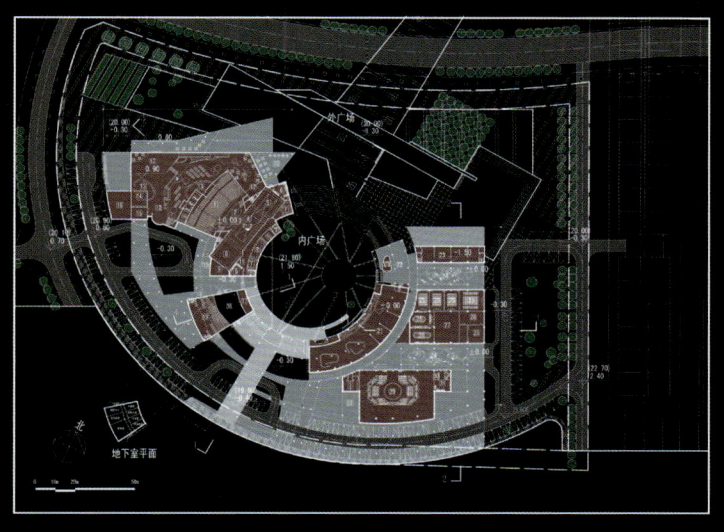
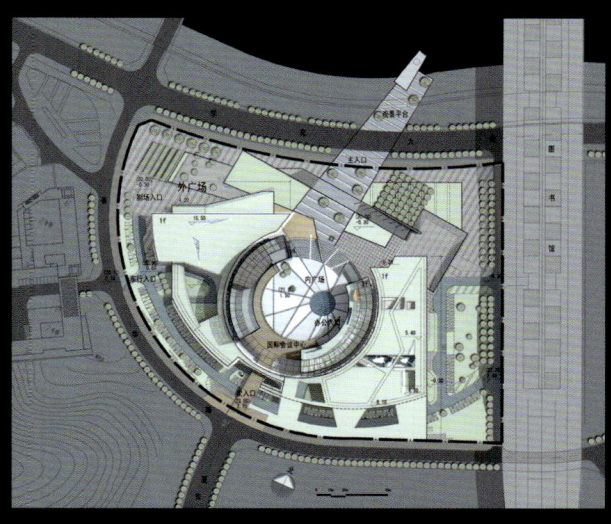

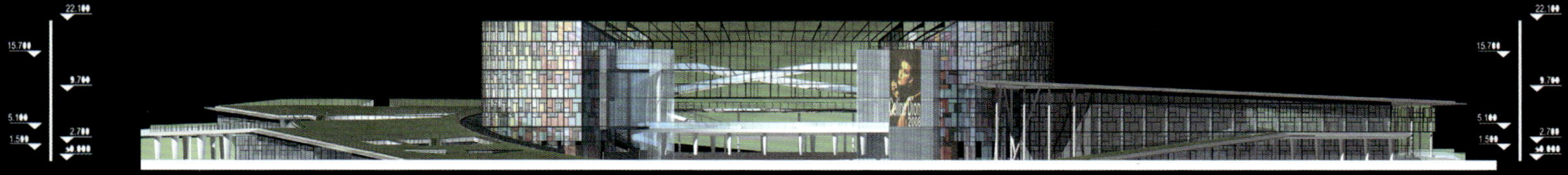

东北立面图

北立面图

房间名称及编号：

1 大会议厅
2 乐池
3 台唇
4 主台
5 排练厅
6 休息室
7 化装间
8 侧台
9 库房
10 货运平台
11 抢妆 候场
12 大堂
13 服务台
14 寄存
15 商务中心
16 银行
17 便利店
18 设备用房
19 放映厅
20 音响设备

21 展厅
22 架空层
23 网络中心\计算机房
24 庭院
25 教室
26 会议室
27 多功能厅（80人）
28 多功能厅（40人）
29 多功能厅
30 服务间
45 机房
46 器材室
47 控制室
48 资料室
49 打印室

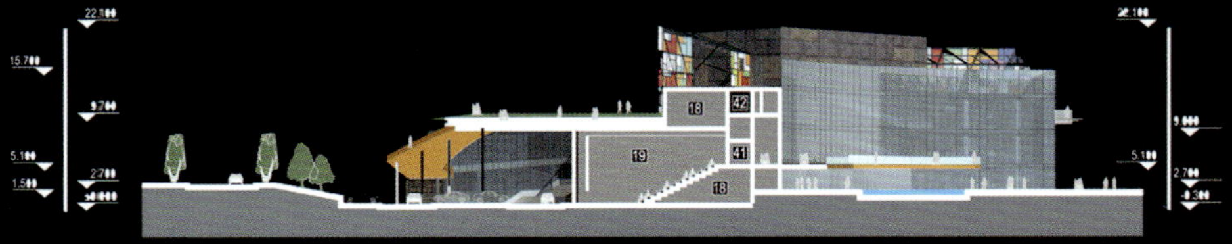

31 接待厅
32 主台上空
33 排练厅上空
34 侧台上空
35 库房上空
36 抢妆 候场上空
37 检修坡道
38 上空
39 景观平台
40 景观坡道
41 放映室
42 管理办公
43 大空间办公
44 VOID 办公

4-4 剖面图

2-2 剖面图

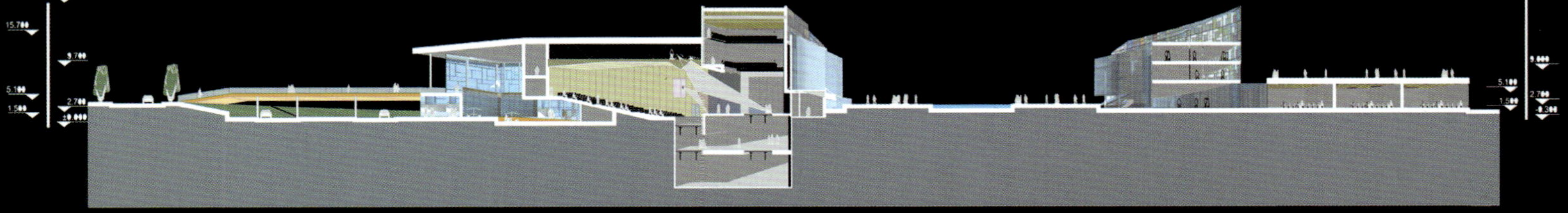

This is a rebirth of an old school. We have contending passions for the reconstruction of the school. We strive to resume its former culture, tradition and honor.

这是一所老学校的新生，对于推倒重建，我们怀着矛盾的心情，力求在这里还能寻回原有的文脉、传统和荣誉。

Ezhou No. 8 Middle School • Ezhou

Ezhou | Project Location
2007 | Project Date
22,300m² | Project Scale
conceptual design | Design Phase
construction in progress | Project Status

鄂州第八中学·鄂州

中国，湖北省，鄂州市 | 项目地点
2007年 | 设计时间
2.23万m² | 项目规模
方案设计 | 设计阶段
建成 | 项目现状

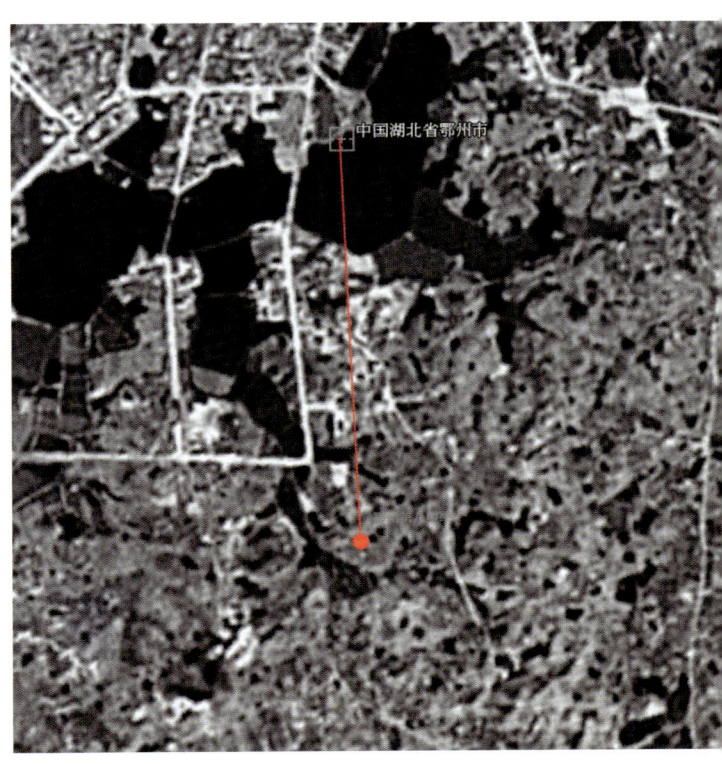

借鉴传统书院的型制,力图体现"唯楚有才"的文化底蕴。

By referring to the layout of traditional Chinese academies, the designers try to embody the profound culture of traditional Chinese academies.

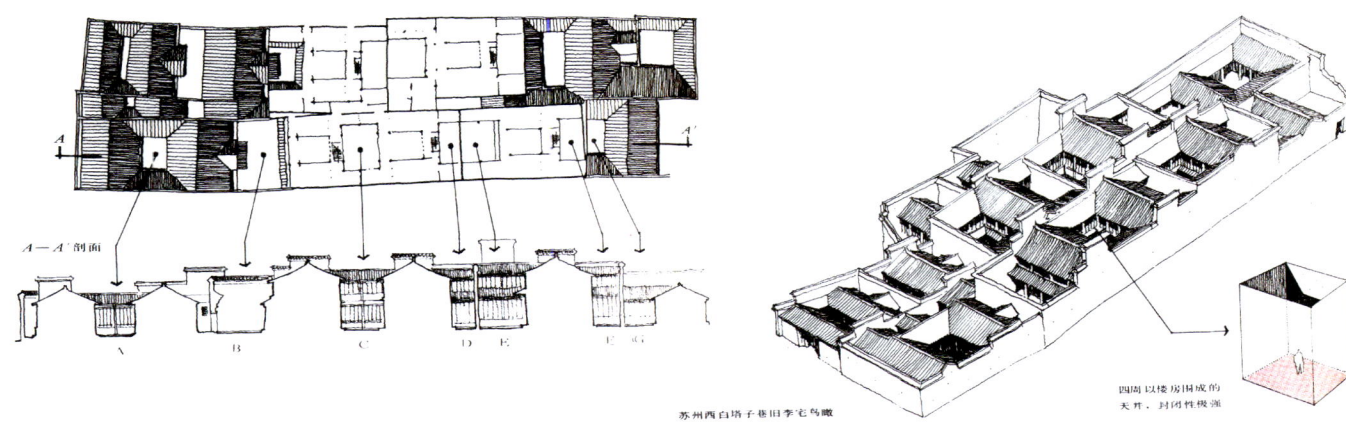

一些民居建筑,往往沿着一条轴线设置多进厅堂,而每个厅堂之前都有相应的天井,这样就可以形成一系列大小不等的天井。

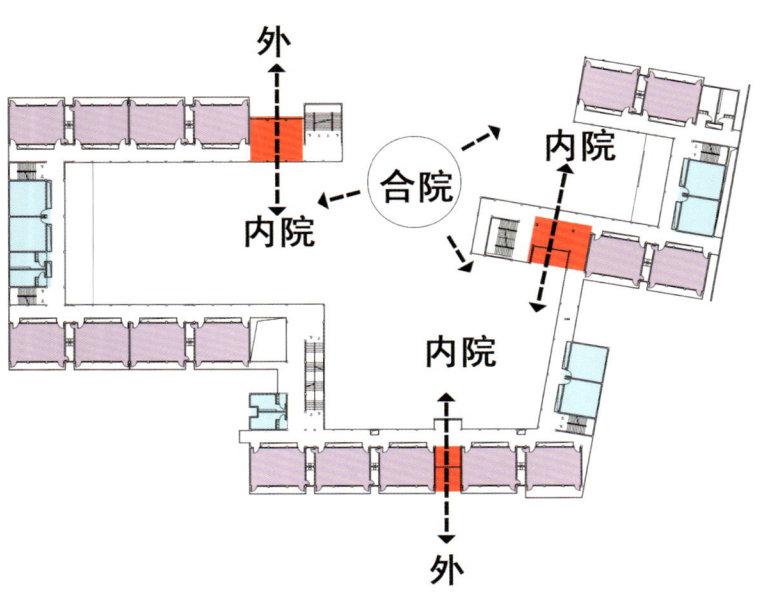

所有的教学楼和办公楼下架空为停车和活动空间

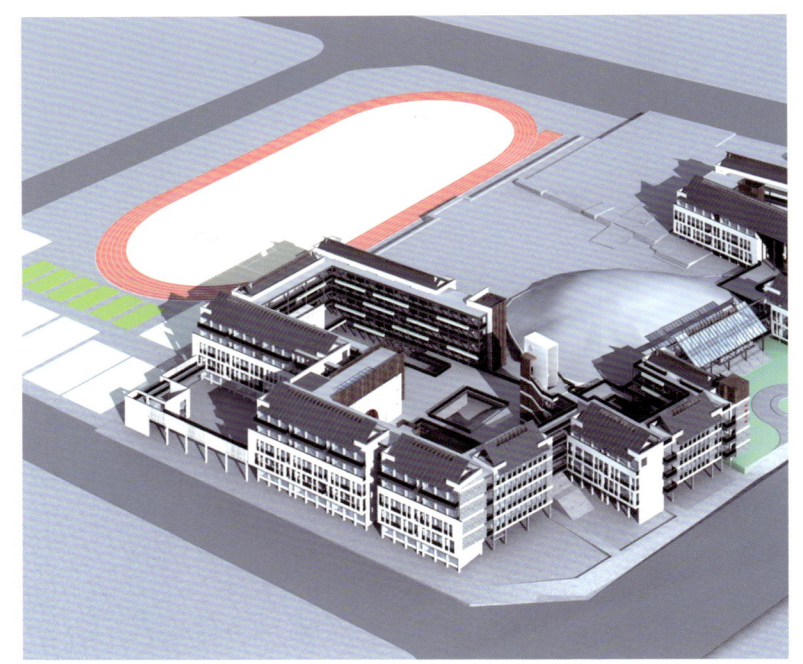

剖面图

总平面　　　　　　　　　　　立面图

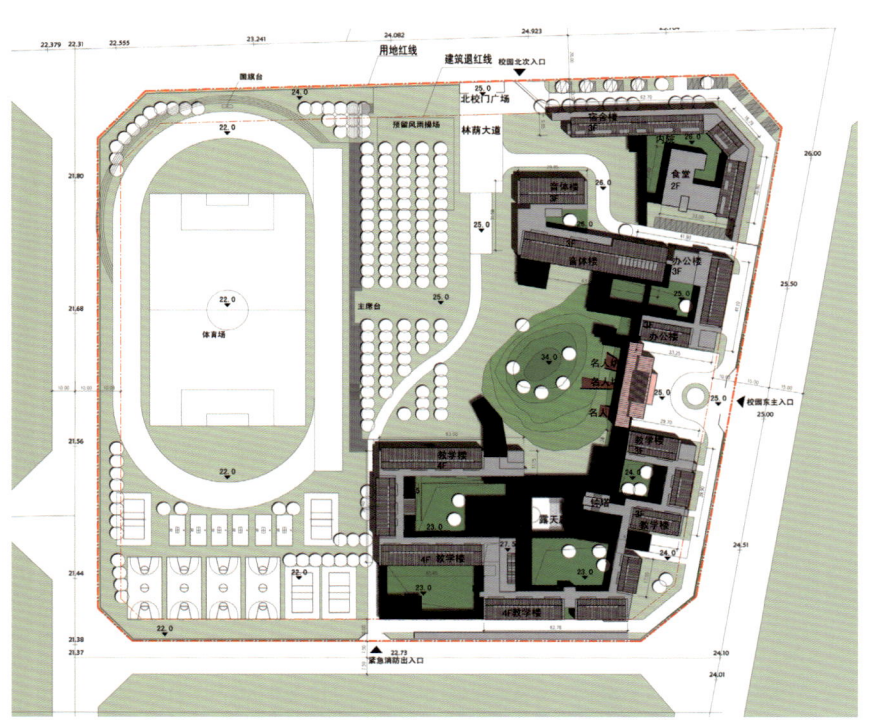
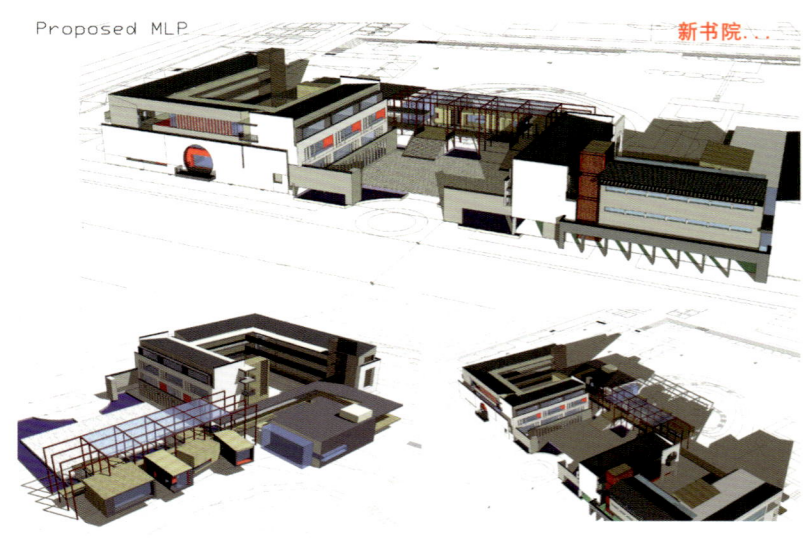

整个校园由相互联系、尺度不同的四大半开放式院落组成，通达的长廊与架空层贯穿整个校园，为学生与老师们提供了更灵活多样的活动空间。

The campus consists of four half-opened courtyards of different sizes, which link with each other. A long corridor and the empty space run through the campus, providing more flexible and diversified spaces for the activities of the students and teachers.

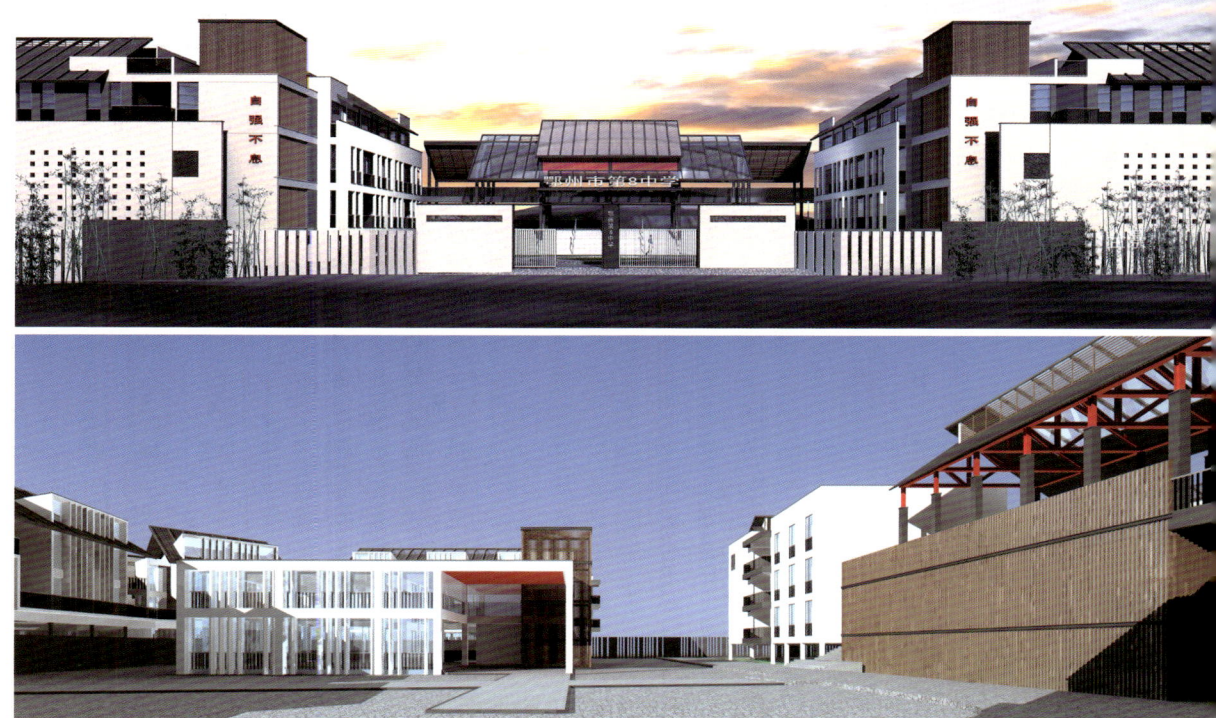

通过对教学单元进行模块化的研究，将主教学区分为3个院落，每个院落东西向都为教学辅助用房。在每个院落的标准层上设计了联系校园内外院的"空中花园"，使之同时成为学生们学习交流的空间。项目一改传统教室模块紧凑、走廊空间窄小的缺点，增加阳台数量，扩大走廊尺寸，保持走廊与教室间距1米，减少走廊对教学的影响。

After a study of modularization of the teaching units, the main teaching area is divided into three courtyards in the design. Auxiliary rooms for teaching are located in the east and west part of each courtyard. A hanging garden, which connects inside and outside courtyards of the campus, is designed on the typical floor of each courtyard. It is also a space for the students to study and communicate with each other. Unlike traditional schools with compact classroom modules and limited corridor space, this school has more balconies and larger corridors beside classrooms. The space between the corridor and the classroom is one meter, thus the influence of the corridor on teaching is reduced.

The design ideas derive from the landform of this plot. The landform provides us many usable conditions.

设计思路来源于这块用地的地形,它为我们提供了很多可利用的条件。

Longgang Pingshan Gymnasium • Shenzhen

Shenzhen | Project Location
2004 | Project Date
12,000m² | Project Scale
conceptual design | Design Phase
construction under planning | Project Status

龙岗坪山体育馆·深圳

中国，广东省，深圳市 | 项目地点
2004年 | 设计时间
1.2万m² | 项目规模
方案设计 | 设计阶段
未建 | 项目现状

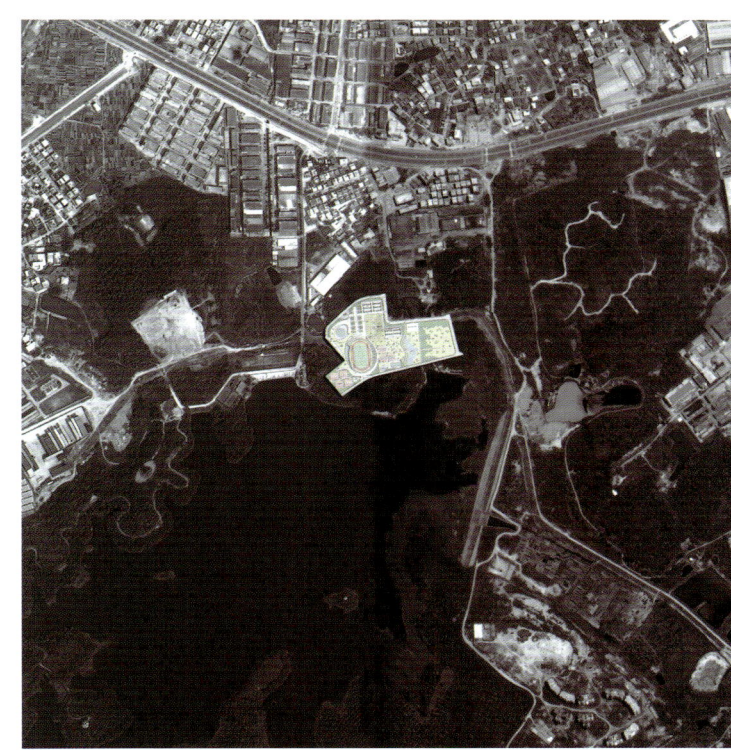

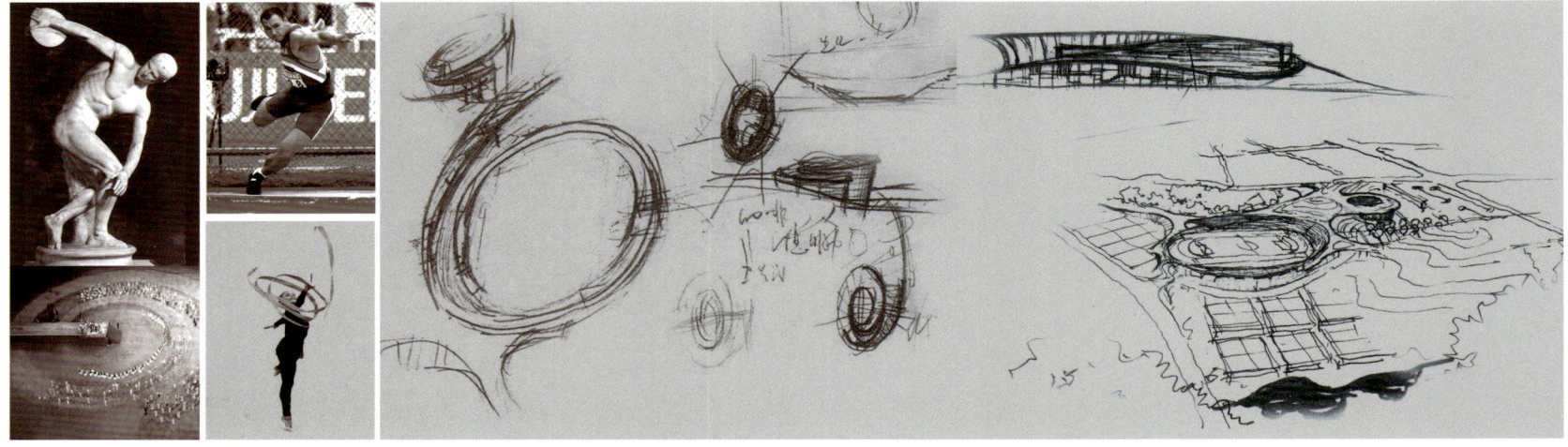

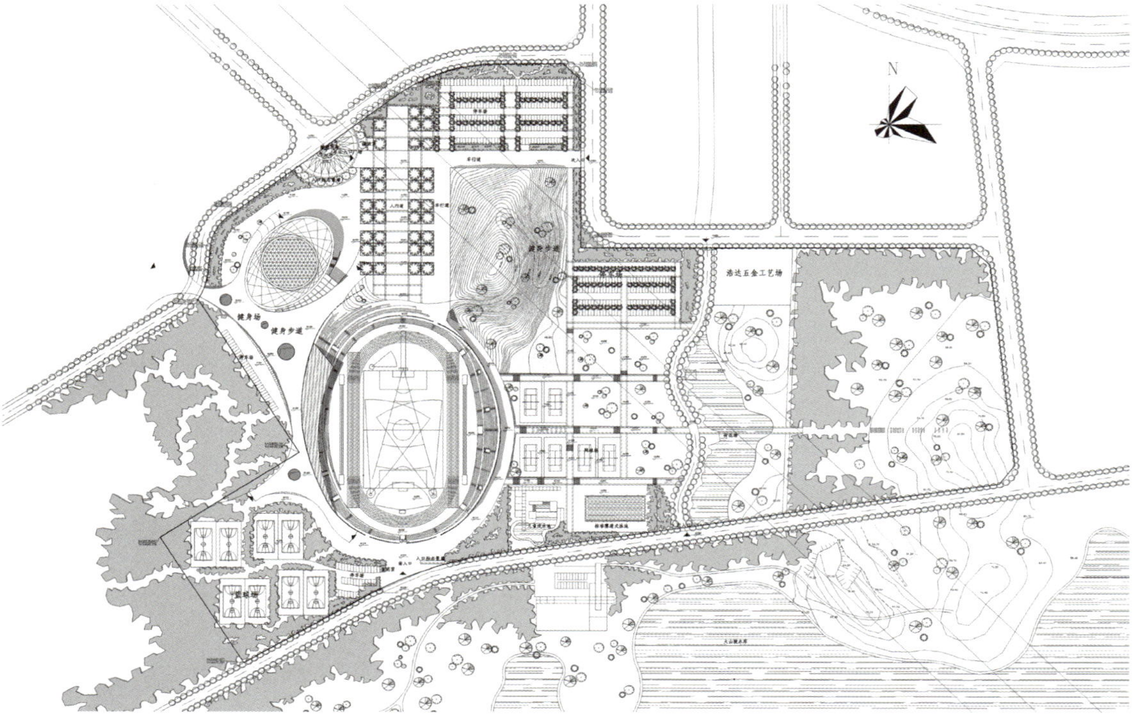

项目总览

用地原为起伏的山地，最大高低差20多m。基地上空的高压走廊斜穿而过，南面为水库，东面有自然积留而成的水体，周边绿化植被良好，是理想的健身运动场所。体育场位于体育中心南部，占地约25,000m²，可容纳观众约10,000人。体育馆位于体育中心西北部，占地约4,200m²，容纳观众约3,000人。

Project Overview

The site is in an uneven hilly region, with a maximum height difference of more than 20m. An overhead high-voltage line run across the base site aslant. In the south is a reservoir. In the east is a naturally developed body of water. Covered by good vegetation, the site is an ideal place for physical exercises and sports. Located in the south of the sports center, the stadium covering a land area about 25,000 square meters can hold about 10,000 audiences. Located in the northwest of the sports center, the gymnasium covering a land area about 4,200 square meters can hold about 3,000 audiences.

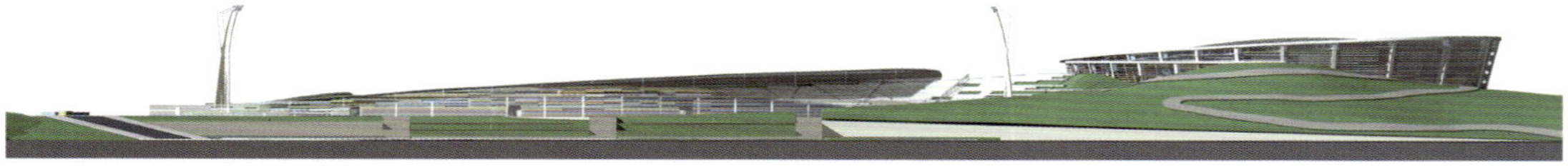

卫生间
直道外边缘线
13150
5673
950
2471
3209
14250
7650
7650
11920
可辩认视线区
清晰视线区

7865
直道外边缘线
商铺
道路
通道
通道
2570
950
11920
6000
11250
清晰视线区

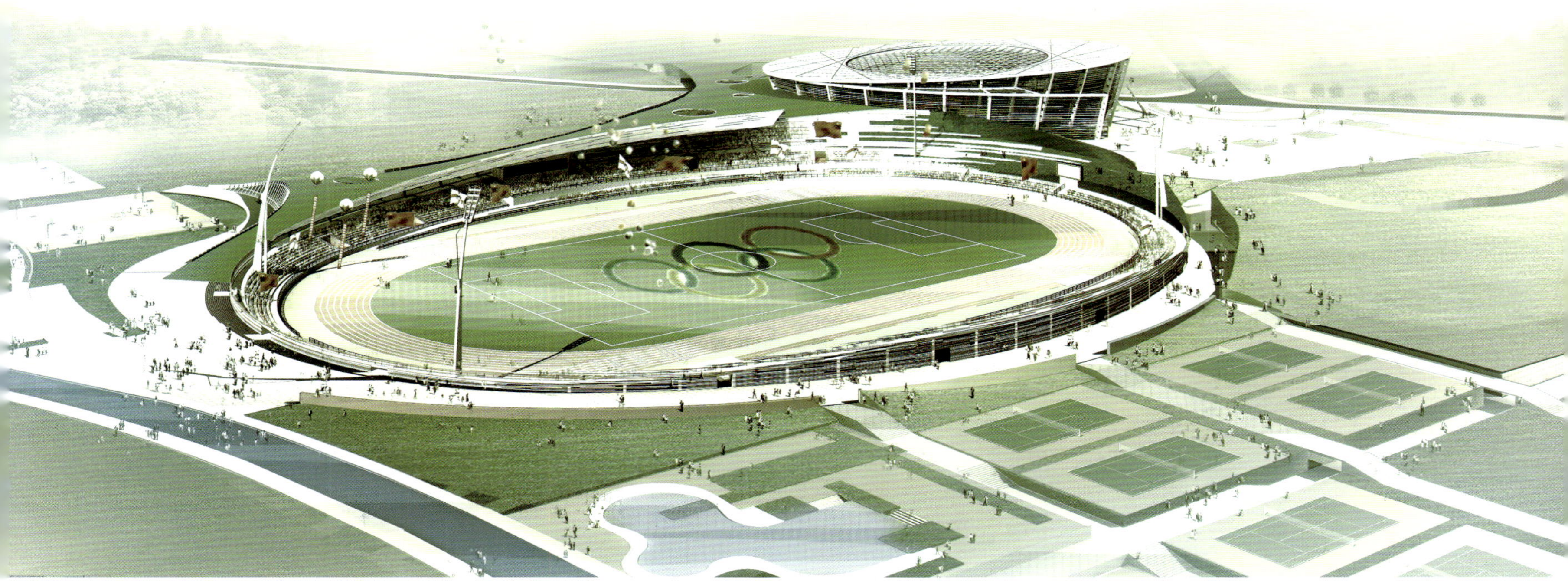

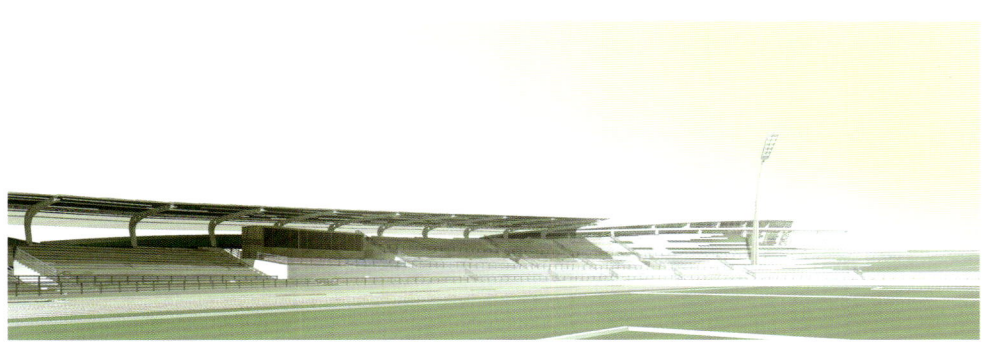

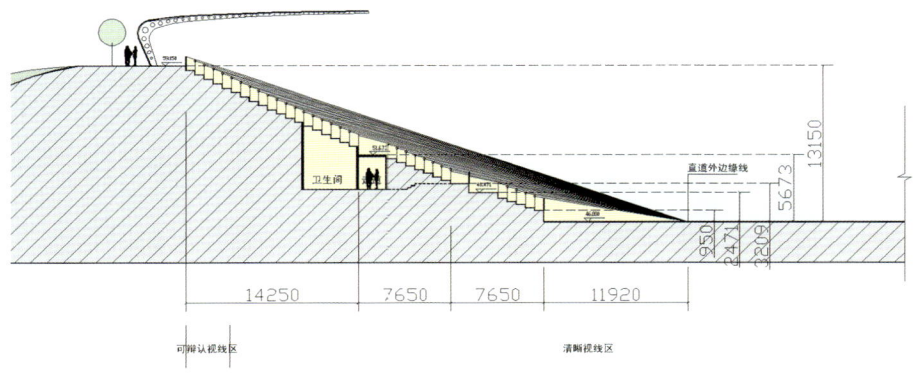

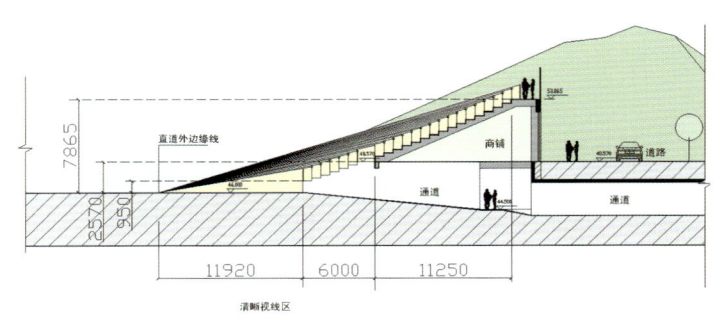

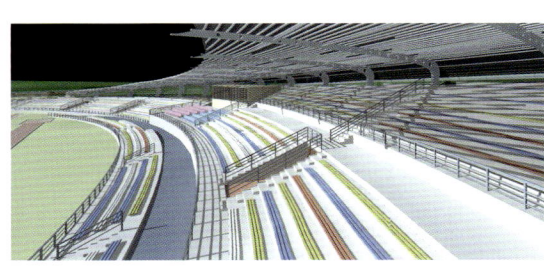

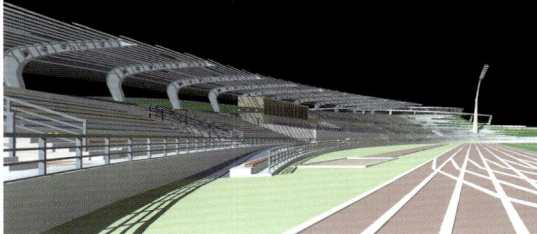

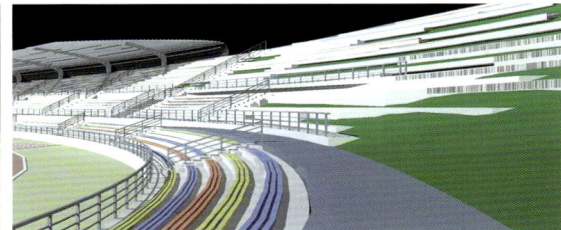

造型与空间

设计构思源于掷铁饼力量从凝聚到迸发过程中的运动轨迹，建筑的造型和空间也延续了对铁饼螺旋线的诠释。

不同起点的两段依次展开的弧形台阶，依附椭圆体量盘旋而上。阳光通过光廊顶部和侧墙格栅在台阶上投下斑驳的阴影，彩色水平向密置的铝合金格栅在光廊侧墙和体育馆外墙上闪烁，与整体连续流动的曲面体量一同在绿色山的背景中熠熠生辉，像旋转飞驰的流星。屋顶中心的圆形天棚采用特氟隆包层的非织性玻璃纤维膜，具有透光性和柔光性，阳光可透入体育馆内，但不造成耀眼的眩光，阴影柔和，没有强烈对比，适合各种室内比赛。

Shape and Space

The design derives from the motion trail of a discus in the process from the time when the thrower's strength is concentrated to the time when the discus is released. The architectural form and space are designed to be like the spiral line of a discus to be thrown.

Two sections of arc-shaped steps spread in sequence from different starting points and spiral up around an elliptical structure. After passing through the top of the corridor and the grids on the wall, sunlight fells upon the steps, forming blocky shadows. The horizontally-distributed colorful aluminum alloy grids twinkle on the side wall of the corridor and exterior wall of the gymnasium, shining along with the continuously-flowing curve mass in the background of green hills, like flying meteors. The round awning in the center of the roof is made of Teflon-coated unwoven glass fiber membrane. It transmits light and softens light. Sunlight can enter the gymnasium but no glare will be generated. The shadow is soft with no strong comparison, so the gymnasium is suitable for all kinds of indoor sports.

243

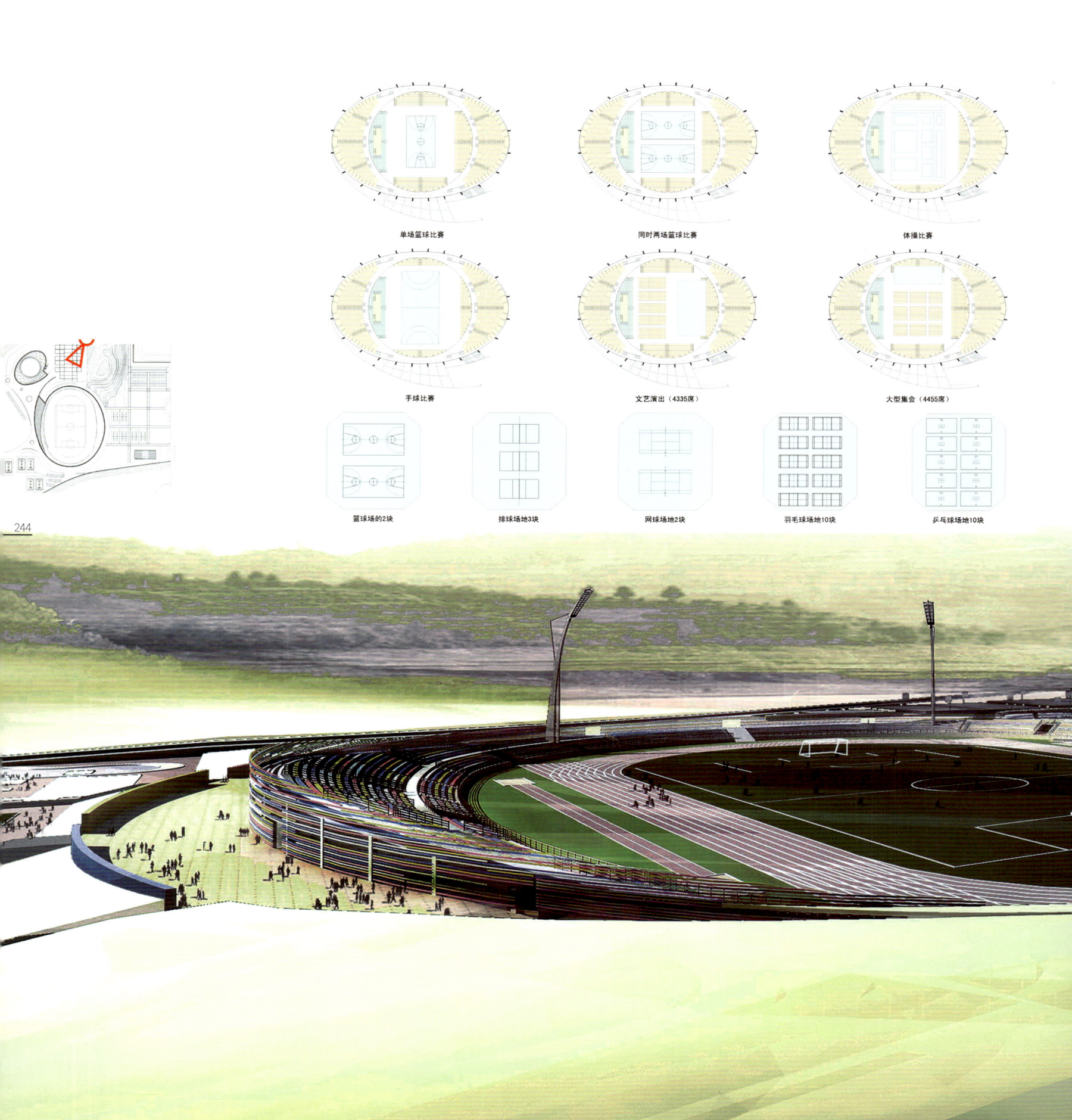

| 单场篮球比赛 | 同时两场篮球比赛 | 体操比赛 |
| 手球比赛 | 文艺演出（4335席） | 大型集会（4455席） |

篮球场的2块　　排球场地3块　　网球场地2块　　羽毛球场地10块　　乒乓球场地10块

景观设计

1、项目运用自然的形态和植被，保持自然山形地貌和水体，创造起伏多变、生动有趣的生态公园。
2、项目对基地自然资源进行整合利用。运用轴线、网状步道、不同标高穿插的环路、视线通廊等动线将山、水、场、馆连成一个整体。

Landscape Design

1. By applying natural forms and vegetation, the natural hilly landform and water body are kept to create a lively and interesting eco-park with a rolling downy landscape.
2. The natural resources of the site are integrated and utilized in this project. Axes, walkway network, ring road with different elevations, and channels allowing sight line are used to connect the hills, water body, stadium and gymnasium together to become an integral whole.

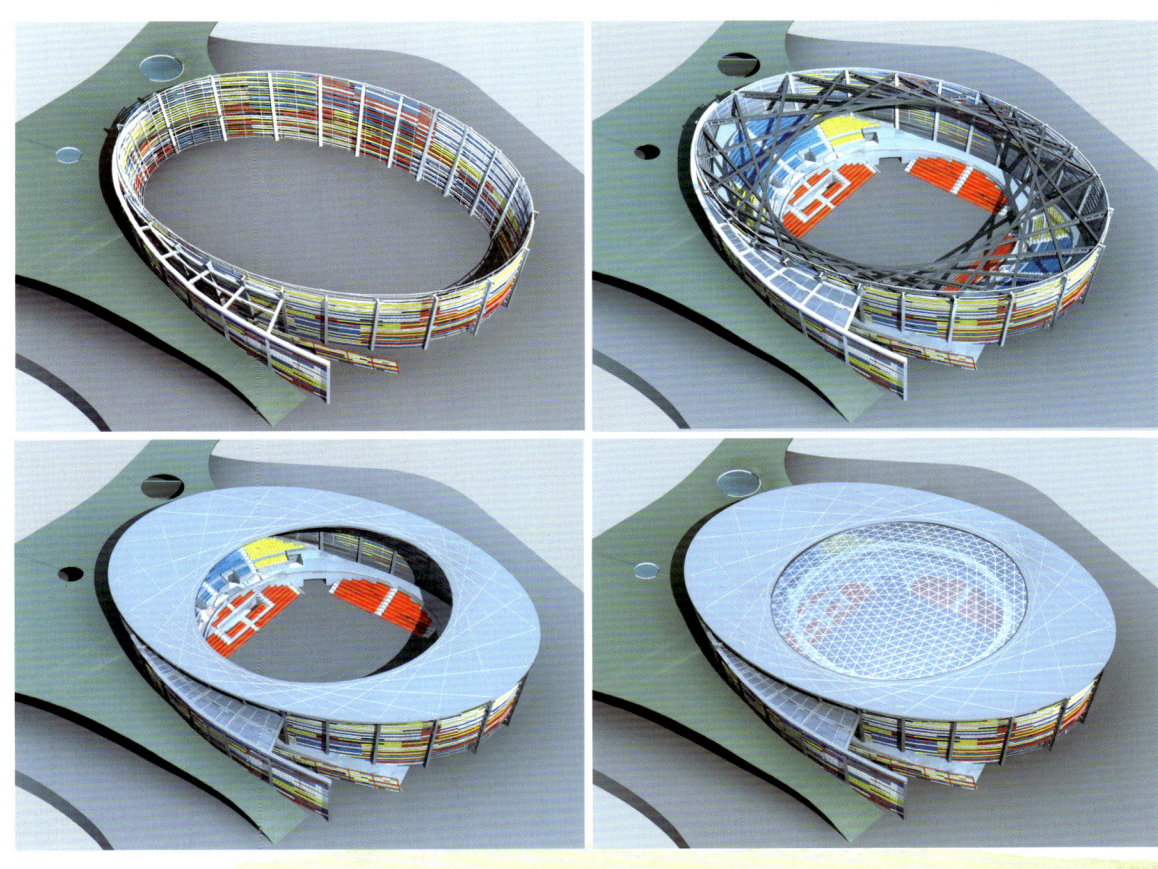

Numerous mountains and waters in Guangxi are worthy of painting.

画不尽的广西山山水水。

Guangxi Art Gallery • Nanning

Nanning | Project Location
2009 | Project Date
36,000m² | Project Scale
conceptual design | Design Phase
construction not started | Project Status

广西美术馆·南宁

中国，广西省，南宁市 | 项目地点
2009年 | 设计时间
3.6万m² | 项目规模
方案设计 | 设计阶段
未建 | 项目现状

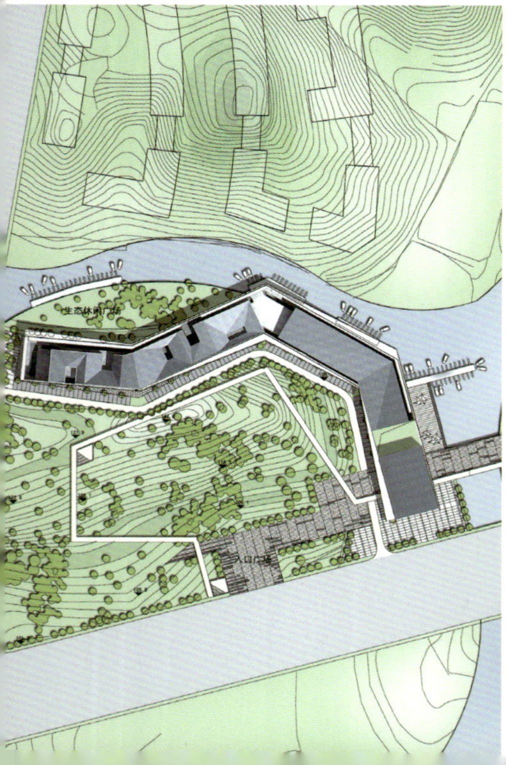

248

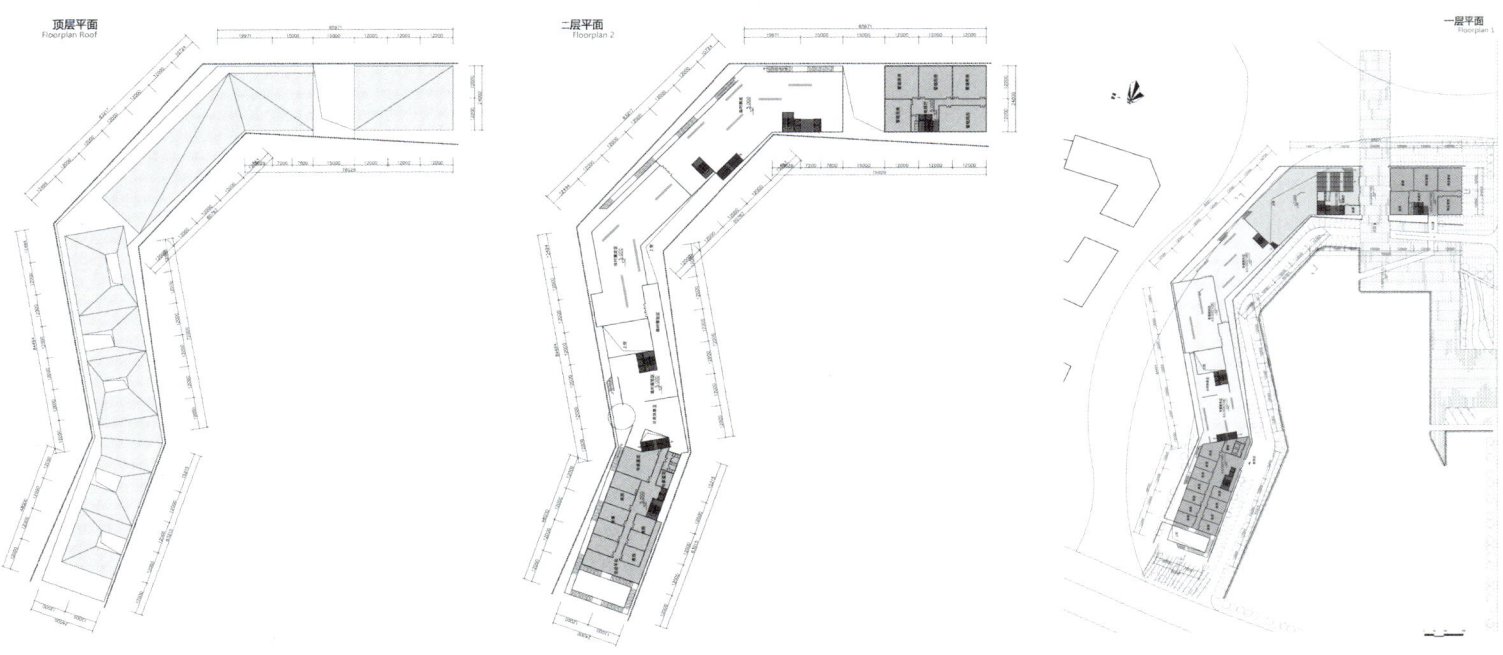

项目总览

美术馆设计在平坦的河滩旁,紧邻水系,结合亲水平台,宛如河畔侗寨,背面水,与自然和谐共生。

Project Overview

Located on a flat flood land, the art gallery is adjacent to water body. With a water-viewing platform, it looks like a riverside village of the Dong people. Fronting water body and with hills on the back, it coexists in perfect harmony with nature.

顶层平面 Floorplan Roof 二层平面 Floorplan 2 一层平面 Floorplan 1

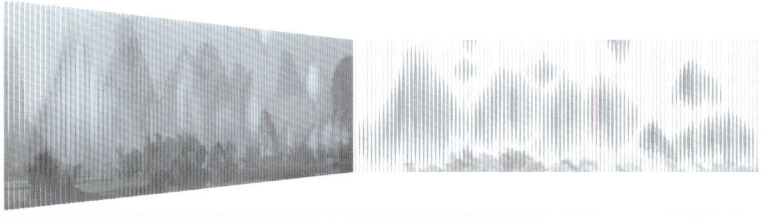

建筑表皮

建筑外立面的设计充分融合了广西独特山水景色与壮民族的文化精髓。

美术馆以壮锦、鼓纹、编织物等饰纹作为形式来源，经几何化、抽象化的提炼，并形成独特的建筑语汇。赋予建筑之外表皮，并结合参数化设计，形成独特的印象。

整个建筑如同巨幅写意山水画一般，随着游览者的脚步而逐渐打开，以此衬托出设计的主题："画不尽的山山水水，赏不完的袅袅炊烟"。

Facade Concept

In the design of building facades, Guangxi's unique natural landscapes and the essence of Zhuang people's culture are taken into account.

The art gallery is decorated with the patterns and ornamentation used on Zhuang brocade, drums, and knit fabrics, which are geometrized and abstracted to form unique architectural vocabularies and cover the buildings with a unique facade. Parametric design is applied to create a unique impression.

Like a large landscape painting of freehand brushwork style, the whole building seems spreading gradually as a visitor walks beside it, embodying the design theme of "numerous mountains and waters for painting, wisps of smoke curling upwards from the village chimneys for appreciation".

西立面 WEST ELEVATION

北立面 NORTH ELEVATION

南立面 SOUTH ELEVATION

东立面 EAST ELEVATION

美术馆顺应山势、水势呈折线自由生长。共同形成"山村水寨"的空间概念，以此取得建筑环境之间的呼应。

The art gallery goes well with the landform of hills and water body to present polygonal lines in its shape. Thus a perception of "mountain & water village" on the space is generated and the building fits the environment quite well.

方案充分汲取壮寨民居与聚落的特点，依山傍水，顺应地形，依山就势，沿阶而下。结合自然界独特的梯田景观，使建筑与山水之间形成良好的互动，密不可分，共同形成了融洽共生的原始村寨形态。

With features of folk houses in Zhuang village, the art gallery is backed by hills and nears water body. It is built in line with the landform. With the unique terrace landscape as the background, the building interacts well with the hills and water body to become a whole and present a harmonious relationship between them, like a primitive village.

Provide an open stage to the city.

为城市提供一个开放的舞台。

Beirut Culture and Arts Center • Beirut (Lebanon)

Beirut | Project Location
2008 | Project Date
15,000m² | Project Scale
conceptual design | Design Phase
construction not started | Project Status

贝鲁特文化艺术中心·贝鲁特(黎巴嫩)

黎巴嫩,贝鲁特市 | 项目地点
2008年 | 设计时间
1.5万m² | 项目规模
方案设计 | 设计阶段
未建 | 项目现状

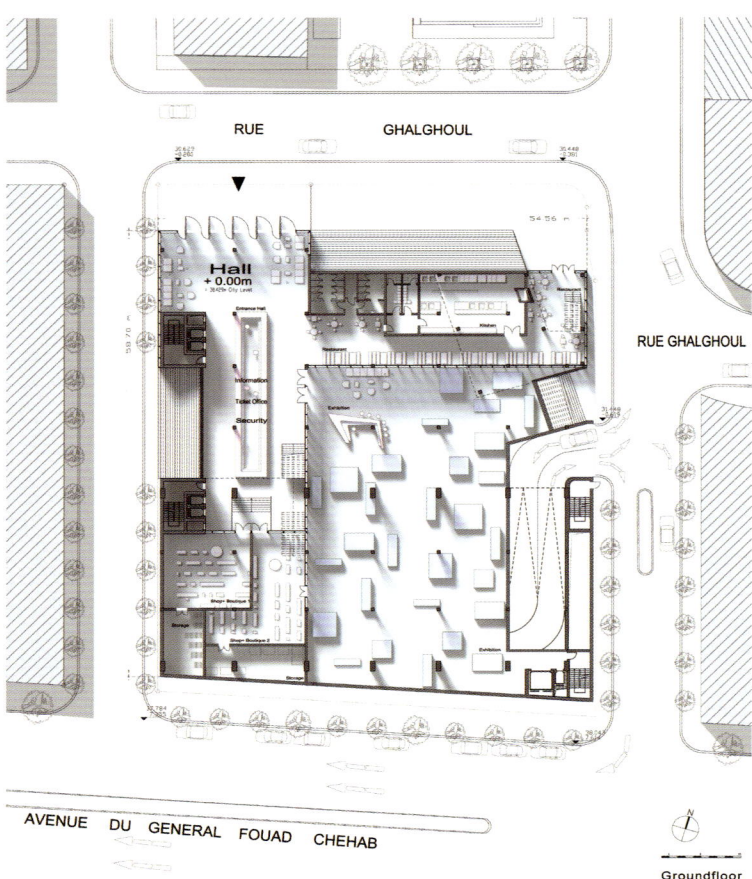

剖面图 C1-C1

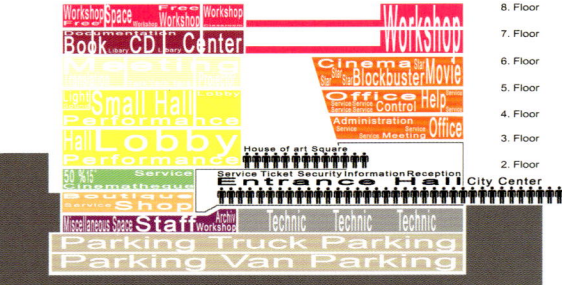

剖面图 C2-C2

剖面图 C3-C3

剖面图 C4-C4

剖面图 C5-C5

项目总览

本建筑设计主要是围绕着一个中央庭院来完成的。它主要有三面面向城市，通向中央大厅的是一个巨大的玻璃门，可以让过往者通过玻璃门看到大厅内的活动情况。贯穿东西的一个楼梯形成了一个通向中央庭院的通道，供参观者通行。入口大厅座落于西北面，方便参观者与工作人员通行，所采用的规格是此类建筑所惯用的。南面以通透的立面向城市展示它的风采。本建筑以其视角细节的表现向路人以及这个城市展现城市地标的魅力。

这个文化中心的目的是提高人们的生活质量，希望这个舞台能够让人们更加紧密地联系在一起。创造一个和谐的社会圈子，让这里成为我们文化与艺术交流之所吧！

Project Overview

The design focuses on a central courtyard, with three sides facing the city. A large glass door leads to the central hall. People outside the hall can see clearly the activities in the hall through the glass door. A staircase, running from east to west, forms a passageway leading to the central courtyard. The entrance hall is located in the northwest, where can be accessed easily by visitors and staff. Its specification is the same as that of similar buildings. In the south is a transparent facade, which faces the city and shows its graceful bearing.

The purpose of this cultural center is to improve people's life quality. Hope this center can bring people together more closely and create a harmonious community. Let this center become a space for people to exchange ideas on culture and arts.

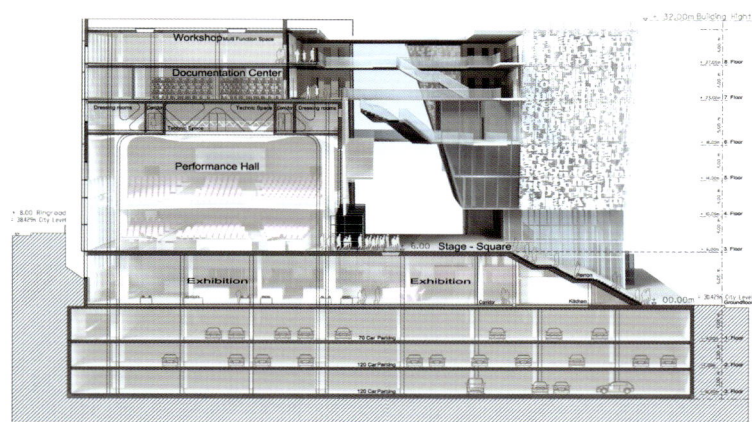

Section A-A
A-A剖面

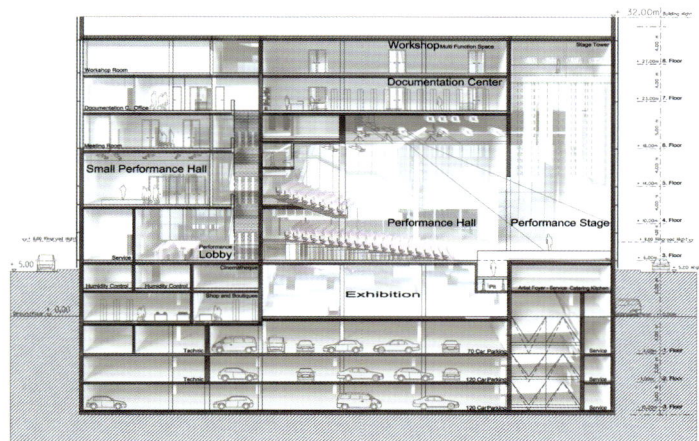

Section B-B
B-B剖面

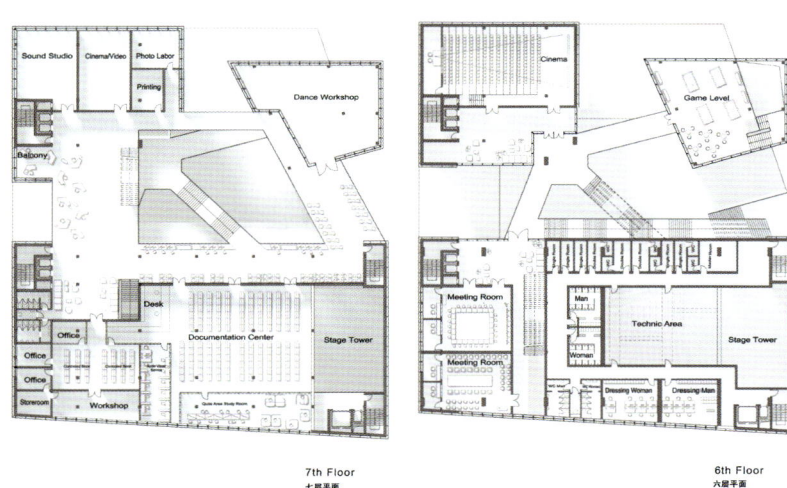

7th Floor
七层平面

6th Floor
六层平面

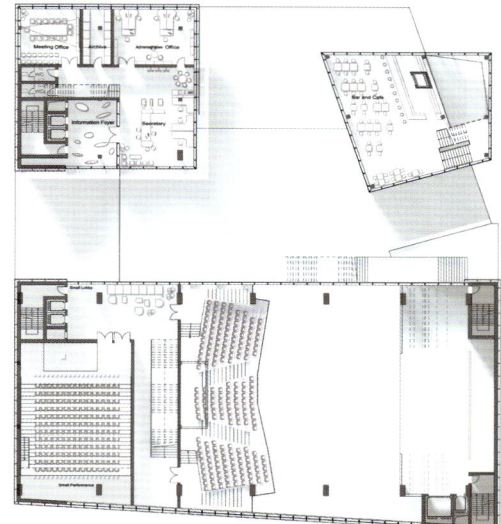

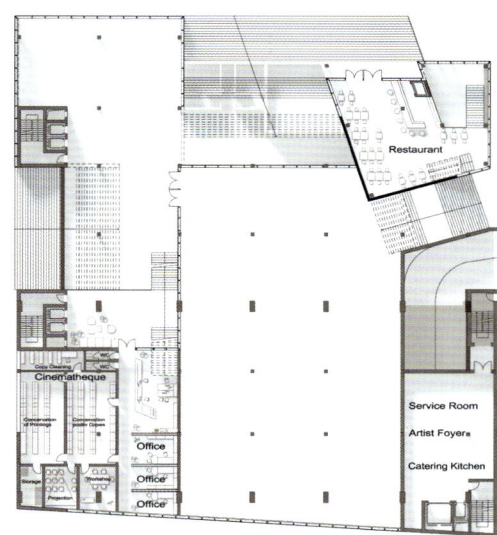

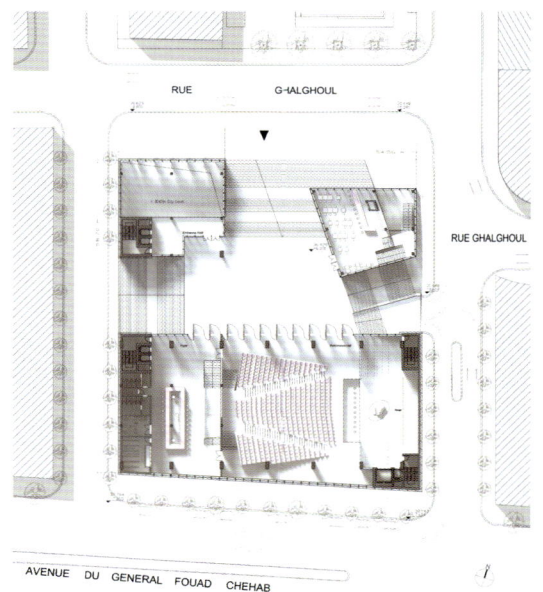

RUE G-IALGHOUL

RUE GHALGHOUL

AVENUE DU GENERAL FOUAD CHEHAB

Traditional monuments only focus on the information conveyed by their appearance while this project tries to treat the internal space of the building as a vehicle for commemoration and also gives the monument an epochal character.

传统的纪念碑仅关注直接由外表传达的信息,而这一设计力求将建筑所形成的内部空间作为表达纪念碑的载体,同时也赋予该纪念碑崭新的时代性。

Xinhai Revolution Monument (Tower) • Wuhan

Wuhan | Project Location
2009 | Project Date
1,000m² | Project Scale
conceptual design | Design Phase
construction under planning | Project Status

辛亥革命纪念碑（塔）·武汉

中国，湖北省，武汉市 | 项目地点
2009年 | 设计时间
1,000m² | 项目规模
方案设计 | 设计阶段
拟建 | 项目现状

项目总览

建筑整体布局方面，我们将广场一并纳入了设计范畴，希望在游人踏上广场的那一刻就开始纪念之旅。

参观路线由地下开始，走过黑暗的通道后，来到宁静的地下大厅。电梯观光面朝向封闭的电梯井，进入电梯后顺势直上，井壁上各种景象再现了当年革命的场景。随着电梯的上升，上部光线渐强，走出电梯后便是位于碑顶的天台。

材料方面，裂开的大地裂缝采取抛光大理石，广场采用锈蚀大理石，玻璃体采用钢化玻璃，坡道采用红色钢化玻璃，底部可加金属网，碑体采用花岗岩或素混凝土。

Project Overview

For the overall layout, the square is also included in the design, with a hope that every visitor can start a journey of commemoration at the moment when he sets foot on the square.

The visiting route starts from underground. After passing a dark passageway a visitor reaches a quiet underground hall. The transparent side of the sightseeing elevator faces the closed elevator shaft. Visitors going up in the elevator can see all kinds of revolutionary scenes on the shaft wall. As the elevator rises, it becomes brighter and brighter. When the elevator reaches the very top, a visitor stepping out of the elevator is on a platform located at the top of the monument.

With respect to materials, the ground crack is made of polished marble, the square made of rusted marble, the glass structure made of toughened glass, the slope made of red toughened glass (a metal net may be added on the base), and the monument made of granite or plain concrete.

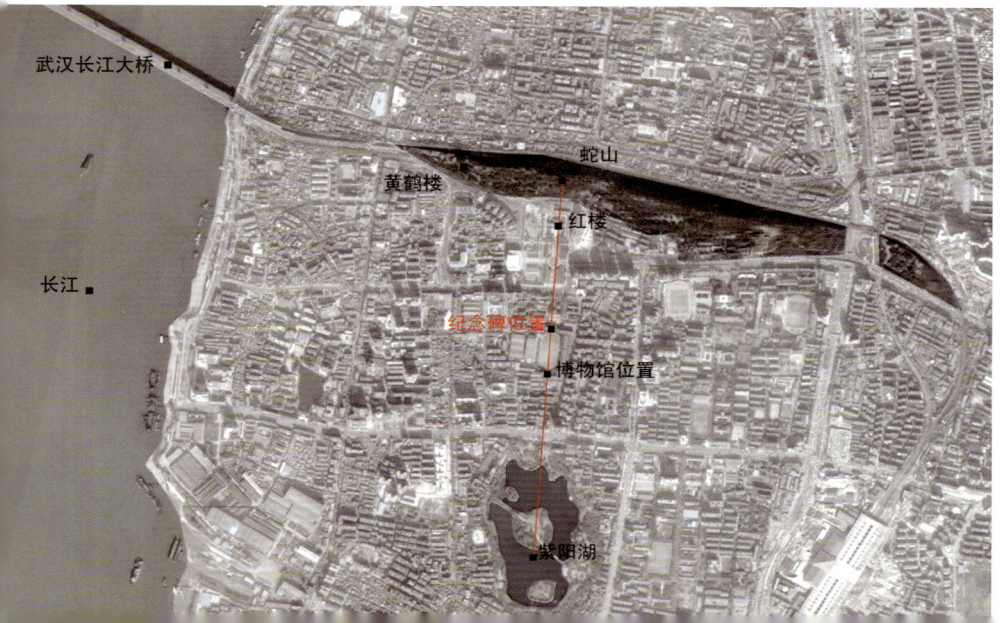

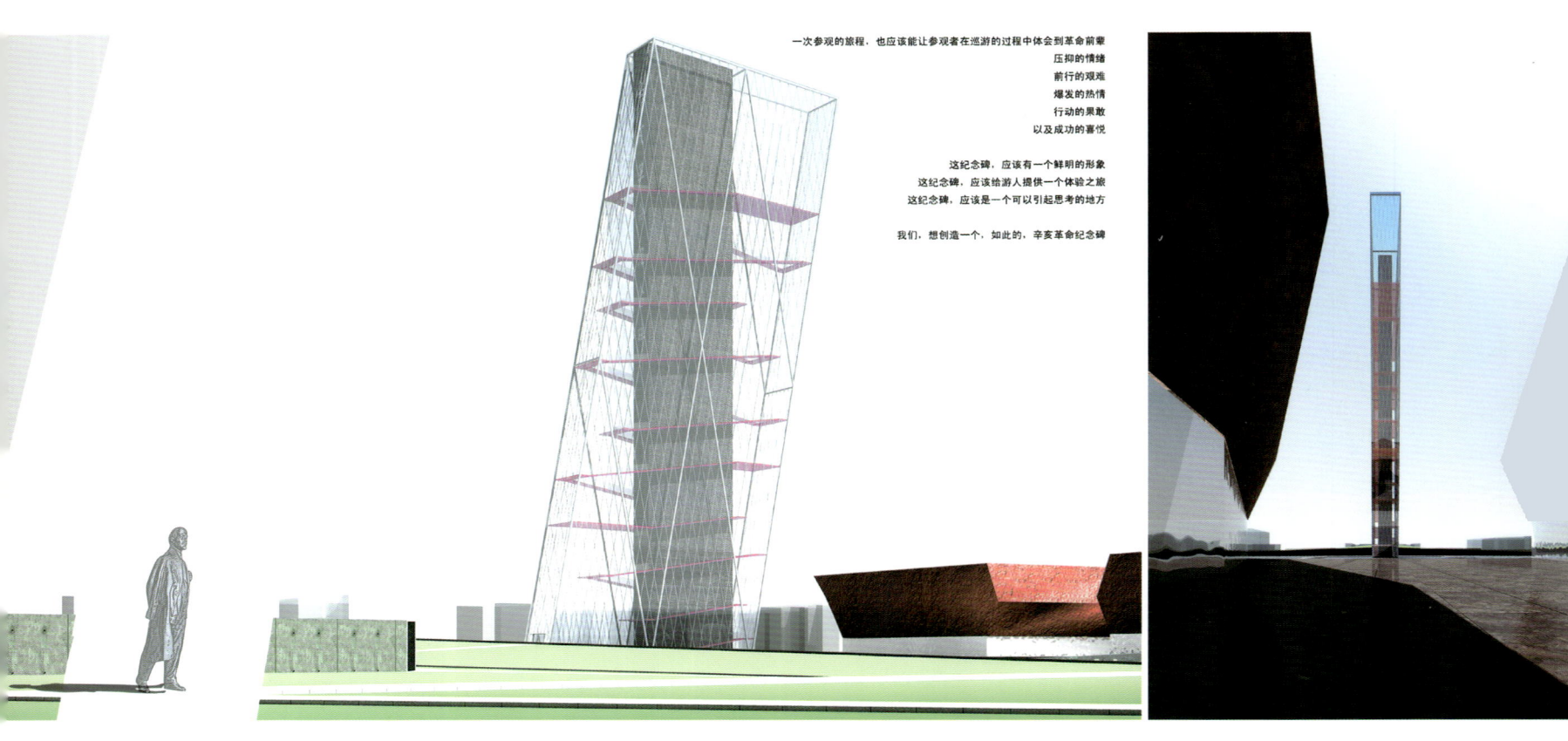

我们选择了一个不平衡的稳定体作为设计母体,建筑本身就犹如一个革命者,手指前方。碑体是它的重心,维持着前行这一不平衡状态的稳定。碑体本身、缠绕着碑体的坡道,还有毫不遮挡视线却保护着碑体及游人的玻璃体,构成了革命纪念碑的整体。

An unbalanced stabilized model is chosen as the design matrix. The building looks like a revolutionary, with a hand pointing forwards. The monument is the center of gravity of the building, maintaining the stability of the building like a man moving forward. The whole revolutionary monument is made up of the monument body, a slope around the monument, and a glass structure which does not hinder sight line but protects the monument and the visitors.

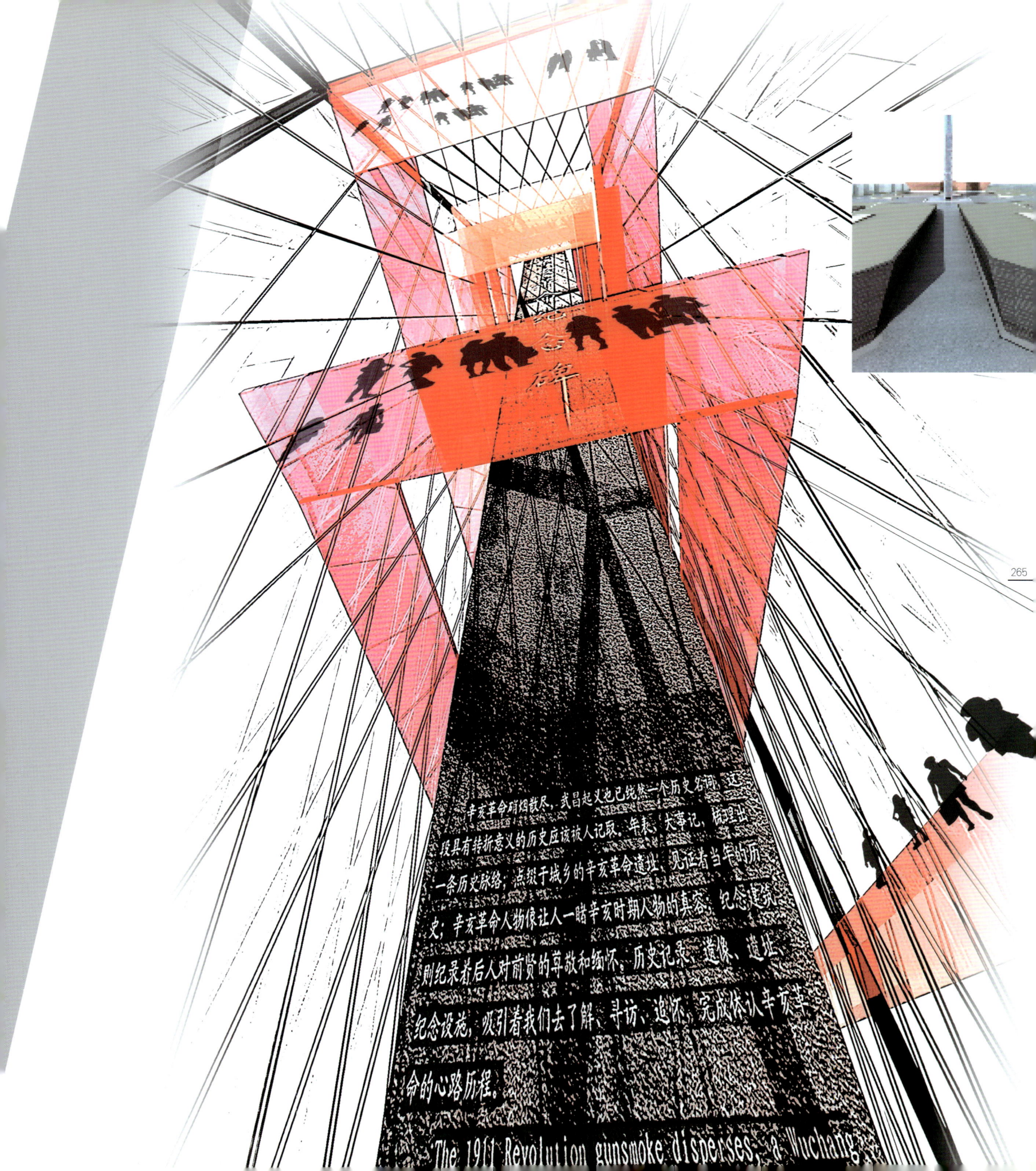

我们希望纪念碑不再是一个孤零零的个体,也不希望游人满腹热情地来,却只能在碑内简单地徘徊,让观光变得乏味肤浅。这是一处会勾起人思念的地方,是一处会让游人逗留、遐想、思考的建筑。一次参观的旅程,也应该能让参观者在参观的过程中体会到革命前辈压抑的情绪、前行的艰难、爆发的热情、行动的果敢以及成功的喜悦。

We hope the monument is not just a single structure. We do not hope that tourists, who come with passion, can only wander around the monument and thus their sightseeing is simple and boring. This place should make visitors recall the history and stay for some time to think and imagine. The visit should also arouse the visitors to feel and imagine the revolutionaries' depressed emotion, difficulties in their revolutionary career, enthusiasm, courage and resolution, and joys of success.

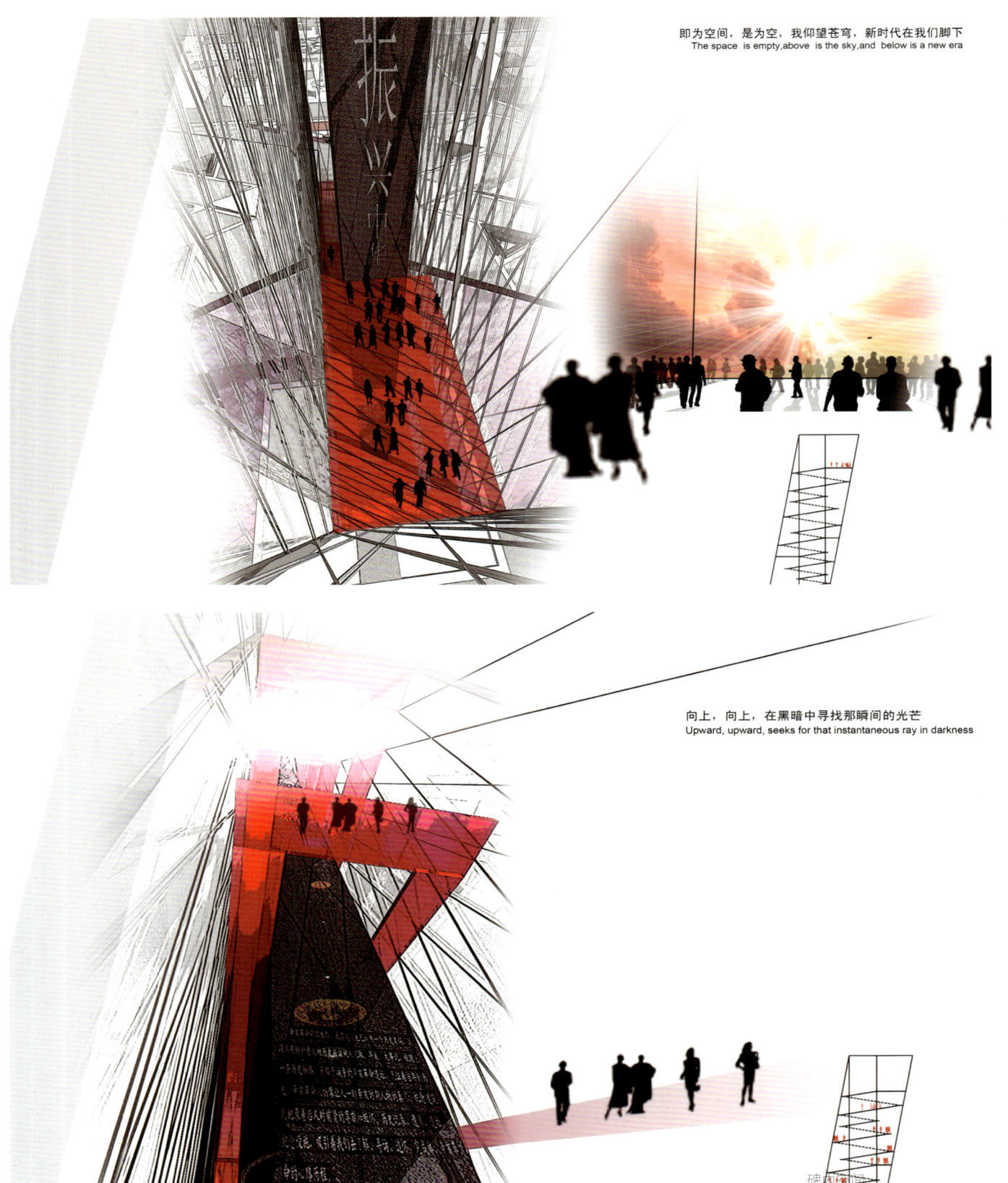

即为空间,是为空,我仰望苍穹,新时代在我们脚下
The space is empty, above is the sky, and below is a new era

向上,向上,在黑暗中寻找那瞬间的光芒
Upward, upward, seeks for that instantaneous ray in darkness

二十年革命路,弹指间过,留下不朽的丰碑。
The 20-year revolution passed quickly and now only an everlasting monument stands erectly.
纪念碑的内空间,观光的内井,时空的遂道。
The monument has an internal space, internal well for sightseeing, and a tunnel.

This accomplished building improves the urban space.

伴随着建筑的建成,也完善了城市的空间。

Guangxi City Planning & Construction Exhibition Hall • Nanning

Nanning | Project Location
2009 | Project Date
40,000m² | Project Scale
conceptual design | Design Phase
construction not started | Project Status

广西城市规划建设展览馆·南宁

中国,广西省,南宁市 | 项目地点
2009年 | 设计时间
4万m² | 项目规模
方案设计 | 设计阶段
未建 | 项目现状

项目总览

广西城市规划建设展示馆位于南宁市五象岭新区，为地下一层，地上五层的会展建筑。项目总建筑面积约44,000m²。会议中心2,500m²，要求单独开放，其它建筑面积为41,500m²。规划展示馆地上部分的空间结构较为复杂，主要由矩形160m×50m（仅二层），矩形190m×45m（仅三、四层，下面架空2层）及矩形150m×55m（仅五层，下面架空4层）三个矩形沿高度依次首尾连接组成，平面投影呈三角形。

Project Overview

Located in new Wuxiangling District, Nanning, the Guangxi City Planning & Construction Exhibition Hall has one underground floor and five aboveground floors. The total building area is about 44,000 square meters. The conference center, with a building area of 2,500 square meters, opens separately to the public. Other buildings covers a building area of 41,500 square meters. The aboveground spatial structures are complicated, consisting of three rectangular structures, i.e. a 160m×50m rectangle (2F), a 190m×45m rectangle (3F & 4F; two floors under them are empty space), and a 150m×55m rectangle (5F; four floors under it are empty space). The three rectangles connects end to end vertically and their planar projection is a triangle.

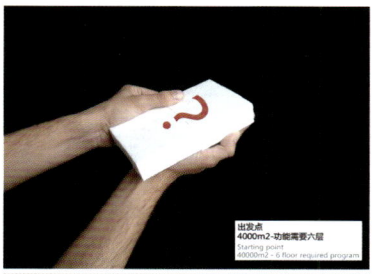
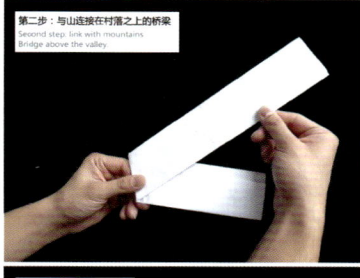
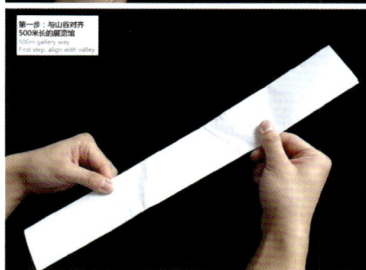
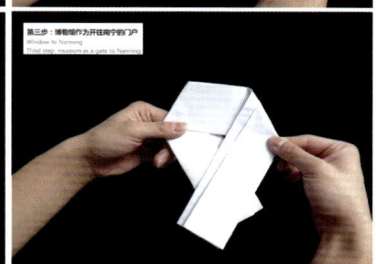

概念

设计概念是建立六个主要参数化步骤：
1) 建筑必须融入现有的景观；
2) 建筑为整个地区提供公共服务；
3) 不同的功能相互关联又相对独立，但从属于同一家博物馆；
4) 博物馆的功能组织应该在一个平台——并主要是在一层；
5) 提供灵活的博物馆展览空间；
6) 新的地标将成为入口大门以及南宁的城市之窗。

Concept

The design idea is established by six major parameters.
1) Architecture must integrate in the existing landscape.
2) Building provides public service for the whole area.
3) The different functions can be accessed and worked independently but are still one museum.
4) The museum function should organize in flat – mainly in one floor.
5) Offer flexible museum exhibition space.
6) The new landmark will become an entrance and the window to Nanning.

概念分析

功能分布

设计提供两条公共面：

第一，（红色）穿过建筑及其内院，联系山、水及城市主轴。
第二，（黄色）作为临时展厅，面向商业街，并形成独立内院，为建筑提供休憩场地。同时，兼当VIP入口，两面均有山水景观。

Layout of Functional Areas

Two public walkways are designed:

The first one (red) passes through buildings and the courtyard, connecting hills, water body, and main axis of the city.
The second one (yellow), as a temporary exhibition gallery, faces the shopping street. An independent inside courtyard is used as a rest area and also VIP entrance. At both sides are mountain and water landscapes.

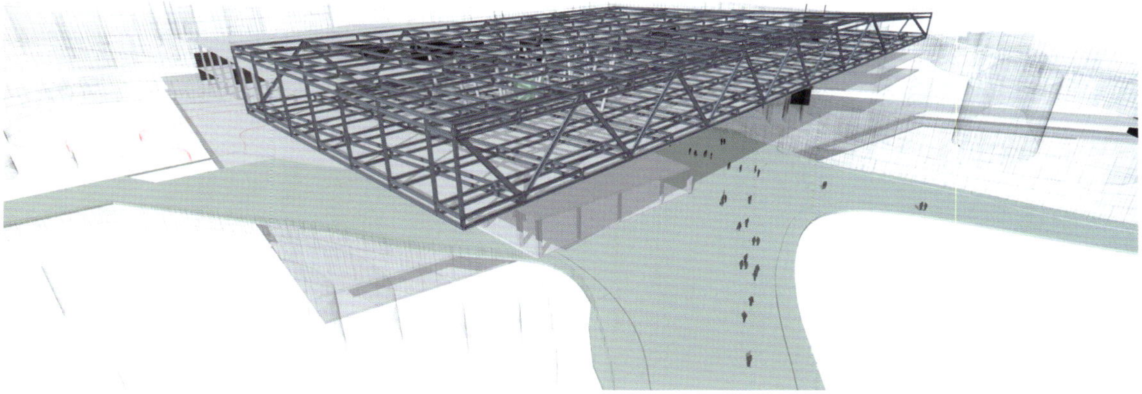

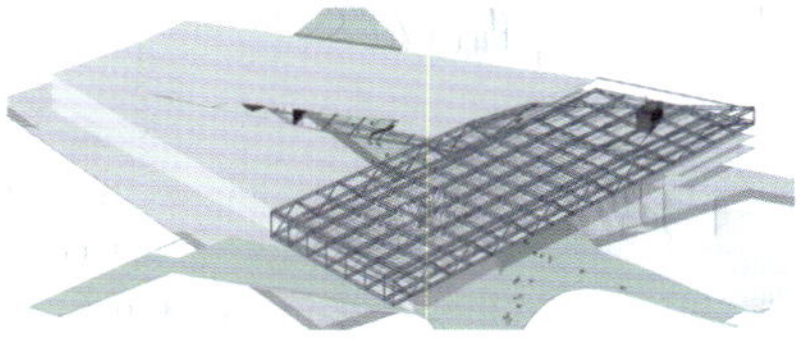

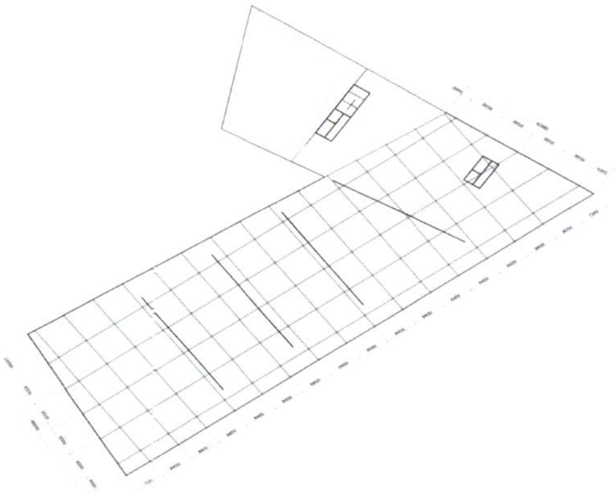

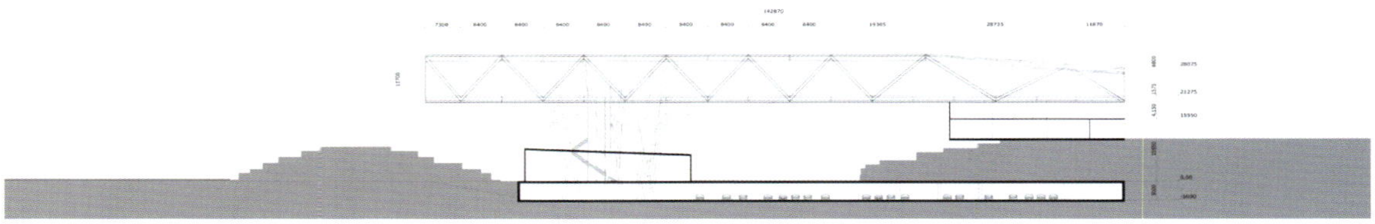

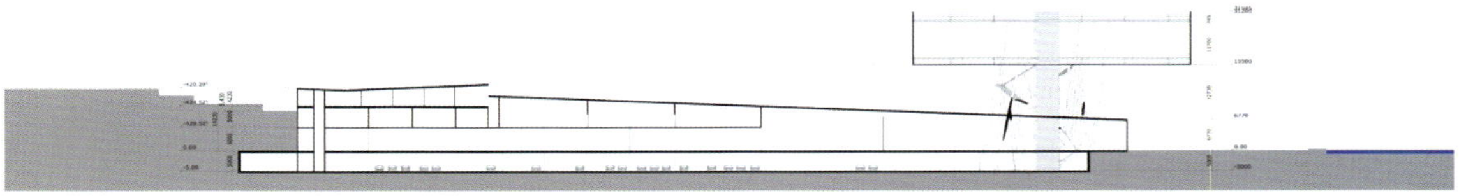

设计为南宁新区创造了一扇"门"。

Building Gate creates an opening to the urban space of the new developing district of Nannnig.

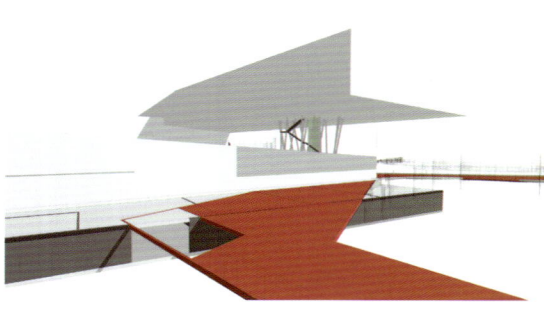

西南广场同时作为VIP入口，可用于特殊节日的庆祝场地。

临时展厅与主要城市街道通过广场相连，为临时展厅的单独使用提供了可能性。作为各大功能区主入口的公共广场。

As a VIP entrance, the southwest plaza may be used as a place for celebration during special holidays.
Since temporary exhibition halls connect major streets cross the plaza, it is possible that they are used separately. They are a public plaza leading to main entrances of all functional areas.

共能布局

怎样把众多功能融合到一起，是设计要解决的首要任务。设计分为三大功能区，三大部分各自独立，又相互融为一体，同时赋予建筑体永恒的地标性特征。三大功能块既可统一使用，又各自相互独立。合理清晰的入口设计，为建筑空间的弹性使用提供了可能性。

Function Layout

The organization and the development of different functions in one building was a main task of the design process. The inner functions are divided in three main parts. Temporary Exhibition, Conference Center and Urban Museum can be separate developed. Also there is a possibility to reorganized the building during the lifetime. Floor plans are designed under permission of flexibility and an easy use. Different entrances make a independent development for different users possible.

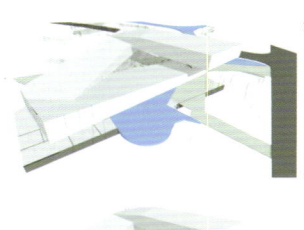

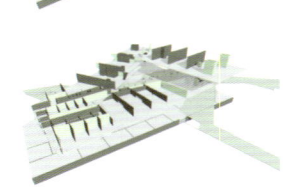
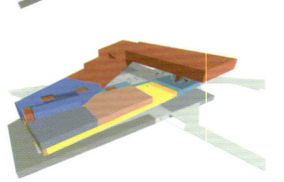
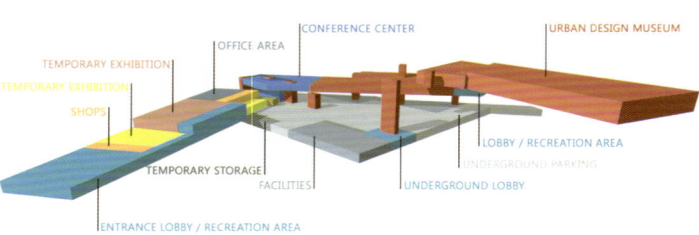

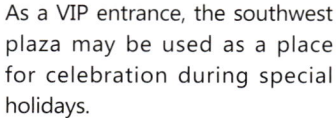

功能分布
Function Layout

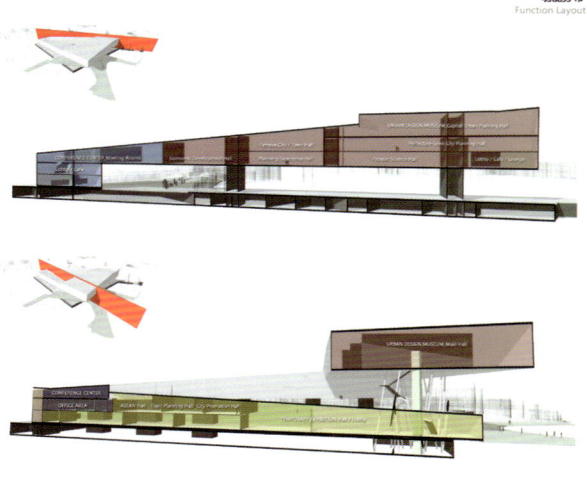

纳米技术创造的新特性
New properties created by NANO technology

媒体立面—发光二极管
Media façade - LED

太阳能-可持续的照明和空调系统
Self-sustainable solar energy for lighting and air conditioning

材料·耐久、耐候性
Material - weather resistant and durable

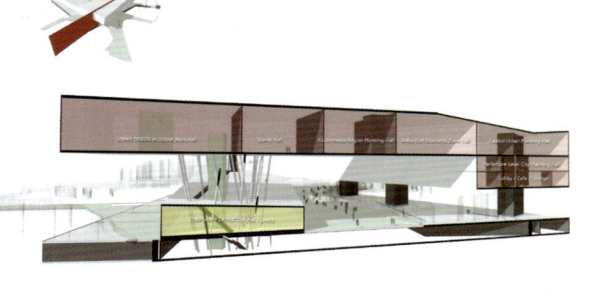

立面——表皮
建设未来

建筑物的外观具有重大意义。除了美观需求，提供内部空间的保护，立面在生态概念中也发挥着重要组成作用。

思小梦大

纳米技术是神奇的，可以赋予建筑立面各种功能。这是喷涂于建筑表面的一个小于0.1mm的薄层。建筑立面就像一个人的皮肤——保护，呼吸，排汗。

该建议认为：

— 金属表面
— 照明（LED）
— 可再生能源
— 耐久性

Build For The Future

The façade of the building is of major importance. Beside the apparent aesthetic purpose and provides protection of its interior, the façade also plays an important part of its ecological concept.

Think Small, Dream Big

Nanotechnology is the magic that can load the elevation with various functions. It is a thin layer, less than 0.1mm print on the main surface. The elevation become a skin like a human being – protect, breath, sweat.

The proposal considers:

- Metallic surface
- Lighting (LED)
- Renewable energy
- Durability

区域位置

南宁市位于广西的西南部，毗邻粤港澳，背靠大西南，面向东南亚，是连接东南沿海与西南内陆的重要枢纽。

Site Location

Nanning, located in the southwest of Guangxi, nears Guangdong, Hong Kong and Macau, and is a important city on the way to southwest China and Southeast Asia. It is an important hub connecting southeast costal regions and scuthwest inland.

This is the presentation of the spirits of a traditional university in another city. The values and features of the university should be integrated in the buildings.

这是一所具有传统精神的学校在另一个城市的代表,需要体现出这个学校的价值观和特点。

Shenzhen Production, Teaching and Research Base of Huazhong University of Science and Technology • Shenzhen

Shenzhen | Project Location
2007 | Project Date
45,024m² | Project Scale
conceptual design | Design Phase
construction in progress | Project Status

华中科技大学深圳产学研基地·深圳

中国，广东省，深圳市 | 项目地点
2007年 | 设计时间
45,024m² | 项目规模
方案设计 | 设计阶段
在建 | 项目现状

方案一

方案二

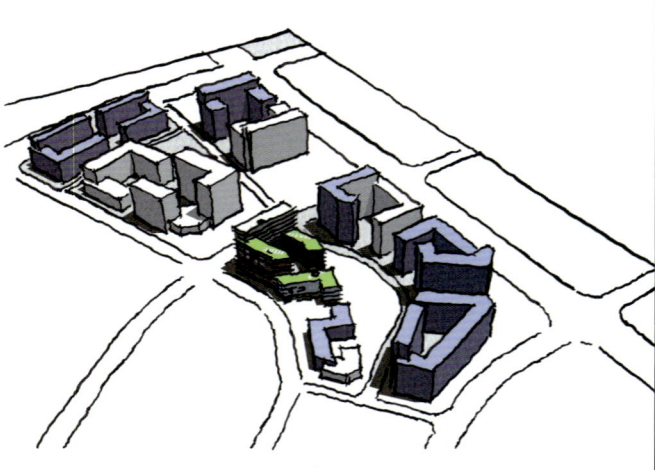

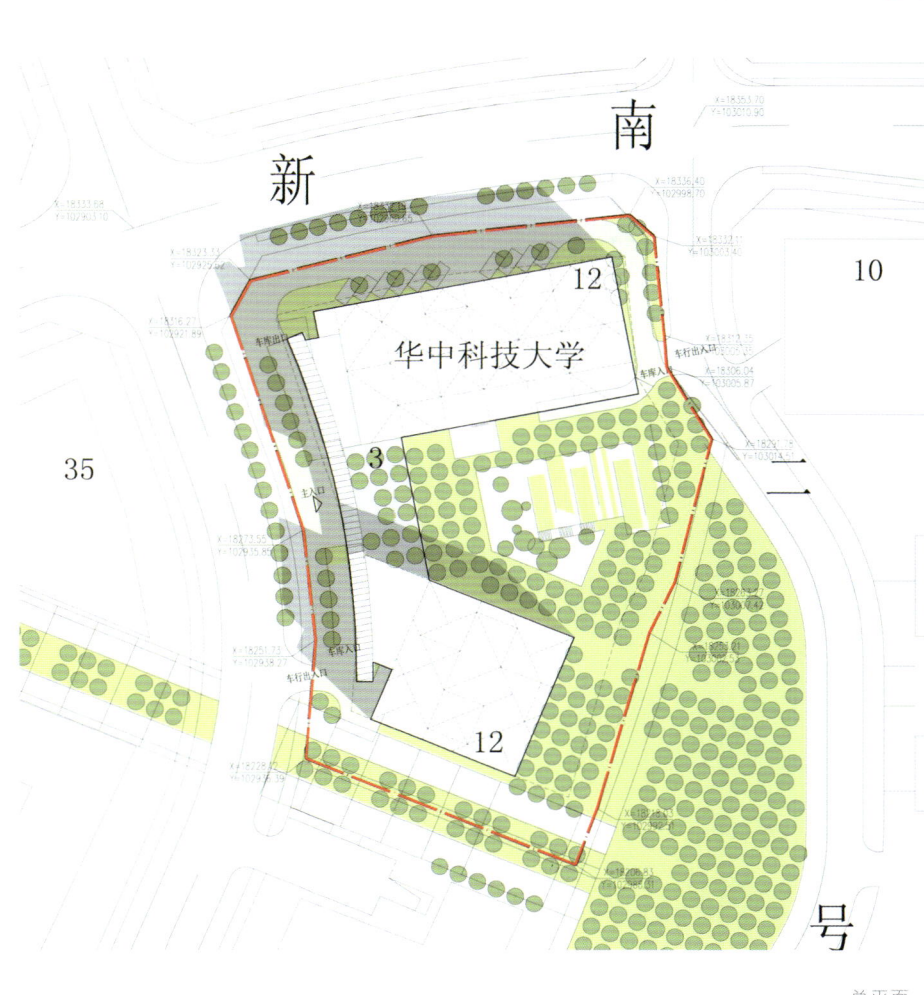

总平面　　方案三

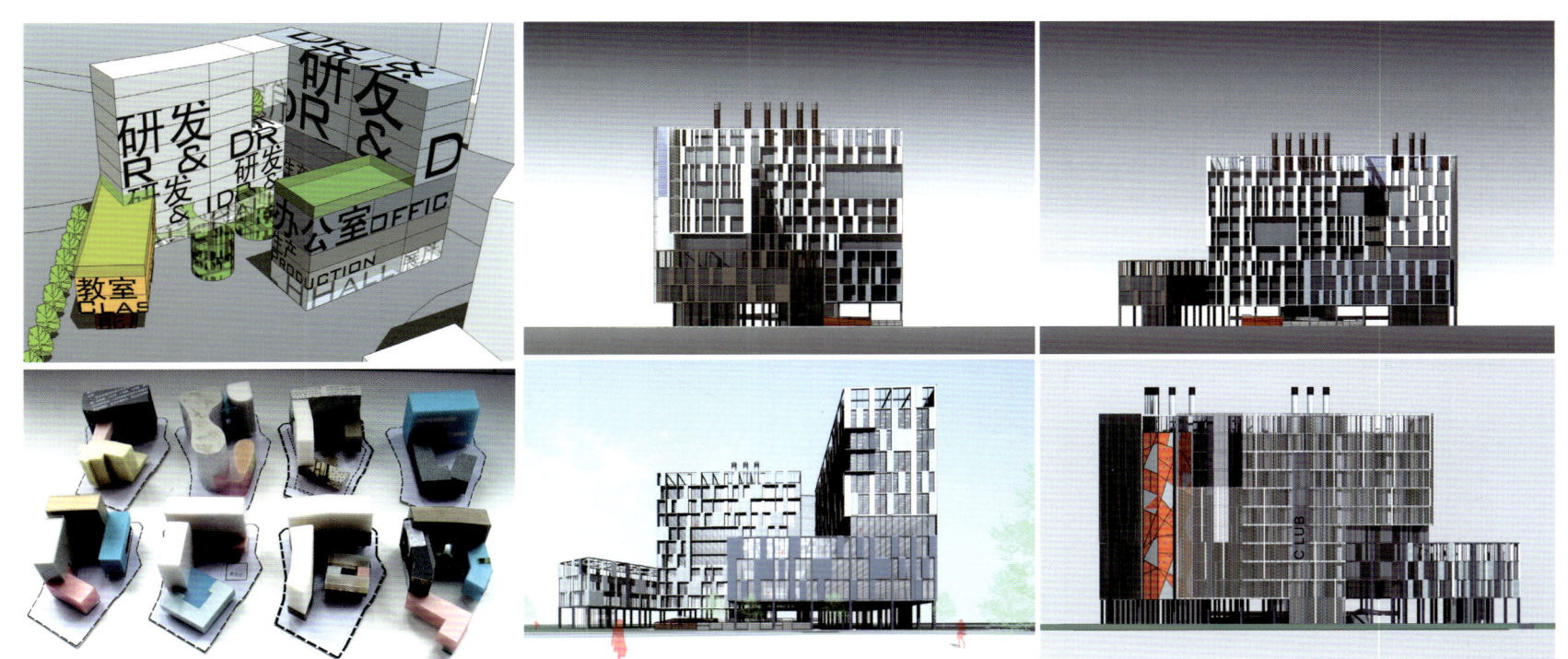

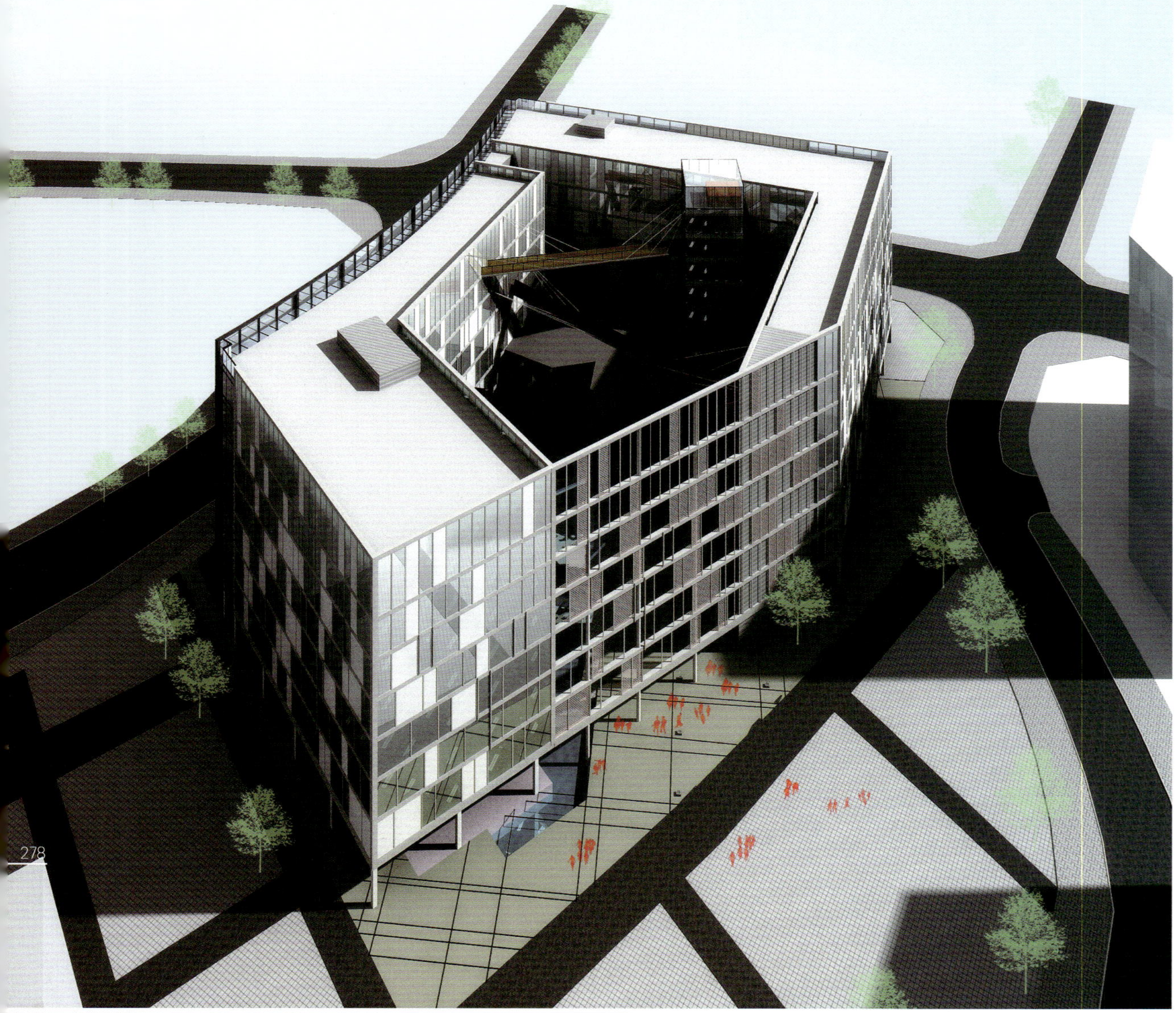

项目总览

本项目研发生产空间有别于传统的办公空间,追求舒适是我们的出发点。预留两层高的花园,为各个孵化器单元提供交流休息的区域,同时提高了办公区的整体环境。

项目还引入传统民居中的节能经验,如热通道、格栅遮阳、拔风中庭,力求使之成为一座低耗能的后现代建筑,适应未来。

Project Overview

This research, development and production space of the base is different from conventional office space. The starting point in design is to create comfort. A two-storey-high garden is designed to provide a rest and communication area to all incubator units and improve the general environment of the office area.

Energy saving ways used in traditional folk houses are introduced in the design of this project, such as heat channels, grating awning, and atrium with draft function. The design aims to build a low-energy-consumption postmodern architecture to adapt to the future.

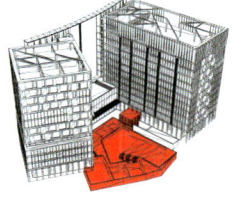 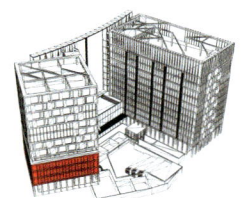 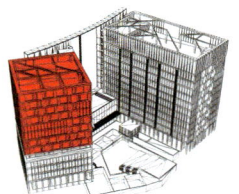
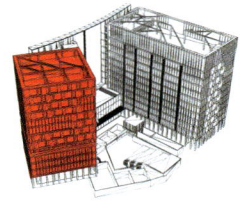 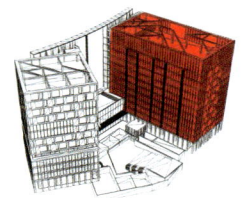 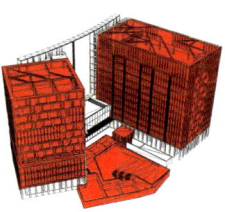

The provision of diversified architectural forms and spaces is good for embodying the history and traditions of the college.

提供建筑与空间的多样性将有助于大学历史和传统的的沉淀。

Foshan Polytechnic College · Foshan

Sanshui, Foshan | Project Location
2009 | Project Date
230,000m² | Project Scale
conceptual design | Design Phase
construction not started | Project Status

佛山职业技术学院·佛山

中国，广东省，佛山市，三水区 | 项目地点
2009年 | 设计时间
23万m² | 项目规模
方案设计 | 设计阶段
未建 | 项目现状

项目总览

本项目旨在建设一个让校友引以自豪、自然和谐的校园。在基地竖向研究，充分考虑到上方平衡，集约开放。我们依山就势提取了一个10m标高的平台，作为校园规划的主结构线。校园建筑在此轴线上自由散落分布。

Project Overview

The design aims to build a natural and harmonious campus of which all alumni are proud. In the vertical design, the balance of the upper structures is emphasized and intensive and open spaces are created. A 10m-high platform is designed as the main structure line for the campus planning. Buildings in the campus are distributed freely along this axis line.

基地原生态美和谐建构了环境生态网络，同步营造校园分共生活空间，使建筑和人以恰当的方式介入并与之结合。

An ecological network is built to live in perfect harmony with the ecosystem of the base. At the same time, a campus space is created to make buildings and people blend in the environment in an appropriate way.

绿色、宁静而富含生机的山水生态校园容纳各种生活形式，具有层次丰富的场所。

The green, quiet, and vigorous landscaped campus contains different life styles and has places of different levels.

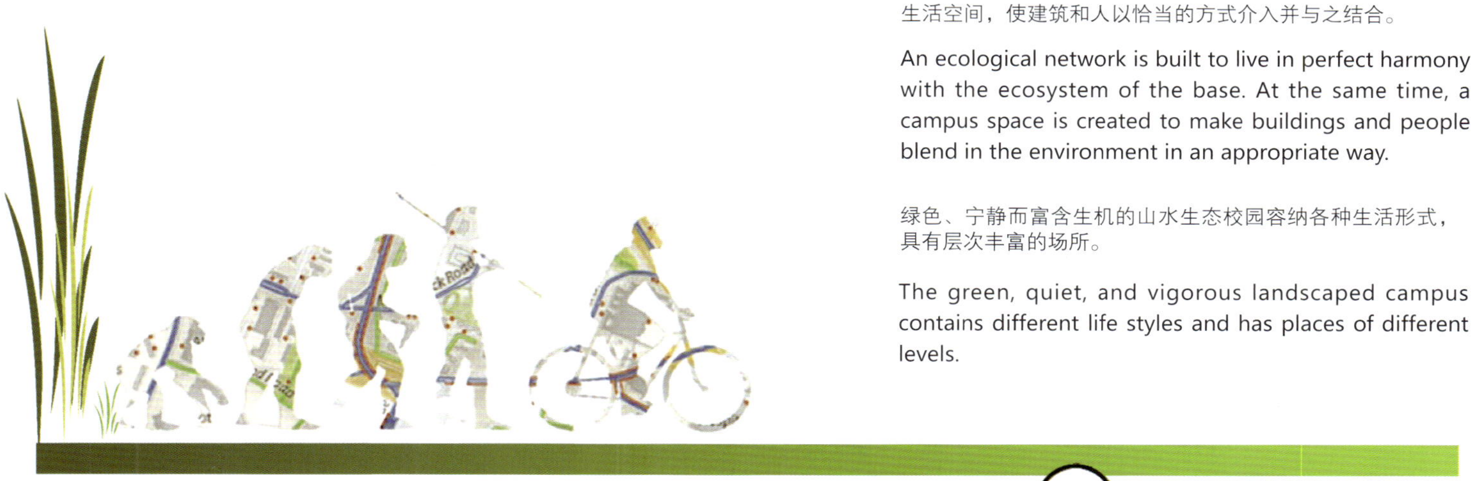

基地红线范围面积
587,332 sq.m
base area with redline
587,332 sq.m

除去农田保护区
remove nature protected area

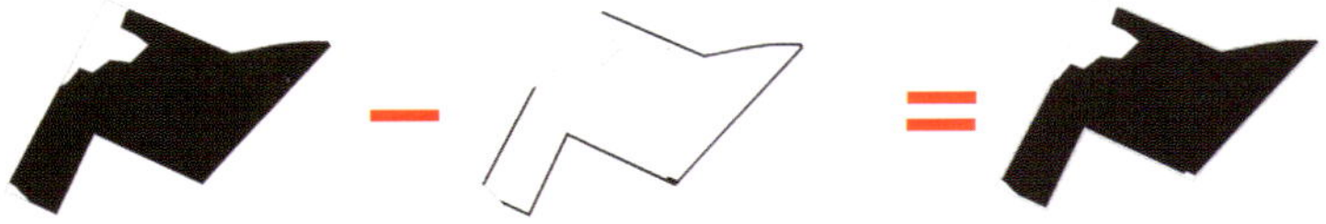

除去城市规划规定退距
remove backway as urban planning regulations

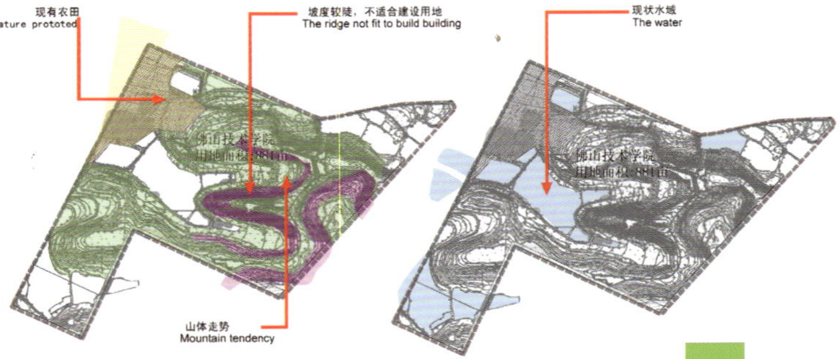

现有农田 Nature protected
坡度较陡，不适合建设用地 The ridge not fit to build building
现状水域 The water
山体走势 Mountain tendency

基地研究：
校园特色的形成不仅在于建筑空间的特色，更来自基地环境的影响和对环境景观的表现。在基地竖向研究，充分考虑土方平衡，集约开放，我们依山就势提取了一个 **10米标高的平台**，作为校园规划的**主结构线**，校园建筑在此轴线上**自由散落分布**。

10m标高平台
The flot roof on 10 meter's high

| 篮球场 / Basketball court |
| 网球场 / Tennis court |
| 排球场 / Volleyball court |
| 体育馆 / Gymnasium |
| 游泳池 / Swimming pool |
| 体育场 / Stadium |
| 实习实训及附属用房 / Campus school |
| 专用实验楼 / Laboratory |
| 教学中心 / Teaching centre |
| 图书馆 / Library |
| 校史馆 / History hall |
| 行政楼 / Administration office |
| 招待所 / Rest house |
| 后勤综合楼 / Logistical building |
| 医务所 / Doctor's office |
| 学生活动中心 / Student centre |
| 食堂 / Canteen |
| 学生宿舍 / Students dormitory |
| 教师宿舍 / Teacherage |

亲切是我们对校园景观尺度的追求。在这次的设计中，我们觉得场地最关键的特性是视线与人的行走路径。在这里对建筑起主导作用的因素是尺度，而不是风格。贯穿校区的步行道路用起伏的坡地、台阶和折形坡道进行组合。

Affability is our pursuit for campus landscaping. We think the most crucial feature of the site in the design is line of vision and pedestrian paths. It is dimension, instead of style, that plays a dominant role in the buildings. Pedestrian roads, which run through the campus, consist of undulating slopes, steps, and winding ramps.

学校不是闭门造车，不应成为一座缺乏交流的孤岛。
职业学院培养的是知识与能力兼备的，且能迅速适应现代化城市要求的社会型人才。我们在校园规划中引入了"学院街"的概念，带入城市规划的思想，以一条不规则的街带串联校园空间。在校园中丰富的视线，令人惊奇的角落以及所倡导的交流场所成为与传统学校不一样的校园空间，拥有城市空间和自然环境的学校充当学生间充分交往、交流的介质和溶剂。街巷式的城市空间组织方式成为本方案的特点，设计中的学院街、街与街交叉的广场、保留的山体森林公园散落在绿化中具有城镇型的学院空间，这正是我们所向往的。

A school should not indulge in fantasy and become an isolated island lack of communication.
A vocational college brings up compound talents who have both knowledge and ability and can meet urban demands quickly. In the campus planning, we introduce a concept of "College Street" as well as city planning ideas and design an irregular street belt to connect campus spaces. Rich lines of vision, amazing corners, and communication places form a campus space different from that of traditional schools. With au urban space and natural environment, the school becomes a medium and solvent for students to communicate with each other. Street & lane-style urban space organization is a feature of this project. The college street, plaza with crossing streets, preserved forest park scatter among green belts to show an urban-style campus space, to which we are looking forward.

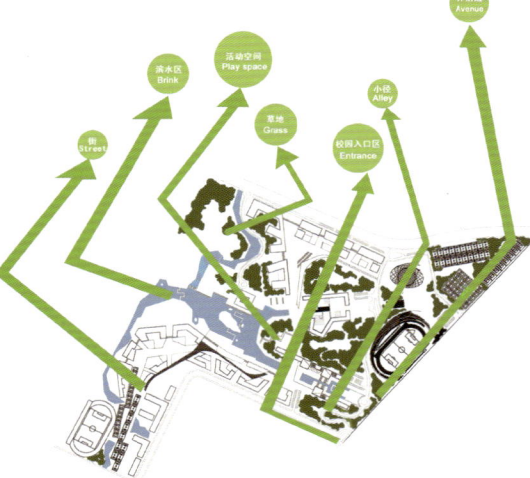

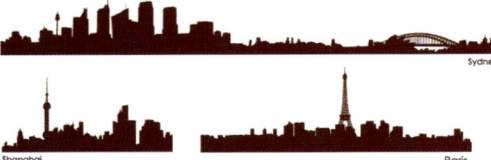

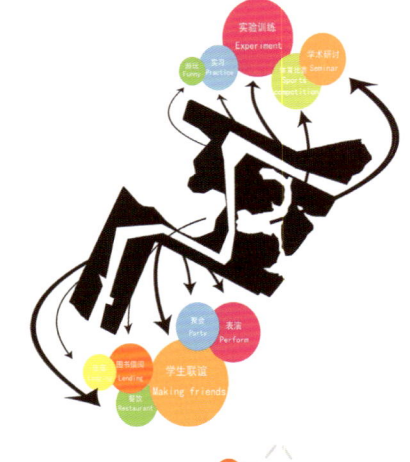

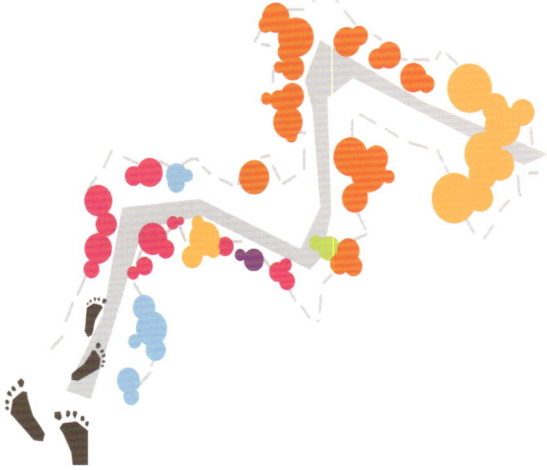

学院街
concept 2

大学需要表达的是一种自由的精神，空间一旦形成，就会有活力。
—— 汉宝德

A college needs to express a kind of liberal spirit. Once the space comes into being, its vitality exists.

-- Pao-Teh Han

台阶和坡道形成景观节点，突出集散功能，创造浓郁氛围。
The landscaping nodes resulting from steps and ramps show a distributing function and create a dense atmosphere.

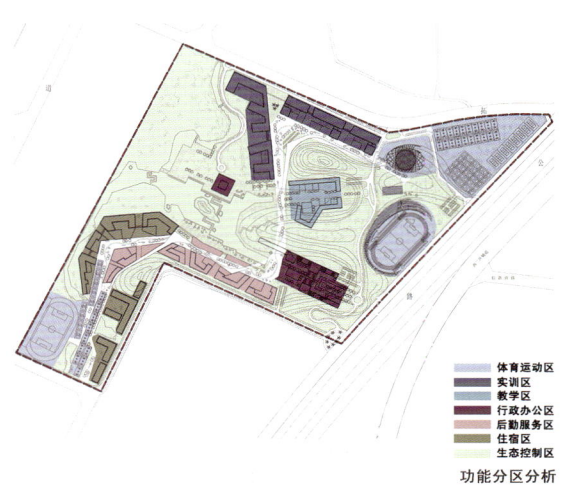

功能分区分析

竖向分析

项目在校园规划中引入了"山水校园"和"学院街"的概念

1. 学院街：项目在校园规划中引入了"学院街"的概念，带入城市规划的思想，以一条不规则的街带串联校园空间。在校园中丰富的视线、令人惊奇的角落以及所倡导的交流场所成为与传统学校不一样的校园空间。

2. 山水校园：新校园规划的关键不在于建筑和空间的刻意表现，而在于首要建构一个生态环境的网络，尊重原生态的同时致力于校园公共生活的营造。这是一个绿色、宁静而富含生机的校园，容纳着各种校园活动，拥有层次丰富的场所。

The concepts of "landscaped campus" and "academic street" are introduced into the campus planning

1. Academic Street: the concept of "academic street" is introduced into the campus planning, including city planning ideas. The irregular street connects the spaces of campus. Plenty of channels allowing sight lines, amazing corners, and communication places create campus spaces different from that of traditional schools.
2. Landscaped Campus: the key of new campus planning is not purposeful expression by means of architectures and spaces but building a networked ecological environment featuring native ecosystem, and facilitateing pubic activities in the campus. This is a green, quiet and dynamic campus allowing all sorts of student activities and a place with rich spaces.

Some buildings are treated as "city walls" of the campus. Quiet and peaceful woods are created in the campus to provide adequate green space and quiet environment to children.

将部分建筑处理成校园的"城墙",在校园内营造出一片宁静、平和的森林,给孩子们充足的绿地和宁静的天空。

Shenzhen Bay Middle School · Shenzhen

Shenzhen | Project Location
2005 | Project Date
70,000m² | Project Scale
conceptual design | Design Phase
construction not started | Project Status

深圳湾中学 · 深圳

中国，广东省，深圳市 | 项目地点
2005年 | 设计时间
7万m² | 项目规模
方案设计 | 设计阶段
未建 | 项目现状

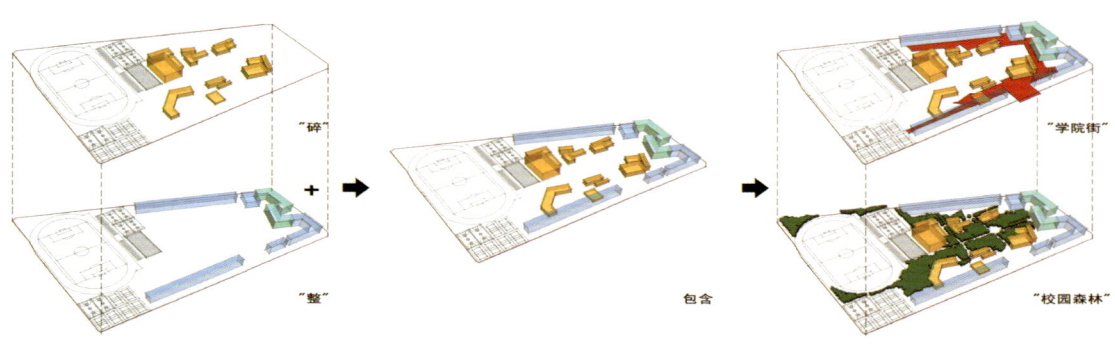

项目总览

本项目旨在为校园营造一片净土，给孩子们充足的绿地和开阔的天空。在校园这片树荫下，孩子们可以不受干扰地学习、活动，充分发展自己的个性。

Project Overview

The purpose of this project is to create a clean campus so as to offer students a broad space with adequate green area. Under the shelter of the campus, the students may study and play freely and develop their personalities.

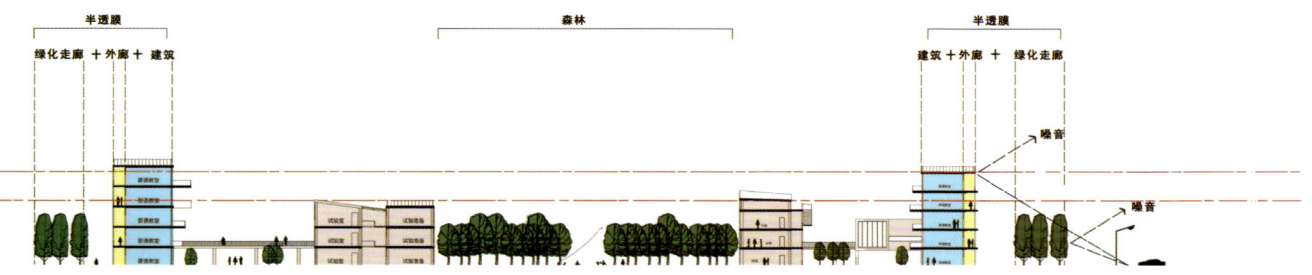

模型图

设计在基地周围形成18m高、"整"的外皮体量,内部是零碎、丰富、小体量的公共建筑。它们之间以街道和广场相联系,形成亲切、轻松的校园学习环境。18m高的建筑和外围的绿化走廊形成一层"半透膜"。这层"半透膜"阻挡了道路的噪音和灰尘。公共建筑围合的内部包含大片的校园森林,与体育场连成一体,成为孩子们开阔的活动空间。

A building with the height of 18 meters is designed around the site. Inside the campus are some dispersedly-distributed small public buildings, which are linked together by streets and squares to form an amiable and relaxed study environment. The 18m-high building and external green corridor constitute a layer of "semipermeable membrane", which prevents noises and dust from outside roads. Inside the space encircled by the public buildings is a large area campus woods, which joins up with the stadium and is a broad space for the activities of the students.

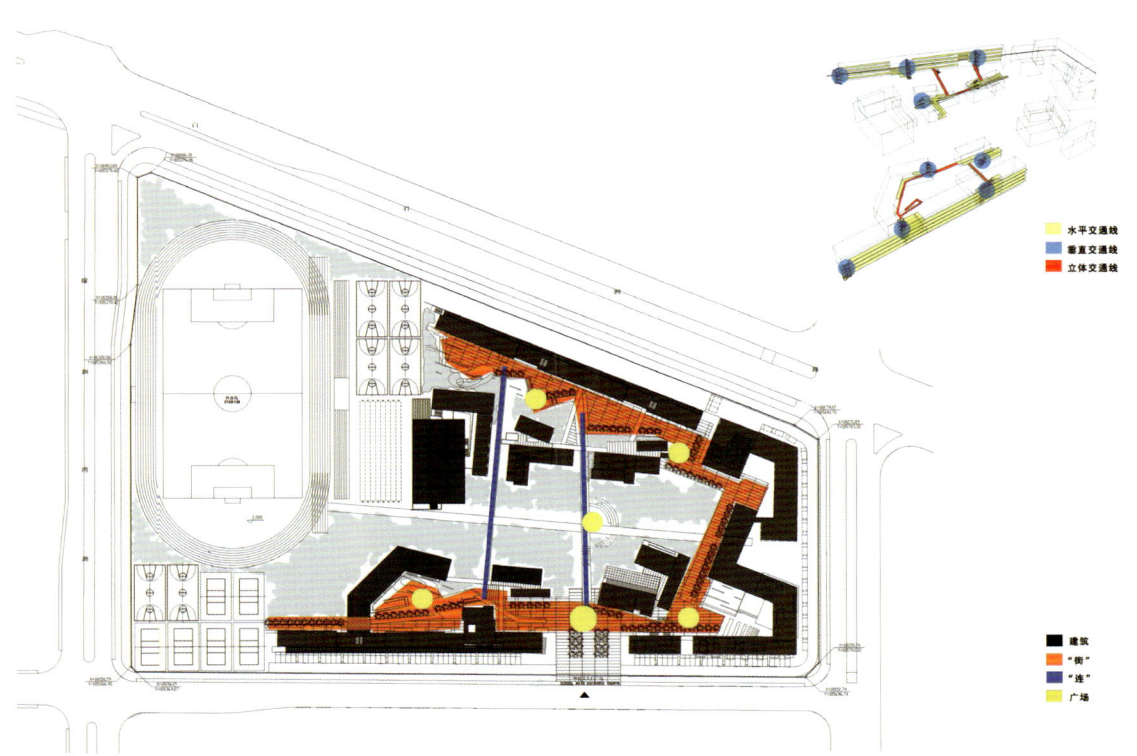 

总平面

分区明确

阅览室

阶梯教师

书库

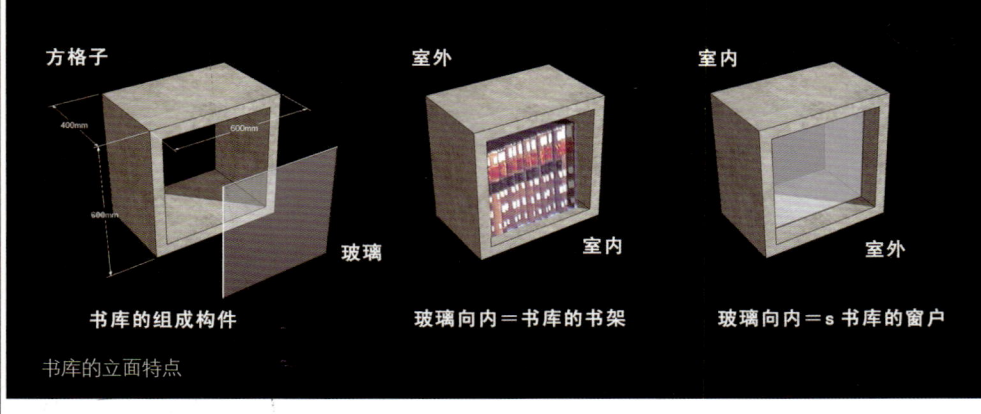

| 方格子 | 室外 | 室内 |

书库的组成构件   玻璃向内＝书库的书架   玻璃向内＝s书库的窗户

书库的立面特点

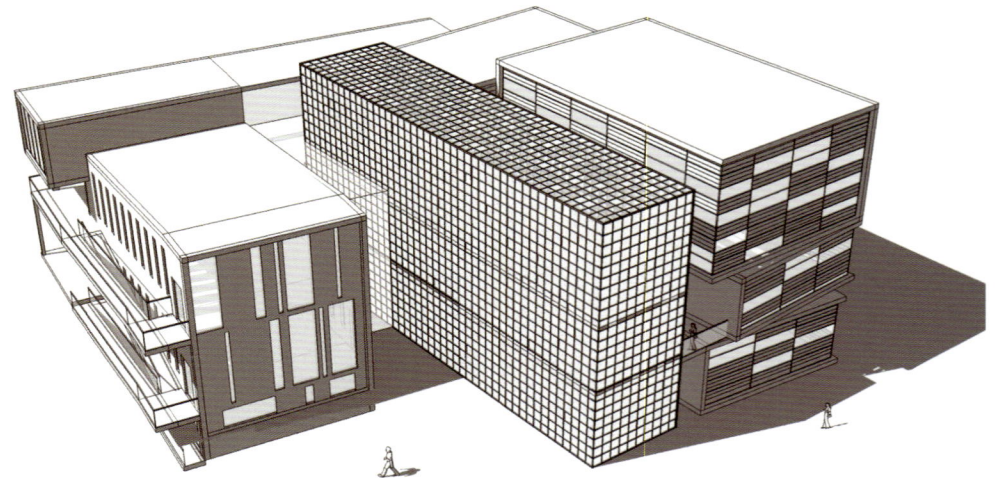

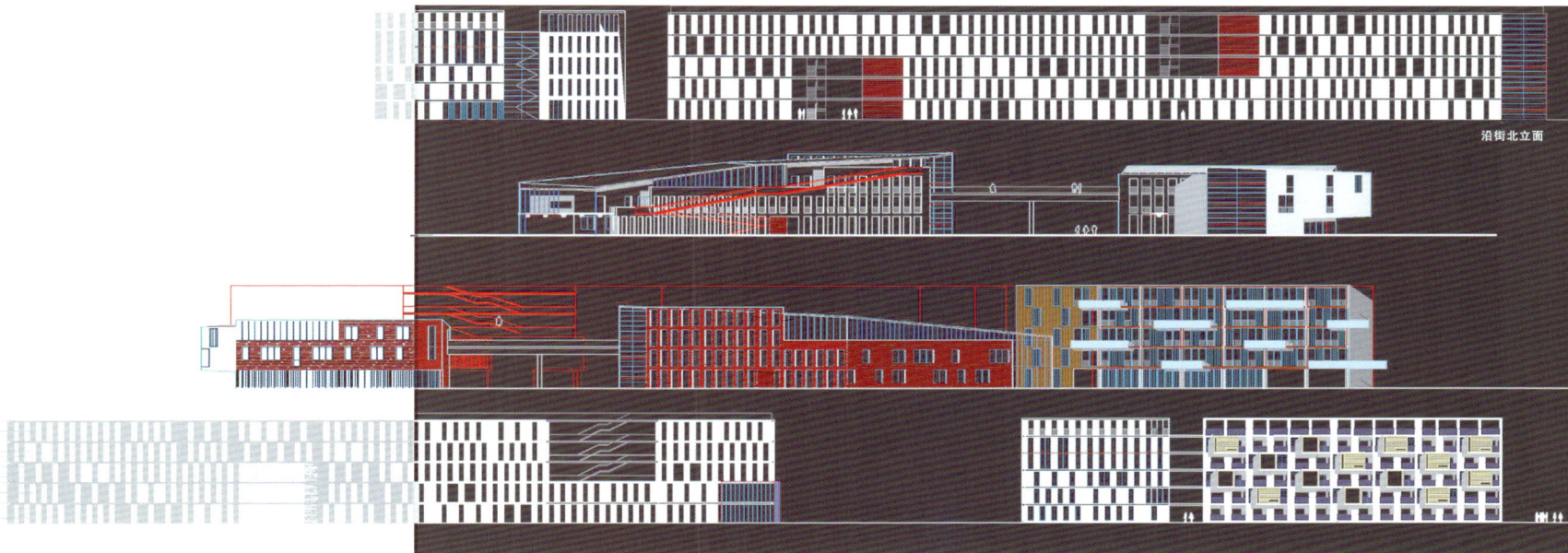

立面图1

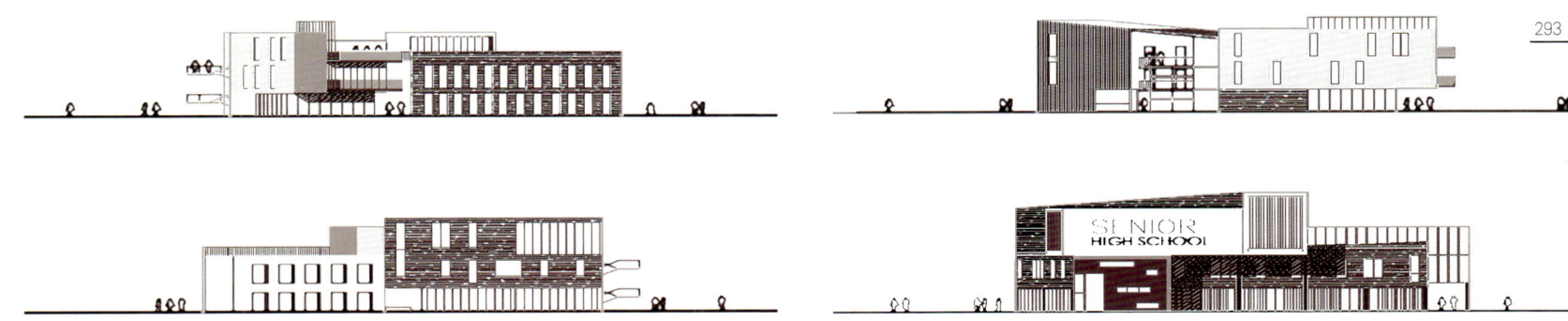

立面图2

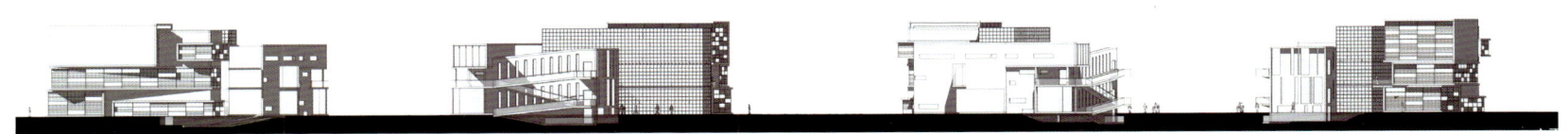

立面图3

Hotels......

Hotels

酒店

In our mind, a hotel is a building designed for short-time living. Therefore we focus our attention on studies of this behavioral pattern. This requires that our design not only dwells on studies of different hotel standards, how to improve hotel use and increase its application efficiency, how to set its background, service and maintenance space, etc., but also show our design concepts by studying what atmosphere the living space offered to customers is put.
Two kinds of contradictive ideas become the starting point of our design. We strive to pursue their balance.
1. How to make them "used" like a customer's home, natural, warm, and related, without strangeness and inconvenience.
2. How to make people become curious about them and impress upon them and make them become a brand new opportunity through which people feel different regions, cultures or special hotel cultures.

在我们的心目中，酒店是为短暂的居住活动而设计的建筑。因此，我们主要着眼于对这一行为模式的研究。这将要求我们在设计时不能仅仅停留在研究具体酒店不同标准的层面，除了考虑到如何提高其使用和实用的效率，它的后台、服务及维护空间如何设置等，还要将设计概念体现在为顾客提供的生活空间究竟浸润在一个什么样的氛围当中。
两种似乎有些矛盾的思路往往会成为我们设计的出发点，我们也会力求追求它们之间的平衡。
1. 如何让它们"用起来"更像是顾客的家，自然、温馨、放松，不会感觉陌生或不便。
2. 如何让人们对其产生一种猎奇的心态，从而给人留下深刻的印象。"看起来"充满新鲜感，让人们可以通过它们感受到不同的地域、文化或者享有酒店特别的人文机会。

A characteristic pure land is created in Cairo, Egypt.

在埃及开罗辟出了一隅独具特色的净土。

Egyptian Hotel • Cairo (Egypt)

Cairo | Project Location
2009 | Project Date
30,000m² | Project Scale
conceptual design | Design Phase
construction not started | Project Status

埃及酒店·开罗(埃及)

埃及,开罗市 | 项目地点
2009年 | 设计时间
3万m² | 项目规模
方案设计 | 设计阶段
未建 | 项目现状

项目总览

历史悠久而富有神秘色彩的埃及是最热门的旅游度假国家之一,她的首都开罗是一个拥有18,000,000人口的商业和经济中心城市,这座要建的酒店是商务与休闲结合的新型酒店。

设计灵感来源于当地的特殊环境。埃及是沙漠之城,绿色的尼罗河从中静静穿过。从GOOGLE地图上可看到开罗由大块的灰色与黄色组成,宛如一处沙漠,建筑用地正位于沙漠中的一片绿洲。酒店凭依其特殊环境,因地制宜。在沙漠的包围下,一座引人注目的独特地标性酒店即将在城市中拔地而起。

Project Overview

Egypt, with a long and mysterious history, is one of the hottest tourist destinations. Its capital, Cairo, is Egypt's commercial and economic center, with a population of 18,000,000. The hotel to be built is a new-type hotel combining the functions of business and leisure.

Design inspiration comes from the local special environment. Egypt is a country with many deserts, through which the green Nile River flows. From the GOOGLE map, we can see that Cairo consists of large grey and yellow blocks, looking like a desert. The plot used for the hotel is located on an oasis in the desert. The hotel is built by making use of the local special conditions. Encircled by a desert, an eye-catching unique landmark hotel will spring up soon in the city.

酒店设计布局是由5个建筑体围合而成的一个院落，中间有一个巨大的湖。建筑底层架空，花园像尼罗河一样淌入酒店，滋润着每个角落。庭院设计采用创造宽广的空间及自然通风条件，营造自然舒适的环境。酒店周围设计了绿化带，酒店每个房间的窗口都能欣赏到绿色的园林。这座酒店为埃及开罗辟出了一隅独具特色的净土。

The hotel is designed as a courtyard enclosed by five buildings, with a large lake in the center. The first floor of buildings is on stilts. A garden, like Nile River, extends into the hotel to moisten every corner. The courtyard has a broad space and natural ventilation condition, creating a natural and comfortable environment. Green belts are designed around the hotel. A green garden can be viewed from windows of each room in the hotel. This hotel creates a unique pure land for Cairo, Egypt.

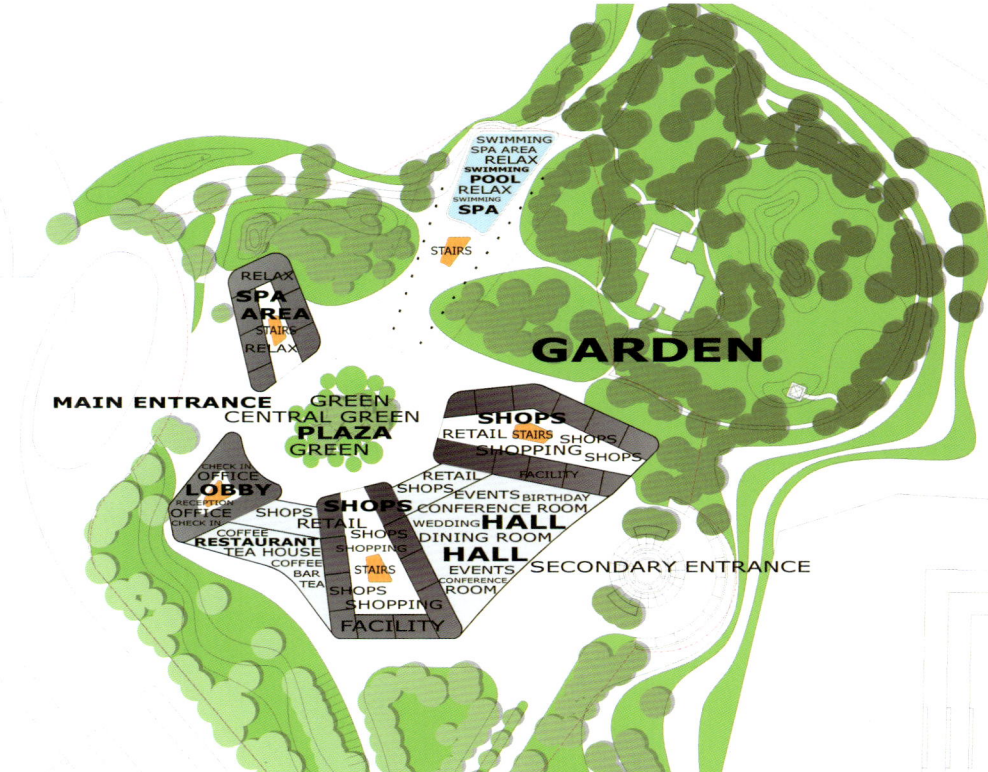

GROUNDFLOOR SCHEME

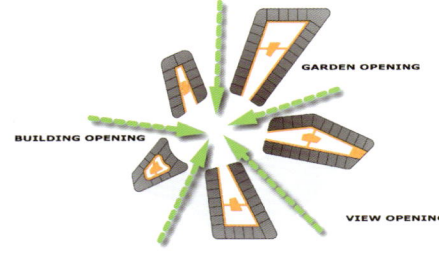

BUILDING ANALYSE

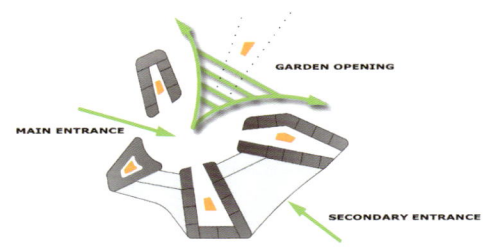

BUILDING ANALYSE

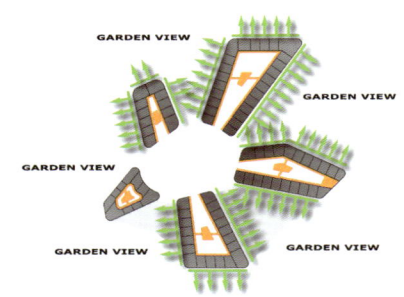

BUILDING ANALYSE

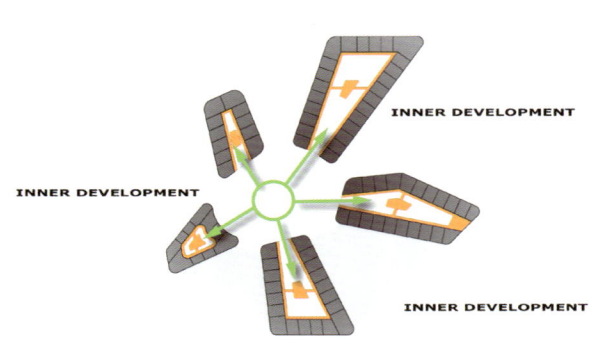

BUILDING ANALYSE

REGULARFLOOR SCHEME UPPERFLOOR SCHEME

The introduction of exotic styles makes every user have extraordinary experience of different times and places.

异域风情的引入让每位使用者都有异时异地的非凡体验。

Luneng Sanya Hotel • Sanya

Sanya | Project Location
2009 | Project Date
140,000m² | Project Scale
conceptual design | Design Phase
construction not started | Project Status

鲁能三亚酒店·三亚

中国,海南省,三亚市 | 项目地点
2009年 | 设计时间
14万m² | 项目规模
方案设计 | 设计阶段
未建 | 项目现状

项目总览

本项目位于海南省三亚市三亚湾新城片区新城路与海虹交界处，西面和南面毗邻城市干道，地势较为平坦。用地西、南部均与城市干道相衔接，交通十分便利。酒店以失落的文明为主题，凸显神秘，有意想不到的视觉刺激。

酒店共有客房548间，最小使用面积39m²，最大面积套房可达250m²。客房的内部装饰设计色彩鲜明、主题突出，浓烈的异域风情能让旅客的眼球得到极大满足，奢华而不失优雅。无论是舒适度或是美观度，都做到了独具一格。

Project Overview

This hotel is located at the intersection between Xincheng Road and Haihong Road in Sanya Bay New Town Region, Sanya, Hainan Province. Its west and south sides are adjacent to trunk roads. The plot is very flat. The western and southern parts of the plot have access to the trunk roads of the city. Therefore, the hotel is easily accessible. "Lost civilization" is the theme of the hotel, highlighted by mysterious and unexpected visual stimulation.

The hotel has 548 guest rooms. The smallest one is 39 square meters in area while the largest suite is 250 square meters. The decoration in all guest rooms features bright color and a definite theme. All rooms are luxurious and elegant, with strong exotic design that is really eye-catching. These rooms are really of unique style in terms of comfort and appearance.

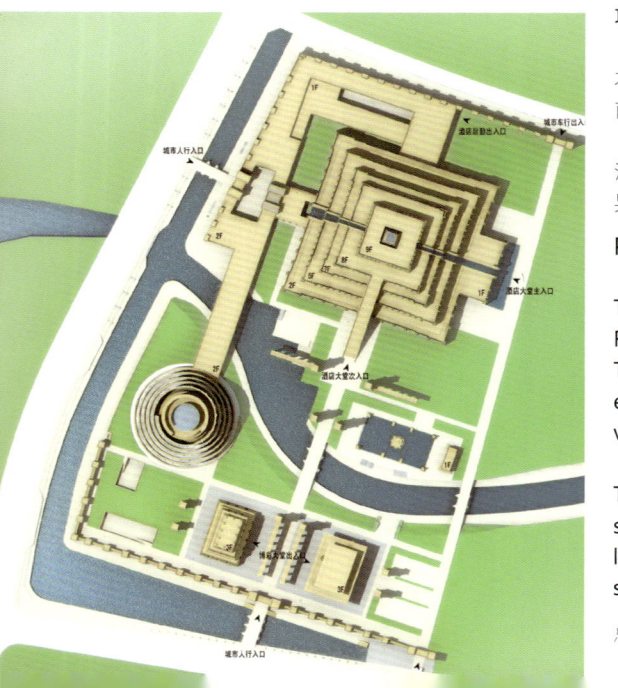

总平面

场所环境与交通条件

博彩酒店的客源大多为外地客人,如果没有发达的枢纽型机场、航运、铁路和高速公路,博彩酒店是无法生存的。

1. 场地环境:位于海南省三亚市三亚湾新城片区新城路与海虹交界处,西面和南面毗邻城市干道,地势较为平坦。
2. 交通条件:用地西、南部均与城市干道相衔接,交通十分便利。

Environment and traffic condition

Most of customers of a gaming hotel come from other places. If there is not developed hub-type airport, waterway, railway and freeway, the gaming hotel cannot survive.

1. Environment: the hotel is located at the intersection between Xincheng Road and Haihong in the new Sayawan district, Sanya, Hainan Province, neighboring trunk roads to the west and south. The land is very flat.
2. Traffic condition: both western and southern parts of the land link up with trunk roads. Therefore the project has an easy access to transportation facilities.

梦幻巴比伦

以古老失落的巴比伦文明为文化主题,并以此为背景从中挖掘出相应的创作素材,其中包括残缺的通天塔、传说中的空中花园、巴比伦运河、芦苇船以及充满异域风情的城墙、建筑群、雕塑群、雕塑和巨柱等,以再现曾经"富饶、瑰丽"的美索不达米亚文明。

Fantastic Babylon

Focusing on a cultural theme of lost Babylon civilization, this project uses corresponding elements, such as incomplete Tower of Babel, Hanging Garden, Babylon Canal, Reed ship, and exotic walls, building complex, sculptures, and giant columns, etc., so as to represent the fertile and charming Mesopotamian civilization.

308

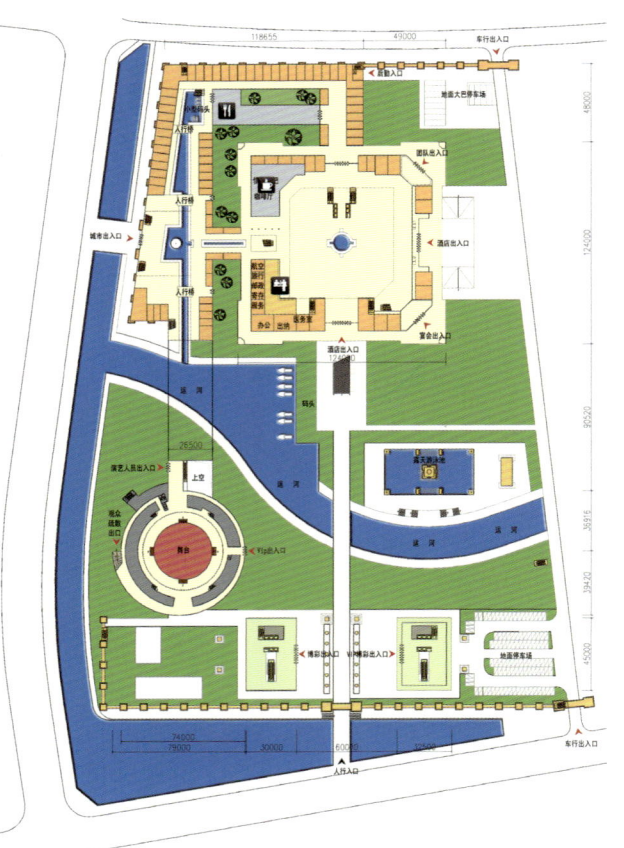

一层平面

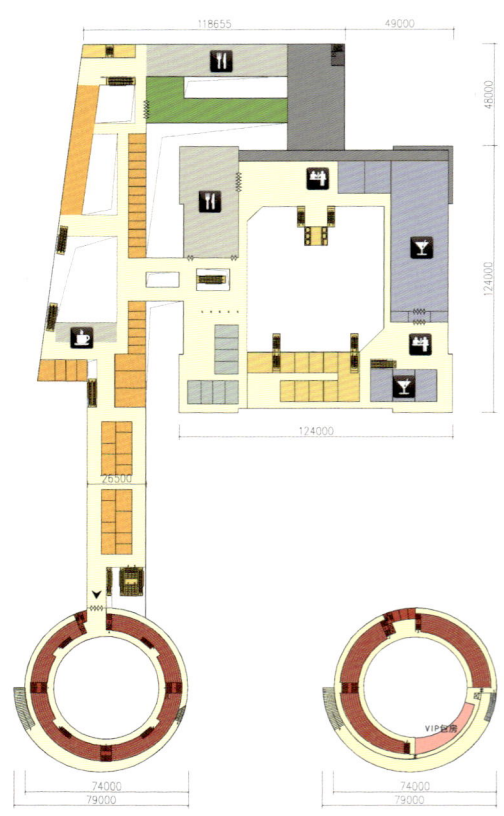

二层平面　　演会中心三层平面

We have made a sufficient research based on the conservation and positive reconstruction of natural environment.

基于对自然环境的保护和积极的再造，我们做了充分的研究。

Guanling Hotel, Beihai, Guangxi • Beihai

Beihai, Guangxi | Project Location
2009 | Project Date
57,100m² | Project Scale
conceptual design | Design Phase
construction not started | Project Status

广西北海冠岭酒店·北海

中国，广西省，北海市 | 项目地点
2009年 | 设计时间
5.71万m² | 项目规模
方案设计 | 设计阶段
未建 | 项目现状

项目总览

北海冠岭五星级酒店位于北海市冠头岭海滨，总建筑面积约3.4万m²，拥有300间标准客房。建筑为一平面"S"形面海排开的8层分级跌落板式建筑，竖向稍有外倾。建筑设有二层面海开敞的半地下室，建筑高度约27.6m。

设计将海浪和贝壳这两种独具海滩风情的元素运用到建筑形体中，与整个周边环境融为一体，和谐共生，让建筑富有浓烈的地域色彩。

Project Overview

The five-star Guanling Hotel is located at the Guantouling seashore in Beihai. With 300 standard guestrooms, the hotel has a total building area of around 34,000 square meters. With an S-shape perspective plane, the 8-floor slab-type cascading building faces the sea and overhangs slightly. It has a two-floor half open basement, about 27.6m in height, facing the sea.

Two characteristic beach elements, wave and shell, are used in the design of the architectural shape, which integrate well with the surrounding environment. Thus the building is of strong regional feature.

方案一

方案二

方案三

概念分析

海之贝，闪耀着光芒夺目的自然元素，且具有海边独特的海岸风情。放射状的建筑形态隐约地让人遐想到连绵蜿蜒的海岸，令人记忆犹新，具有浓烈的地域色彩。

Conceptual Analysis

Seashore shells are a natural element with bright light and particular seashore customs. The ray-like buildings make people compare them with winding seacoast, giving people a strong impression and showing dense regional features.

概念分析图

总平面

概念分析

海浪是天然的行为艺术，忽起忽伏、连绵不断的形态充满了内敛的力量与艺术张力。将这些特点提炼到建筑形体中，与整个周边环境、大海的谦逊融为一体，和谐共生。

Conceptual Analysis

Wave is a kind of natural performance art. Its undulating and unending form is full of deep strength and artistic tensile force. These features are used for the architectural form and blend in and live in perfect harmony with surrounding environment and sea.

The building coexists with mountains and the sea. The roofs with different elevation look like sea waves.

建筑是山海共生的，建筑的起伏就如同海浪的波动。

Dameisha Hotel (Liantai Club) • Shenzhen

Shenzhen | Project Location
2004 | Project Date
11,000m² | Project Scale
conceptual design | Design Phase
construction in progress | Project Status

大梅沙酒店（联泰会所）• 深圳

中国，广东省，深圳市 | 项目地点
2004年 | 设计时间
1.1万m² | 项目规模
方案设计 | 设计阶段
在建 | 项目现状

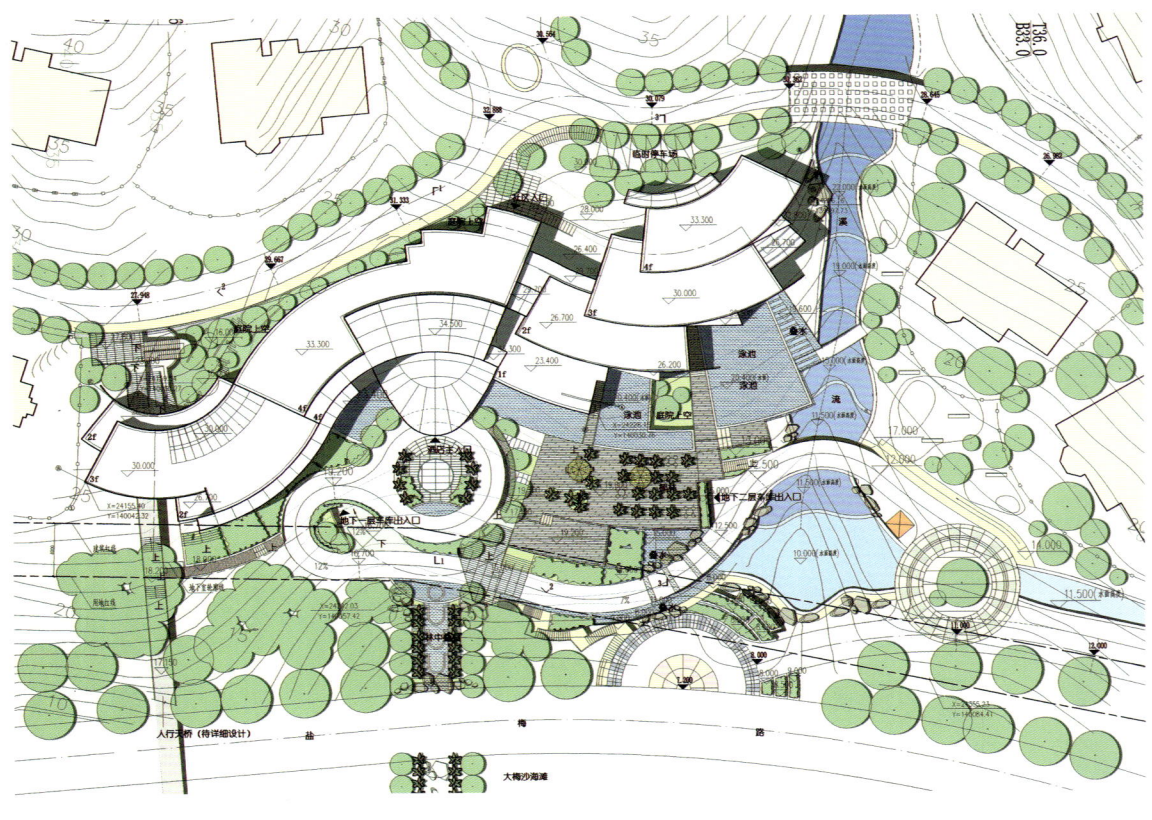

总平面

项目总览

联泰大梅沙酒店项目拟建于人气极旺的大梅沙海滨公园南侧山的半山腰上，属于联泰大梅沙别墅用地的一部分，依山傍水，风景秀丽。酒店定位为一个全海景，以度假为主、商务为辅的山地小型高档酒店。别墅区中需解决的重要问题是在满足酒店功能要求及客房海景要求的前提下，不遮挡任何一栋别墅看海。因此建筑结合地形顺山顺势采用台阶状叠落式的设计，同时也为酒店提供了丰富的户外活动空间。

Project Overview

Liantai Dameisha Hotel is to be built at a hillside in the south of Dameisha Seashore Park, which attracts lots of visitors. The site used for the hotel is a part of the plot used for Liantai Dameisha Villa. Since the hotel is backed by hills and is close to water body, it boasts beautiful sceneries. The hotel is positioned as a small high-end hotel in a mountainous region overlooking full seaview for guests spending holidays here as well as business guests. Since this hotel is built in a villa area, it should not hinder the sight line of any occupant in the villa to enjoy the seaview, but the hotel functions and the requirement that a guest in any guestroom can see the seaview must be satisfied. Therefore, a cascading building is designed, which is built on the hill slope without making change to the original landform. Such design also makes the hotel have a plentiful outdoor space.

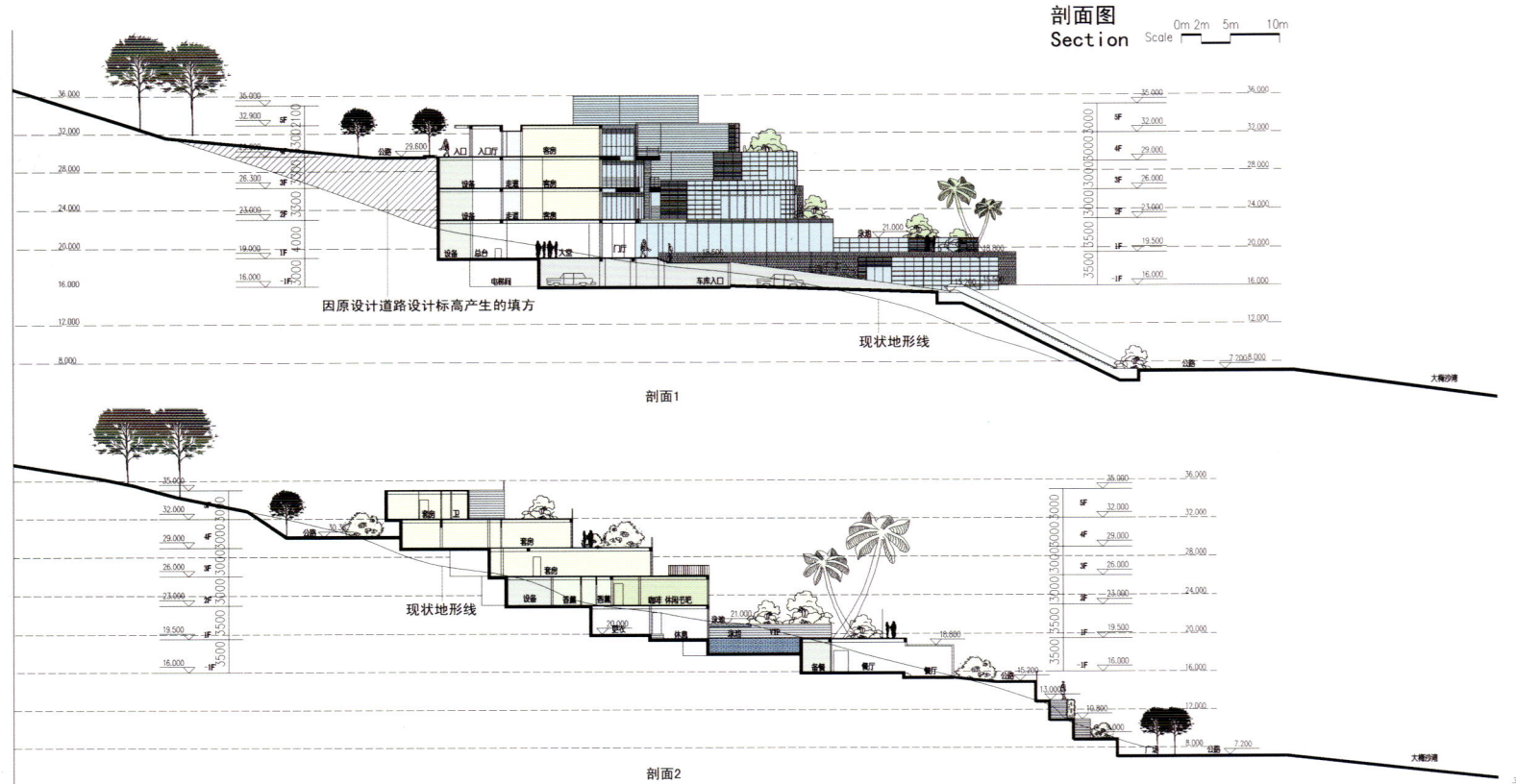

剖面图

321

A landmark needs not only a prominent image but also a particularity.

地标不仅仅需要有突出的建筑形象,也需要营造场所的独特性。

Dongjiu Building • Yixing

Yixing | Project Location
2008 | Project Date
100,000m² | Project Scale
conceptual design | Design Phase
construction not started | Project Status

东汶大厦·宜兴

中国，江苏省，宜兴市 | 项目地点
2008年 | 设计时间
10万m² | 项目规模
方案设计 | 设计阶段
未建 | 项目现状

项目总览

江南水乡狭窄的街道，曲折的河水，处处充满韵味。但是随着现代城市的发展，越来越多的江南水乡旧城就像现代钢筋混凝土丛林中的盆景，失去了往日的生气与活力。然而，人们却依旧向往那个充满诗情画意的江南小镇。本方案凭借地块的优越条件，将现代的建筑形式以及传统的构造手法糅合起来，重现这美好的一幕。

Project Overview

Narrow streets and zigzag rivers in towns in the south of the lower reaches of the Yangtze River present lingering charm. With the development of modern cities, more and more high-rise buildings and skyscrapers have been built in the cities, so the old town areas just like bonsais in a modern reinforced concrete jungle, losing the past vitality. However, people still prefer poetical and picturesque old towns. Modern architectural forms and traditional construction methods are combined in the design, making full use of the predominant conditions in the site, so as to reproduce the features of such old towns in the south of the lower reaches of the Yangtze River. One-storey houses are mixed with multi-storey buildings, with diversified gable walls.

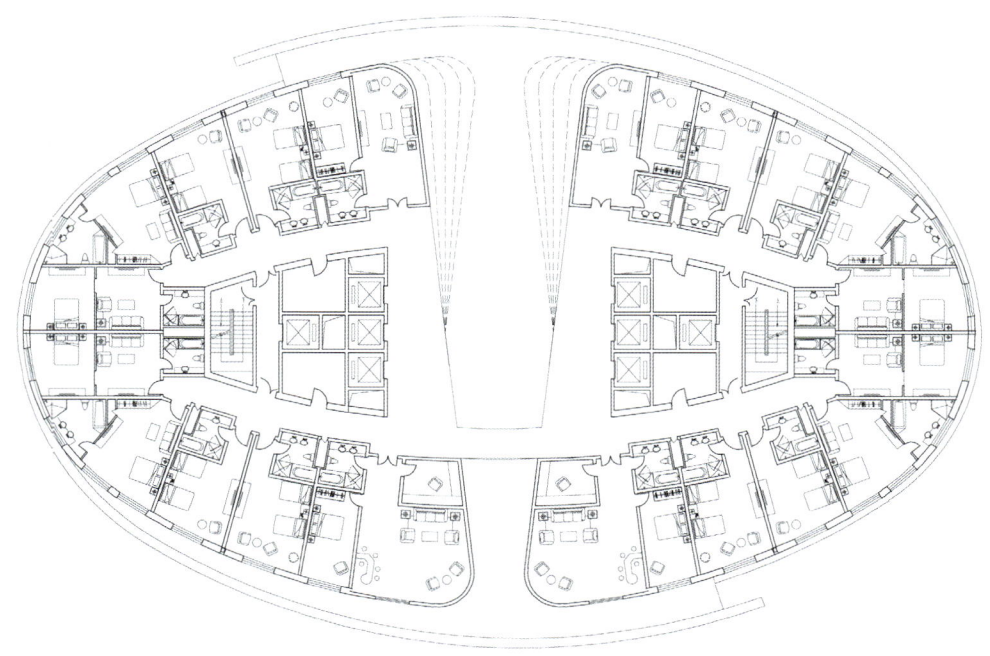

塔楼标准层平面

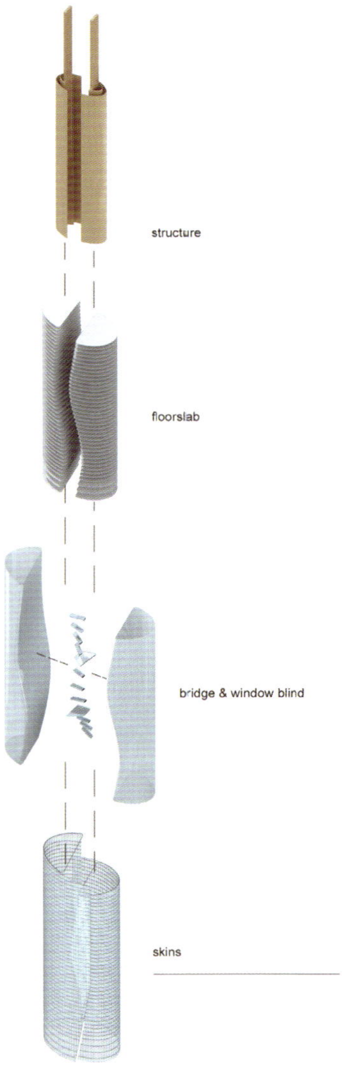

塔楼整体结构 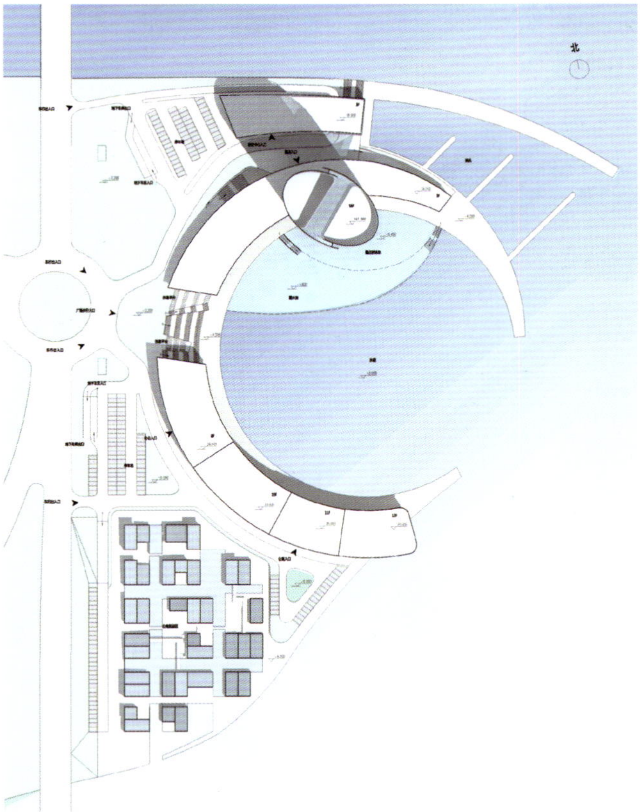　　　　　总平面

塔楼设计1号

在反复比较之后，方案选择了宜兴的一个主要文化符号——"梁祝"作为主题，通过对男女形象的抽象化处理，得到了一条美学上可表述的阴阳曲线。用这条曲线切分一个完整的椭圆柱体，最终形成了一对看似一体，实则分离的塔楼形象。

Tower Design

After repeated comparison, "The Butterfly Lovers", a major cultural sign in Yixing is chosen as the theme of this building. Through abstraction of male and female images, a curve, which can present Yin and Yang aesthetically, is created. This curve is used to divide a complete elliptical column so as to form a pair of towers, which looks like a whole but separates from each other actually.

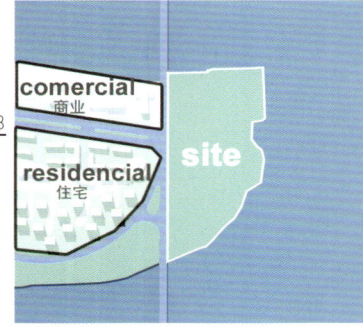

original conditions 原始条件

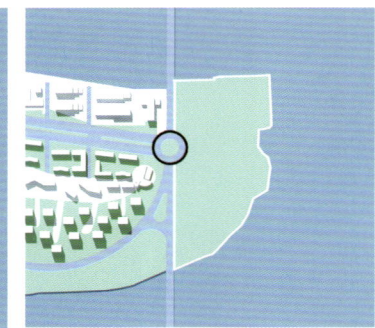

Step 1: 设计了一个转盘，作为道路的节点以及城市景观的终端。

Step 1: A roundabout is designed as a road node and urban landscape terminal.

Step 2: 顺延左侧的规划，将基地上下分区，即公共区及私密区。公共区包含了酒店、办公、商业等，私密区包含了公寓及高档住宅。

Step 2: the leftward planning divides the site into a pubic area and private area. The public area covers hotels, office buildings, and business facilities, etc. while the private area covers apartments and high-end residences.

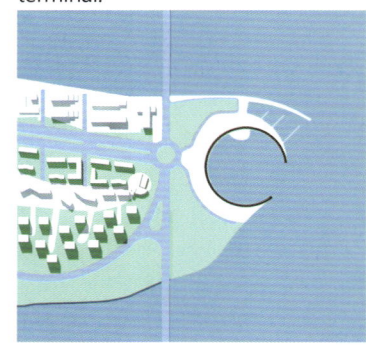

Step 3: 创造了一个弧形内湖，在不干扰内部功能的情况下，为城市提供了景观优美的休闲场所和商业聚集地。

Step 3: An arc lake is built to offer a beautiful recreation place and a venue for business meeting on condition that internal functions are not affected.

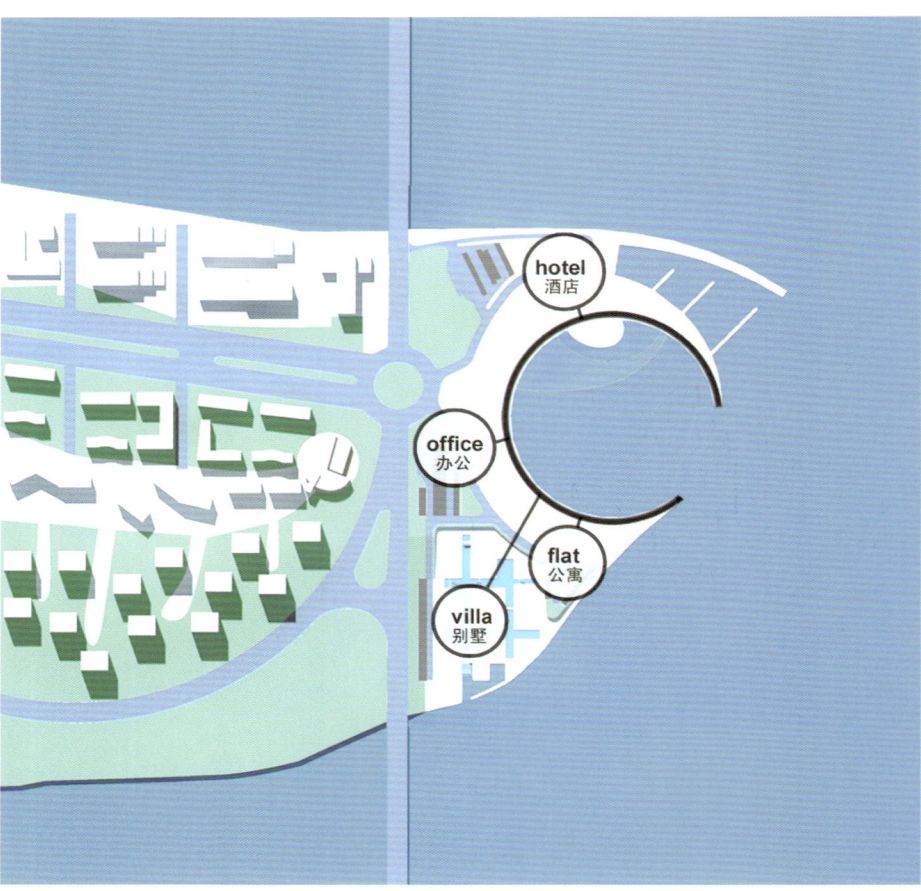

Step 4: 沿内湖岸呈项链状布置各个功能，并利用高差创造丰富的层次。

Step 4: Various functions are arranged along the bank of the lake, looking like a necklace. Diversified levels are created by making use of height difference.

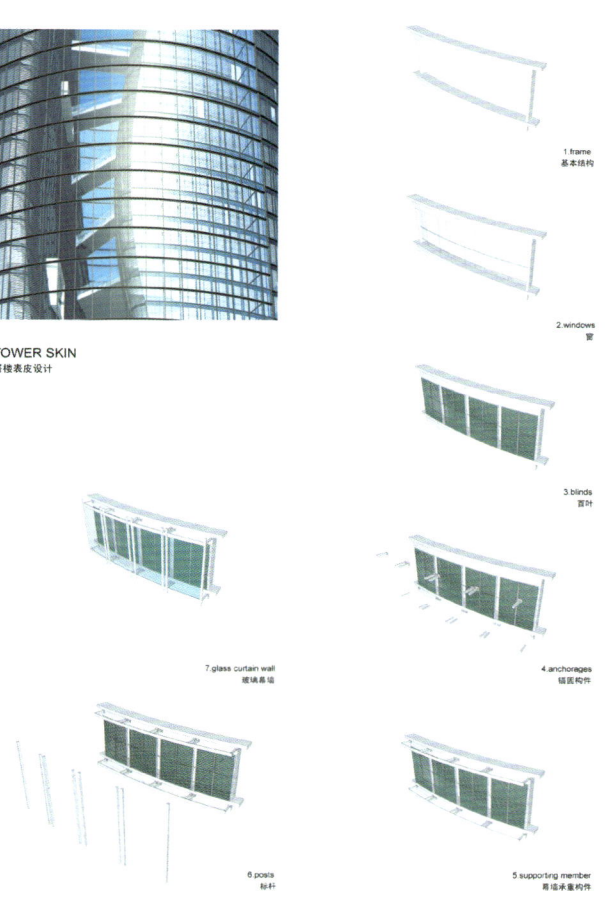

TOWER SKIN
塔楼表皮设计

1. frame 基本结构
2. windows 窗
3. blinds 百叶
4. anchorages 锚固构件
5. supporting member 幕墙承重构件
6. posts 标杆
7. glass curtain wall 玻璃幕墙

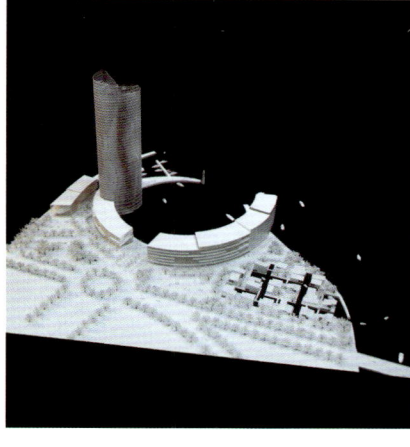
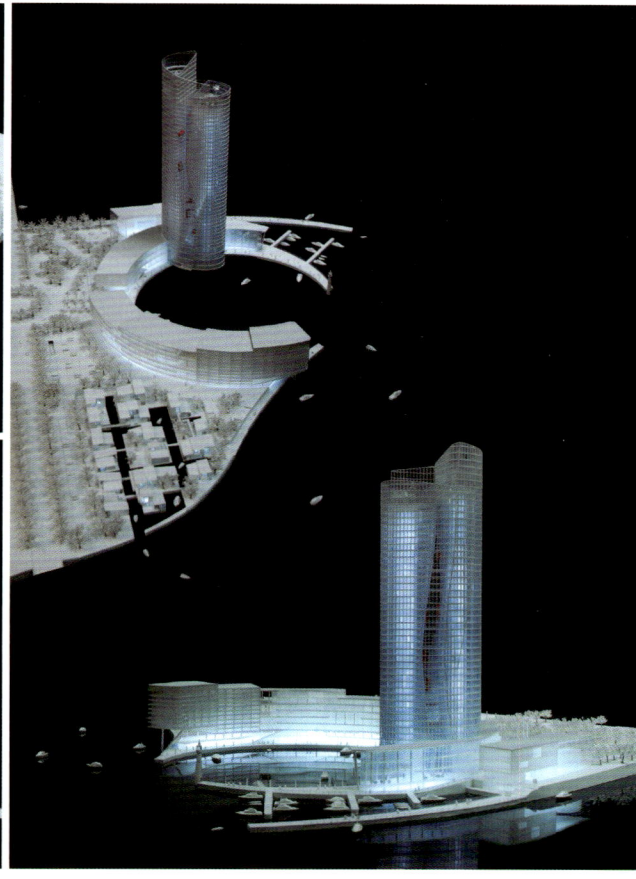

剖面图

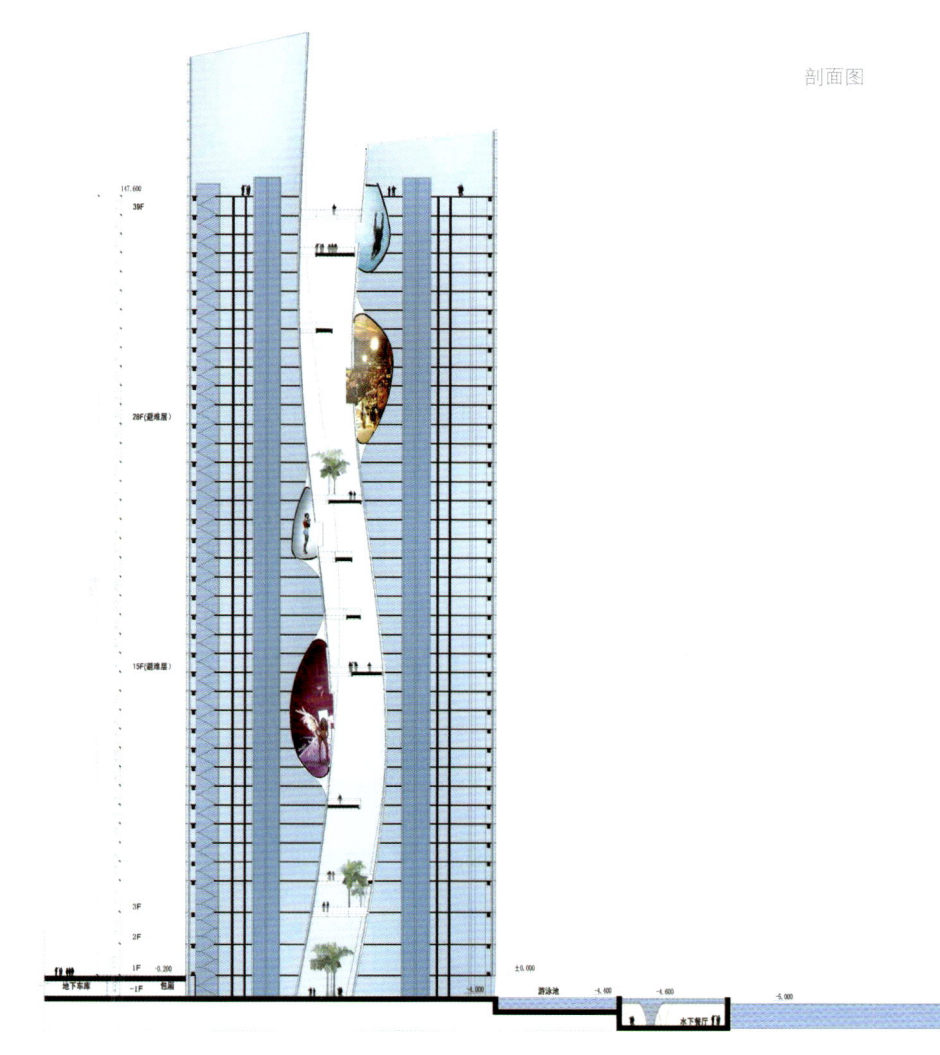

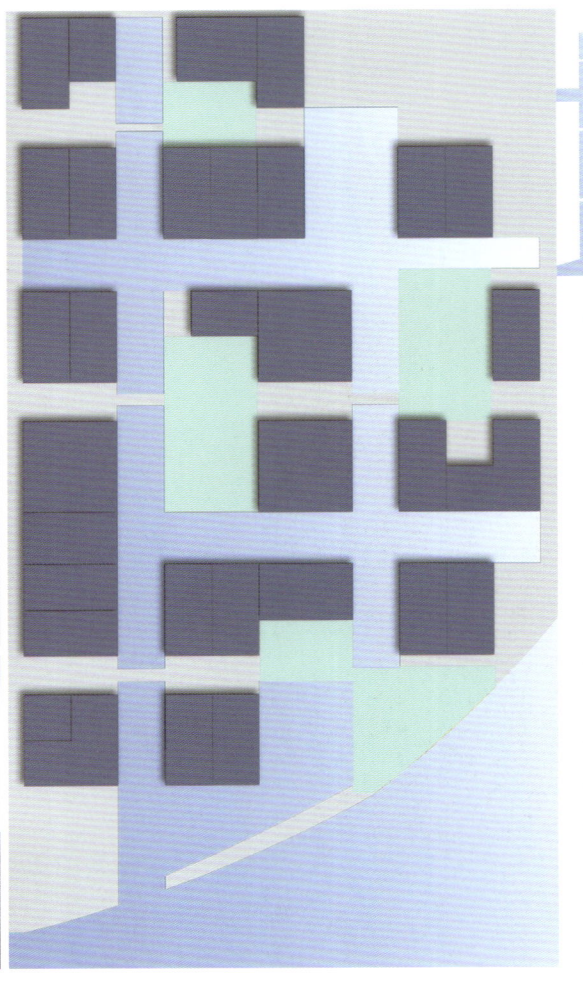
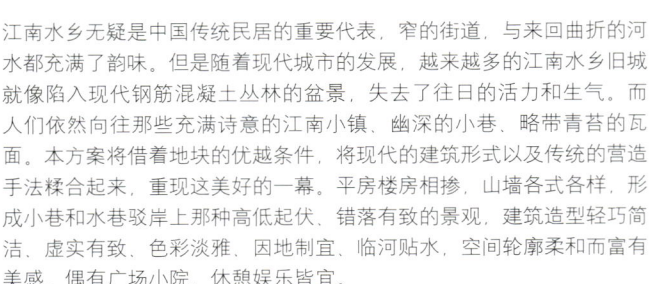

循环水系统　　居住空间　　漫步区域　　空间节点

江南水乡无疑是中国传统民居的重要代表，窄的街道、与来回曲折的河水都充满了韵味。但是随着现代城市的发展，越来越多的江南水乡旧城就像陷入现代钢筋混凝土丛林的盆景，失去了往日的活力和生气。而人们依然向往那些充满诗意的江南小镇、幽深的小巷、略带青苔的瓦面。本方案将借着地块的优越条件，将现代的建筑形式以及传统的营造手法糅合起来，重现这美好的一幕。平房楼房相掺，山墙各式各样，形成小巷和水巷驳岸上那种高低起伏、错落有致的景观，建筑造型轻巧简洁、虚实有致、色彩淡雅、因地制宜、临河贴水，空间轮廓柔和而富有美感。偶有广场小院，休憩娱乐皆宜。

With respect to high-end waterside apartment design, undoubtedly folk houses in the south of the lower reaches of the Yangtze River are important representatives of traditional Chinese folk houses. Both narrow streets and roundabout river water are characteristic. However, with the development of modern cities, more and more folk houses become bonsais in the reinforced concrete jungles and lose their vitality. People still long for poetical towns in the south of the lower reaches of the Yangtze River, long and quiet lanes, and tiles with few mosses in the towns because a particular life scene can be seen everywhere in those towns. Relying on this plot's superior conditions, we will combine modern architectural form with tradition construction methods to make the beautiful scene reappear. Bungalows stand together with storied buildings. Various walls make up undulating and orderly landscapes. Buildings have a light and concise shape and are lightly colored. All of them are close to water. Their contour is soft and beautiful. Squares and courtyards are interspersed among them. Thus people can have a rest or amuse themselves there.

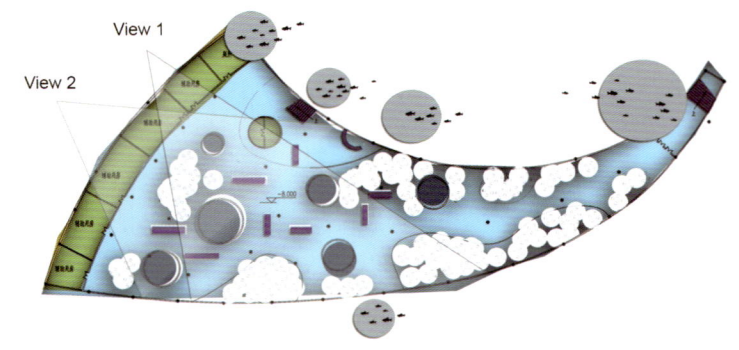

Urban Design and Landscape Planning
......

Urban Design and Landscape Planning

城市设计与景观规划

The compounding nature of the urban functions brings about the vitality for the region; more and more attention is being paid to creating the urban life of foot in the urban development. In the designing of the large-scale group architectures integrating residence, commerce, entertainment, office and hotel etc., we highlight the relationship between the landmark feature underlined by the developers and the urban culture by way of introducing the open urban space of different categories and layers, advocating the foot-based traffic and revivification of community life, establishing three-dimensional urban traffic system and internal movement system thus creating the city form with sustainable development.

城市功能的复合性带来区域的活力,城市的开发也越来越注重朝创造步行城市生活的方向发展。在集居住、商业、娱乐、办公、酒店等功能为一体的大型群体建筑的设计环境中,我们注重协调处理开发商所强调的地标性和城市文脉之间的关系,引入不同类型和层次的开放城市空间,倡导步行交通及还原街区生活,建立立体化城市交通体系及内部动线体系,建立可持续发展的城市形态。

As China's No. 1 electronic street, Huaqiangbei is not only a vehicle of technology but also a recorder of urban changes and development.

华强北作为"中国电子第一街",不仅仅是技术的载体,还是一个城市变迁与发展的记录者。

Huaqiangbei Three-dimensional Street and City Design • Shenzhen

Futian District, Shenzhen | Project Location
2009 | Project Date
city planning | Design Phase
construction not started | Project Status

华强北立体街道城市设计·深圳

中国，广东省，深圳市，福田区 | 项目地点
2009年 | 设计时间
城市规划 | 设计阶段
未建 | 项目现状

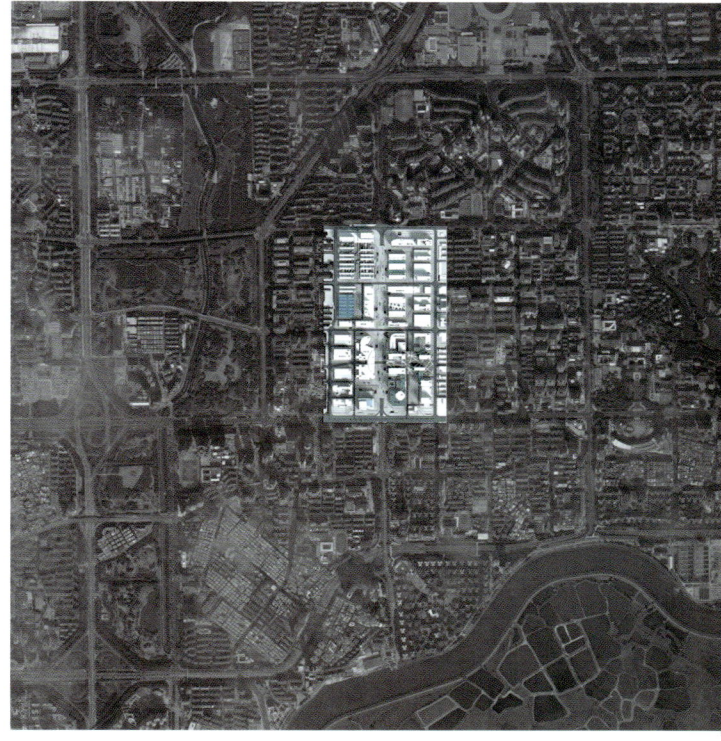

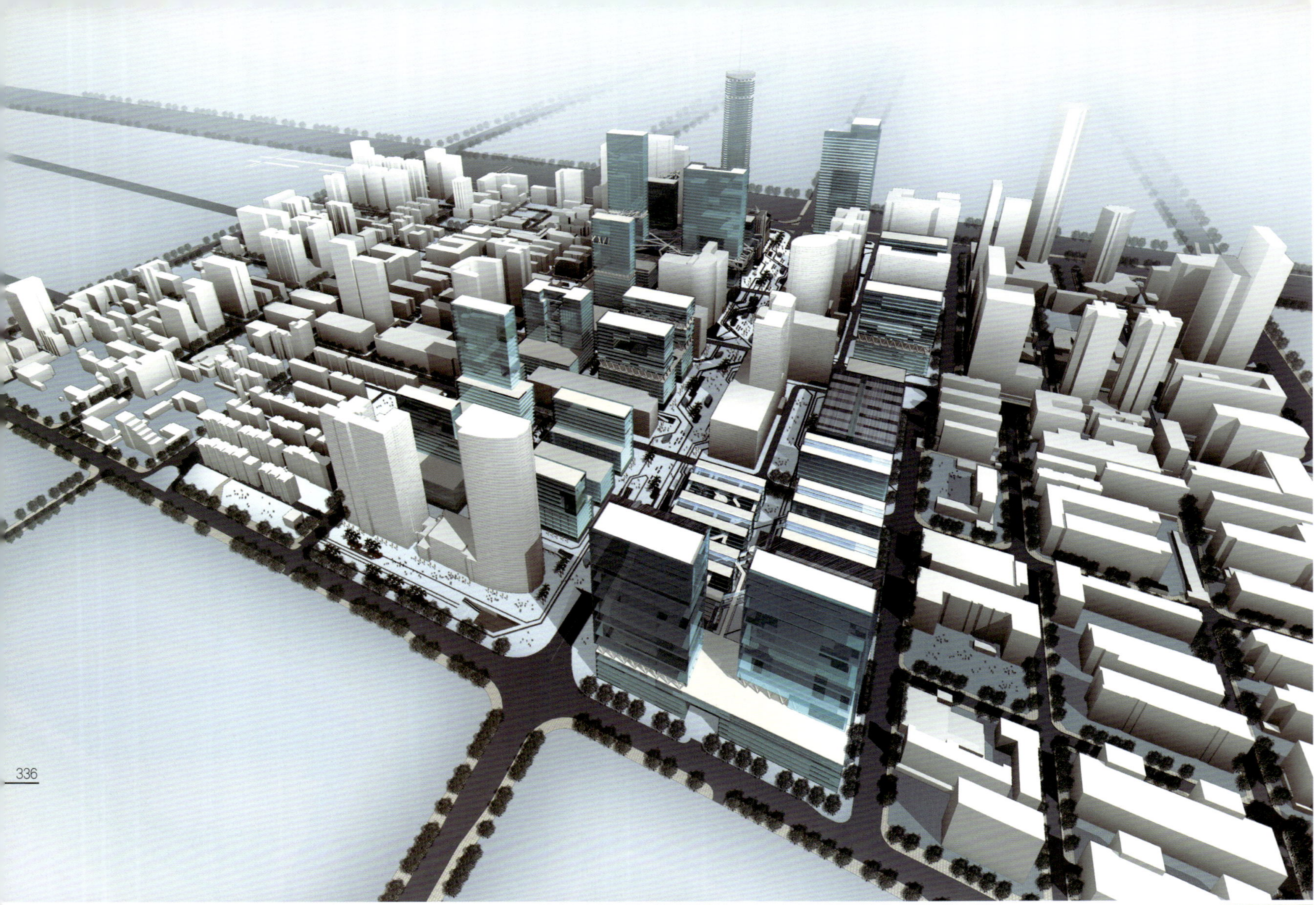

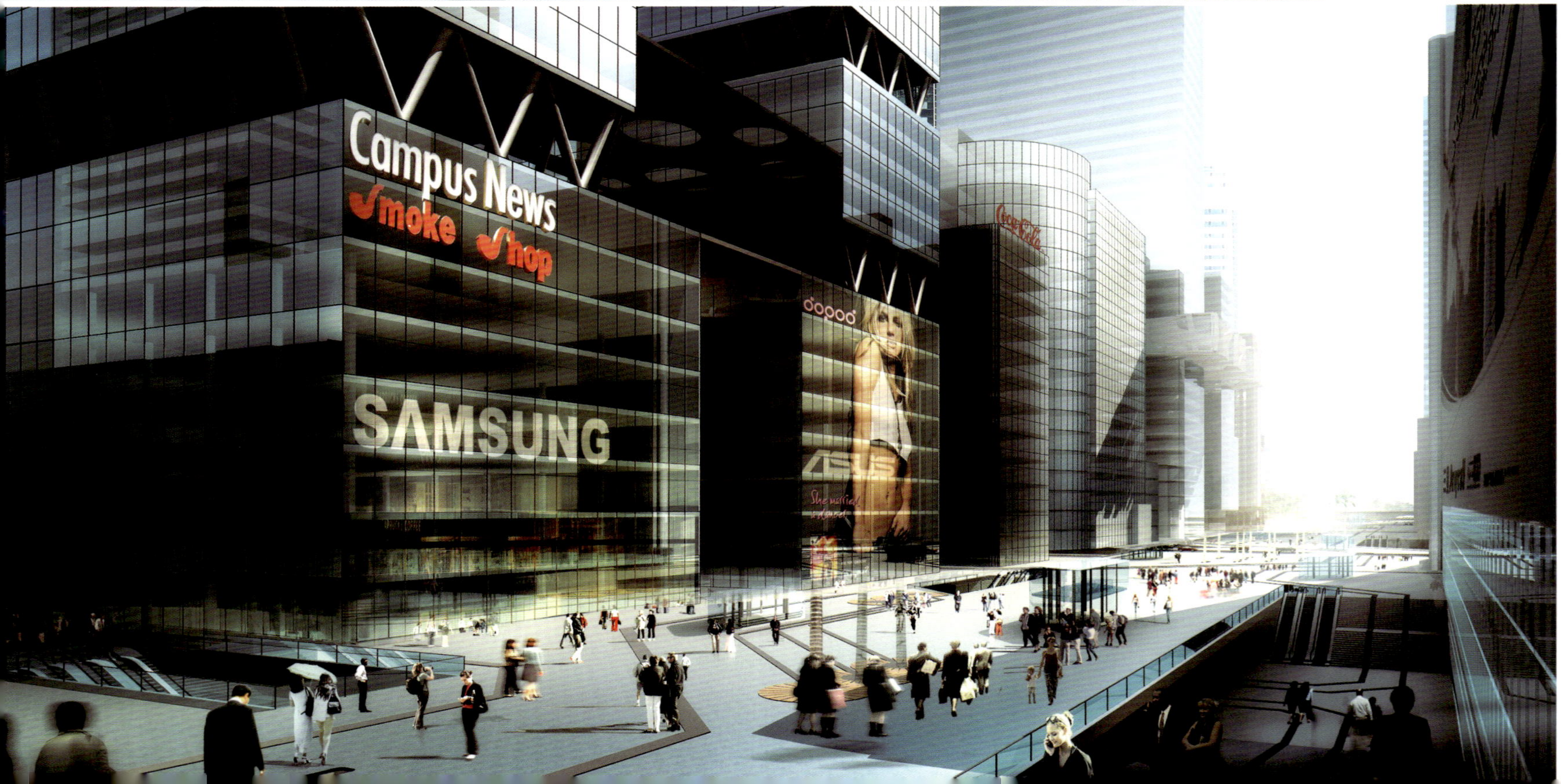
Campus News
Smoke Shop
SAMSUNG
dopod

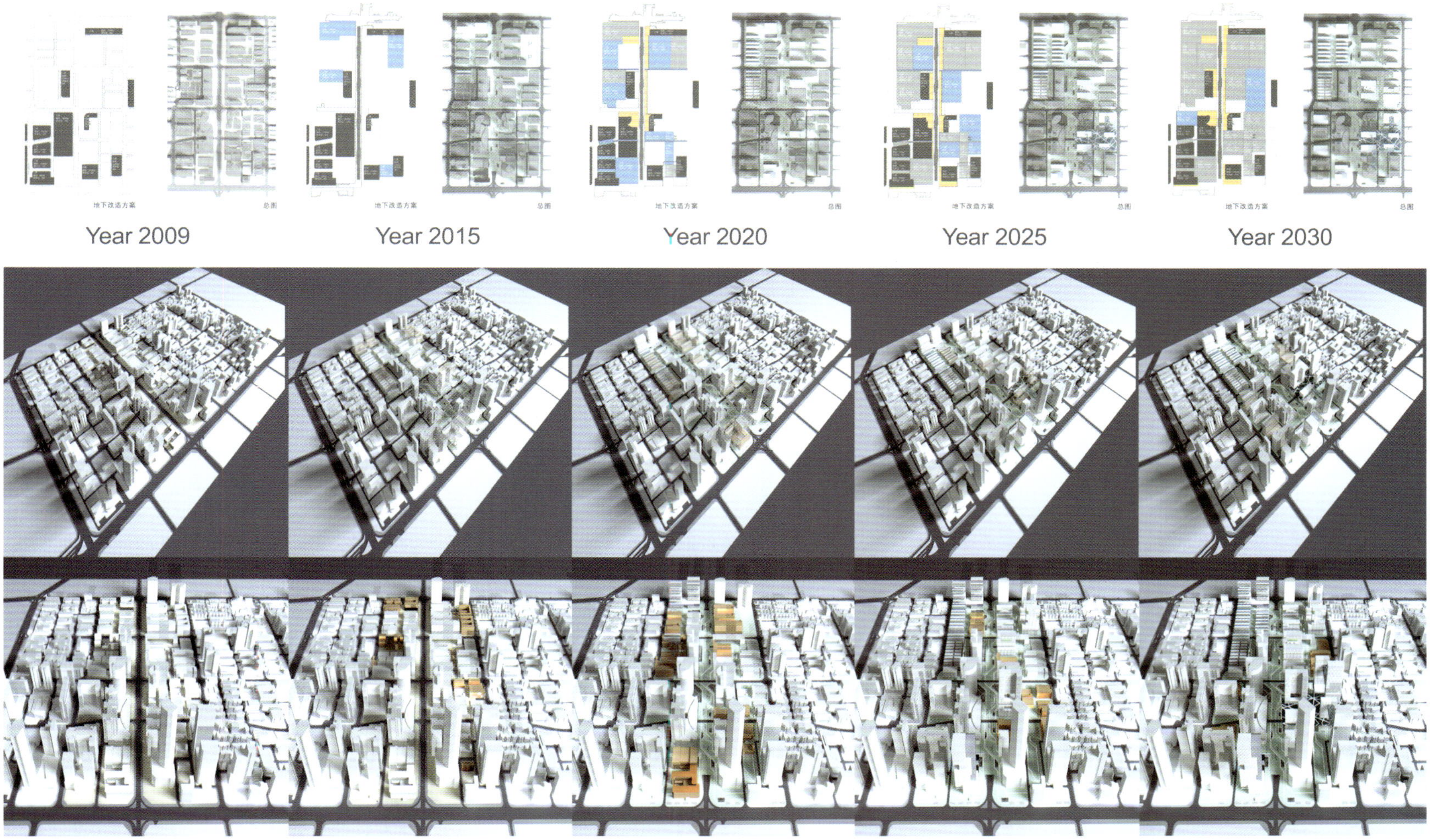

Year 2009　　　Year 2015　　　Year 2020　　　Year 2025　　　Year 2030

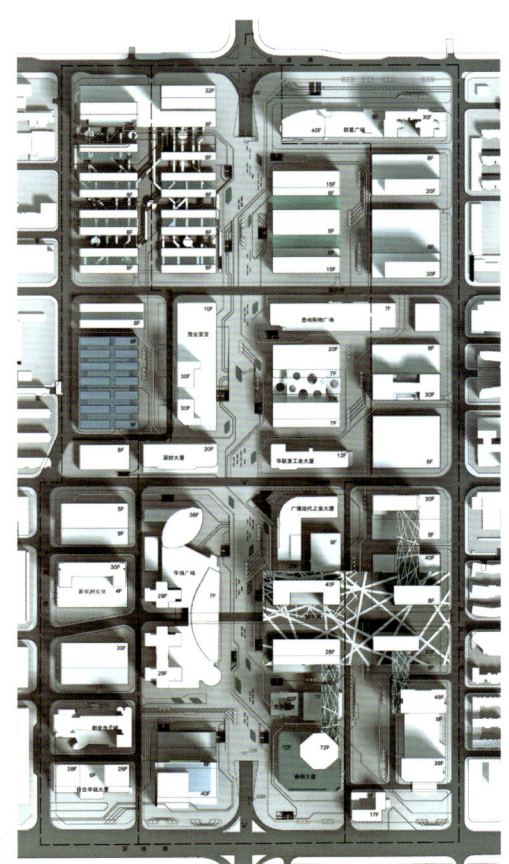

项目总览

作为"中国电子第一街"的华强北，高速的发展带来的巨大人流和车流是其持续兴旺的根本，但同时也为这个从上世纪80年代走来，脱胎于旧工厂区系统格局的片区带来前所未有的压力。

面对2010年多条地铁线路即将开通的黄金发展契机，多功能混合发展已经成为一个健康城市地区的必要发展策略。配备公共交通节点的多功能混合区将在市中心地区越来越多地被使用。底层商铺，上层居住的商务住楼模式的多功能混合准则已成为常见的实现模式。零售商将得到楼上及周边的固定客源，居民的利益则体现在他们能够通过短距离的步行买到食品或是看场电影。

而我们的设计不仅仅是回答一个结果，展示一张蓝图，重要的是展现其可持续发展的过程。

Project Overview

Huaqiangbei is China's No. 1 street for trading electronic products. Its rapid development attracts a huge stream of people and traffic flow, which is the fundamental reason for Huaqiangbei's continuous prosperity but also bring an unprecedented pressure to this region which had been developed from an old factory area in 1980s.

Since several metro lines will be put into use around this area in 2010, development of multiple functions has become an essential development strategy for building a healthy urban area. More and more multifunctional areas with public transportation will be established in the downtown area. Business-living buildings (i.e. first floor for stores and higher floors for residence) are common now. Retailers can attract regular customers who live in and around such building while residents of such building can buy foods or see a movie within a short distance.

Our design is not only to give an answer and present a blueprint but also to show the whole process of sustainable development.

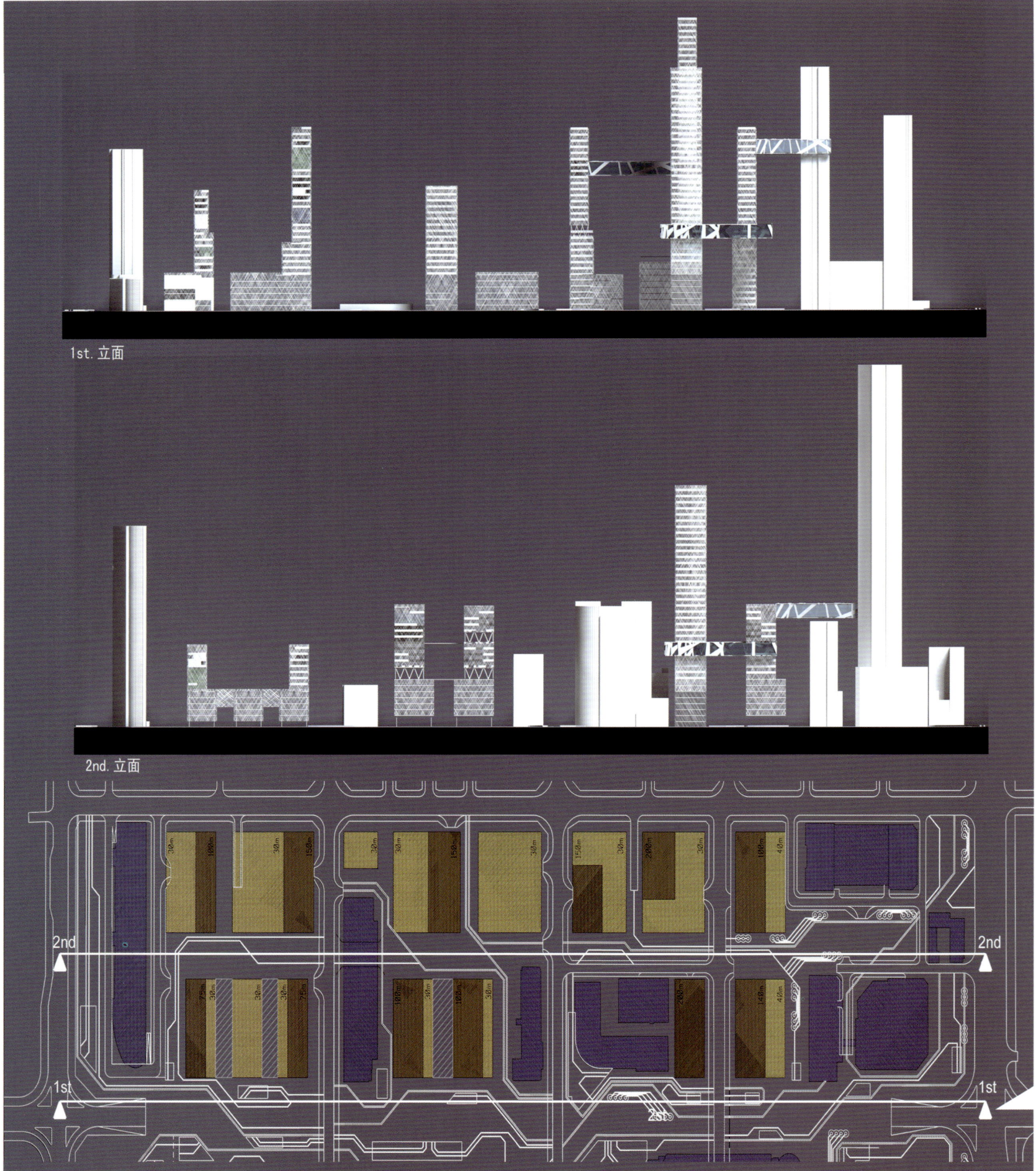

多功能混合发展　Mixed-Zone Development

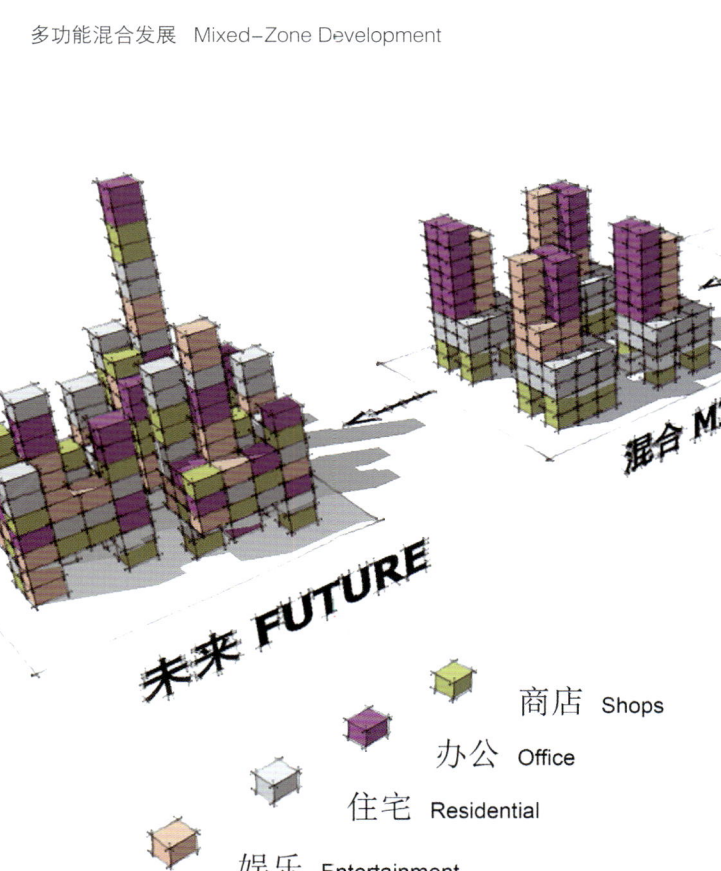

商店 Shops
办公 Office
住宅 Residential
娱乐 Entertainment

在整个20世纪后期，对于许多城市规划人员和其它专业人员而言，城市多功能混合发展的许多优点日益明显。在这里，这种概念将被再一次提出。随着城市的工业化发展，将住宅与复杂的工厂区分离开来已经变得不那么重要。完全独立的分区把各类型的发展变成"孤岛"。在大多数情况下，汽车已经成为大范围内住宅区、独立商业区以及办公区之间一个重要的运输桥梁，这就造成了城市对汽车的依赖。同时，多功能混合发展已经成为一个健康城市地区必要的发展策略。

分区法已作相应修改，城市越来越多地试图通过使用多功能混合使用分区来解决这些问题，配备公共交通节点多功能混合区将在"市中心"地区越来越多地被使用。底层商铺、上层居住的商住楼模式的多功能混合准则常见于实现模式。零售商将得到楼上及周边的固定客源，而居民的利益则体现在他们能够通过短距离的步行就能买到食品或是看场电影。

For many urban planners and other professionals, multifunctional mixed development showed many strong points increasingly in the late 20th century. Here this concept is mentioned again. With the industrialized development in cities, it is not so important to separate residences from factory areas because such separation may change various types of development into "an isolated island". In most cases, automobile has become an important transportation bridge between residential quarters, independent business districts, and office areas. Thus urban development relies on automobiles. Meanwhile, multifunctional mixed development has become a necessary development strategy for a healthy urban area.

The zoning method has been modified accordingly. A city is using more and more mixtures of multiple functions to solve these problems. More and more multi-functional mixed areas with public traffic nodes will appear in "downtown". The multifunctional mixing rule of business-living building mode characterized by "shops on the ground floor and residences on upper floors" has been in common use. Retailers can attract customers living upstairs and in surrounding areas while residences can buy foods or go to cinema after walking a short distance.

On the basis of without changing many things, we want to make this region change qualitatively.

我们非常想在不改变很多的情况下，使该区域发生根本的质变。

Overseas Chinese Town
LOFT • Shenzhen

Futian District, Shenzhen | Project Location
2002 | Project Date
56,000m² | Project Scale
conceptual design | Design Phase
construction completed | Project Status

华侨城LOFT · 深圳

中国，广东省，深圳市，福田区 | 项目地点
2002年 | 设计时间
5.6万m² | 项目规模
方案设计 | 设计阶段
未建 | 项目现状

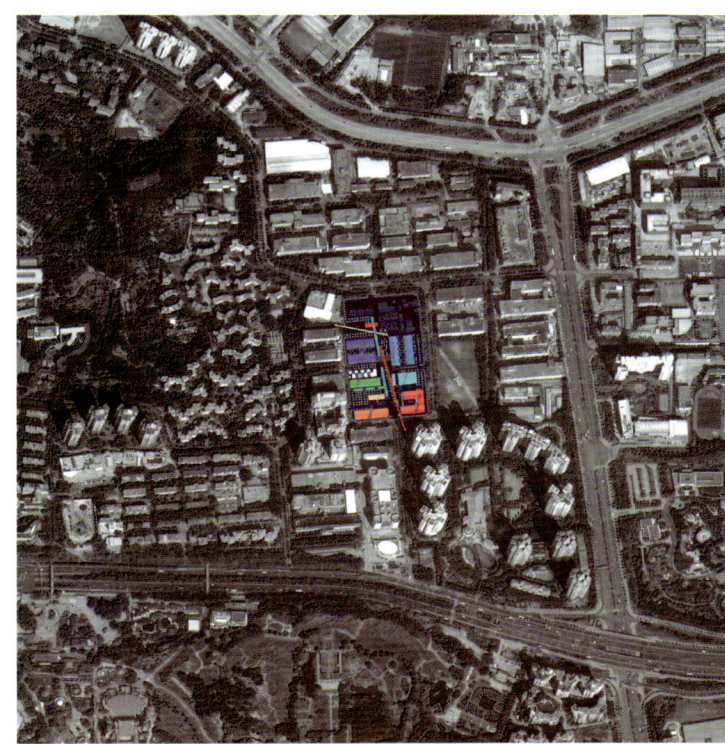

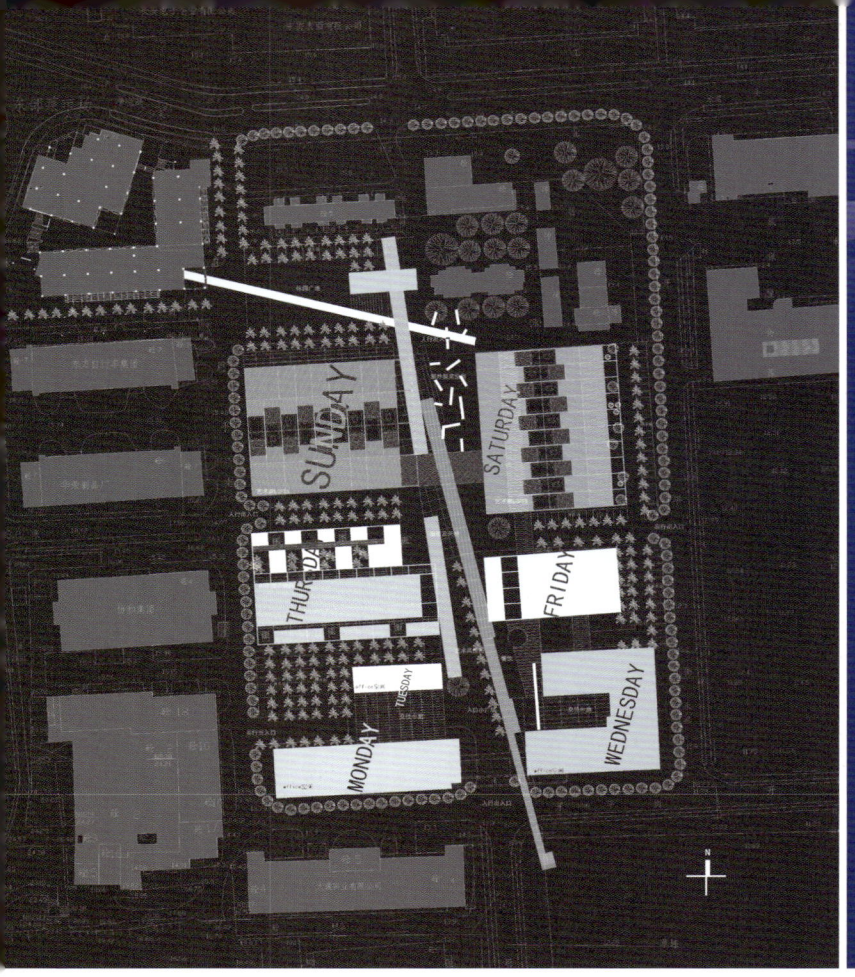

总平面

项目总览

这是一个旧工业区及厂房的更新和改造。我们的重点放在重塑建筑群内在的肌理和联系，建立起贯穿南北的通道。同时，通过镶嵌新建功能模块，使原来相当消极的空间产生新的活力。我们深信能理解和延续这一区的历史和文脉环境是渐成的，而不是建成的。

Project Overview

This is a project for renovating and reconstructing an old industrial zone and factory buildings. We pay attention to rebuild the intrinsic textures and connections of the buildings, design a passageway running from south to north, and embed new function modules to bring new vitality to the original inactive space. We firmly believe that understanding and extending the historical and cultural background in this area needs time but not depending on once and for all change through construction.

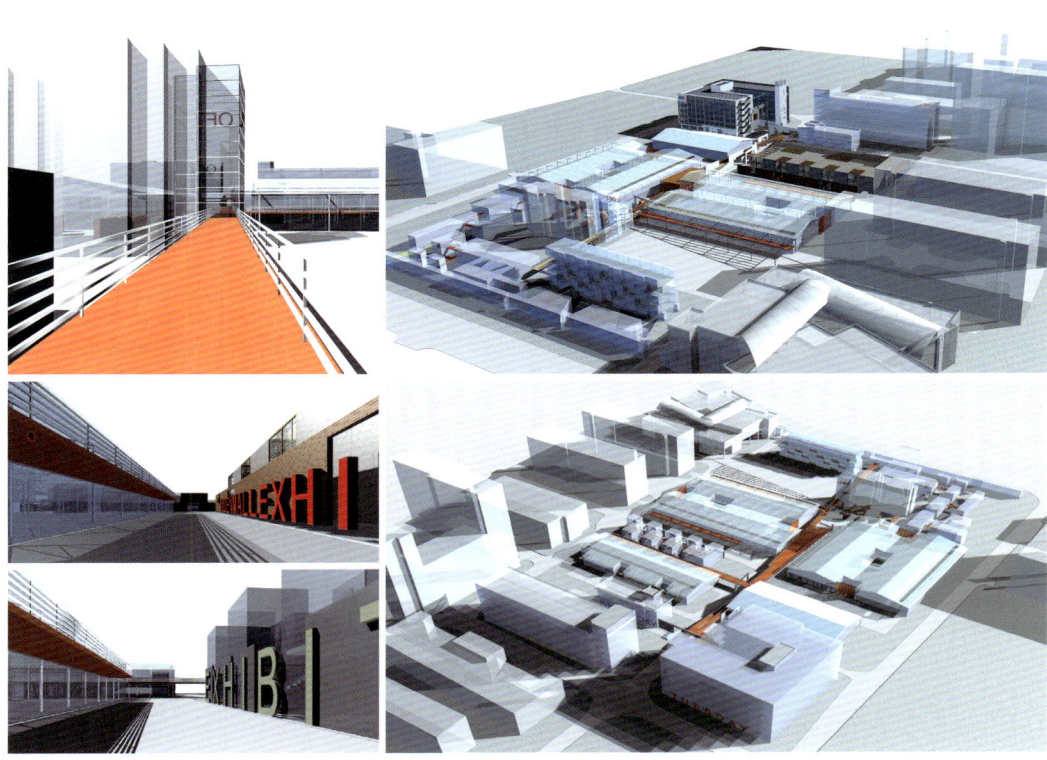

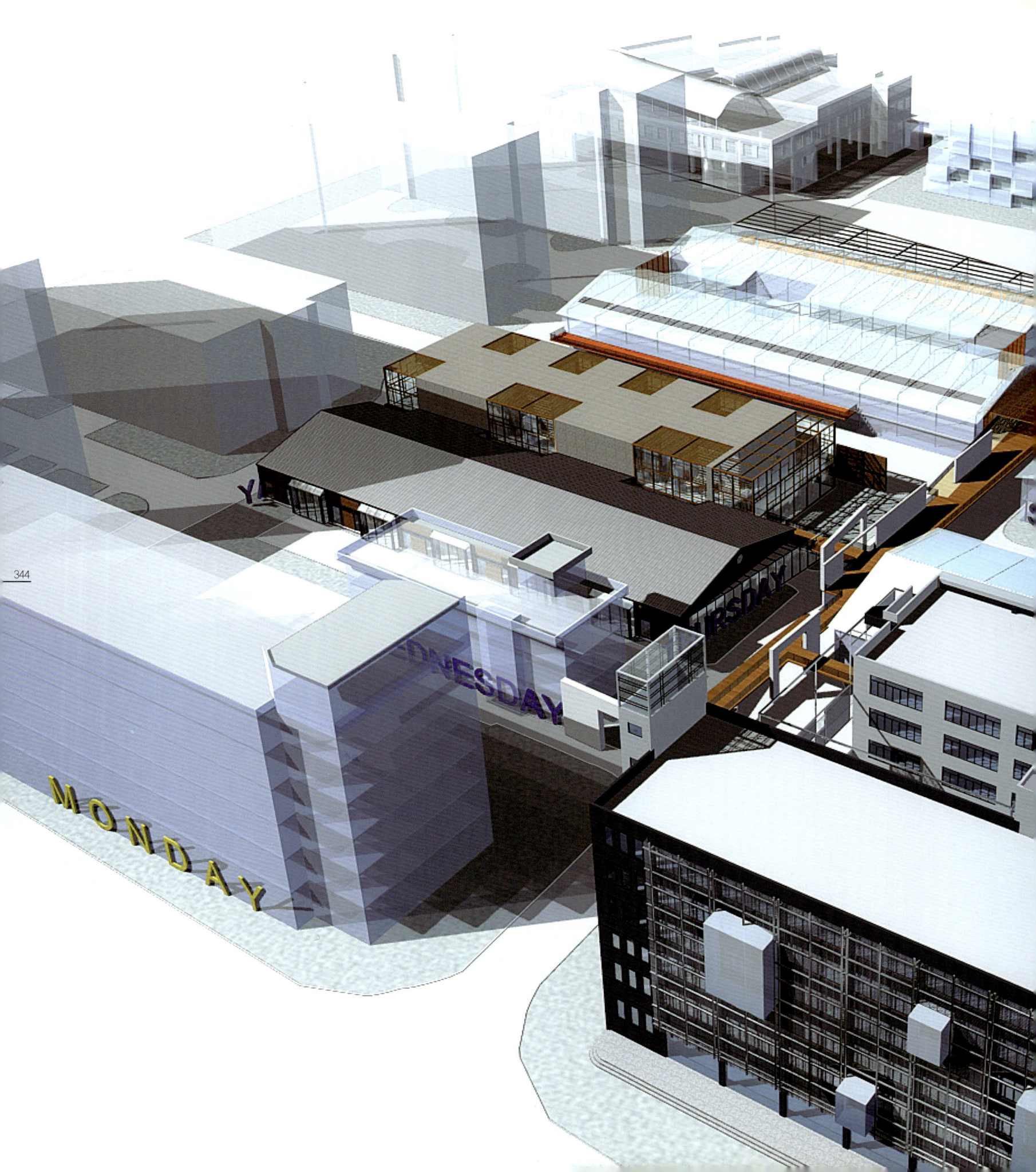

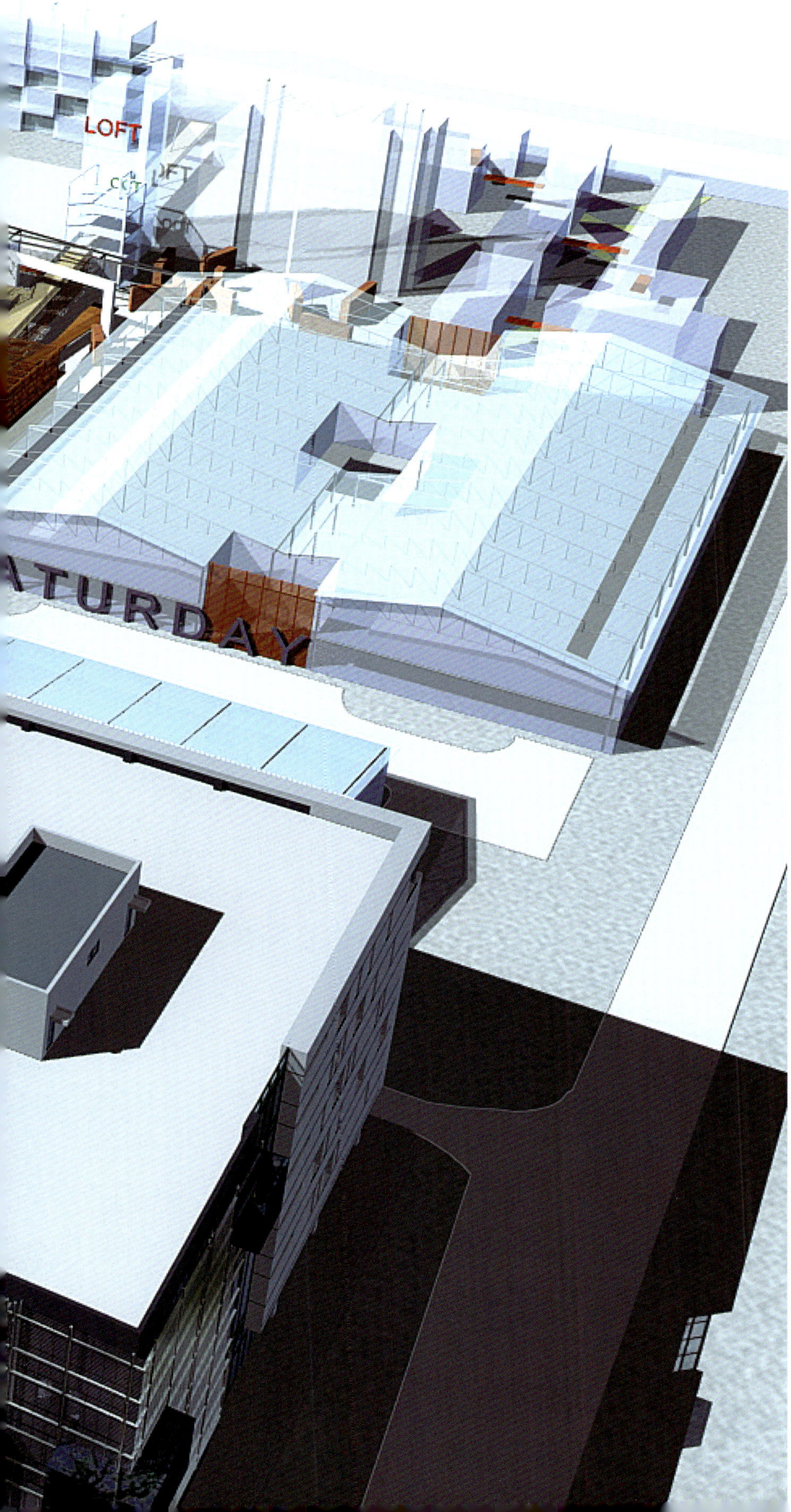
LOFT

THURSDAY

THURSDAY

THURSDAY

THURSDAY

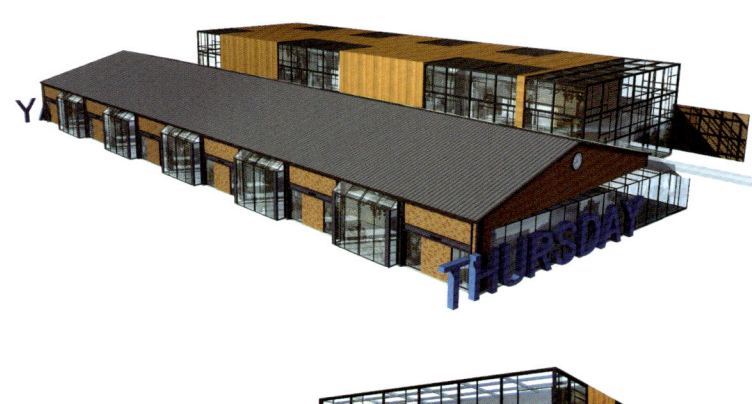

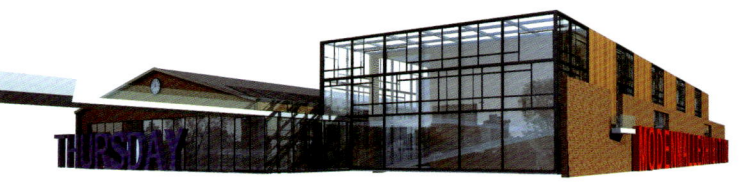

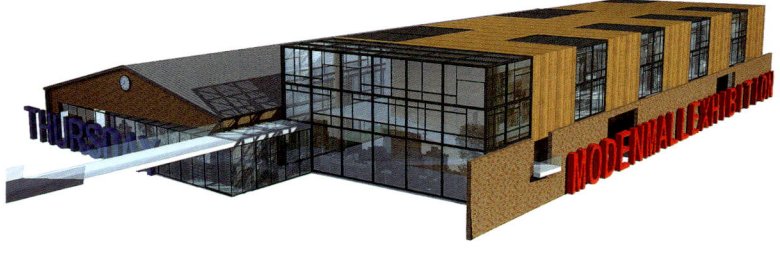

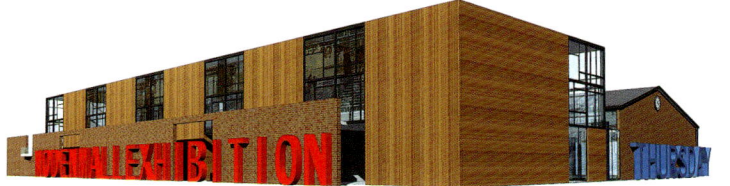

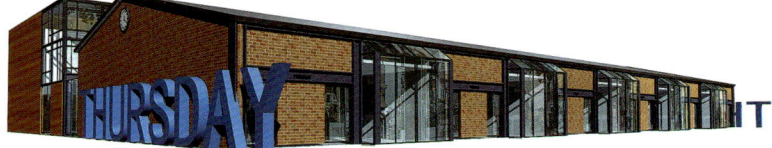

It is a supporting role but it will not only be a place where people can travel and play but also an important viewing point for the conference and exhibition center.

它是一个配角,既要成为人们能游玩的点,更重要的是还需要为会展中心提供一个重要的观景点。

Eastern Square of Nanning Conference & Exhibition Center • Nanning

Nanning | Project Location
2004 | Project Date
28,000m² | Project Scale
conceptual design | Design Phase
construction not started | Project Status

南宁会展中心东广场·南宁

中国，广西省，南宁市 | 项目地点
2004年 | 设计时间
2.8万m² | 项目规模
方案设计 | 设计阶段
未建 | 项目现状

EXHIBITION AND CONVENTION SQUARE   NANNING CHINA

方案二阶段总平面图

项目总览

本项目旨在以最简明、经济的方式创造出多功能、多层次的场所。项目设计要求体现展望未来、多元文化、五彩缤纷和媒介桥梁的特征。

Project Overview

This design aims at creating a multifunctional and multi-level place in the simplest and most economical way. It is required that the project design must consider future demands, show multiculture, present multiple colors, and reflect the features of a media bridge.

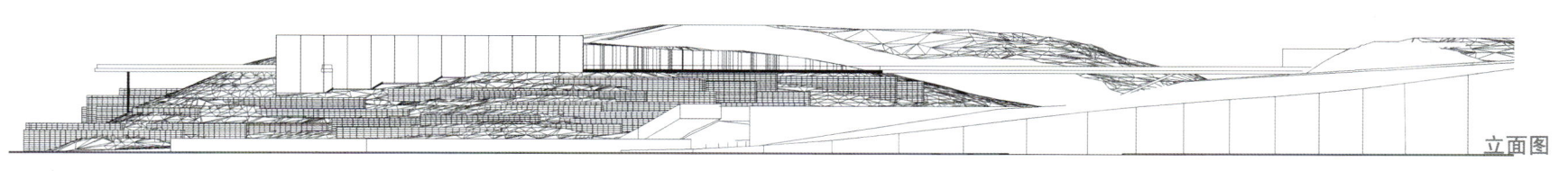

立面图

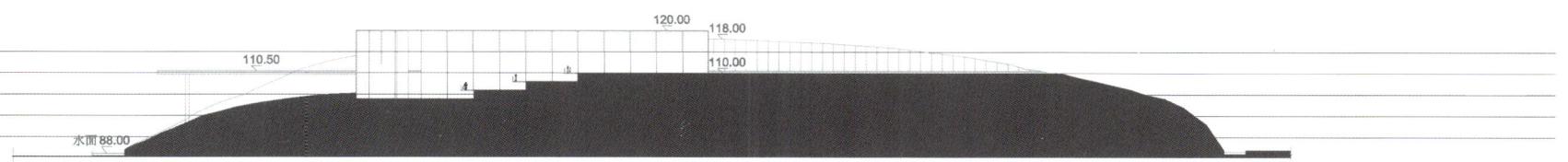

剖面图

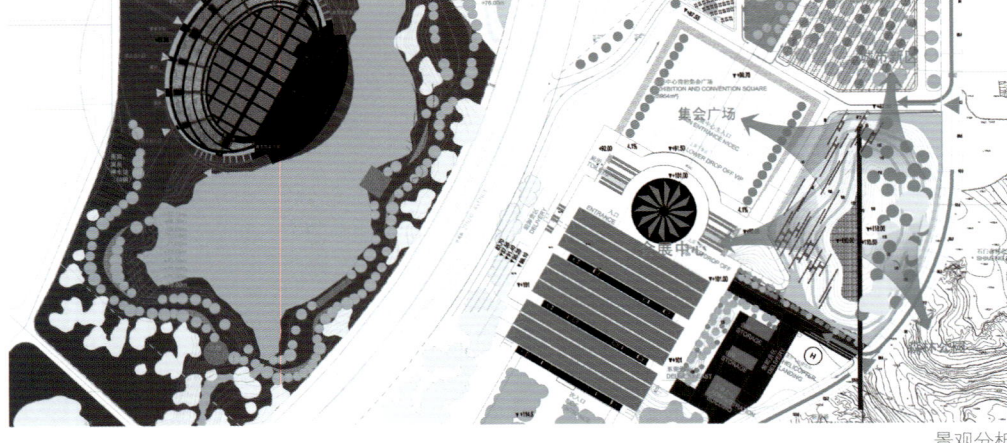

景观分析

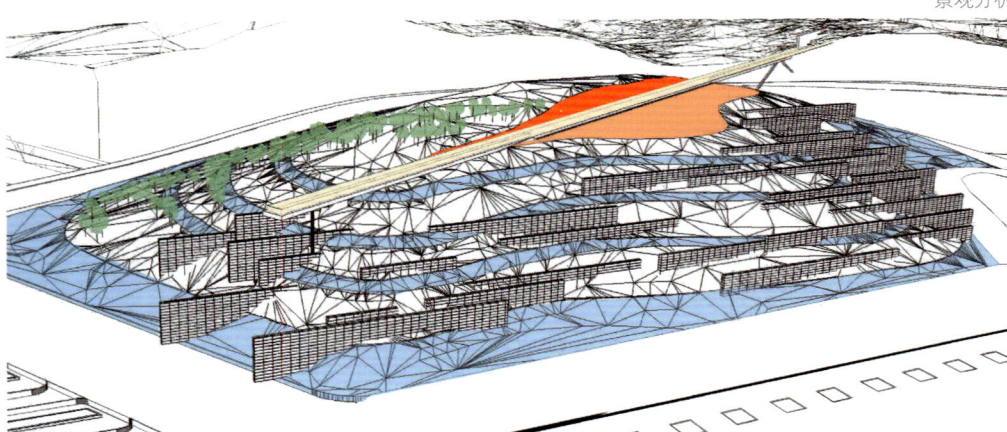

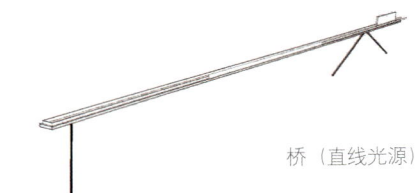 桥（直线光源）

 山顶广场（面光源）

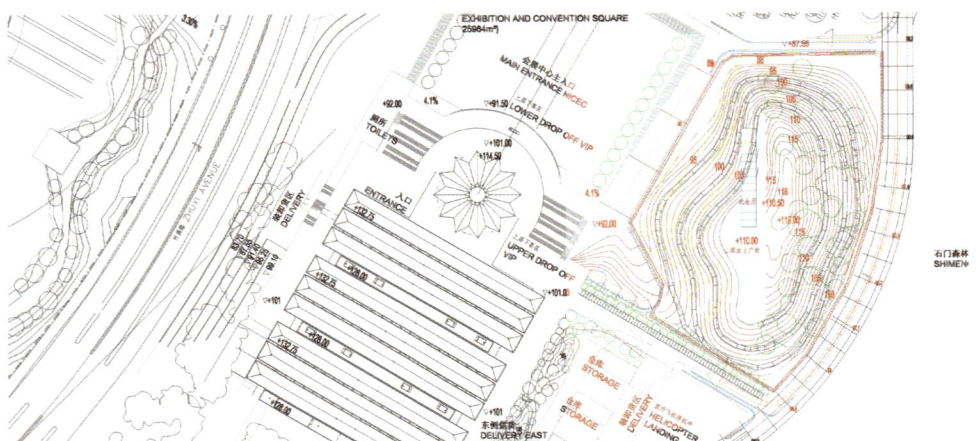

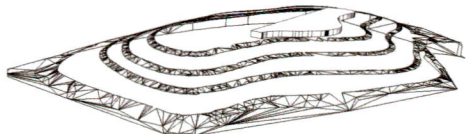 河流（曲线光源）

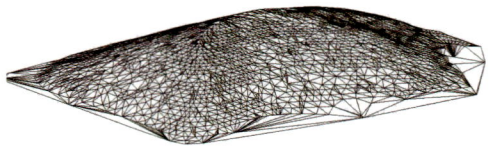 山体

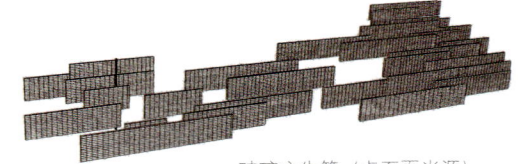 玻璃广告箱（点至面光源）

设计元素

本案中设计了一座桥，连接森林公园和广场，向会堂方向延伸，隐喻是"10+1相聚在南宁"，即博览会的桥梁紧密联系着中国与东盟十国的友好合作。

建筑整体造型

A bridge is designed to connect the forest park and the square. It extends toward the direction of the conference hall, implying that "10+1 countries" get together in Nanning. The exposition's bridge seems a close link between China and ten ASEAN countries for friendly cooperation.

新丝绸之路：一条曲水从山顶环绕山体缓缓流下，形成一条五彩缤纷的绚丽彩带，象征着对中国及东盟传统文化继承和发展的辉煌与灿烂。

New Silk Road – a zigzag river flows down slowly from the top of the hill, forming a colorful ribbon representing the outstanding achievement in carrying on and developing China and ASEAN's culture.

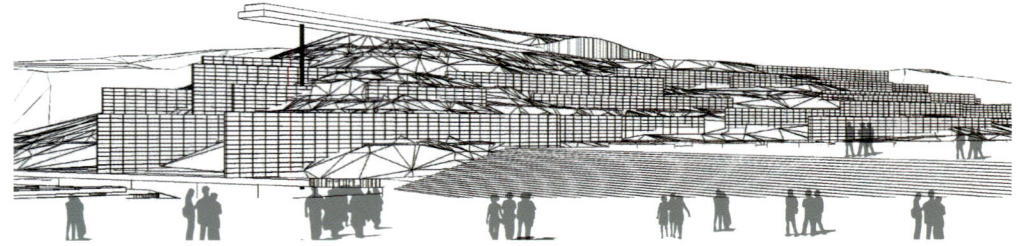

Our design aims at creating publicness, share, harmonious relationship, and driving force of people's life. Criticizing modern features does not mean negating modern features. The yearning for traditional streets and blocks is not returning to Middle Ages but to facilitate the restoration of human nature and a supplementation to modernization.

我们的设计目标是共有、共享与和谐相处，并积极推动人的生活。对于现在的批判，决不是对现代的否定，对于街和街区的向往不是要促成中世纪的再现，而是以人为本的回归和补充。

Design for the Region around the Sci-Tech Park in New Guangming District · Shenzhen

New Guangming District, Shenzhen | Project Location
2008 | Project Date
4,750,000m² | Project Scale
conceptual design | Design Phase
construction not started | Project Status

深圳光明新区科技公园周边地区城市设计·深圳

中国，广东省，深圳市，光明新区 | 项目地点
2008年 | 设计时间
475万 m² | 项目规模
方案设计 | 设计阶段
未建 | 项目现状

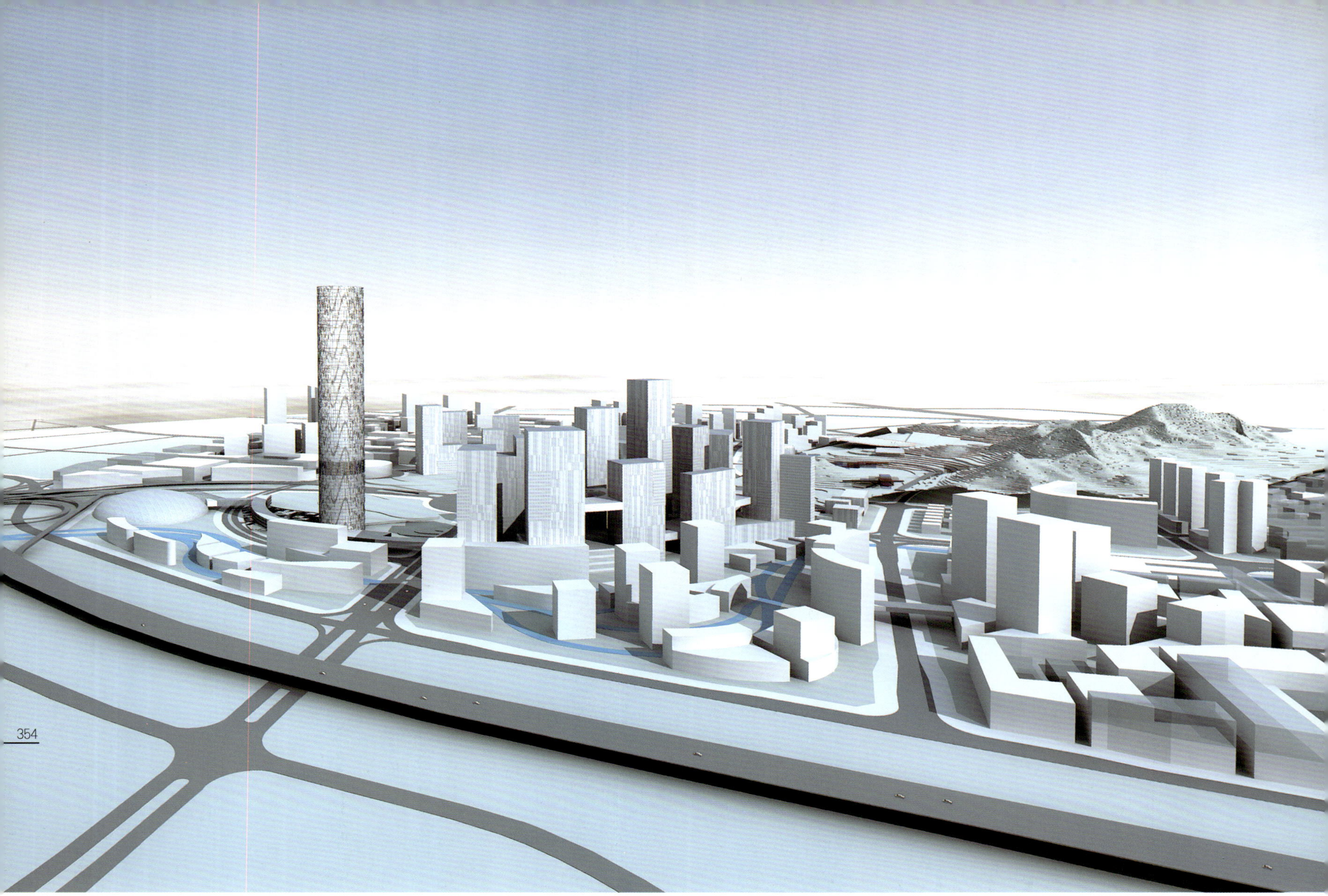

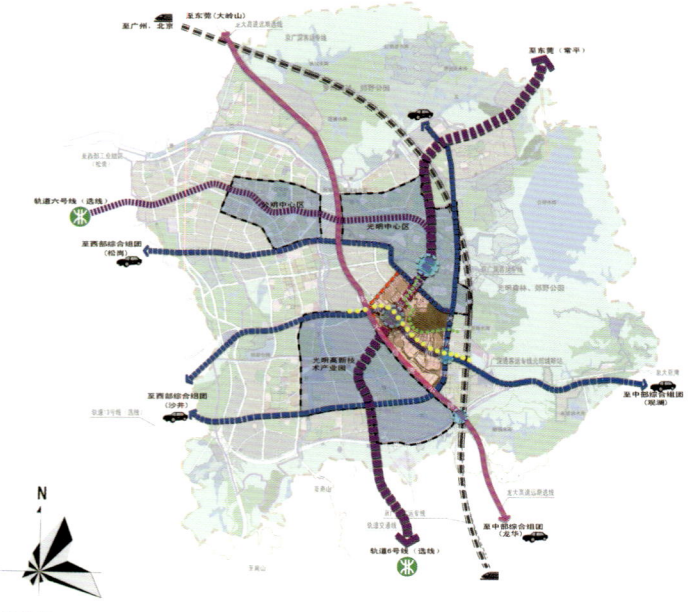

设计理念及构思

在规划方案在光明大道与观光大道交接处引入了一个环形超级城市集落,这里将链状串联行政、办公、文化、商业、科研、展览、酒店、公寓等多个集落,这一"集落群——巨型环"将成为这一片区的中心区；而巨型环的中心720m直径的中心城市花园。为整个规划片区甚至周边片区提供服务,而其自身也能形成一个高度集约的生活体系。

Design Ideas and Conceptions

A ring-shaped super-large urban community is planned near the intersection between Guangming Avenue and Guanguang Avenue, where many functional areas, such as administration, office, culture, business, research, exhibition, hotel, and apartment, are connected in series, looking like a chain. This "group of functional areas – a giant ring" will become the central part of this region. In the center of the giant ring is a central city garden, 720m in diameter. The community serves the people in the planned region or even surrounding regions, and itself is a function-intensive living system.

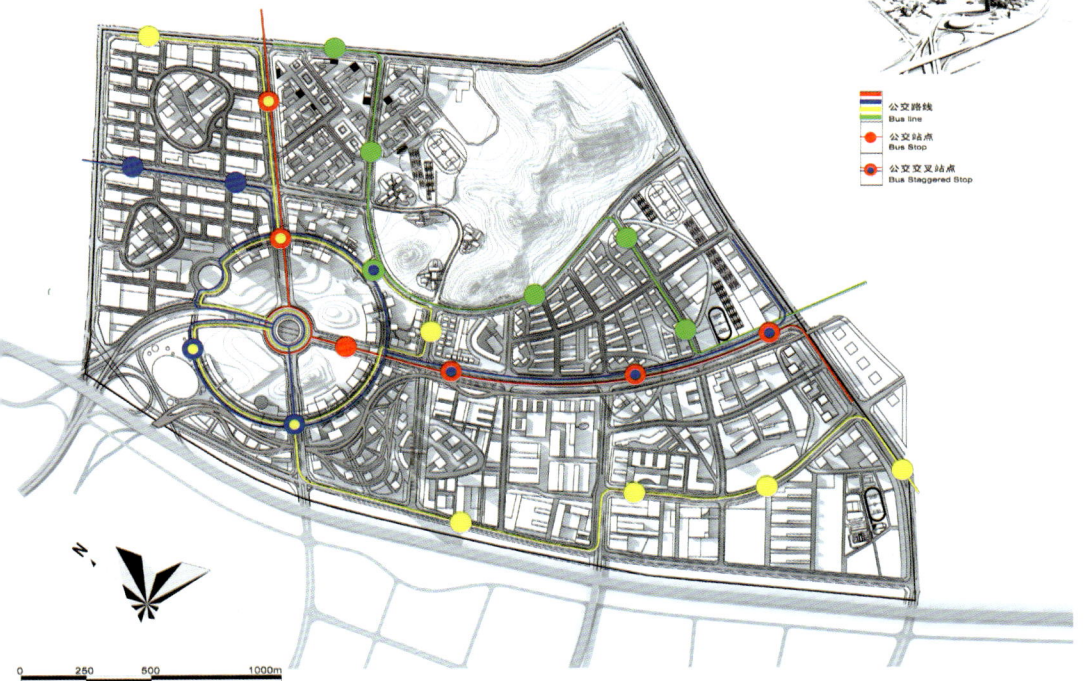

总平面

城市的形态、空间、尺度、交通等都需有张有弛，动态意味着活力，平衡代表着健康，我们设计的是弘法自然的"太极城市"。城市的优良性能最重要的是，它需具备自我调节、生长的能力；动态平衡城市是城市可持续性发展的前提。

The form, space, scale and transportation of a city should create a relaxing and active environment. Dynamic means vitality while balance stands for health. We design a "Tai Chi city" preaching on the principles of nature. For excellent performance of a city, the most important thing is that the city must have a capacity to regulate and develop itself. Dynamical balance is essential for a city's sustainable development.

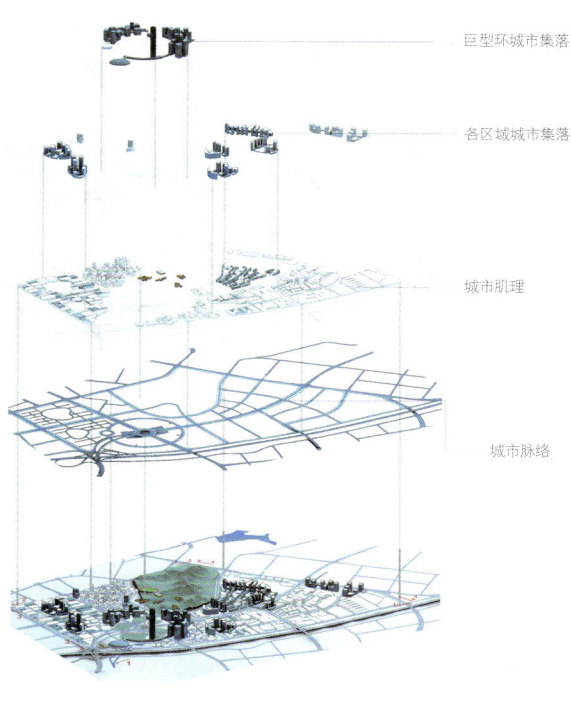

巨型环城市集落
各区域城市集落
城市肌理
城市脉络

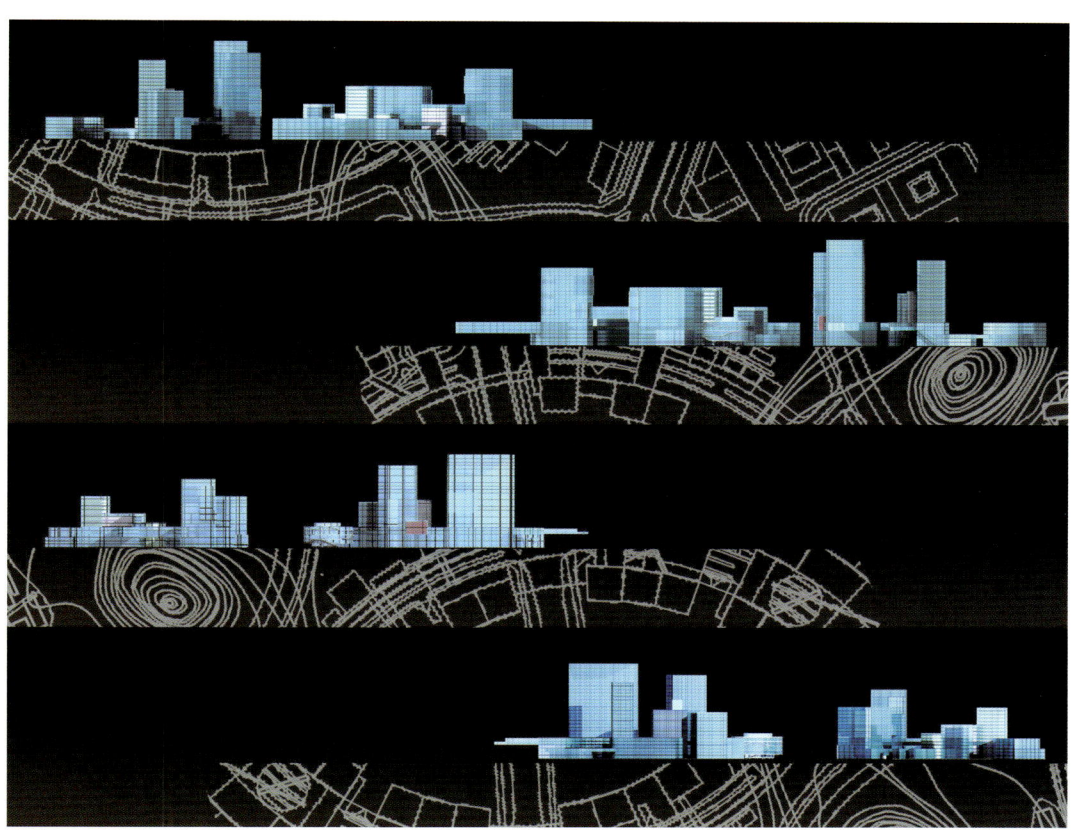

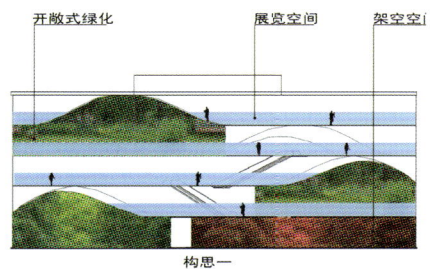

开敞式绿化　展览空间　架空空间

构思一

透视图

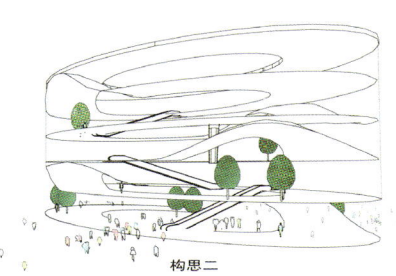

构思二

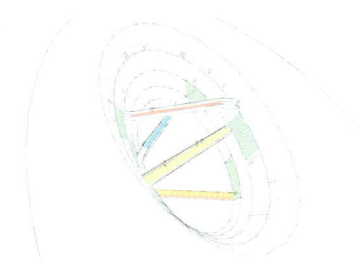

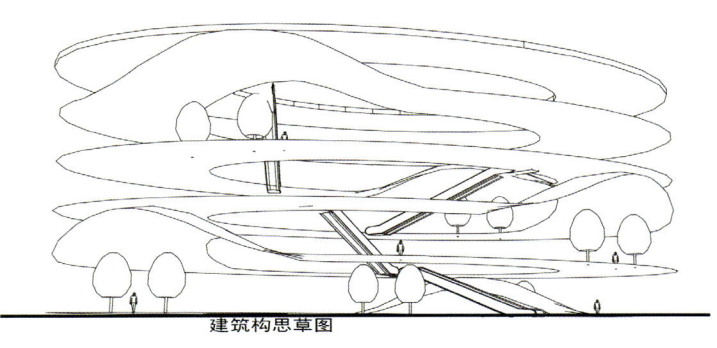

建筑构思草图

作为整个光明新区的重要组成部分，设计通过引入750m直径的中央公园，将位于用地东部的光明科技公园和以茅洲河为线索的城市中央绿轴横向相连接起来，完善了整体光明新区的城市花园体系。城市设计通过不同层次的多元的带状公园线性体系，形成由微观到宏观的城市绿化体系。

A central park, 750m in diameter, is introduced in the design. As an important integral part of the new Guangming District, this park connects Guangming Sci-Tech Park in the east of the community and the city's central green axis with Maozhou River as a main part. Thus the urban garden system in the whole new Guangming District is improved. A linear system consisting of different levels of pluralistic ribbon-shaped parks constitute a whole urban greening system with well-arranged parts.

功能分析图

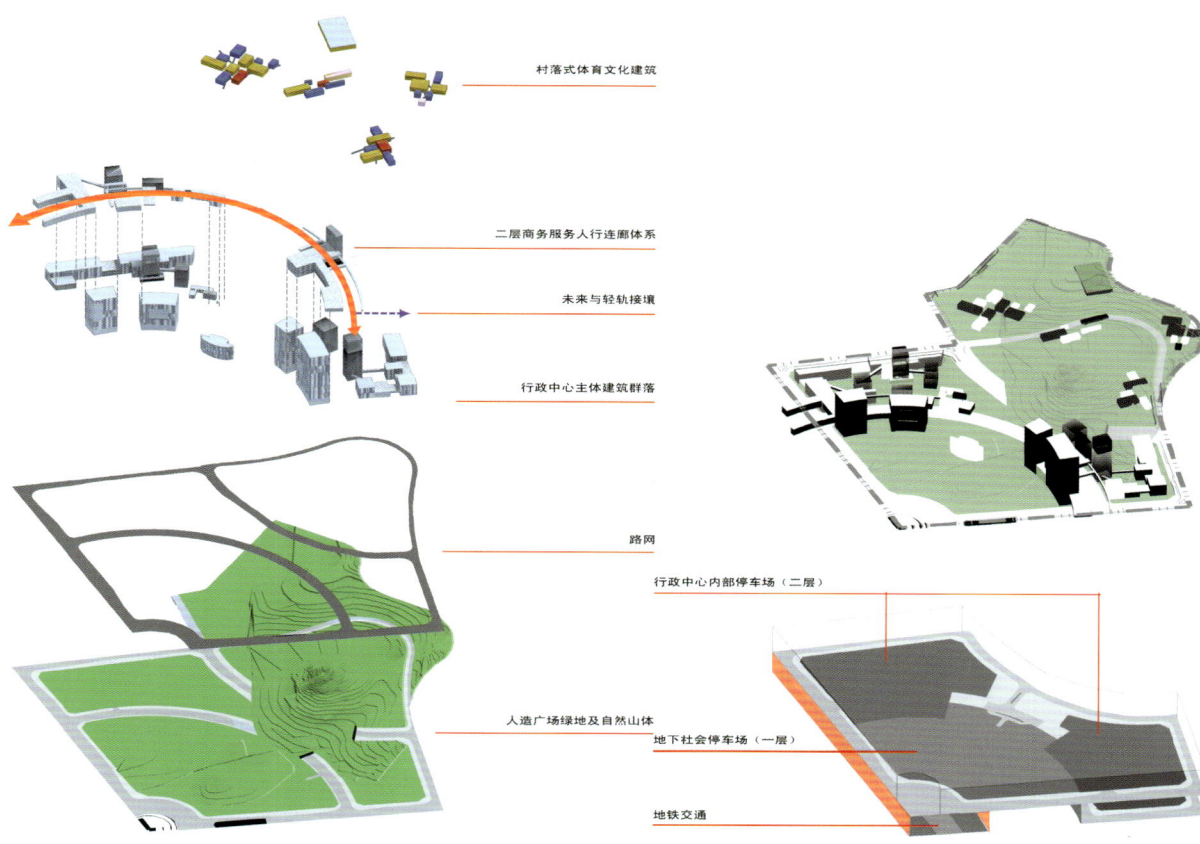

村落式体育文化建筑

二层商务服务人行连廊体系

未来与轻轨接驳

行政中心主体建筑群落

路网

行政中心内部停车场（二层）

人造广场绿地及自然山体

地下社会停车场（一层）

地铁交通

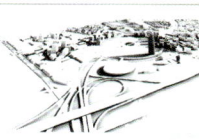

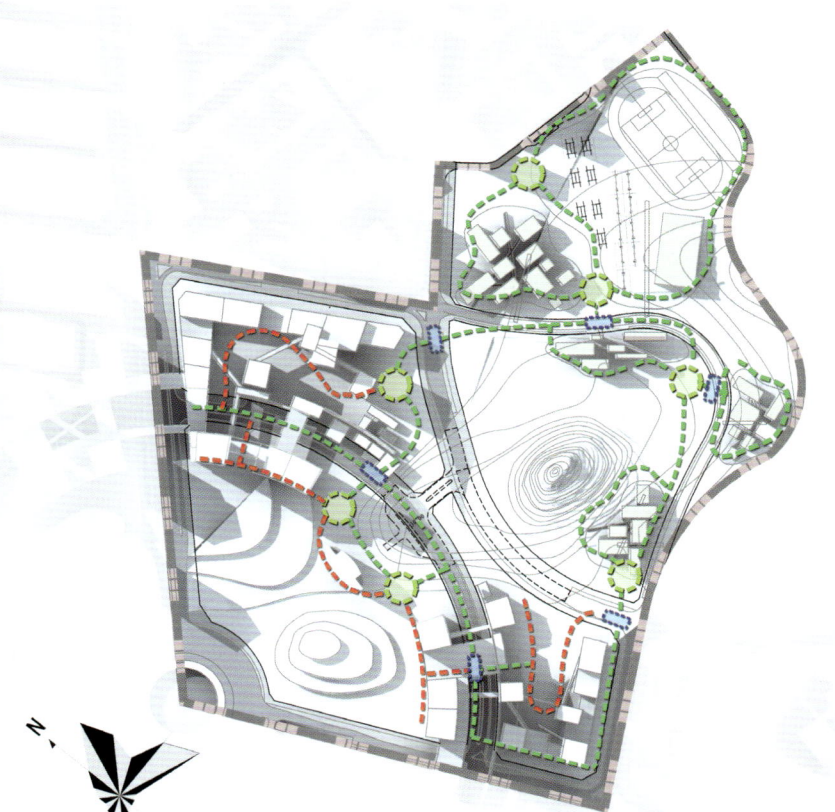

公共人行道
内部人行道
主要步行节点
主要步行道路交汇处

立面分析图

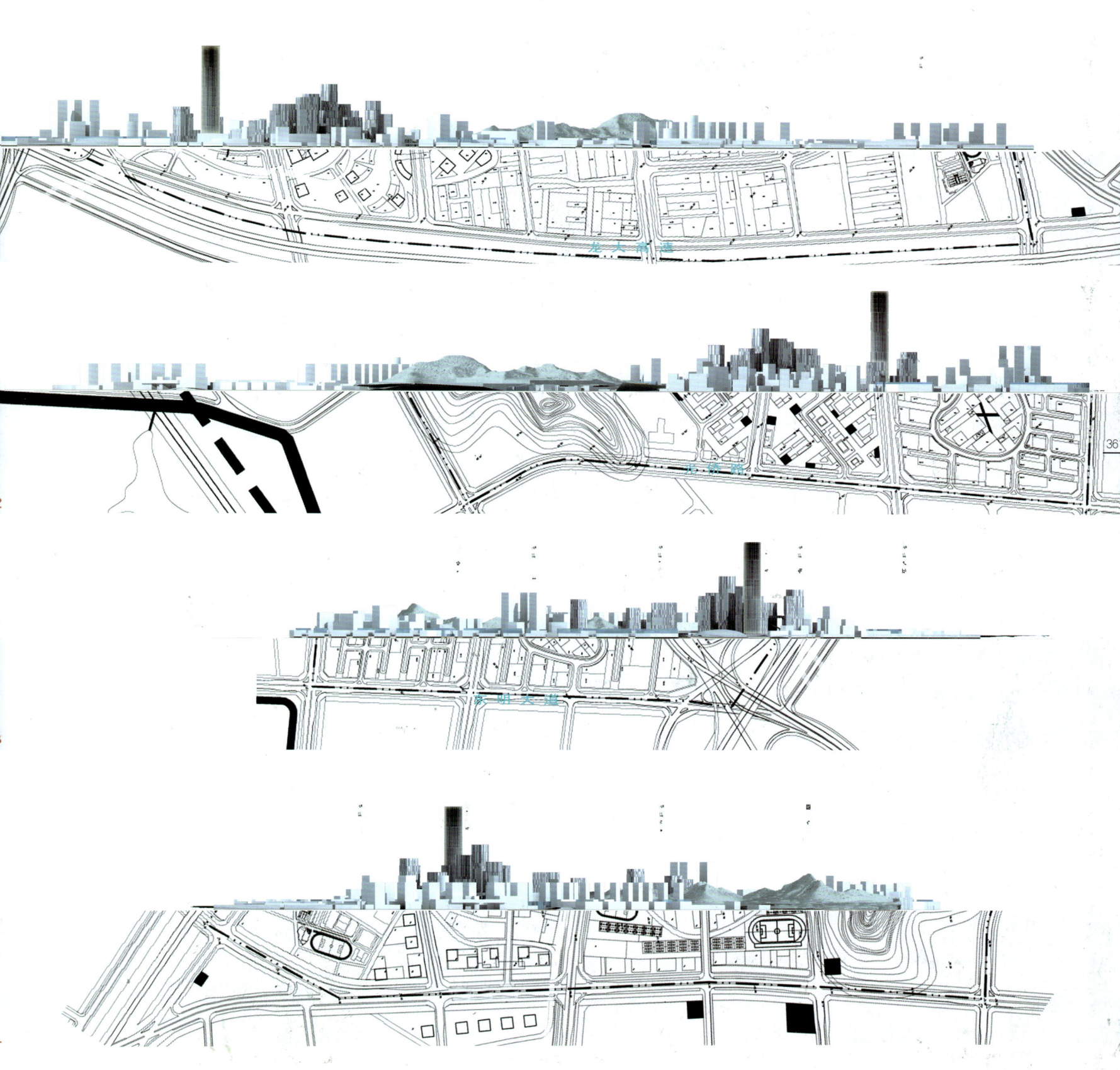